Laboratory Manual for Principles of General Chemistry

Laboratory Manual for Principles of General Chemistry

Seventh Edition

J. A. Beran
Texas A & M University—Kingsville

John Wiley & Sons, Inc.

The author of this manual has outlined extensive safety precautions in each experiment. Ultimately, it is your responsibility to practice safe laboratory guidelines. The author and publisher disclaim any liability for any loss or damage claimed to have resulted from, or been related to, the experiments.

ACQUISITIONS EDITOR Jennifer Yee
PRODUCTION EDITOR Sarah Wolfman-Robichard
MARKETING MANAGER Robert Smith
COVER DESIGN: Dawn Stanley
ILLUSTRATION EDITOR Gene Aiello
PHOTO EDITOR Tara Sanford
FRONT COVER PHOTO © Michael Abbey/Photo Researchers, Inc.

This book was set in 10/12 Times Roman by GGS Book Services, Atlantic Highlands and printed and bound by Courier Westford. The cover was printed by Phoenix.

This book is printed on acid-free paper. ♾

ISBN: 0-471-21498-1

Printed in the United States of America

10 9 8 7 6 5 4

Preface

The goal for writing the 7th edition of the laboratory manual was to provide a laboratory experience that was informative and challenging, yet enjoyable for its users—the students. The student should find rewards in the hands-on experience of meticulously collecting information that can be calculated, plotted, analyzed, and, as a consequence, finalized into a postulate that is scientifically logical. A sense of pride and confidence in understanding chemical principles first hand should be the result of the laboratory experience.

Based upon the comments from the large number of users of the previous six editions of this manual, we know that we are meeting many of the objectives of general chemistry laboratory programs. Scientifically speaking, of course, our goal to provide the "perfect" manual will never be attainable. However, we feel that we are closer in reaching that goal in the 7th edition.

For the 7th edition, we listened and responded! We incorporated many of the suggestions expressed by users of the 1st–6th editions and potential users of the 7th edition. The 7th edition is the best yet!

Breadth (and Level) of the 7th Edition

This manual covers two semesters (or three quarters) of a general chemistry laboratory program. A student may expect to spend three hours per experiment in the laboratory; limited, advanced preparation and/or extensive analysis of the data may lengthen this time. The experiments were chosen and written so that they may accompany any general chemistry text.

Features of the 7th Edition

Techniques. The explicit presentation of basic laboratory techniques that students must learn in general chemistry continues to be emphasized in the 7th edition. Two kinds of icons are used in the manual—those that are instructional and those that identify a laboratory technique to better complete the Experimental Procedure.

- The numbered icons, along with a detailed description of the corresponding techniques, are found in the laboratory techniques section of the manual beginning on page 11.
- Icons for the techniques used in each experiment are identified before the Introduction for the purpose of preparing students for the Experimental Procedure. Additionally, the technique icons are repeated in the margin of the Experimental Procedure to remind students of the time and situation for the correct use of the appropriate technique.

Users of the manual, the students, prompted most of the changes that were implemented to enhance the presentation of laboratory techniques.

Experimental Procedure. For clarity and safety, a word or short phrase introduces each paragraph of the Experimental Procedure to focus on its objective or goal of accomplishment.

Safety and Disposal. Icons that cite safety cautions and the safe disposal of test solutions are located at appropriate positions in the Experimental Procedure. In addition, a Caution statement and a Disposal statement are placed in the text of the Experimental Procedure to ensure that students are aware of immediate laboratory safety procedures.

Reduced Amounts of Chemicals. During the preparation of the 7th edition, the author listened to and concurred with reviewers, and *especially* students, who were users of the previous editions: *test tubes are the best apparatus* to observe chemical phenomena in the general chemistry laboratory! The contents of test tubes are easily observed at various angles, readily heated, and quickly set aside for reference without potential contamination of nearby chemicals or chemical reactions.

Consequently, "small" test tubes have replaced most "well plate" applications in the Experimental Procedures. Since the volume of a well in a 24-well plate approximates that of a 75-mm (3-inch) test tube, the small quantities of solutions used and suggested for the experiments have not been changed and, consequently, the disposal problems remain the same as in the 6th edition.

Figures and Illustrations. Many photographs and line drawings have been added and/or replaced with new and more detailed arrangements and labeling.

Web Site. http://www.wiley.com/college/chem/beran

The web site provides additional support for faculty and students who use this laboratory manual. Additionally, the web site includes downloadable Excel files of helpful **Report Sheet Templates** that can be used by students and/or instructors for experiments in which a numerical analysis of data is required.

New to the 7th Edition

Content and Organization Nearly 90% of the Prelaboratory Assignment questions and Laboratory Questions are new and/or revised from the 6th edition. The **Prelaboratory Assignment** questions were designed to emphasize the chemical principles presented in the Introduction of the experiment. Representative calculations and data analysis and occasional "points of emphasis" required for the completion of the Experimental Procedure are also addressed in the Prelaboratory Assignment.

Emphasis of the **Laboratory Questions** has focused on the "what if" of nearly every part of the Experimental Procedure. These questions seek to employ a thorough understanding of the "depths" of the experiment and just where the chemical principles can be used for interpreting unexpected data or "turning points" in the experiment.

A number of new experiments and revisions of others (see below) has introduced a degree of "relevance" to the laboratory experience, the relevance being associated with environmental and health issues. Even the qualitative analysis experiments (Experiments 34–37) provide the option for students to analyze one of their "own" samples for various ions.

New and Revised Experiments. The author has attempted to solve at least one foreseen problem associated with the general chemistry laboratory . . . "boring!" Three new experiments and several revisions of 6th edition experiments have been added with the intent that the experiments are more interesting, but without reducing the level of expectations and value of the laboratory experience.

Early in the sequence of experiments (Experiment 3) **Water Analysis: Solids** has been included in the 7th edition. Students analyze *any* water sample for total, dissolved, and suspended solids, observing some of the physical properties of matter and learning several valuable techniques. These analyses are also conducted at municipal drinking water laboratories and thus lend a meaning to the data that are collected.

An oxidation-reduction experiment typically completed in an analytical chemistry laboratory (permanganate-iron(II) analysis in the 6th edition) has been replaced by **Vitamin C Analysis** (Experiment 16), also an oxidation-reduction analysis but with a twist of interest to students. So much is said of dietary supplements, labeling of the RDA levels on nearly all food items, conflicting newspaper articles touting what is good and what is bad, . . . that this experiment perks students' interest in the results from their data analysis. The chemical properties of potassium permanganate, however, are not lost in the 7th edition—the temperature effect on the kinetics of the permanganate-oxalate reaction is observed in **Factors Affecting Reaction Rates** (Experiment 22).

A third experiment again focuses on the importance of our world's more precious natural resource—water. **Hard Water Analysis** (Experiment 30) combines not only the chemical principles of acid-base chemistry, buffering action, and complex formation but also on the comprehensive technique of a titrimetric analysis. Beyond that, the experiment brings forth the importance, sources, and levels of hard water used for drinking water and industrial processes. Additionally, this experiment directly compliments the water studies of Experiments 2 and 19, after a varied series of experiments on acid-base chemistry and chemical equilibrium.

The **Determination of a Rate Law** (Experiment 23) has been expanded to include the determination of the activation energy for the 6th edition's hydrogen peroxide/iodide ion reaction. This addition makes the kinetic studies in the 7th edition *complete*. While the time constraints for completion of the experiment may be jeopardized (>3 hours), most all of the data analysis can be completed outside of the laboratory period. Students learn of activation energy and of all the "straight lines" for kinetic data in the lecture, but the inclusion of an actual activation energy determination in the laboratory helps to clarify the study of kinetics.

For additional flexibility in laboratory assignments, a synthesis of a tin oxide has been added to **Empirical Formulas** (Experiment 7). Students often know in advance what the empirical formula for magnesium oxide and the calcium oxide/carbon dioxide ratio should be. However, with the two common oxidation states of tin, the "expected" empirical formula for the tin oxide may not be so certain. The completion of the experiment will provide a solution in which students have to believe in their data.

Good Reviews. The author's new and revised experiments that were a part of the 6th edition drew many favorable responses. The thermodynamics experiment (**The Thermodynamics of the Dissolution of Borax,** Experiment 31) took students to a higher level of involvement and understanding at an appropriate time in the general chemistry program—many of the laboratory techniques and chemical principles that had been introduced in earlier experiments were brought together in this experiment. The complexity of the data analysis for this and other experiments has been simplified with the development of the Report Sheet Templates available from Wiley.

Similarly the response to the revised dry lab on **Atomic and Molecular Structure** (Dry Lab 3) had only positive feedback from students and colleagues. For once, a vision of the "unseen (atoms and molecules)" was coordinated into a single, complimentary "sit down and figure it out" study.

A Few Parts Not Included in the 7th Edition. In the interest of laboratory safety and the disposal of chemicals, the synthesis of chrome alum from **Synthesis of an Alum** (Experiment 17), the substitution of hydrogen peroxide for potassium dichromate in **Oxidation-Reduction Reactions** (Experiment 14), the detection of the bromide ion from the qualitative analysis of the **Common Anions** (Experiment 34), and the confirmatory test for barium ion as the chromate salt in the **Qual III Group** (Experiment 37) are not a part of the 7th edition. Additionally, repetitive experiments illustrating the common-ion effect and the temperature effect on a chemical equilibrium in **LeChâtelier's Principle; Buffers** (Experiment 24) are also not included.

On-the-Shelf Experiments. Several experiments, a dry lab, and an appendix appearing in the 6th edition, do not appear in the 7th edition. If these omissions are critical to your program, they can be ordered as a "lab separate" from Wiley.

Three experiments were omitted from the 6th edition. **The Identification of a Compound: Physical Properties** was omitted for two reasons: (1) working with the many suggested organic compounds (along with the associated lab safety and disposal problems) can be a threatening experience in the early part of a laboratory experience and (2) the experiment is not an experience that stimulates any early excitement for chemistry in the laboratory. In effect, **Water Analysis: Solids** replaces this experiment. Second, the **Stoichiometric Analysis of a Redox Reaction** is replaced with **Vitamin C Analysis** for the reasons cited earlier. Third, the **Aluminum Analysis** is a gas-collection over water experiment, similar to, but less complete, than the gas-collecting experiment of **Calcium Carbonate Analysis: Molar Volume of Carbon Dioxide.** Thus Aluminum Analysis is omitted to make room for the **Hard Water Analysis** experiment.

Also not included in the 7th edition is the dry lab, **Oxidation-Reduction Equations,** a detailed exercise for balancing a wide array of redox equations. Additionally, the appendix entitled **Glassworking** from the 6th edition has been omitted. This very valuable technique seems to have lost its intrigue in general chemistry—none of the experiments in the 7th edition really *require* this technique and therefore is omitted. The latter omissions were prompted by comments from reviewers and users of previous editions.

Laboratory Equipment. Simple laboratory glassware and equipment, shown in the first sections of the manual, are necessary for completing most experiments. Where appropriate, the apparatus or technique is shown in the experiment with a line drawing or photograph. In addition to analytical balances, spectrophotometers (Experiments 25), pH meters (Experiment 27), and ammeters (Experiment 33) are suggested; however, if this instrumentation is unavailable, these experiments can be modified or omitted without penalizing students.

Contents of the Manual

The manual has five major sections:

- *Laboratory Safety and Guidelines.* Information on self-protection, what to do in case of an accident, general laboratory rules, and work ethics in the laboratory are presented.
- *Laboratory Data.* Guidelines for recording and reporting data are described. Sources of supplementary data (handbooks and World Wide Web sites) are listed. Suggestions for setting up a laboratory notebook are presented.
- *Laboratory Techniques.* Seventeen basic laboratory techniques present the proper procedures for handling chemicals and apparatus. Techniques unique to qualitative analysis are presented in Dry Lab 4.
- *Experiments and Dry Labs.* Thirty-eight experiments and four "dry labs" are subdivided into 11 principles.
- *Appendices.* Seven appendices include conversion factors, the treatment of data, the graphing of data, names of common chemicals, vapor pressure of water, concentrations of acids and bases, and water solubility of inorganic salts.

Contents of Each Experiment

Each experiment has six sections:

- *Objectives.* One or more statements establish the purposes and goals of the experiment. The "flavor" of the experiment is introduced with an opening photograph.
- *Techniques.* Icons identify various laboratory techniques that are used in the Experimental Procedure. The icons refer students to the Laboratory Techniques section where the techniques are described and illustrated.
- *Introduction.* The chemical principles, including the appropriate equations and calculations that are applicable to the experiment, are presented in the opening paragraphs. New and revised illustrations have been added to this section to further enhance the understanding of the chemical principles that are used in the experiment.
- *Experimental Procedure.* The Procedure Overview, a short introductory paragraph, provides a perspective of the Experimental Procedure. Detailed, stepwise directions are presented in the Experimental Procedure. Occasionally, calculations for amounts of chemicals to be used in the experiment must precede any experimentation.
- *Prelaboratory Assignment.* Questions and problems about the experiment prepare students for the laboratory experience. The questions and problems can be answered easily after studying the Introduction and Experimental Procedure. Approximately 90% of the Prelaboratory questions and problems are new to the 7th edition.
- *Report Sheet.* The Report Sheet organizes the observations and the collection and analysis of data. Laboratory Questions, for which students must have a thorough and in-depth understanding of the experiment, appear at the end of the Report Sheet. Approximately 90% of the Laboratory Questions are new to the 7th edition.

Instructor's Resource Manual

The *Instructor's Resource Manual* (available to instructors from Wiley) continues to be most explicit in presenting the details of each experiment. Sections for each experiment include

- an Overview of the experiment
- an instructor's Lecture Outline
- Teaching Hints for one-on-one instruction (including cautions and disposal procedures)
- representative or expected data and results
- Chemicals Required
- Special Equipment
- Suggested Unknowns
- answers to the Prelaboratory Assignment and Laboratory Questions
- a Laboratory Quiz.

Offered as a supplement to the *Instructor's Resource Manual* is a **Report Sheet Template** for those experiments requiring the numerical analysis of data. The format of the templates is according to Microsoft Excel software and is available from Wiley upon adoption.

The Appendices of the *Instructor's Resource Manual* detail the preparation of all of the solutions, including indicators, a list of the pure substances, and a list of the special equipment used in the manual *and* the corresponding experiment number for each listing. Users of the laboratory manual have made mention of the value of this supplement to the laboratory package.

Reviewers

The valuable suggestions provided by the following reviewers are greatly appreciated:

D. Neal Boehnke
Jacksonville University

Alan Hazari
University of Tennessee

Newton P. Hillard, Jr.
Eastern New Mexico University, Portales

Robert E. Hollins
Chicago State University

Maureen Kendrick Murphy
Huntingdon College

Margaret G. Kimble
Indiana University, Purdue University at Fort Wayne

Paul Popieniek
Sullivan County College

Kyle R. Willian
Tuskegee University

Lynne Zwman
Kirkwood Community College

Acknowledgments

The author thanks Dr. John R. Amend, Montana State University, for permission to use his basic idea in using emission spectra (without the aid of a spectroscope) to study atomic structure; Dr. Gordon Eggleton, Southeastern Oklahoma State University, for encouraging the inclusion of the paper chromatography experiment (Experiment 4); the general chemistry faculty at Penn State University, York Campus for the idea behind the thermodynamics experiment (Experiment 31).

What a staff at Wiley! Thanks to Jennifer Yee, Acquisitions Editor, for her keen insight, helpful suggestions, and unending commitment to see the manual through its birth; Sarah Wolfman-Robichaud, Production Editor, for coordinating the production of the manual; Tara Sanford, Associate Photo Editor at Wiley, for assistance in obtaining the new photographs for this edition; Gene Aiello, Illustration Editor at Wiley, for advancing the quality and detail of the line drawings.

Thanks also to the Chemistry 1111 and 1112 students and laboratory assistants at Texas A&M—Kingsville for their keen insight and valuable suggestions; also to my colleagues and assistants for their valuable comments.

A special note of appreciation is for Judi, who has unselfishly permitted me to follow my professional dreams and ambitions since long before the 1st edition of this manual in 1978. She has been the "rock" in my life. And also to Kyle and Greg, who by now have each launched their own families and careers—a Dad could not be more proud of them and their personal and professional accomplishments. My father and mother gave their children the drive, initiative, work ethic, and their blessings to challenge the world beyond that of our small Kansas farm. I shall be forever grateful to them for giving us those tools for success.

James E. Brady, St. Johns University, Jamaica, NY, who was a co-author of the manual in the early editions, remains the motivator to review and update the manual and to stay at the forefront of general chemistry education. Gary Carlson, my *first* chemistry editor at Wiley, gave me the opportunity to kick off my career in a way I never thought possible or even anticipated. Thanks Jim and Gary.

The author invites corrections and suggestions from colleagues and students.

J. A. Beran
MSC 161, Department of Chemistry
Texas A&M University—Kingsville
Kingsville, TX 78363

Contents

Wearing proper laboratory attire protects against chemical burns and irritations.

Laboratory Safety and Guidelines

The chemistry laboratory is one of the safest environments on campus or in an industrial setting. Every chemist, trained to be aware of the potential dangers of chemicals, is additionally careful in handling, storing, and disposing of chemicals. Laboratory safety should be a constant concern to everyone in the laboratory.

Be sure that you and your partners practice laboratory safety and follow basic laboratory rules. It is your responsibility, *not* the instructor's, to **play it safe.** A little extra effort on your part will assure others that the chemistry laboratory continues to be safe. Accidents do and will occur, but most often they are caused by carelessness, thoughtlessness, or neglect.

On the inside front cover of this book, there is space to list the location of important safety equipment and other valuable reference information that are useful in the laboratory. You will be asked to complete this at your earliest laboratory meeting.

This section of the manual has guidelines for making laboratory work a safe and meaningful venture. Depending on the specific laboratory setting or experiment, other guidelines for a safe laboratory may be enforced. Study the following guidelines carefully before answering the questions on the Report Sheet of Dry Lab 1.

A. Self-Protection

1. Approved safety goggles or eye shields *must be worn* at all times to guard against the laboratory accidents of others as well as your own. Contact lenses should be replaced with prescription glasses. In rare cases where contact lenses must be worn, eye protection (safety goggles) is absolutely necessary. A person wearing prescription glasses must also wear safety goggles or an eye shield. Discuss any interpretations of this with your laboratory instructor.
2. Shoes *must* be worn. Wear only shoes that shed liquids. Sandals, canvas shoes, and high-heeled shoes are *not* permitted.
3. Secure long hair and remove (or secure) neckties and scarves.
4. Wear nonsynthetic (cotton) clothing that is not torn or frayed. Shirts and blouses should not be frilled or flared, and the sleeves should be close-fit. In case of fire, synthetic (e.g., nylon, spandex) clothes may be difficult to remove quickly because they tend to stick to the skin.
5. Clothing should cover the skin from "neck to (below the) knee" and *at least* to the wrist. Discuss any interpretations of this with your laboratory instructor. See the opening photo.
6. Gloves are often worn to protect the hand when transferring corrosive liquids. Always consult with your laboratory instructor.

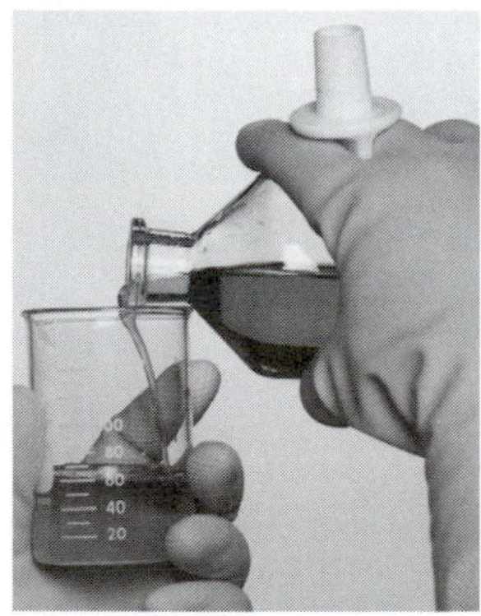

Laboratory gloves protect the skin from chemicals.

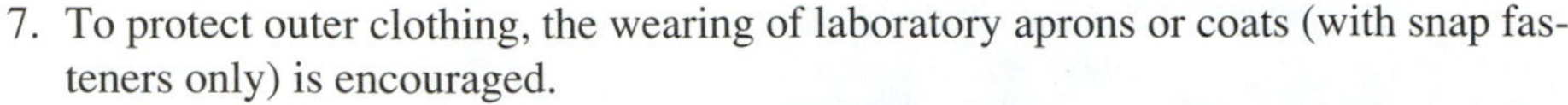

7. To protect outer clothing, the wearing of laboratory aprons or coats (with snap fasteners only) is encouraged.

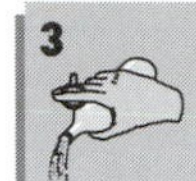

8. *Never* taste, smell, or touch a chemical or solution unless specifically directed to do so (see B.4 below). Individual allergic or sensitivity responses to chemicals cannot be anticipated. Poisonous substances are not always labeled.
9. For additional information on handling chemicals in the laboratory, refer to Technique 3, page 14.

B. In Case of an Accident

1. *Do not panic.* The most important first action after an accident is the care of the individual. ***Alert your laboratory instructor immediately!*** If a person is injured, provide or seek aid *immediately;* clothing and books can be replaced and experiments can be performed again later. Second, take the appropriate action regarding the accident: clean up the chemical (see B.8 below), use the fire extinguisher (see B.6 below), and so on.
2. Even if the accident or injury is regarded as minor, *notify your instructor* at once. A written report of an(y) accident may be required. Check with your laboratory instructor.

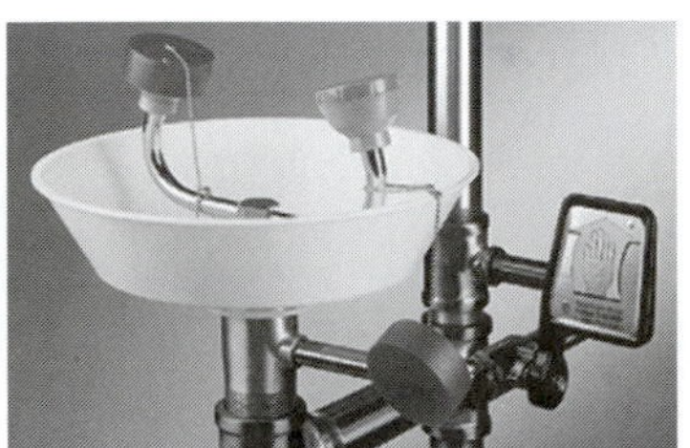

An eye wash can quickly remove chemicals from the eyes.

3. Wash your hands often during the laboratory. Always wash your hands before leaving the laboratory and your arms and face immediately thereafter in the washroom. Toxic or otherwise dangerous chemicals may be inadvertently transferred to the skin and from the skin to the mouth.
4. Whenever your skin (hands, arms, face, etc.) comes into contact with chemicals, quickly flush the affected area for several minutes with tap water followed by thorough washing with soap and water. Use the eyewash fountain to flush chemicals from the eyes and face. *Get help immediately.* Do *not* rub the affected area, especially the face or eyes, with your hands before washing.

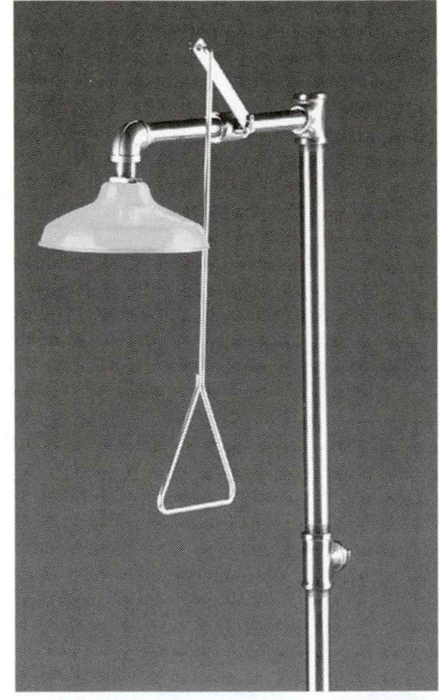

A safety shower can quickly remove chemicals from the body.

5. Chemical spills over a large part of the body require immediate action. Using the safety shower, flood the affected area for at least 5 minutes. Remove all contaminated clothing if necessary. Use a mild detergent and water only (no salves, creams, lotions, etc.). Get medical attention.
6. In case of fire, discharge a fire extinguisher at the base of the flames and move it from one side to the other. Small flames can be smothered with a watchglass (do *not* use a towel, it may catch on fire). Do *not* discharge a fire extinguisher when a person's clothing is on fire—use the safety shower.
7. For abrasions or cuts, flush the affected area with water. Any further treatment should be given only after consulting with the laboratory instructor.

   For burns, the affected area should be rubbed with ice, submerged in an ice/water bath, and/or placed under running water for several minutes to withdraw heat from the burned area.
8. Treat chemical spills in the laboratory as follows:
   - Alert your neighbors and the laboratory instructor.
   - Clean up the spill as directed by the laboratory instructor.
   - If the substance is volatile, flammable, or toxic, warn everyone of the accident.

9. For additional information on the disposal of chemicals in the laboratory, refer to Technique 4, page 15.

C. Laboratory Rules

In addition to the guidelines for self-protection (Part A), the following rules must be followed.

1. *Smoking, drinking, eating, and chewing* (including gum and tobacco) are not permitted at any time because chemicals may inadvertently enter the mouth or lungs. Your hands may be contaminated with an "unsafe" chemical. Do not place any ob-

jects, including pens or pencils, in your mouth during or after the laboratory period. These objects may have picked up a contaminant from the laboratory bench.

2. Do *not* work in the laboratory alone. The laboratory instructor must be present.
3. Inquisitiveness and creativeness in the laboratory are encouraged. However, variations or alterations of the Experimental Procedure are forbidden without prior approval of the laboratory instructor. If your chemical intuition suggests further experimentation, consult with your laboratory instructor.
4. Maintain an orderly, clean laboratory desk and drawer. Immediately clean up all chemical spills, paper scraps, and glassware. Discard wastes as directed by your laboratory instructor.
5. Keep drawers or cabinets closed and the aisles free of any obstructions. Do *not* place book bags, athletic equipment, or other items on the floor near any lab bench.
6. At the end of the laboratory period, completely clear the lab bench of equipment, clean it with a damp sponge or paper towel (and properly discard), and clean the sinks of all debris. Also clean all glassware used in the experiment (see Technique 2, page 13).

7. Be aware of your neighbors' activities; you may be a victim of their mistakes. Advise them of improper techniques or unsafe practices. If necessary, tell the instructor.
8. For all other rules, **listen to your instructor!**

D. Working in the Laboratory

1. Maintain a wholesome, professional attitude. Horseplay and other careless acts are prohibited. No personal audio or other "entertainment" equipment is allowed in the laboratory.
2. Do *not* entertain guests in the laboratory. Your total concentration on the experiment is required for a safe, meaningful laboratory experience. You may socialize with others in the lab, but do not have a party! You are expected to maintain a learning environment.
3. Scientists learn much by discussion with one another. Likewise, you may profit by discussion with your laboratory instructor or classmates, but *not* by copying from them.
4. *Prepare* for each experiment by completing the Prelaboratory Assignment and by studying the Objectives, Techniques, Introduction, and Experimental Procedure *before lab*. Advanced preparation will save you time, reduce the chances of personal injury and damage to equipment, and provide a more meaningful learning experience (and a better lab grade).

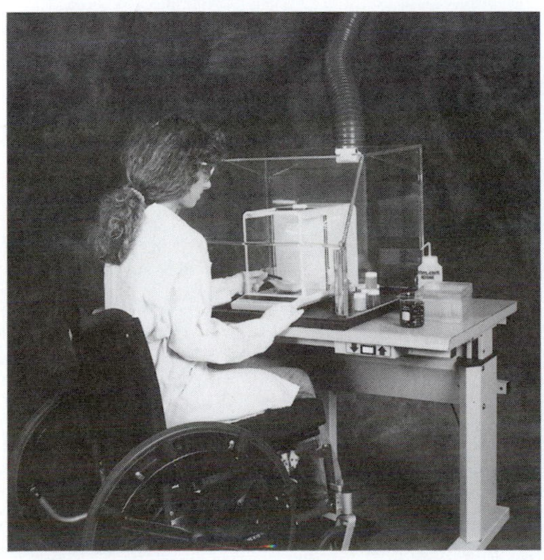

Laboratory facilities must be designed for safety.

5. Are any calculations required before you begin to collect the data for the experiment? Complete those before the laboratory period begins.
6. As a guide to studying the Experimental Procedure, always try to understand the reason for each step and why you are doing it. **Think,** whistle if you like, while working in the laboratory.
7. Note beforehand the need for any extra equipment in the experiment and obtain it all at once from the stockroom.
8. Review the Laboratory Questions at the conclusion of the Report Sheet before *and* as you perform the experiment. These questions are intended to enhance your understanding of the chemical principles on which the experiment is based.

Notes on Laboratory Safety and Guidelines

Laboratory data should be carefully recorded.

Laboratory Data

The lifeblood of a good scientist depends on the collection of reliable and reproducible data from experimental observations and on the analysis of that data. The data must be presented in a logical and credible format, that is, the data must appear such that other scientists will believe in and rely on the data that you have collected.

Believe in your data, and others should have confidence in it also. A scientist's most priceless possession is integrity. **Be a scientist.** A scientist is conscientious in his/her efforts to observe, collect, record, and interpret the experimental data as best possible. Only honest scientific work is acceptable.

You may be asked to present your data on the Report Sheet that appears at the end of each experiment, or you may be asked to keep a laboratory notebook (see guidelines below). For either method, a customary procedure for collecting, recording, and presenting data is to be followed. A thorough preview of the experiment will assist in your collection and presentation of data.

A. Recording Data

1. Record all data entries *as they are being collected* on the Report Sheet or in your laboratory notebook. Be sure to include appropriate units after numerical entries. Data on scraps of paper (such as mass measurements in the balance room) will be confiscated.
2. Record the data *in permanent ink* as you perform the experiment.
3. If a mistake is made in recording data, cross out the incorrect data entry with a *single* line (do *not* erase, white-out, overwrite, or obliterate) and clearly enter the corrected data nearby (see Figure D.1). If a large section of data is deemed incorrect, write a short notation as to why the data is in error, place a single diagonal line across the data, and note where the correct data is recorded.

	Trial 1
Mass of $CaCO_3$ sample, initial	0.218 g
Mass of $CaCO_3$, after heating	~~0.184 g~~ 0.164
Mass of CO_2 in sample	0.054 g

Figure D.1 Procedures for recording and correcting data.

4. For clarity, record data entries of values <1 with a zero in the "one" position of the number; for example, record a mass measurement as 0.218 g rather than .218 g (see Figure D.1).
5. Data collected from an instrument and/or computer printout should be attached to the Report Sheet.

B. Reporting Data

The quantitative data that are collected must reflect the reliability of the instruments and equipment used to make the measurements. For example, most bathroom scales in the United States weigh to the nearest pound (± 1 lb); therefore, reporting a person's weight should reflect the precision of the measurement—a person's weight should be expressed as, e.g., 145 ± 1 pounds and *not* 145.000 . . . pounds! Conversely, if the mass of a substance is measured on a balance that has a precision of ± 0.001 g, the mass of the object should be expressed as, e.g., 0.218 g and *not* as 0.2 g.

Scientists use ***significant figures*** *to clearly express the precision of measurements.* The number of significant figures used to express the measurement is determined by the specific instrument used to make the measurement.

The number of significant figures in a measurement equals the number of figures that are certain in the measurement plus one additional figure that expresses uncertainty. The first uncertain figure in a measurement is the last significant figure of the measurement. The above mass measurement (0.218 g) has three significant figures—the first uncertain figure is the "8" which means that the confidence of the measurement is between 0.219 g and 0.217 g or 0.218 ± 0.001 g.

Rules for expressing the significant figures of a measurement and manipulating data with significant figures can be found in most general chemistry texts.

A simplified overview of the "Rules for Significant Figures" is as follows:

- Significant figures are used to express measurements, dependent on the precision of the measuring instrument.
- All definitions (e.g., 12 inches = 1 foot) have an infinite number of significant figures.
- For the addition and subtraction of data with significant figures, the answer is rounded off to the number of decimal places equal to the *fewest* number of decimal places in any one of the measurements.
- For the multiplication and division of data with significant figures, the answer is expressed with the number of significant figures equal to the *fewest* number of significant figures for any one of the measurements.

Expressing measurements in scientific notation often simplifies the recording of measurements with the correct number of significant figures. For example, the mass measurement of 0.218 g, expressed as 2.18×10^{-1} g, clearly indicates three significant figures in the measurement. Zeros at the front end of a measurement are not significant.

Zeros at the end of a measurement of data may or may not be significant. However, again that dilemma is clarified when the measurement is expressed in scientific notation. For example, the volume of a sample written as 200 mL may have one, two, or three significant figures. Expressing the measurement as 2×10^3 mL, 2.0×10^3 mL, or 2.00×10^3 mL clarifies the precision of the measurement. Zeros at the end of a number *and* to the right of a decimal point are always significant.

In reporting data for your observations in this laboratory manual, follow closely the guidelines for using significant figures for correctly expressing the precision of your measurements and the reliability of your calculations.

C. Accessing Supplementary Data

You will also profit by frequent references to your textbook or, for tabular data on the properties of chemicals, the *CRC Handbook of Chemistry and Physics,* published by the Chemical Rubber Publishing Company of Cleveland, Ohio, or the

Merck Index, published by Merck & Co., Inc., of Rahway, New Jersey. Books are generally more reliable and more complete sources of technical information than are classmates.

The World Wide Web has a wealth of information available at your fingertips. Search the web for additional insights into each experiment. As a beginning you may wish to access the author's personal web page, which is continuously "under construction," at *http://www.tamuk.edu/chemistry/beran.*

Ask your laboratory instructor about his/her favorite web sites.

Other (suggested only) web sites that may enhance your appreciation of the laboratory experience are listed here:

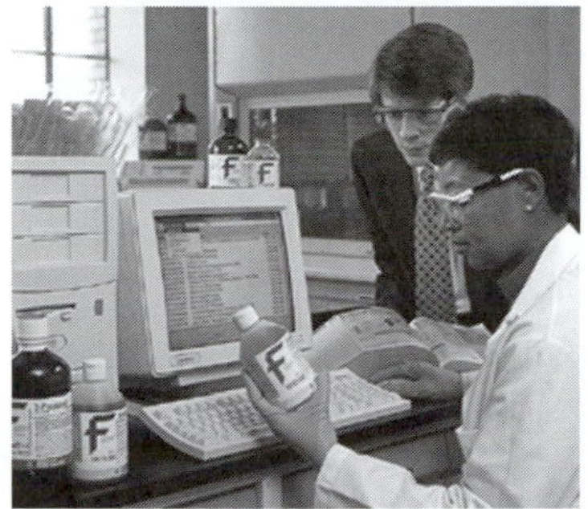

Scientific data can be obtained from the internet or analyzed with appropriate software.

- *http://webbook.nist.gov/chemistry* (database of technical data)
- *http://www.ilpi.com/msds* (MSDS information of chemicals)
- *http://www.yahoo.com* (search Science and Chemistry)
- *http://chem-courses.ucsd.edu* (search undergraduate labs)
- *http://JChemEd.chem.wisc.edu*
- *http://www.chemtutor.com*
- *http://chemfinder.camsoft.com* (information on compounds)
- *http://www.chemistrycoach.com*
- *http://www.chemcenter.org*
- *http://chemistry.about.com*
- *http://www.siraze.net/chemistry*
- *http://bouman.chem.georgetown.edu/genchem*
- *http://physics.nist.gov/cuu* (database of technical data)

D. Laboratory Notebook

The laboratory notebook is a personal, permanent record, i.e., a journal, of the activities associated with the experiment or laboratory activity. The first 3–4 pages of the notebook should be reserved for a table of contents. The laboratory notebook should have a sewn binding and the pages numbered in sequence.

Each experiment in the laboratory notebook should begin on a new page, using only the right-hand pages of the laboratory notebook, and should include the following sections with clear, distinct headings:

- the title of the experiment
- bibliographic source of the experiment
- co-workers for the experiment
- the purpose and/or objective(s) of the experiment
- a brief, but clearly written experimental procedure that includes the appropriate balanced equations for the chemical reactions
- a brief description or sketch of the apparatus
- a section for the data that is recorded (see Recording and Reporting Data above) as the experiment is in progress, i.e., the Report Sheet. This data section must be planned and organized carefully. The quantitative data is to be organized, neat, and recorded with the appropriate significant figures and units; any observed, qualitative data must be written legibly, briefly, and with proper grammar. All data must be recorded in permanent ink. Allow plenty of room for recording observations, comments, notes, etc.
- a section for data analysis that includes representative calculations, an error analysis (see Appendix B), instrument and computer printouts, graphical analyses (see Appendix C), and organized tables. Where calculations using data are

involved, be orderly with the first set of data. Do *not* clutter the data analysis section with arithmetic details.
- a section for results and discussion

The left-hand pages of the laboratory notebook are generally used for quick arithmetical calculations, special instructions or announcements, specific cautions for handling chemicals, etc.

At the completion of each day's laboratory activities, each page of laboratory activity should be dated and signed by the chemist, any co-worker, *and* the laboratory instructor.

The laboratory instructor will outline any specific instructions that are unique to your laboratory program.

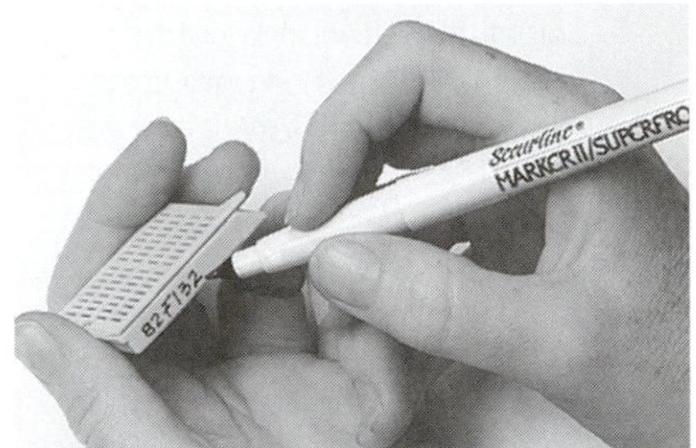

Marking pens help to organize samples.

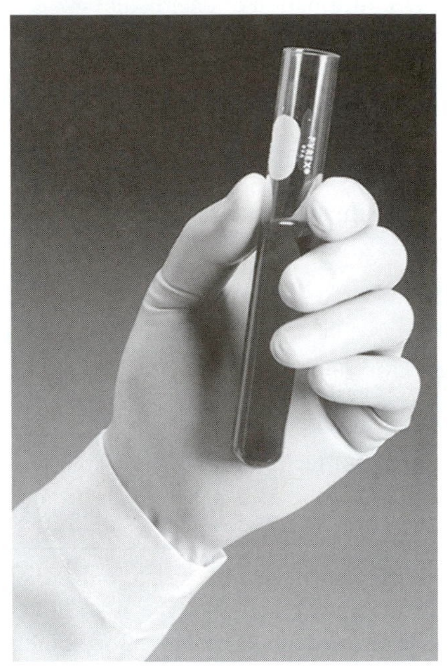

Test tubes are a chemist's companion.

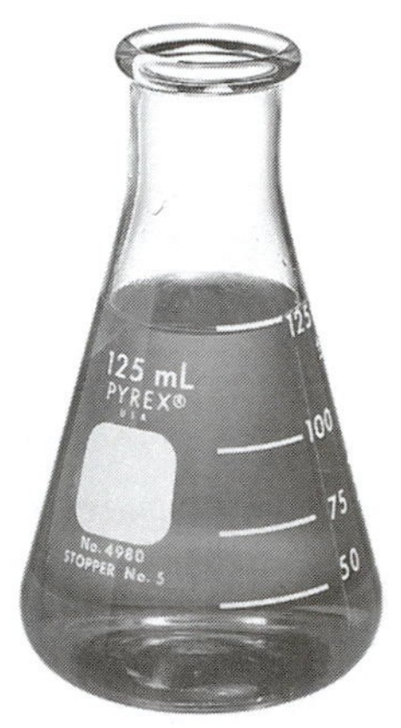

Erlenmeyer flasks are convenient for containing solutions.

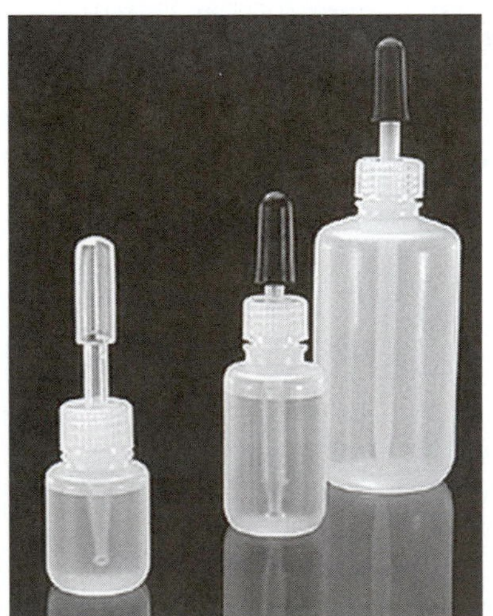

Dropping bottles assist in transferring small volumes of solutions.

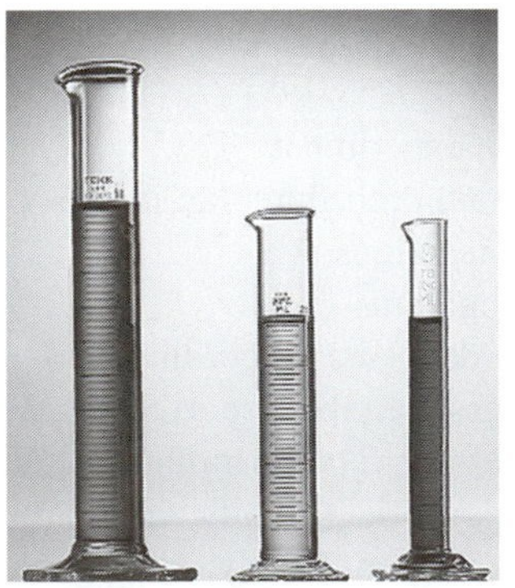

Graduated cylinders measure quantitative volumes of solutions.

A wash bottle containing deionized water must always be handy.

Common Laboratory Desk Equipment Check List

No.	Quantity	Size	Item	First Term In	First Term Out	Second Term In	Second Term Out	Third Term In	Third Term Out
1	1	10-mL	graduated cylinder						
2	1	50-mL	graduated cylinder						
3	5	—	beakers						
4	2	—	stirring rods						
5	1	500-mL	wash bottle						
6	1	75-mm, 60°	funnel						
7	1	125-mL	Erlenmeyer flask						
8	1	250-mL	Erlenmeyer flask						
9	2	25 × 200-mm	test tubes						
10	6	18 × 150-mm	test tubes						
11	8	10 × 75-mm	test tubes						
12	1	large	test tube rack						
13	1	small	test tube rack						
14	1	—	glass plate						
15	1	—	wire gauze						
16	1	—	crucible tongs						
17	1	—	spatula						
18	2	—	litmus, red and blue						
19	2	90-mm	watch glasses						
20	1	75-mm	evaporating dish						
21	4	—	dropping pipets						
22	1	—	test tube holder						
23	1	large	test tube brush						
24	1	small	test tube brush						
	1	—	marking pen						

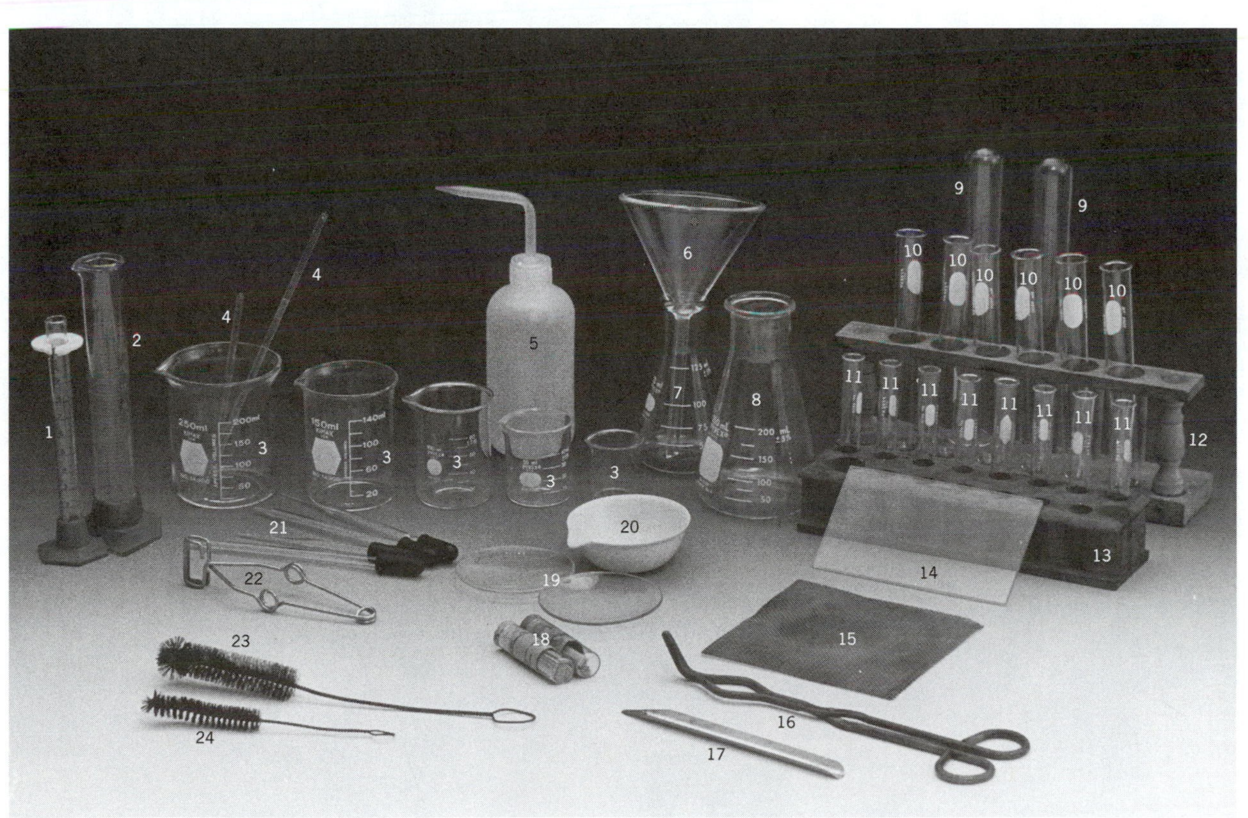

Special Laboratory Equipment

Number	Item	Number	Item
1	reagent bottles	16	porcelain crucible and cover
2	condenser	17	mortar and pestle
3	500-mL Erlenmeyer flask	18	glass bottle
4	1000-mL beaker	19	pipets
5	Petri dish	20	ring and buret stands
6	Büchner funnel	21	clamp
7	Büchner (filter) flask	22	double buret clamp
8	volumetric flasks	23	Bunsen burner
9	500-mL Florence flask	24	buret brush
10	−10°C–110°C thermometer	25	clay pipe-stem triangle
11	100-mL graduated cylinder	26	rubber stoppers
12	50-mL buret	27	wire loop for flame test
13	glass tubing	28	pneumatic trough
14	U-tube	29	rubber pipet bulb
15	porous ceramic cup	30	iron support ring

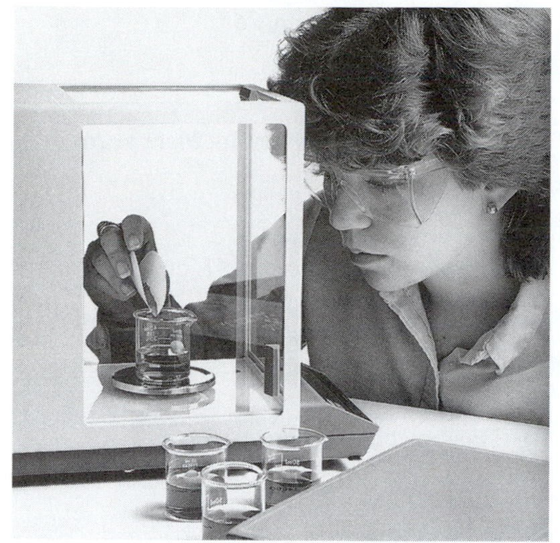

Laboratory Techniques

The Application of proper laboratory techniques improves data reliability.

Scientific data that are used to analyze the characteristics of a chemical or physical change must be collected with care and patience. The data must be accurate; that is, it must be reproducible to within an "acceptable" margin of error. Reproducible data implies that the data collected from an observed chemical or physical change can be again collected at a later date by the same scientist or another scientist in another laboratory.

A scientist who has good laboratory skills and techniques generally collects good, reproducible data (called **quantitative data**). For that reason, careful attention as to the method (or methods) and procedures by which the data are collected is extremely important. This section of the laboratory manual describes a number of techniques that you will need to develop for collecting quantitative data in the chemistry laboratory. You do not need to know the details for all of the techniques at this time (that will come with each successive experiment that you encounter), but you should be aware of their importance, features, and location in the laboratory manual. Become *very* familiar with this section of the laboratory manual! Consult with your laboratory instructor about the completion of the Laboratory Assignment at the end of this section.

In the Experimental Procedure of each experiment, icons are placed in the margin at a position where the corresponding laboratory technique is to be applied for the collection of "better" data. The following index of icons identifies the laboratory techniques and page numbers on which they appear:

Technique 1. Inserting Glass Tubing through a Rubber Stopper ***p. 13***

Technique 2. Cleaning Glassware ***p. 13***

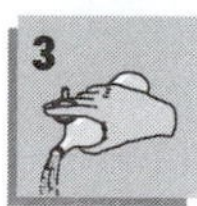

Technique 3. Handling Chemicals ***p. 14***

Technique 4. Disposing of Chemicals ***p. 15***

Technique 5. Preparing Solutions ***p. 15***

Technique 6. Measuring Mass ***p. 16***

Technique 7. Handling Small Volumes ***p. 17***

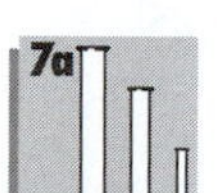

A. Test Tubes for Small Volumes ***p. 17***

B. Well Plates for Small Volumes ***p. 17***

Technique 1. Inserting Glass Tubing Through a Rubber Stopper

Caution: *Perhaps more accidents occur in the general chemistry laboratory as a result of neglect in this simple operation than all other accidents combined. Please review and practice this technique correctly when working with glass tubing. Serious injury can occur to the hand if this technique is performed incorrectly.*

Moisten the glass tubing and the hole in the rubber stopper with glycerol or water (**note:** glycerol works best). Place your hand on the tubing 2–3 cm (1 in.) from the stopper. Protect your hand with a cloth towel (Figure T.1). Simultaneously *twist* and *push* the tubing slowly and carefully through the hole. Wash off any excess glycerol on the glass or stopper with water and dry.

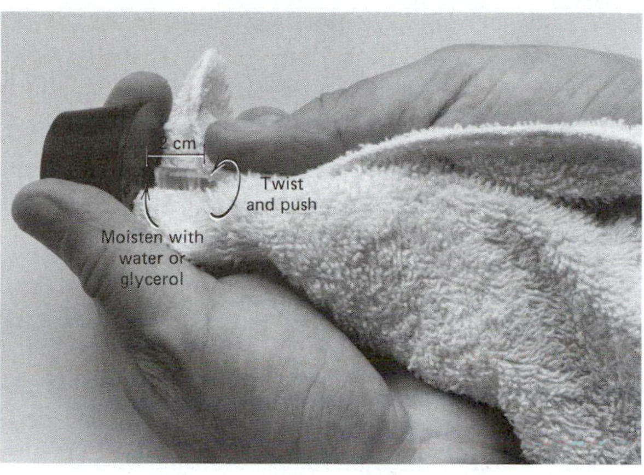

Figure T.1 Inserting glass tubing through a rubber stopper.

Technique 2. Cleaning Glassware

It's like eating from a dirty plate! A chemist is very concerned about contaminants causing errors in experimental data. Cleanliness is extremely important in minimizing errors in the precision and accuracy of data.

Clean all glassware with a soap or detergent solution using *tap* water. Use a laboratory sponge or a test tube, pipet, or buret brush as appropriate. Once the glassware is thoroughly cleaned, first rinse several times with tap water and then once or twice with small amounts of deionized water. Roll each rinse around the entire inner surface of the glass wall for a complete rinse. Discard each rinse through the delivery point of the vessel (i.e., buret tip, pipet tip, beaker spout). Deionized water should never be used for washing glassware; it is too expensive.

Invert the clean glassware on a paper towel or rubber mat to dry (Figure T.2a); do *not* wipe or blow-dry because of possible contamination. Do *not* dry heavy glassware (graduated cylinders, volumetric flasks, or bottles), or for that matter any glassware, over a direct flame.

The glassware is clean if, following the final rinse, no water droplets adhere to the clean part of the glassware (Figure T.2b).

A laboratory detergent.

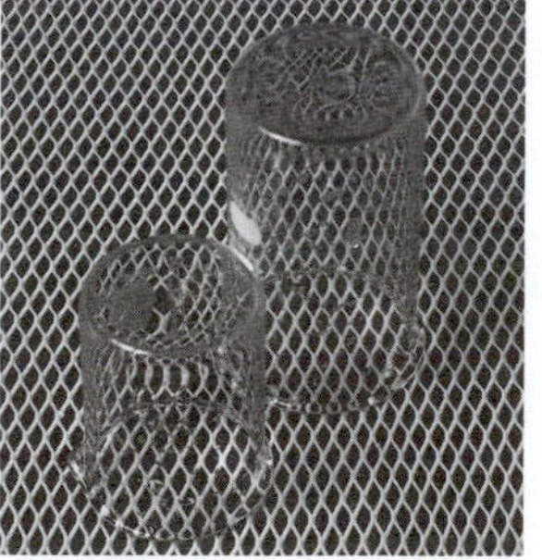

Figure T.2a Invert clean glassware on a paper towel or rubber mat to air-dry.

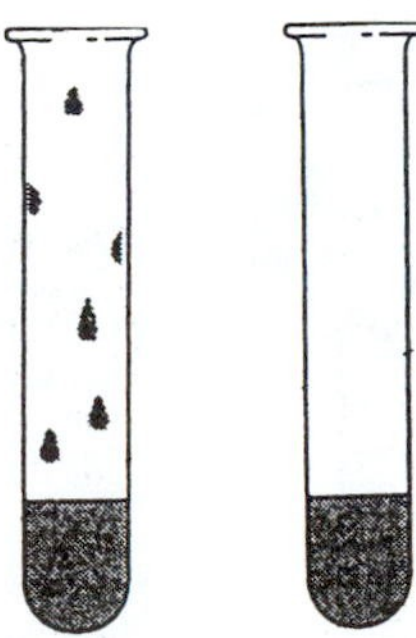

Figure T.2b Water droplets (left) do *not* adhere to the wall of clean glassware (right).

TECHNIQUE 3. HANDLING CHEMICALS

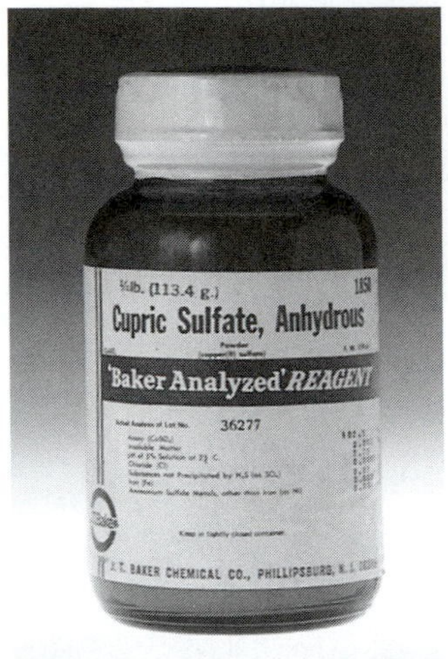

Figure T.3a Chemicals are labeled with systematic names.

Handling chemicals is as safe as handling your dog; if it is done carefully, no injuries will occur. Most techniques for handling chemicals are also regarded as safety guidelines.

- Read the label on a reagent bottle at least *twice* before removing any chemicals (Figure T.3a). The wrong chemical may lead to serious accidents or "unexplainable" results in your experiments (see Dry Labs 2 for an understanding of the rules of chemical nomenclature). Techniques 9 and 10 illustrate the correct procedures for transferring solids and liquid reagents.
- Avoid using excessive amounts of reagents. *Never* dispense more than the experiment calls for. *Do not return excess chemicals to the reagent bottle!*
- *Never* touch, taste, or smell chemicals unless specifically directed to do so. Skin, nasal, and/or eye irritations may result. If inadvertent contact with a chemical does occur, wash the affected area immediately with copious amounts of water and inform your laboratory instructor (see Laboratory Safety B.4, 5).

Chemicals are often labeled according to National Fire Protection Association (NFPA) standards which describe the four possible hazards of a chemical and a numerical rating from 0 to 4. The four hazards are health hazard (blue), fire hazard (red), reactivity (yellow), and specific hazard (white). A photograph of a label is shown in Figure T.3b.

If you wish to know more about the properties and hazards of the chemicals with which you will be working in the laboratory, safety information about the reagents is available in a bound collection of the Material Safety Data Sheets (MSDS). The MSDS collection is also accessible at various sites on the World Wide Web (see Laboratory Data, Part C).

In this manual, the international caution sign (shown at left) is used to identify a potential danger in the handling of a solid chemical or reagent solution or hazardous equipment.

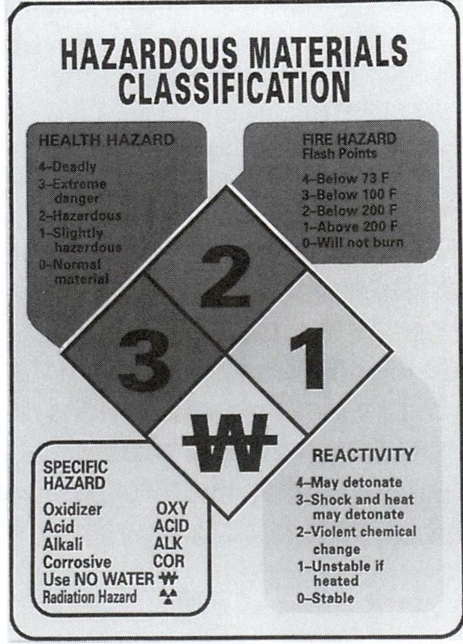

Figure T.3b Hazardous materials classification system.

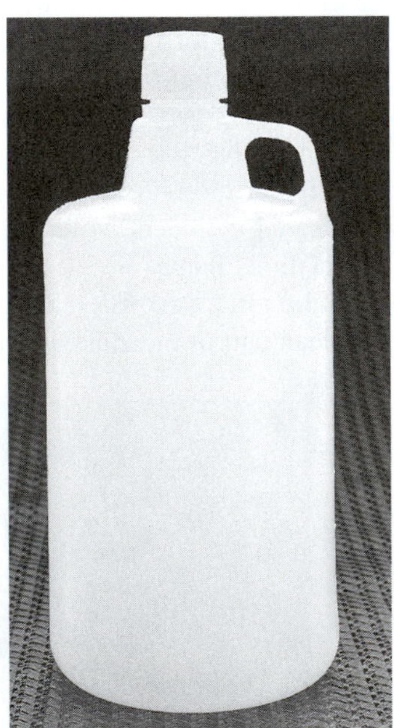

Figure T.4 Waste disposal containers are available in the laboratory.

TECHNIQUE 4. DISPOSING OF CHEMICALS

Disposal of waste chemicals has become an increasingly important concern in recent years. Although most of the chemicals in this manual are considered "safe," any carelessness or abuse in their handling can be dangerous.

You must properly dispose of each test chemical. Assume that *nothing* (besides soap and water!) is to be discarded in the sink. Discard waste chemicals as directed in the Experimental Procedure of each experiment or by the laboratory instructor. There are no "general" waste disposal containers; each is labeled for a specific waste (Figure T.4). Instructions for the disposal of the various test chemicals are listed in the Experimental Procedure of each experiment.

Know the proper procedure for disposal before carelessly discarding your test solution. Double-check all labels before discarding. Improper waste disposal is foolish and potentially disastrous to you, your friends, your apparatus, and the laboratory!

A disposal icon is used throughout the manual to bring attention to the proper disposal of chemicals.

The final disposal of chemicals is the responsibility of the stockroom personnel. Information for the proper disposal of chemicals is available from the MSDS collection or at various sites on the World Wide Web (see Laboratory Data, Part C).

TECHNIQUE 5. PREPARING SOLUTIONS

The preparation of an aqueous solution is often required for an experimental procedure. The preparation begins with either a solid reagent or a solution more concentrated than the one needed for the experiment. At either starting point, the minimum amount (the number of *moles*) of compound required for the experiment is calculated: from a solid, the mass and the molar mass of the compound are needed to calculate the number of moles of compound required for the preparation of the solution; from a more concentrated solution, the concentration and volume (or mass) of the concentrated solution must be known in order to calculate the number of moles of compound needed for the preparation of the aqueous solution. In both cases, the calculated (and then also measured) moles of compound are diluted to final volume. Knowledge of moles and mole calculations is absolutely necessary.

In the laboratory preparation, *never* insert a pipet, spatula, or dropping pipet into the reagent used for the solution preparation. Always transfer the calculated amount from the reagent bottle as described in Techniques 9 and 10.

Solutions are commonly prepared in volumetric flasks (Figure T.5) according to the following procedure:

- Place water (or the less concentrated solution) into the volumetric flask until it is one-third to one-half full.

(a)

(b)

(c)

(d)

Figure T.5 Place water (or the less concentrated solution) into the flask before slowly adding the solid or more concentrated solution. Dilute the solution to the "mark" with water; stopper and invert the flask 10–15 times.

- Add the solid (or add the more concentrated reagent) *slowly, while swirling,* to the volumetric flask. (**Caution:** *Never dump it in!*)
- Once the solid compound has dissolved or the more concentrated solution has been diluted, add water (dropwise if necessary) until the calibrated "mark" etched on the volumetric flask is reached (see Technique 16A for reading the meniscus). While securely holding the stopper, invert the flask slowly 10–15 times to ensure that the solution is homogeneous.

TECHNIQUE 6. MEASURING MASS

The mass measurement of a sample can be completed in two ways. In the traditional method, the mass of weighing paper or a clean, dry container (such as a beaker, watchglass, or weighing boat) is first measured and recorded. The sample is then placed on the weighing paper or in the container and this combined mass is measured. The mass of the weighing paper or container is then subtracted from the combined mass to record the mass of the sample.

On modern electronic balances, the mass of the weighing paper or container can be tared out—that is, the balance can be zeroed again *after* placing the weighing paper or container on the balance, in effect subtracting its mass immediately (and automatically). The sample is then placed on the weighing paper or in the container, and the balance reading *is* the mass of the sample.

For either method the resultant mass of the sample is the same and is called the **tared mass** of the sample.

Tared mass: mass of sample without regard to its container

The laboratory balance is perhaps the most used *and abused* piece of equipment in the chemistry laboratory. Therefore, because of its extensive use, you and others must follow several guidelines to maintain the longevity and accuracy of the balance:

- Handle with care; balances are expensive.
- If the balance is not leveled, see your laboratory instructor.
- Use weighing paper, a watchglass, a beaker, or some other container to measure the mass of chemicals; do *not* place chemicals directly on the balance pan.
- Do *not* drop anything on the balance pan.
- If the balance is not operating correctly, see your laboratory instructor. Do *not* attempt to fix it yourself.
- After completing a mass measurement, return the mass settings to the zero position.
- Clean the balance and balance area of any spilled chemicals.

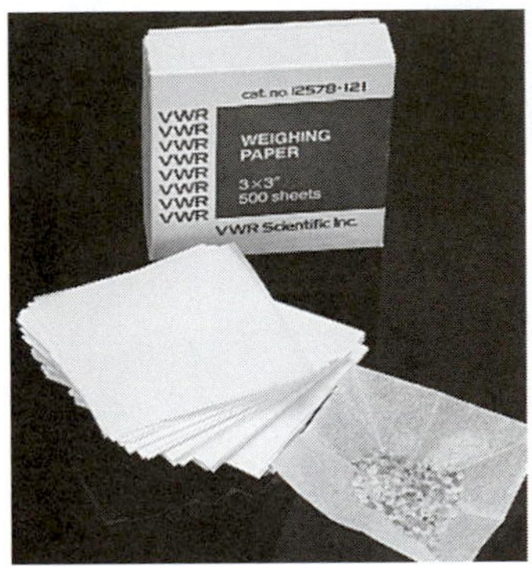

Weighing paper.

Different balances, having varying degrees of sensitivity, are available for use in the laboratory. These are shown in Figures T.6a through T.6d.

Balance	Sensitivity (g)
Triple-beam (Figure T.6a)	±0.01
Top-loading (Figure T.6b)	±0.01 or ±0.001
Top-loading (Figure T.6c)	±0.0001
Analytical (Figure T.6d)	±0.00001

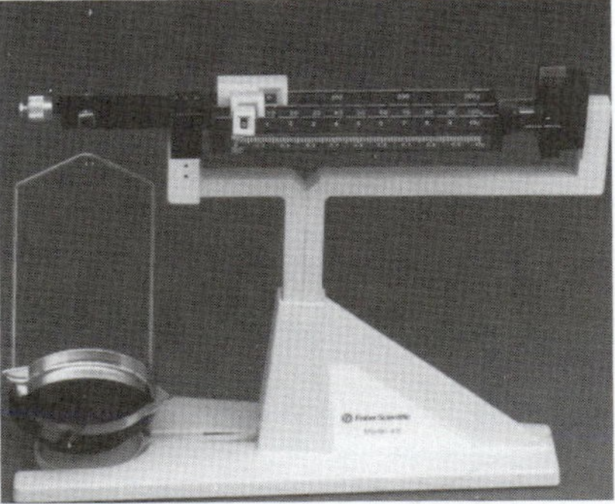

Figure T.6a Triple-beam balance, sensitivity of ±0.01 g.

Figure T.6b Electronic top-loading balance, sensitivity of ±0.01 g and/or ±0.001 g.

Figure T.6c Electronic analytical balance, sensitivity of ±0.0001 g.

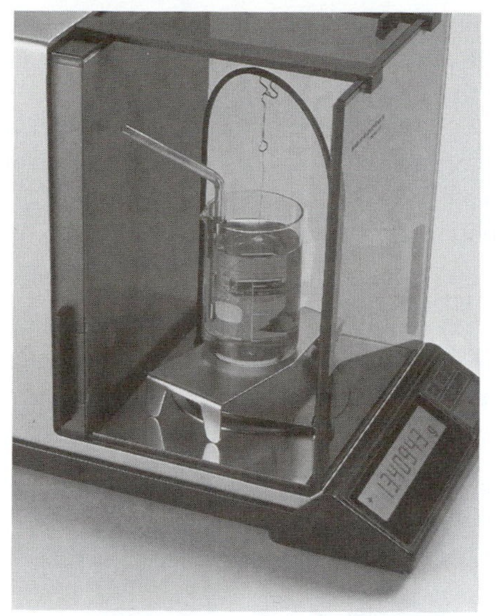

Figure T.6d Electronic analytical balance, sensitivity of ±0.00001 g.

Technique 7. Handling Small Volumes

The use of smaller quantities of chemicals for synthesis and testing in the laboratory offers many safety advantages and presents fewer chemical disposal problems. Many of the experimental procedures in this manual were designed with this in mind. Handling small volumes requires special apparatus and technique.

A. Test Tubes for Small Volumes

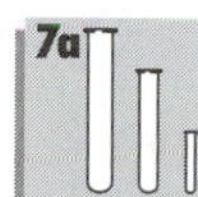

Small test tubes are the chemist's choice for handling small volumes. Common laboratory test tubes are generally of three sizes: the 75-mm (or 3-inch) test tube, the 150-mm (or 6-inch) test tube, and the 200-mm (or 8-inch) test tube (Figure T.7a). The approximate volumes of the three test tubes are as follows:

75-mm (3-inch) test tube	~3 mL
150-mm (6-inch) test tube	~25 mL
200-mm (8-inch) test tube	~75 mL

The 75-mm test tube is often recommended for "small volume" experiments.

B. Well Plates for Small Volumes

Alternatively, a "well plate" can be used for a number/series of reaction vessels (Figure T.7b). The well plate is especially suited for experiments that require observations

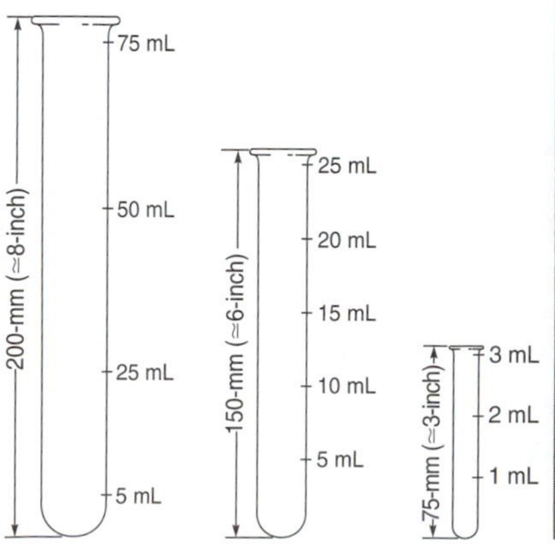

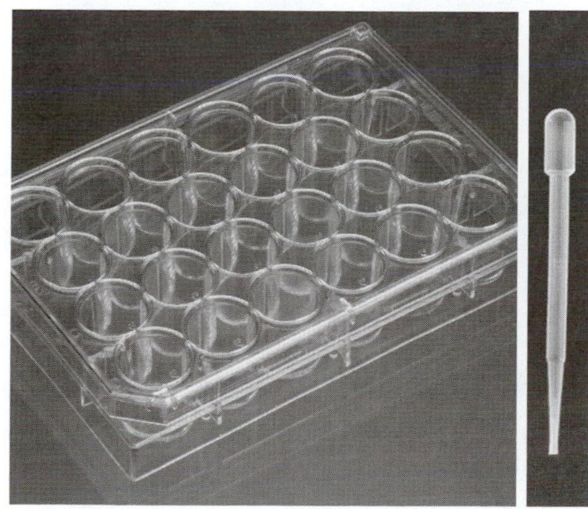

Figure T.7a The three common-sized test tubes for containing reagent solutions.
Figure T.7b A 24-well plate and Beral pipet are used for containing and transferring small quantities of reagent solutions.

from repeated or comparative reactions. The well plate most often recommended is the 24-well plate in which each well has an approximate volume of 3.5 mL (compared to a 3 mL for a small test tube).

For either technique the Beral pipet, a plastic, disposable pipet, or a dropping pipet is often used to transfer small volumes of solutions to and from the test tubes or well plate. The Beral pipet has a capacity of about 2 mL, and some have the volume graduation marks on the stem.

Technique 8. Collecting Gases

The solubility and density of a gas determine the apparatus used for its collection. Water-soluble gases should not be collected over water, but rather by air displacement.

A. Water-Soluble Gases, Air Displacement

Water-soluble gases *more* dense than air are collected by air displacement (Figure T.8a). The more dense gas pushes the less dense air up and out of the gas bottle. Water-soluble gases *less* dense than air are also collected by air displacement (Figure

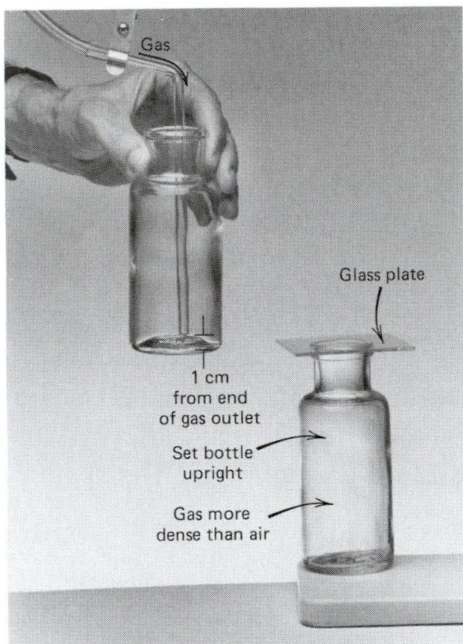

Figure T.8a Collection of water-soluble gases more dense than air.

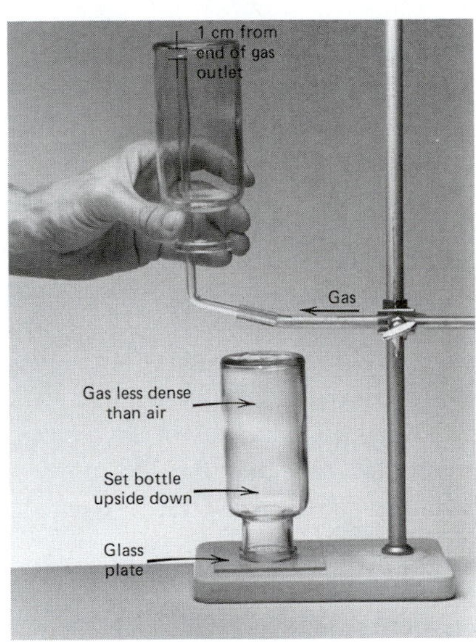

Figure T.8b Collection of water-soluble gases less dense than air.

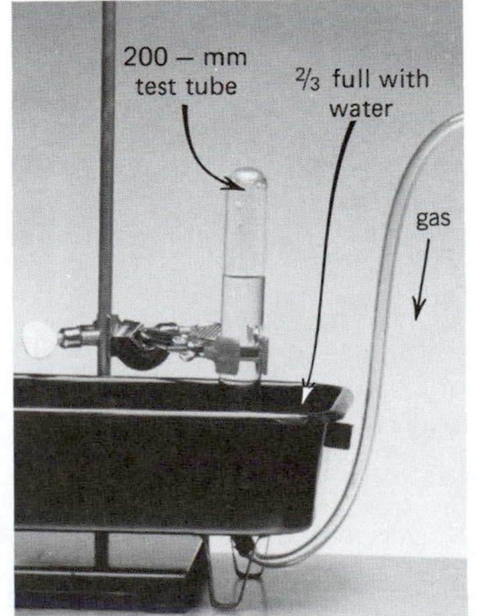

Figure T.8c Collection of water-insoluble gas by the displacement of water.

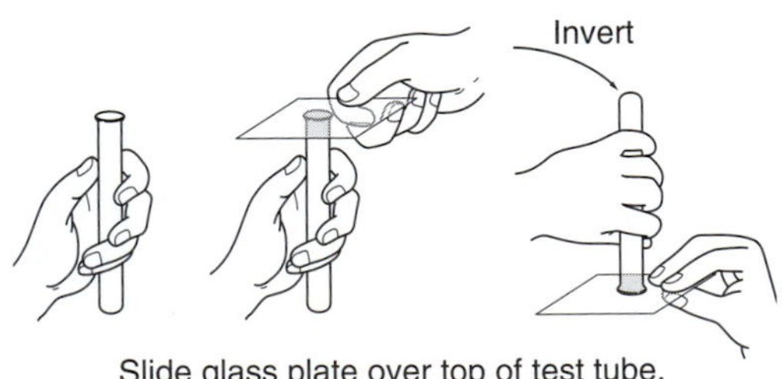

Figure T.8d Inverting a water-filled test tube.

T.8b) except that, in this case, the less dense gas pushes the more dense air down and out of the gas bottle. Note that the gas outlet tube should extend to within 1 cm of the bottom (or top) of the gas bottle.

B. Water-Insoluble Gases, Water Displacement

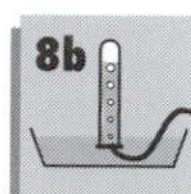

Gases that are relatively insoluble in water are collected by water displacement. The gas pushes the water down and out of the water-filled gas-collecting vessel (Figure T.8c). The gas-collecting vessel (generally a flask or test tube) is first filled with water, covered with a glass plate or plastic wrap (no air bubbles must enter the vessel, Figure T.8d), and then inverted into a deep pan or tray half-filled with water. The glass plate or plastic wrap is removed, and the tubing from the gas generator is inserted into the mouth of the gas-collecting vessel.

Technique 9. Transferring Solids

Before transferring a solid chemical for use in the preparation of a solution or for study of its chemical properties, read the label on the bottle *twice* to be sure it is the correct chemical. For example, is the chemical iron(II) acetate or iron(III) acetate? Is it the anhydrous, trihydrate, or pentahydrate form of copper(II) sulfate?

If the reagent bottle has a hollow glass stopper or if it has a screw cap, place the stopper (or cap) top side down on the bench (Figure T.9a). To dispense solid from the bottle, hold the label against your hand, tilt, and roll the solid reagent back and forth.

- For *larger quantities* of solid reagent, dispense the solid into a beaker (Figure T.9b) until the estimated amount has been transferred. Try not to dispense any more reagent than is necessary for the experiment. Do not return any excess reagent to the reagent bottle—share the excess with another chemist.
- For *smaller quantities* of solid reagent, first dispense the solid into the inverted hollow glass stopper or screw cap. And then transfer the estimated amount of

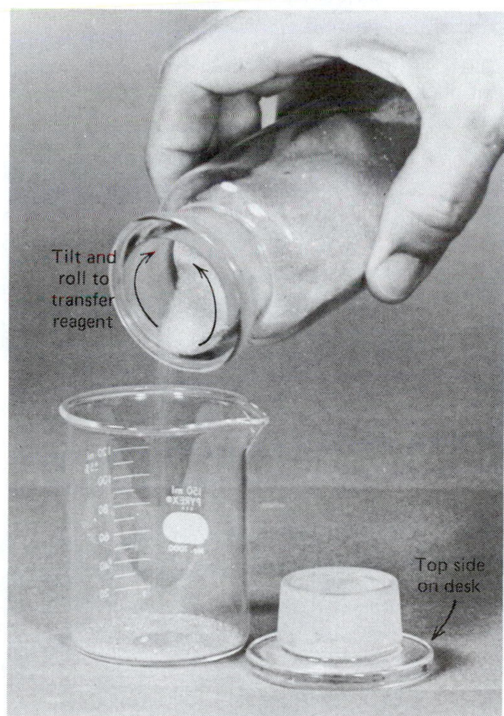

Figure T.9a Transferring a solid chemical from a glass ground reagent bottle. Place the glass stopper top side down.

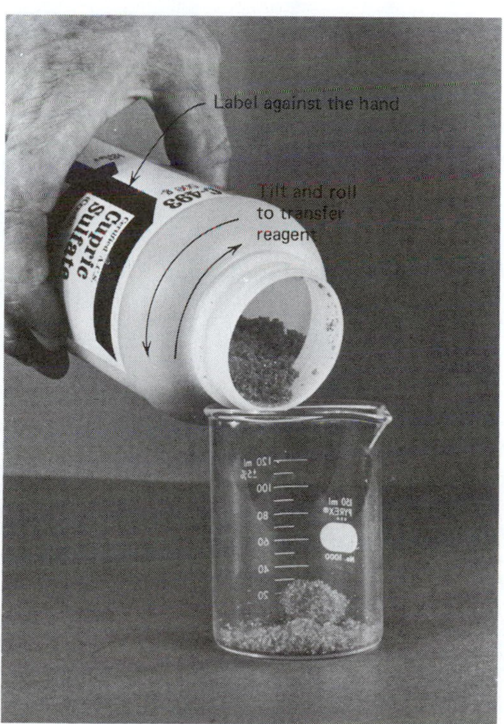

Figure T.9b Tilt and roll the reagent bottle back and forth until the desired amount of solid chemical has been dispensed.

reagent needed for the experiment from the stopper/screw cap to an appropriate vessel. Return the excess reagent in the glass stopper or screw cap to the reagent bottle—in effect, the solid reagent has never left the reagent bottle.

For either situation, *never* use a spatula or any other object to break up or transfer the reagent to the appropriate container unless your laboratory instructor specifically instructs you do to so.

When you have finished dispensing the solid chemical, *recap* the reagent bottle.

Technique 10. Transferring Liquids and Solutions

When a liquid or solution is to be transferred from a reagent bottle, remove the glass stopper and hold it between the fingers of the hand used to grasp the reagent bottle (Figures T.10a, b). Never lay the glass stopper on the laboratory bench; impurities may be picked up and thus contaminate the liquid when the stopper is returned.

To transfer a liquid from one vessel to another, hold a stirring rod against the lip of the vessel containing the liquid and pour the liquid down the stirring rod, which, in

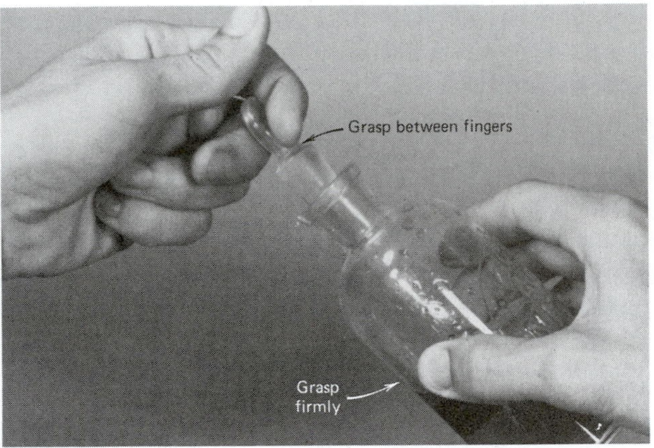

Figure T.10a Remove the glass stopper and hold it between the fingers of the hand that grasps the reagent bottle.

Figure T.10b Transfer the liquid from the reagent bottle with the aid of the stirring rod.

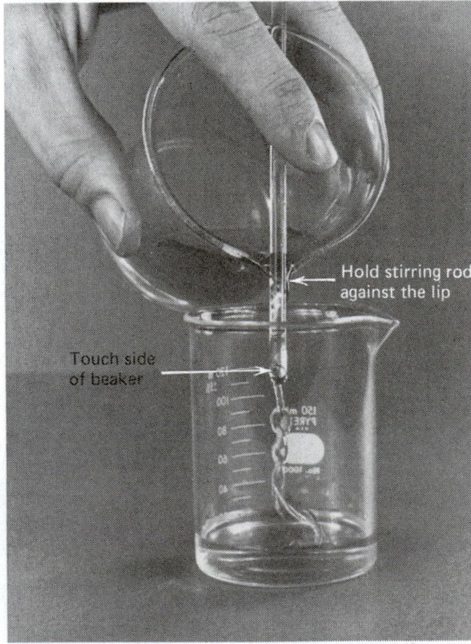

Figure T.10c The stirring rod should touch the lip of the transfer vessel and the inner wall of the receiving vessel.

turn, should touch the inner wall of the receiving vessel (Figures T.10b, c). Return the glass stopper to the reagent bottle.

Do *not* transfer more liquid than is needed for the experiment; do *not* return any excess or unused liquid to the original reagent bottle.

Technique 11. Separating a Liquid or Solution from a Solid

A. Decanting a Liquid or Solution from a Solid

A liquid can be decanted (poured off the top) from a solid if the solid clearly separates from the liquid in a reasonably short period of time. Allow the solid to settle to the bottom of the vessel (Figure T.11a). Transfer the liquid (called the **supernatant**) with the aid of a clean stirring rod (Figure T.11b). Do this slowly so as not to disturb the solid. Review Technique 10 for the transfer of a liquid from one vessel to another.

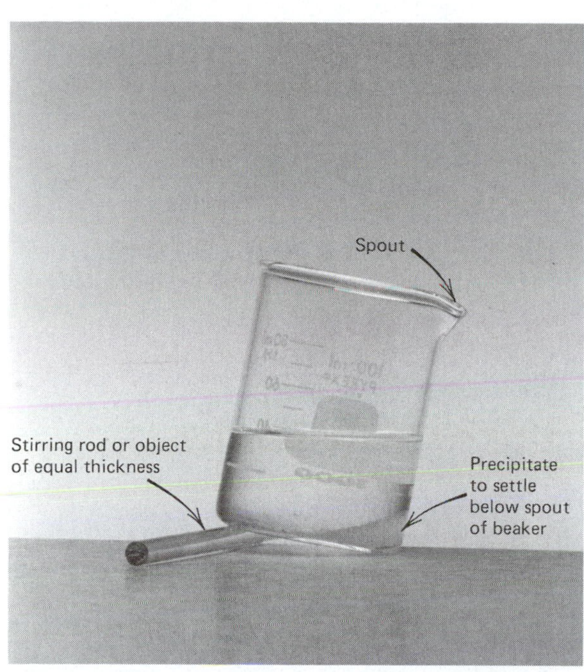

Figure T.11a Tilt the beaker to allow the precipitate to settle at the side. Use a stirring rod or a similar object to tilt the beaker.

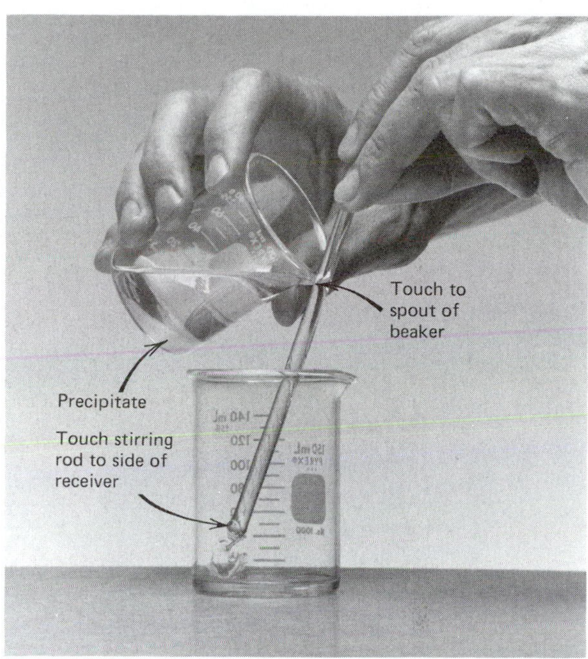

Figure T.11b Transfer the supernatant to a receiving vessel with the aid of a stirring rod.

B. Preparing Filter Paper for a Filter Funnel

If a solid is to be separated from the liquid using a filtering process, then the filter paper must be properly prepared. For a gravity filtration procedure, first fold the filter paper in half (Figure T.11c), again fold the filter paper to within about 10° of a 90° fold, tear off the corner of the outer fold unequally, and open. The tear enables a close seal to be made across the paper's folded portion when placed in a funnel.

Place the folded filter paper snugly into the funnel. Moisten the filter paper with the solvent of the liquid/solid mixture being filtered (most likely this will be deionized water) and press the filter paper against the top wall of the funnel to form a seal. Support the funnel with a clamp or in a funnel rack.

C. Gravity Filtration

Filtrate: the solution that passes through the filter in a filtration procedure

Transfer the liquid as described in Technique 10 (Figure T.11d). The tip of the funnel should touch the wall of the receiving beaker to reduce any splashing of the **filtrate.** Fill the bowl of the funnel until it is less than two-thirds full with the mixture. Always keep the funnel stem full with the filtrate; the weight of the filtrate in the funnel stem creates a slight suction on the filter in the funnel, and this hastens the filtration process.

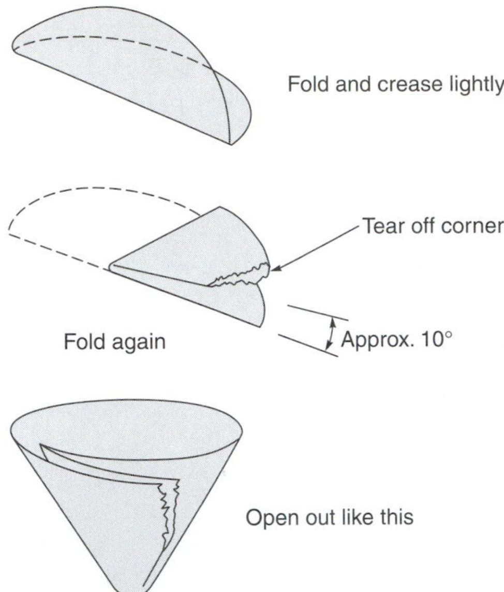

Figure T.11c The sequence of folding filter paper for a filter funnel in a gravity filtration procedure.

Figure T.11d The tip of the funnel should touch the wall of the receiving flask, and the bowl of the funnel should be one-half to two-thirds full.

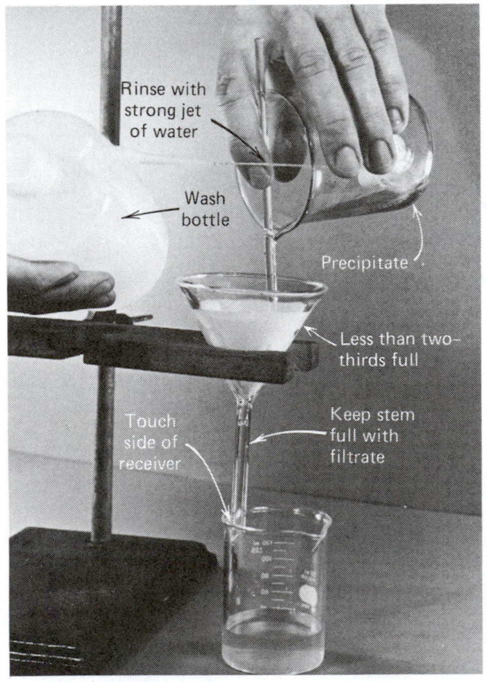

Figure T.11e Flushing the precipitate from a beaker with the aid of a "wash" bottle.

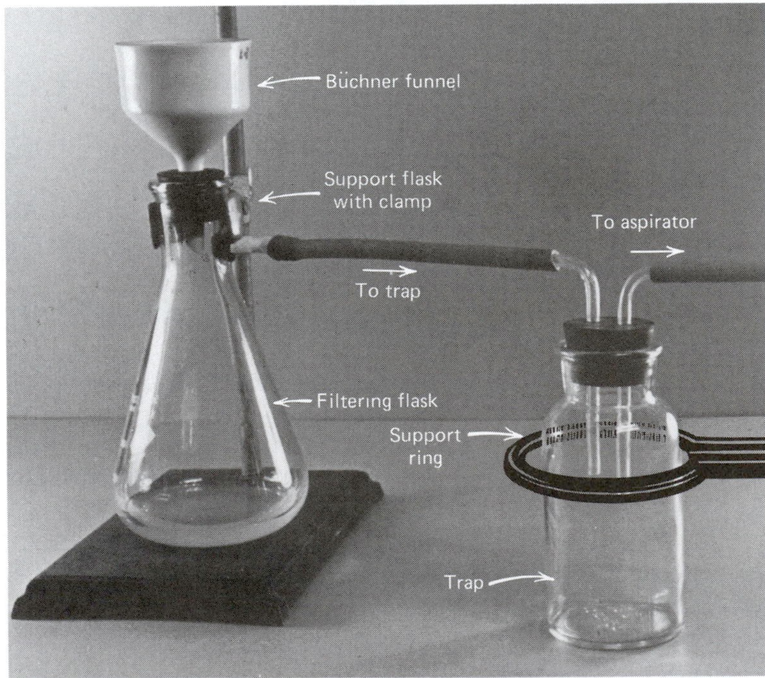

Figure T.11f The aspirator should be fully open during the vacuum filtering operation.

D. Flushing a Precipitate from the Beaker

Flush the precipitate from a beaker using a wash bottle containing the mixture's solvent (usually deionized water). Hold the beaker over the funnel or receiving vessel (Figure T.11e) at an angle such that the solvent will flow out and down the stirring rod.

E. Vacuum Filtration

Set up the vacuum filtration apparatus as shown in Figure T.11f. Although a regular filter funnel can be used, the apex of the filter paper is easily ruptured when a vacuum is applied. A Büchner funnel (a disk of filter paper fits over the flat, perforated bottom of the funnel) set into a filter flask connected to a water aspirator is the apparatus normally used for vacuum filtration. Seal the disk of filter paper onto the bottom of the funnel by applying a light suction to the filter paper while adding a small amount of solvent.

Once the filter paper is sealed, turn the water faucet attached to the aspirator completely open to create a full suction. Transfer the mixture to the filter (Technique 10) and wash the precipitate with an appropriate liquid. To remove the suction, *first* disconnect the hose from the filter flask, and then turn off the water.

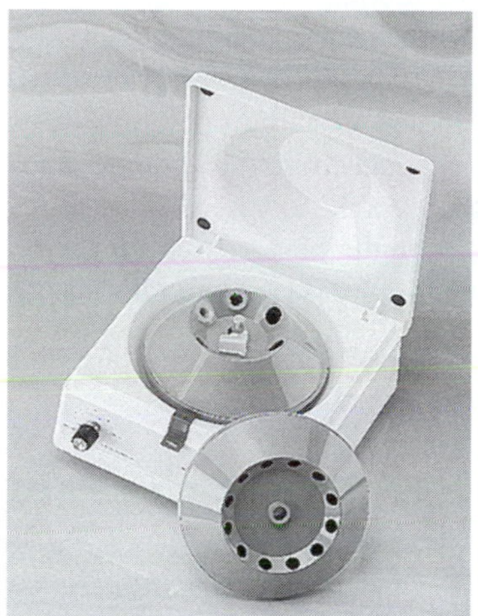

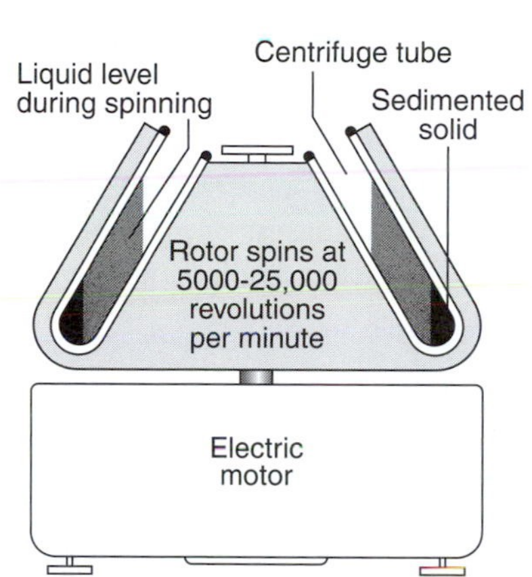

Figure T.11g A laboratory centrifuge forces the precipitate to the bottom of the centrifuge tube.

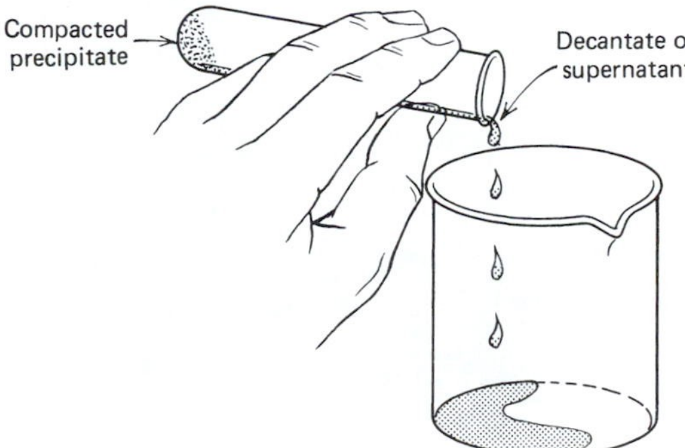

Figure T.11h Decant the supernatant from the compacted precipitate.

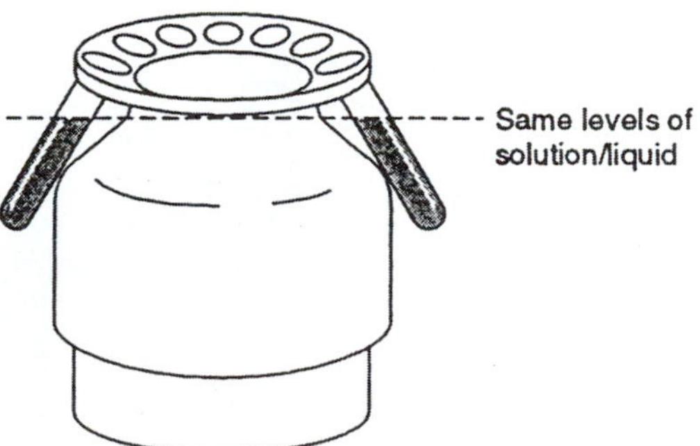

Figure T.11i Balance the centrifuge by placing tubes with equal volumes of liquid opposite each other inside the metal sleeves of the rotor.

F. Centrifugation

Supernatant: the clear liquid covering a precipitate

A centrifuge (Figure T.11g) spins at velocities of 5000 to 25,000 revolutions per minute! A solid/liquid mixture in a small test tube or centrifuge tube is placed into a sleeve of the rotor of the centrifuge. By centrifugal force the solid is forced to the bottom of the test tube or centrifuge tube and compacted. The clear liquid, called the **supernatant,** is then easily decanted without any loss of solid (Figure T.11h). This quick separation of liquid from solid requires 20–40 seconds.

Observe these precautions in operating a centrifuge:

- Never fill the centrifuge tubes to a height more than 1 cm from the top.
- Label the centrifuge tubes to avoid confusion of samples.
- *Always* operate the centrifuge with an *even* number of centrifuge tubes containing equal volumes of liquid placed opposite one another in the centrifuge. This *balances* the centrifuge and eliminates excessive vibration and wear. If only one tube needs to be centrifuged, balance the centrifuge with a tube containing the same volume of solvent (Figure T.11i).
- *Never* attempt to manually stop a centrifuge. When the centrifuge is turned off, let the rotor come to rest on its own.

Technique 12. Venting Gases

Removing "undesirable" gases from a chemical reaction should be accomplished in a fume hood (Figure T.12a). Locate the fume hood in your laboratory.

On occasion the space in the fume hoods is not adequate for an entire class to perform the experiment in a timely manner. With the *approval of your laboratory instructor,* an improvised hood (Figure T.12b) can be assembled. For the operation of an improvised hood, a water aspirator draws the gaseous product from above the reaction vessel—the gas dissolves in the water. To operate the "hood," completely open the faucet that is connected to the aspirator in order to provide the best suction for the removal of the gases. But, as a reminder, *never* substitute an improvised hood for a fume hood if space is available in the fume hood.

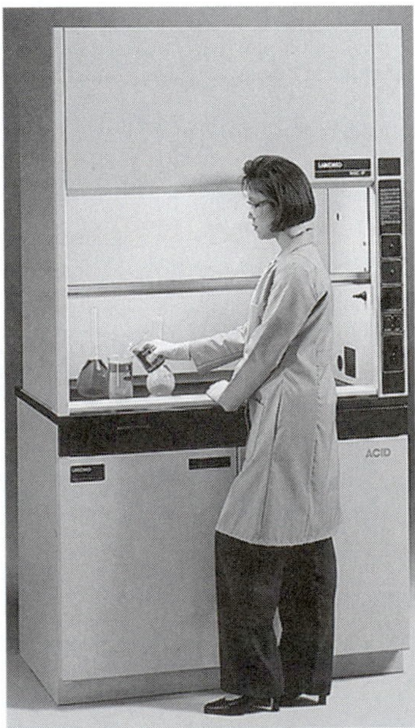

Figure T.12a A modern laboratory fume hood.

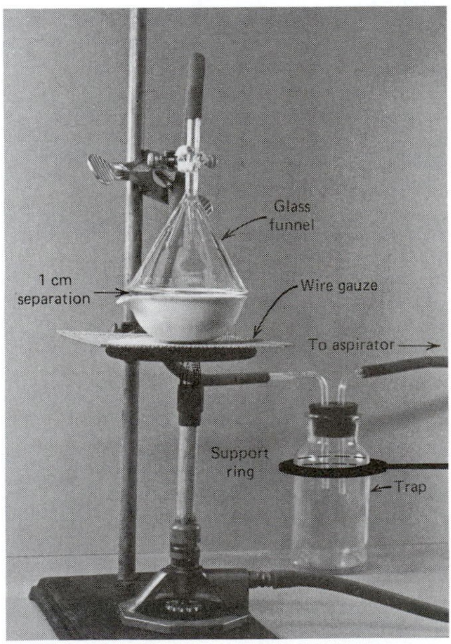

Figure T.12b Position a funnel, connected to a water aspirator, over the escaping gases.

TECHNIQUE 13. HEATING LIQUIDS AND SOLUTIONS

Liquids and solutions are often heated, for example, to promote the rate of a chemical reaction to or hasten a dissolution or precipitation, in a number of different vessels. **Caution:** *Flammable liquids should **never** be heated (directly or indirectly) with a flame. Always use a hot plate—refer to Techniques 13C and 13D where hot plates are used.*

Hot liquids and solutions can be cooled by placing the glass vessel either under flowing tap water or in an ice bath.

A. Test Tube Over a "Cool" Flame

A **cool flame** is a nonluminous flame supplied with a reduced supply of fuel. In practice, the rule of thumb for creating a cool flame for heating a liquid in a test tube is as follows: *if you can feel the heat of the flame with the hand that is holding the test tube clamp, **the flame is too hot!***

For heating a liquid in a test tube, the test tube should be less than one-third full of liquid. Hold the test tube with a test tube holder at an angle of about 45° with the flame. Move the test tube circularly and continuously in and out of the cool flame, heating from top to bottom, mostly near the top of the liquid (Figure T.13a). **Caution:** *Never fix the position of the flame at the base of the test tube, and never point the test tube at anyone; the contents may be ejected violently if the test tube is not heated properly.*

See Technique 13D for heating a solution in a test tube to a specified elevated temperature; the hot water bath in Technique 13D is a safer, but slower, procedure.

B. Erlenmeyer Flask Over a "Cool" Flame

An Erlenmeyer flask less than one-fourth full of liquid may be heated *directly* over a cool flame (see also Technique 13C). Hold the flask with a piece of tightly folded paper or flask tongs (*not* crucible tongs) and gently and continuously swirl the flask in and out of the flame (Figure T.13b). Set the flask on a wire gauze for cooling; do *not* place the hot flask directly on the laboratory bench.

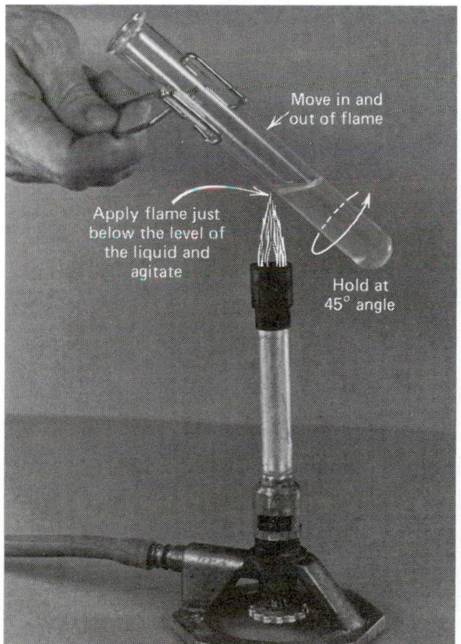

Figure T.13a Move the test tube circularly in and out of the *cool* flame, heating the liquid or solution from top to bottom.

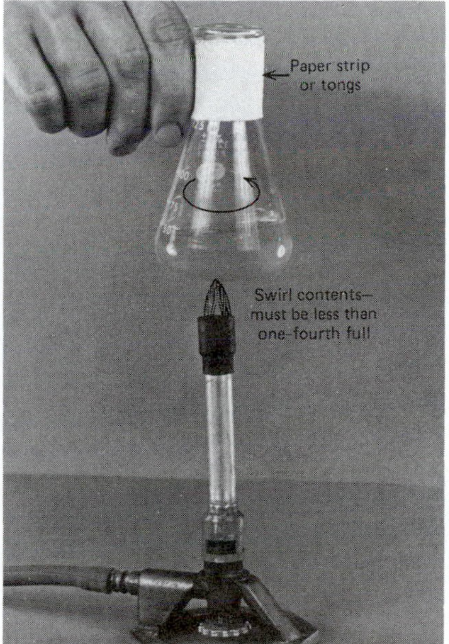

Figure T.13b Swirl the flask, less than one-fourth full, in and out of the cool flame.

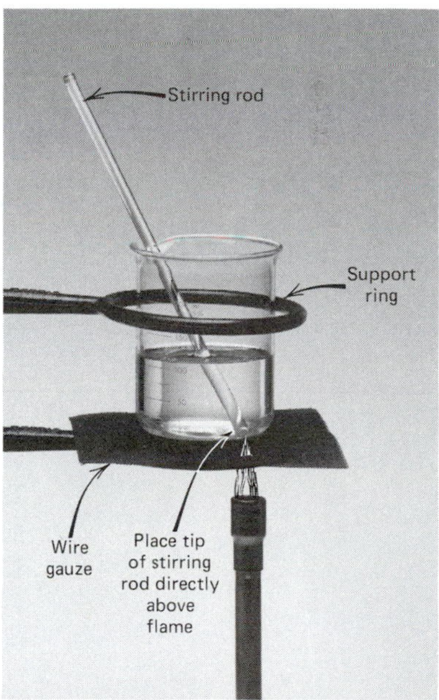

Figure T.13c Place the flame directly beneath the tip of the stirring rod in the beaker. Boiling chips may also be placed in the beaker to avoid "bumping."

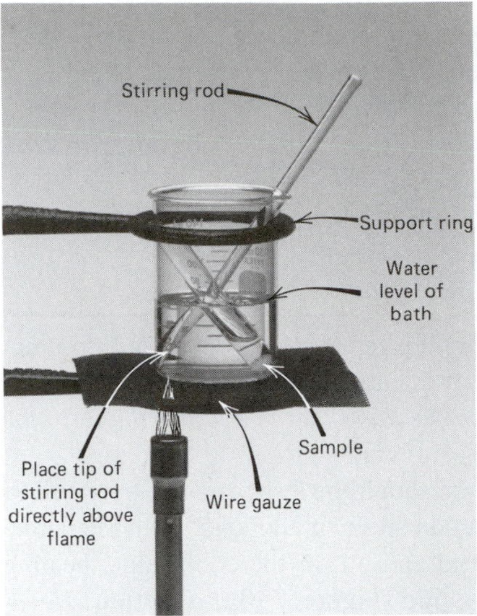

Figure T.13d A hot water bath may be used to maintain solution in test tubes at a constant, elevated temperature for an extended time period.

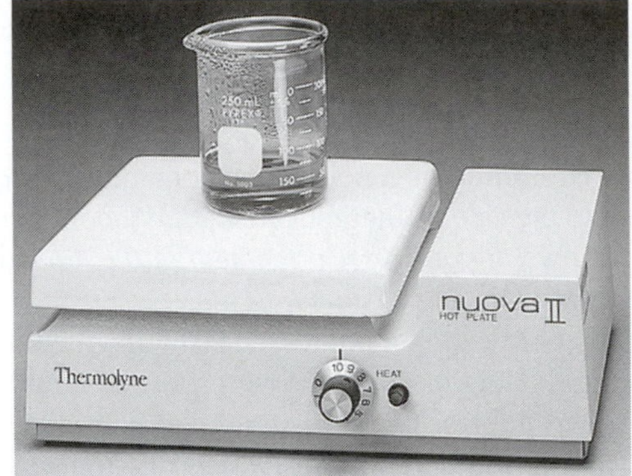

Figure T.13e A hot plate may be used to maintain solutions in a beaker or flask at a constant, elevated temperature for an extended time period.

C. Beaker (or Flask)

Nonflammable liquids in beakers or flasks that are more than one-fourth full can be heated directly using either a direct flame or a hot plate, provided a safe apparatus is assembled. If a flame is to be used, support the beaker (or flask) on a wire gauze that is centered over an iron ring; for a hot plate, merely center the beaker (or flask) on the surface.

Place a support (iron) ring around the top of the beaker (or flask) to prevent it from being accidentally knocked off. To avoid the problem of bumping (the sudden formation of bubbles from the superheated liquid), place a glass stirring rod and/or several **boiling chips** in the beaker. For heating with a flame, position the flame directly beneath the tip of the stirring rod (Figure T.13c). For either method of heating, heat slowly—occasional agitation with a stirring rod allows for more uniform heating of the liquid and minimizes bumping.

Boiling chips.

Boiling chips (also called ***boiling stones****): small, porous ceramic pieces—when heated, the air contained within the porous structure is released, gently agitating the liquid and minimizing bumping. Boiling chips also provide nucleation sites on which bubbles can form.*

D. A Hot Water Bath or Hot Plate

Small quantities of liquids in test tubes that need to be maintained at a constant, elevated temperature over a period of time can be placed in a hot water bath (Figure T.13d). The heat source may be a hot plate or direct flame, depending on the chemicals being used. The setup is the same as that for heating a liquid in a beaker (Technique 13C).

If the liquid is in a beaker or Erlenmeyer flask instead of a test tube, place the beaker or flask on a hot plate and *slowly* heat the liquid (Figure T.13e).

Technique 14. Evaporating Liquids

To remove a liquid from a vessel by evaporation, the flammability of the liquid must be considered. This is a safety precaution.

Use a fume hood or an improvised hood (Technique 12) as recommended to remove irritating or toxic vapors.

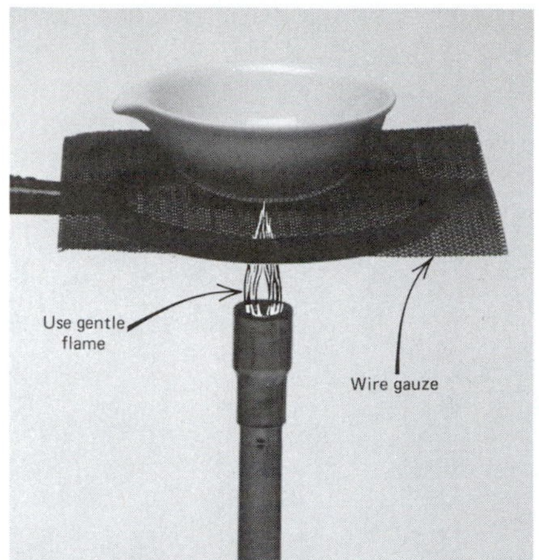

Figure T.14a Evaporation of a nonflammable liquid over a low, direct flame. A hot plate may be substituted for the flame.

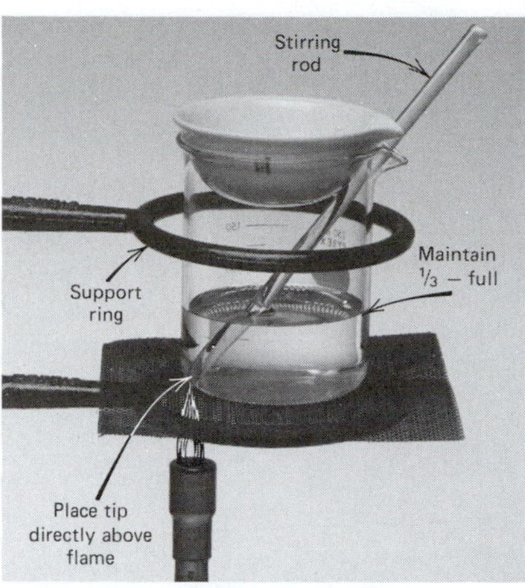

Figure T.14b Evaporation of a nonflammable liquid over a steam bath. A hot plate may be substituted for the flame.

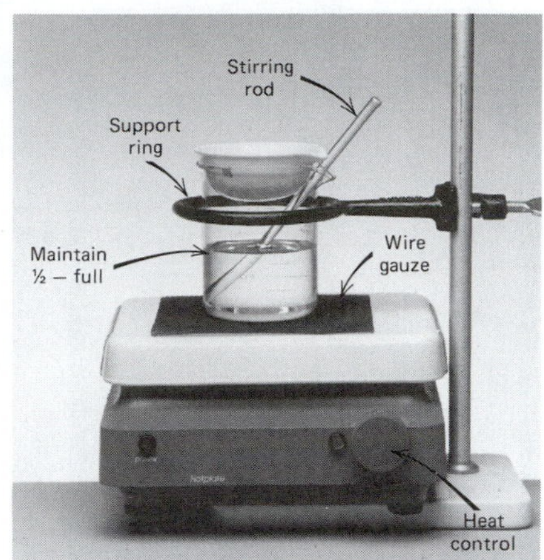

Figure T.14c Evaporation of a flammable liquid over a steam bath using a hot plate for the heat source.

A. Nonflammable Liquids

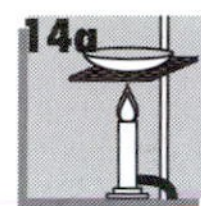

A nonflammable liquid can be evaporated either with a direct flame (Figure T.14a) or over a steam bath (Figure T.14b). If a direct flame is used, place the liquid in an evaporating dish centered on a wire gauze and iron ring. Use a gentle, "cool" flame to slowly evaporate the liquid.

If a steam bath is used, place the liquid in an evaporating dish on top of a beaker set up according to Figure T.14b. Gentle boiling of the water in the beaker is more efficient than rapid boiling for evaporating the liquid. Avoid breathing any vapors.

B. Flammable Liquids

No open flames should be near flammable liquids. Substitute a hot plate for the open flame in Figure T.14b (Figure T.14c). The use of a fume hood or an improvised hood (Technique 12) is suggested if large amounts are evaporated in a laboratory with inadequate ventilation. Consult your laboratory instructor.

Technique 15. Heating Solids

Solids are heated to dry them or to test their thermal stability. A drying oven is often used for low temperature heating, and porcelain crucibles are used for high temperature heating. Beakers and test tubes can be used for moderately high temperature heating.

A. Heating in a Drying Oven

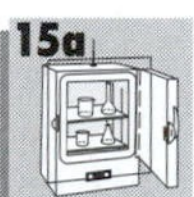

When solid chemicals are left exposed to the atmosphere they often absorb moisture. If an exact mass of a solid chemical is required for a solution preparation or for a reaction, the absorbed water must be removed before the mass measurement is made on the balance. The chemical is often placed in an open container (usually a Petri dish or beaker) in a drying oven (Figure T.15a) set at a temperature well above room temperature (most often at ~110°C) for several hours to remove the adsorbed water. The container is then removed from the drying oven and placed in a desiccator (Technique 15B) for cooling to room temperature. **Caution:** *Hot glass and cold glass look the same—the container from the drying oven is hot and should be handled accordingly.* See your laboratory instructor.

B. Cooling in a Desiccator

When crucibles, crucible lids, and solid chemicals are cooled in the laboratory, moisture tends to condense on the outer surface, adding to the total mass. To minimize this mass error, and for quantitative work, substances and mixtures that may tend to be hygroscopic are placed into a desiccator (Figure T.15b) until they have reached ambient temperature.

Figure T.15a A modern laboratory drying oven.

Figure T.15b A simple laboratory desicooler (left) or a glass desiccator (right) contains a desiccant (usually anhydrous $CaCl_2$) to provide a dry atmosphere.

A desiccator is a laboratory apparatus that provides a dry atmosphere. A desiccant, typically anhydrous calcium chloride, absorbs the water vapor from within the enclosure of the desiccator. The anhydrous calcium chloride forms $CaCl_2{\cdot}2H_2O$; the hydrated water molecules can be easily removed with heat, and the calcium chloride can be recycled for subsequent use in the desiccator.

C. Using a Crucible

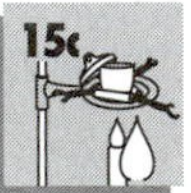

For high temperature combustion or decomposition of a chemical, porcelain crucibles are commonly used. To avoid contamination of the solid sample, thoroughly clean the crucible (so it is void of volatile impurities) prior to use. Often-used crucibles tend to form stress fractures or fissures. Check the crucible for flaws; if any are found return the crucible to the stockroom and check out and examine a second crucible.

1. **Drying and/or Firing the Crucible.** Support the crucible and cover on a clay triangle (Figure T.15c) and heat in a hot flame until the bottom of the crucible glows a dull red. Rotate the crucible with crucible tongs to ensure complete "firing" of the crucible, i.e., the combustion and volatilization of any impurities in the crucible. Allow the crucible and cover to cool to room temperature while on the clay triangle or after several minutes in a desiccator (Technique 15B). If the crucible still contains detectable impurities, add 1–2 mL of 6 *M* HNO_3 (**caution:** *avoid skin contact, flush immediately with water*) and evaporate *slowly* to dryness in the fume hood.
2. **Igniting Contents in the Absence of Air.** To heat a solid sample to a high temperature but *not* allow it to react with the oxygen of the air, set the crucible upright in the clay triangle with the cover in place (Figure T.15d). Use the crucible tongs to adjust the cover.
3. **Igniting Contents for Combustion.** To heat a solid sample to a high temperature and allow it to react with the oxygen of the air, slightly tilt the crucible on the clay triangle and adjust the cover so that about two-thirds of the crucible remains covered (Figure T.15e). Use the crucible tongs to adjust the cover.

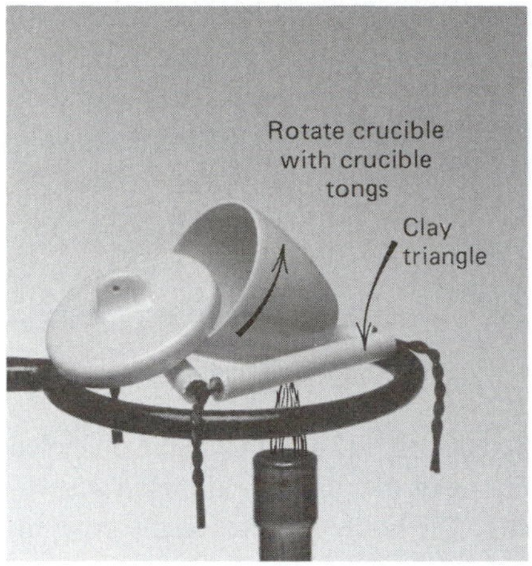

Figure T.15c Drying and/or firing a crucible and cover.

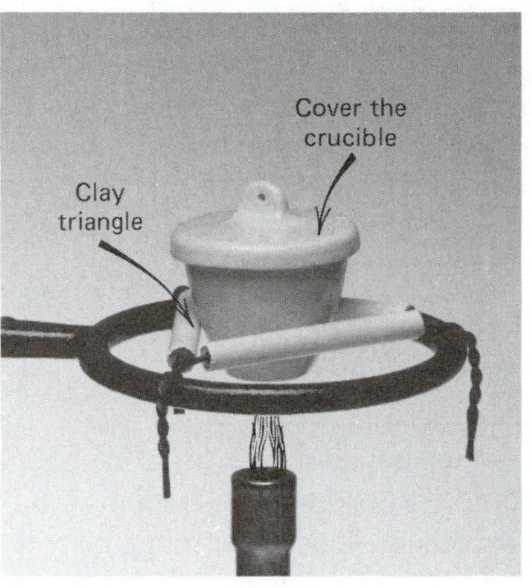

Figure T.15d Ignition of a solid sample in the absence of air.

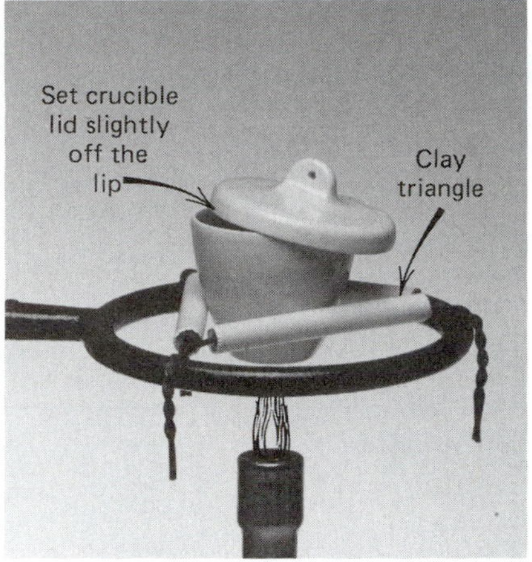

Figure T.15e Ignition of a solid sample in the presence of air for complete combustion.

Technique 16. Measuring Volume

The careful measurement and recording of volumes of liquids are necessary to obtain quantitative data for a large number of chemical reactions that occur in solutions. Volumes must be read and recorded as accurately as possible to minimize errors in the data.

A. Reading and Recording a Meniscus

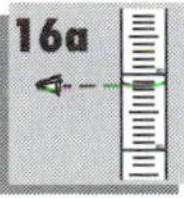

1. **Reading a Meniscus.** For measurements of liquids in graduated cylinders, pipets, burets, and volumetric flasks, the volume of a liquid is read at the *bottom of its meniscus*. Position the eye horizontally at the bottom of the meniscus (Figure T.16a) to read the level of the liquid. A clear or transparent liquid is read more easily, especially in a buret, by positioning a black mark (made on a white card) behind or just below the level portion of the liquid. The black background reflects off the bottom of the meniscus and better defines the level of the liquid (Figure T.16b). Substituting a finger for the black mark on the white card also helps in detecting the bottom of the meniscus but is not as effective.

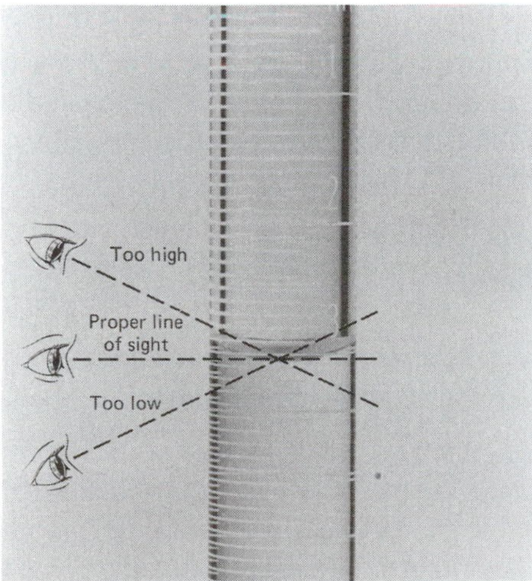

Figure T.16a Read the volume of a liquid with the eye horizontal to the bottom of the meniscus.

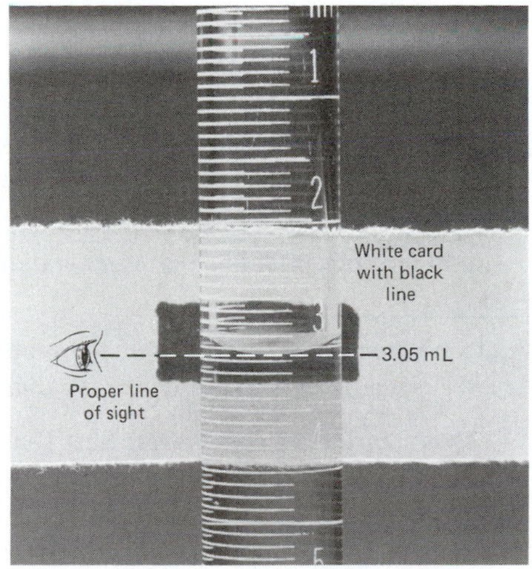

Figure T.16b Use a black line drawn on a white card to assist in pinpointing the location of the bottom of the meniscus.

Volumetric glassware: glassware that has a calibration mark(s) that indicates a calibrated volume, as determined by the manufacturer

2. **Recording a Volume.** Record the volume of a liquid in **volumetric glassware** using all certain digits (from the labeled calibration marks on the glassware) *plus* one uncertain digit (the last digit which is the best estimate between the calibration marks). In Figure T.16b, the volume of solution in the buret is between the calibration marks of 3.0 and 3.1; the "3" and the "0" are certain, the "5" is the estimate between 3.0 and 3.1. The reading is 3.05 mL. Be aware that all volume readings do *not* end in "5!"

This guideline for reading volumes also applies to reading and recording temperatures on a thermometer.

B. Pipetting a Liquid

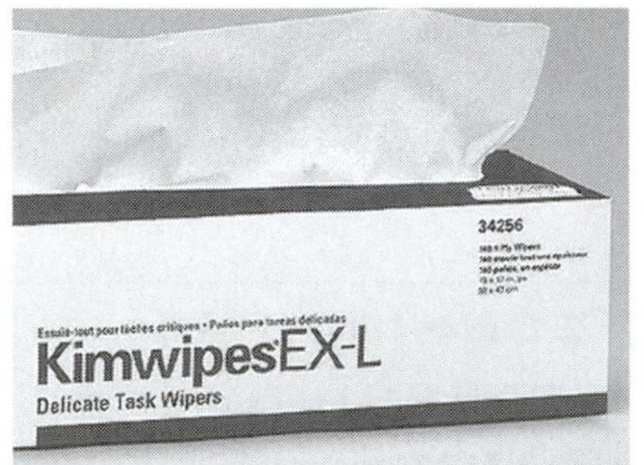

Dust/lint-free tissue.

The most common type of pipet in the laboratory is labeled TD at 20°C. A pipet labeled TD at 20°C (to deliver at 20°C) means that the volume of the pipet is calibrated according to the volume it delivers from gravity flow only. The liquid retained in the pipet tip is considered in the calibration of the pipet and therefore is *not* to be blown out.

A clean pipet in conjunction with the proper technique for dispensing a liquid from a pipet are important in any quantitative determination.

1. **Preparation of the Pipet.** See Technique 2 for cleaning glassware. A clean pipet should have no water droplets adhering to its inner wall. Inspect the pipet to ensure it is free of chips or cracks. Transfer the liquid that you intend to pipet from the reagent bottle into a clean, dry beaker; do *not* insert the pipet tip directly into the reagent bottle. Dry the outside of the pipet tip with a clean, dust-free towel or tissue (e.g., Kimwipe). Using the suction from a collapsed rubber (pipet) bulb, draw several 2- to 3-mL portions into the pipet as a rinse. Roll each rinse around in the pipet to make certain that the liquid washes the entire surface of the inner wall. Deliver each rinse through the pipet tip into a waste beaker. Discard the rinse as directed in the experiment.
2. **Filling of the Pipet.** Place the pipet tip well below the surface of the liquid in the beaker. Using the collapsed pipet bulb (or a pipet pump—*never* use your mouth!), draw the liquid into the pipet until the level is 2–3 cm above the "mark" on the pipet (Figure T.16c). Remove the bulb and quickly cover the top of the pipet with your index finger (*not* your thumb!). Remove the tip from the liquid and wipe off the pipet tip with a clean, dust-free towel or tissue. Holding the pipet in a vertical position over a waste beaker, control the delivery of the excess liquid until the level is "at the mark" in the pipet. Practice (Figure T.16d)! Read the meniscus correctly. Remove any drops suspended from the pipet tip by touching it to the wall of the waste beaker.
3. **Delivery of the Liquid.** Deliver the liquid to the receiving vessel (Figure T.16e) by releasing the index finger from the top of the pipet. The pipet tip should touch the wall of the receiving vessel to avoid splashing. Do *not* blow or shake out the last bit of liquid that remains in the tip; this liquid has been included in the calibration of the pipet . . . remember thin is a TD at 20°C pipet!

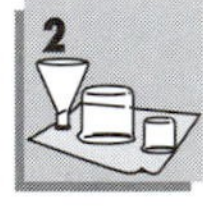

4. **Cleanup.** Once it is no longer needed in the experiment, rinse the pipet with several portions of deionized water and drain each rinse through the tip.

C. Titrating a Liquid (Solution)

A clean buret in conjunction with the proper technique for measuring and dispensing a liquid from a buret is important in any quantitative analysis determination.

Titrant: the solution in the buret

1. **Preparation of the Buret.** See Technique 2 for cleaning glassware. If a buret brush is needed, be careful to avoid scratching the buret wall with the wire handle. Once the buret is judged to be "clean," close the stopcock. Rinse the buret with several 3- to 5-mL portions of water and then **titrant.** Tilt and roll the barrel of the buret so that the rinse comes into contact with the entire inner wall. Drain each rinse through the buret tip into the waste beaker. Dispose of the rinse as advised in the experiment. Support the buret with a buret clamp (Figure T.16f).

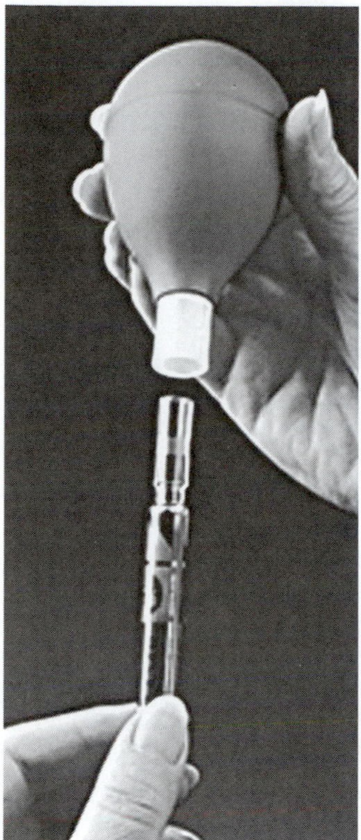

Figure T.16c Draw the liquid into the pipet with the aid of a rubber pipet bulb (*not* the mouth!).

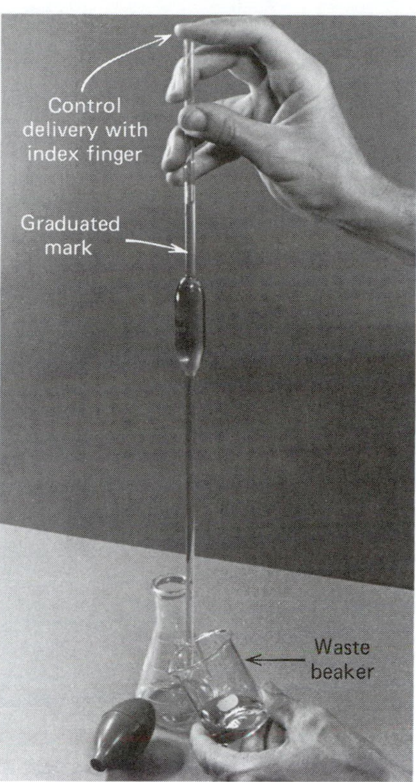

Figure T.16d Control the delivery of the liquid from the pipet with the forefinger (*not* the thumb!).

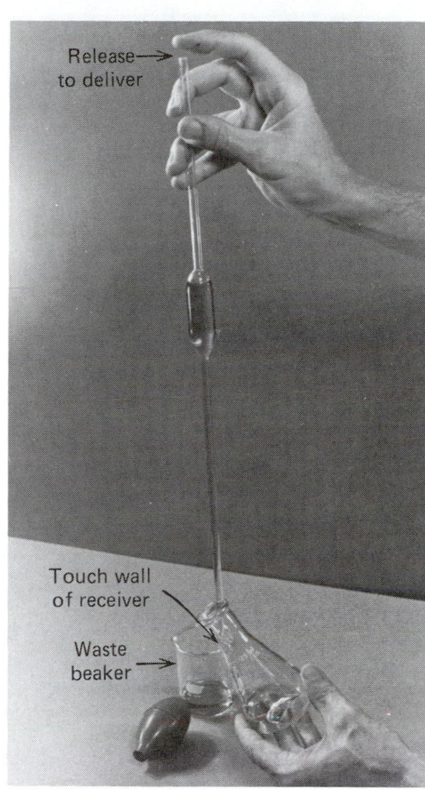

Figure T.16e Deliver the liquid from the pipet with the tip touching the wall of the receiving flask.

2. **Preparation of the Titrant.** Close the stopcock. With the aid of a *clean* funnel, fill the buret with the titrant to just above the zero mark. Open the stopcock briefly to release any air bubbles in the tip *and* allow the meniscus of the titrant to go below the uppermost graduation on the buret. Allow 10–15 seconds for the titrant to drain form the wall; **record the volume** (Technique 16A) of titrant in the buret. Note that the graduations on a buret *increase* in value from the top down (Figure T.16g).

Record the volume: read the volume in the buret using all certain digits (from the labeled calibration marks on the buret) plus one uncertain digit (the last digit which is the best estimate between the calibration marks)

3. **Operation of the Buret.** During the addition of the titrant from the buret, operate the stopcock with your left hand (if right-handed) and swirl the Erlenmeyer flask with your right hand (Figure T.16h). This prevents the stopcock from sliding out of its barrel and allows you to maintain a normal, constant swirling motion of the reaction mixture in the receiving flask as the titrant is added. The opposite procedure, of course, is applicable if you are left-handed (Figure T.16i). Fill the buret after each analysis. Use an Erlenmeyer flask as a receiving flask, rather than a beaker, unless the solution is to be stirred with magnetic stirrer.
4. **Addition of Titrant to Receiving Flask.** Have a white background (a piece of white paper) beneath the receiving flask, generally an Erlenmeyer flask, so that you can better see the endpoint of a titration (the point at which the indicator turns color). If the endpoint is a change from colorless to white, a black background is preferred. Add the titrant to the Erlenmeyer flask as described above; periodically stop its addition and wash the wall of the flask with the solvent (generally deionized water) from a wash bottle (Figure T.16j). Near the endpoint (slower color fade of the indicator, Figure T.16k), slow the rate of titrant addition until a drop (or less) makes the color change of the indicator persist for 30 seconds. **Stop,**

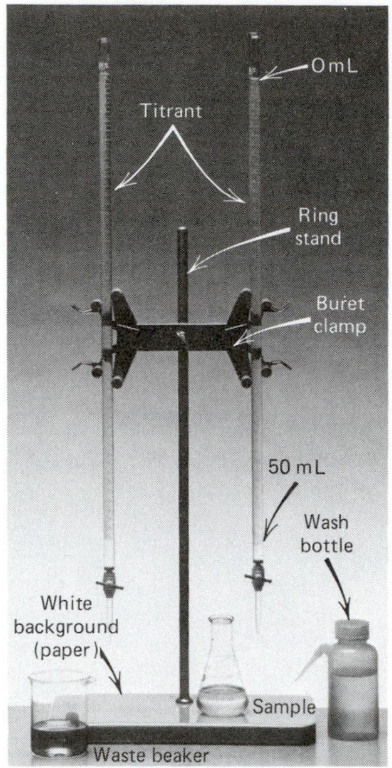

Figure T.16f Setup for a titration analysis.

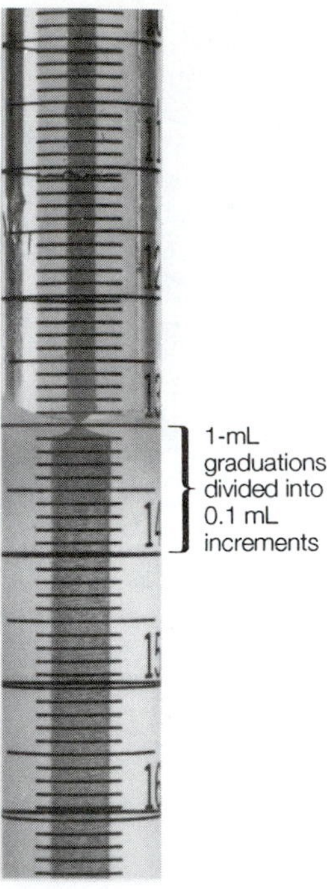

Figure T.16g A 50-mL buret is marked from top to bottom, 0 to 50 mL, with 1-mL gradations divided into 0.1-mL increments.

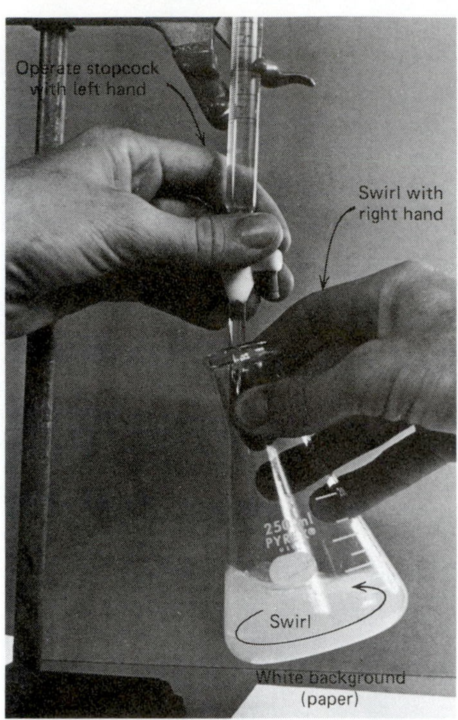

Figure T.16h Titration technique for right-handers.

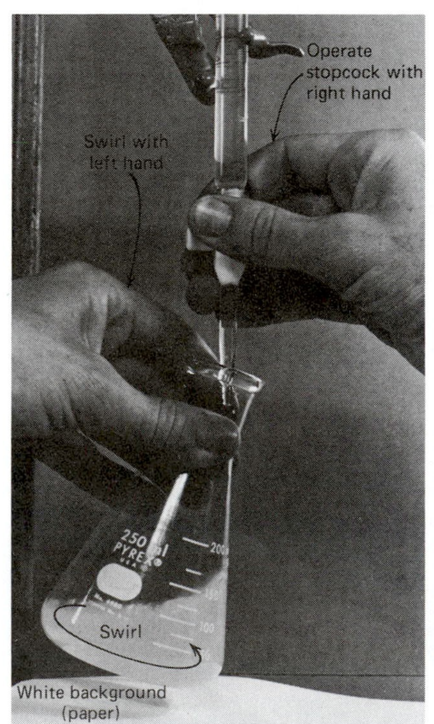

Figure T.16i Titration technique for left-handers.

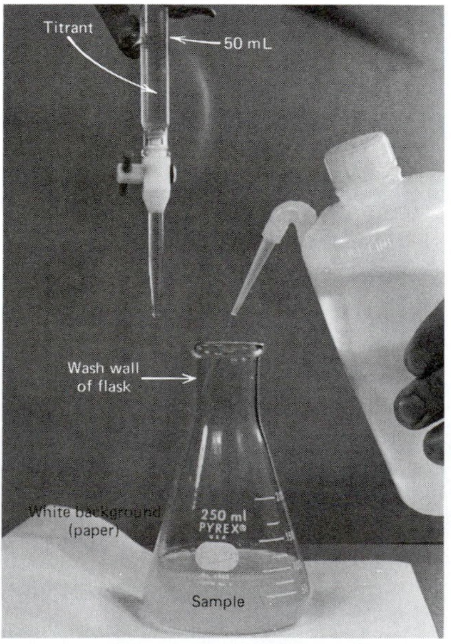

Figure T.16j Place a white background beneath the receiving flask and wash the wall of the receiving flask periodically during the titration.

Figure T.16k A slow color fade of the indicator occurs near the endpoint in the titration.

allow 10–15 seconds for the titrant to drain from the buret wall, read, and record the volume in the buret (Technique 16A).

To add less than a drop of titrant (commonly referred to as a "half-drop") to the receiving flask, suspend a drop from the buret tip, touch it to the side of the receiving flask, and wash the wall of the receiving flask (with deionized water).

5. **Cleanup.** After completing a series of titrations, drain the titrant from the buret, rinse the buret with several portions of deionized water, and drain each rinse through the tip. Discard the excess titrant and the rinses as advised in the experiment. Store the buret as advised by your laboratory instructor.

TECHNIQUE 17. QUICK TESTS

A. Testing for Odor

An educated nose is an important and very useful asset to the chemist. Use it with caution, however, because some vapors induce nausea and/or are toxic. *Never* hold your nose directly over a vessel. Fan some vapor toward your nose (Figure T.17a). *Always* consult your laboratory instructor before testing the odor of any chemical.

B. Testing for Acidity/Basicity

To test the acidity/basicity of a solution with test paper, insert a *clean* stirring rod into the solution, withdraw it, and touch it to the test paper (Figure T.17b). For litmus paper acidic solutions turn blue litmus red; basic solutions turn red litmus blue. *Never* place the test paper directly into the solution.

Other paper-type indictors, such as pHydrion paper (Figure T.17c), are also used to gauge the acidity/basicity of a solution.

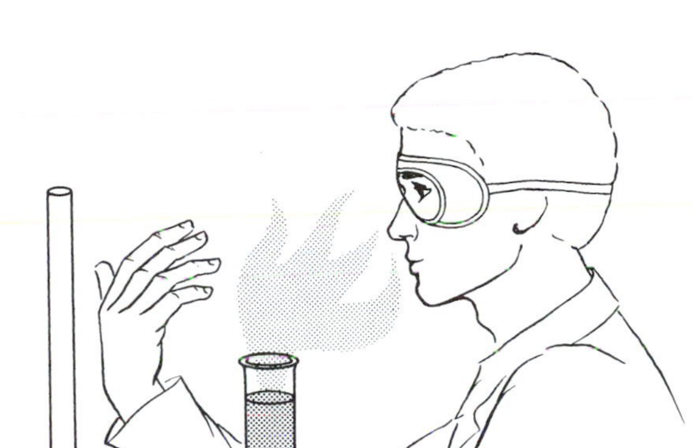

Figure T.17a Fan the vapors gently toward the nose.

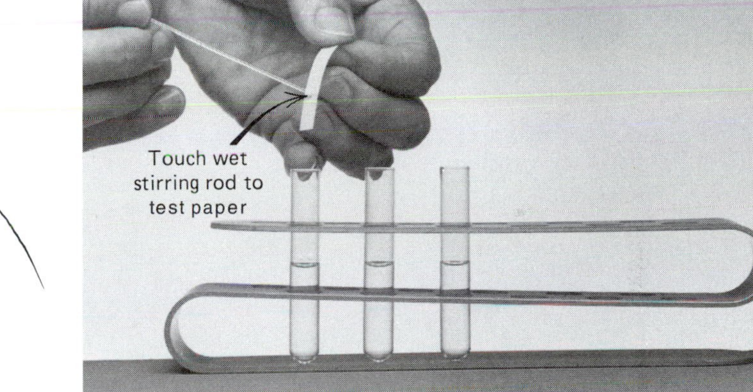

Figure T.17b Test for acidity/basicity.

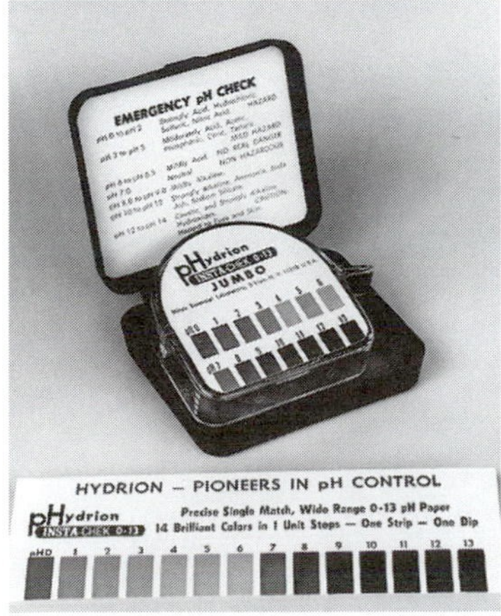

Figure T.17c Test papers impregnated with a mixture of acid-base indicators can be used to measure the approximate pH of a solution.

Disclaimer: The material contained in the Laboratory Safety and Laboratory Techniques sections of this manual has been compiled from sources believed to be reliable and to represent the best opinions of safety in the laboratory. This manual is intended to provide basic guidelines for safe practices in the undergraduate chemistry laboratory. It cannot be assumed that all necessary warning and precautionary measures are contained in this manual, or that other or additional information or measures may not be required.

Further discussions of these and other laboratory techniques can be found on the World Wide Web. Refer to Laboratory Data, Part C.

Notes on Laboratory Techniques

Laboratory Techniques

Date __________ Lab Sec. ______ Name ____ Desk No. __________

Complete with the correct word(s), phrase, or value(s).

Ask your instructor to identify the questions you are to complete.

1. Technique 1. To insert glass tubing through a rubber stopper, first moisten the glass with ______________ or ______________.
2. Technique 2. Glassware should first be washed with ______________ water and a detergent solution followed by final rinses with ______________ water.
3. Technique 2. Glassware is clean when ____.
4. Technique 4. Information on the properties and disposal of chemicals can be found in the stockroom or online from the ______________ collection.
5. Technique 5. Most solutions used for quantitative work are prepared in flasks called ______________________ flasks.
6. Technique 6. The mass of a sample measured on a balance without regard to its container is called its ______________ mass.
7. Technique 6. After completing a mass measurement, the mass settings are to be set to ______________.
8. Technique 7. A well of a 24-well plate has the same approximate volume that is slightly larger than a ______________ mm test tube.
9. Technique 8. To collect a water-soluble gas ______________ dense than air, the mouth of the gas-collecting flask should be pointed downward.
10. Technique 9. Do not use a ______________________ to transfer a solid from its reagent bottle.
11. Technique 10. Transfer liquids or solutions from the reagent bottle to a beaker with the aid of a ________________.
12. Technique 11c. The bowl of the funnel should be less than ______________ full when gravity filtering a mixture.
13. Technique 11f. A centrifuge should be balanced with ______________ numbers of test tubes containing ______________ volumes of solution.
14. Technique 12. If a noxious or nauseating gas is evolved from a reaction mixture, it is good advice to perform the reaction ____________________________ or ____________________________.
15. Technique 13a. A nonluminous flame with a reduced supply of fuel is called a ______________________. Such a flame is most critical when heating liquids and solutions contained in a ______________________.
16. Technique 13c. A nonflammable liquid in a flask or beaker that is *greater than* one-fourth full, can be heated using the laboratory setup shown in Figure ______________.
17. Technique 13d. Solutions in test tubes can be maintained at a constant "higher" temperature with the use of a ______________________.
18. Technique 15a, b. Solids are commonly heated to dryness in a drying oven and then cooled (ideally) in a ______________.
19. Technique 16a. The volume of a liquid should be read at the ______________ of the meniscus.
20. Technique 16b. The volume of a liquid in a pipet should be controlled with the ______________ finger.
21. Technique 16c. A buret should be rinsed with several 3- to 5-mL portions of ______________ before being filled.

22. Technique 16c. During the titration procedure, the stopcock of the buret should be controlled with the ______________ hand for those chemists who are left handed.

23. Technique 16c. The color change of the indicator at the endpoint should persist for ______________ seconds.

24. Technique 17. The acidity or basicity of a solution is easily and quickly tested with ______________ or ______________.

25. All techniques. Technique advisories appear in the margin of Experiment Procedure for each experiment by the presence of an appropriate ______________.

True or False

Ask your instructor to identify the questions you are to complete.

_________ **1.** Dry all clean glassware with a clean towel.

_________ **2.** While cleaning glassware, discard all washes and rinses from the delivery point of the glass vessel.

_________ **3.** To avoid waste in the use of chemicals, always return the unused portion directly to the original reagent bottle.

_________ **4.** Never touch, taste, or smell a chemical unless specifically told to do so.

_________ **5.** Most all chemicals used in experiments can be discarded into the sink.

_________ **6.** The mass of a dry solid can be measured directly on the balance pan.

_________ **7.** If a 3-inch test tube has a volume of 3 mL, then an 8-inch test tube must have a volume of 8 mL.

_________ **8.** To transfer a solution, a stirring rod touches the delivery point of the reagent vessel and the wall of the receiving vessel.

_________ **9.** Only 1 mL of a liquid mixture should be present in a centrifuge tube when placed in the centrifuge.

_________ **10.** A test tube should be less than one-third full when heating with a direct flame.

_________ **11.** If the heat of a flame is felt with the hand holding the test tube clamp (holding the test tube), the flame is a cool flame.

_________ **12.** Blow out the solution remaining in the pipet tip after the solution has drained from the pipet.

_________ **13.** A buret must *always* be filled to the top (the zero mark) before every titration procedure.

_________ **14.** The volume of solution in a buret should be read and recorded 10–15 seconds after completing the titration.

_________ **15.** It is possible to add a half-drop of solution from a buret.

_________ **16.** To test the acidity of a solution with litmus paper, place the litmus paper directly in the solution.

_________ **17.** The odor of a chemical should not be tested unless specifically instructed to do so. The vapors of the chemical should be fanned toward the nose.

_________ **18.** A "blue 3" on a reagent bottle indicates the reagent poses an extreme danger.

Summarize the "Disclaimer" in your own words.

A set of standard SI mass units.

Dry Lab 1

The Laboratory and SI

OBJECTIVES

- To check into the laboratory
- To become familiar with the laboratory and the laboratory manual
- To learn laboratory safety rules and the necessity of practicing these rules in the laboratory
- To develop skills in the use of *Le Système International d'Unités* (SI Units)

INTRODUCTION

All chemical principles, tools, and techniques are developed in the laboratory. The experience of observing a chemical phenomenon and then explaining its behavior is one that simply cannot be gained by reading a textbook, listening to a lecturer, or viewing a video. It is in the laboratory where chemistry comes alive, where chemical principles are learned and applied to the vast natural "chemistry laboratory" that we call our everyday environment. The objectives of a laboratory experience are to design and build apparatus, develop techniques, observe, record and interpret data, and deduce rational theories so that the real world of science is better explained and understood.

In the laboratory, you will use common equipment and safe chemicals to perform experiments. You record your experimental observations and interpret the data on the basis of sound chemical principles. A good scientist is a thinking scientist trying to account for the observed data and rationalize any contradictory data. Cultivate self-reliance and confidence in your data, even if "the data do not look right." This is how many breakthroughs in science occur.

In the first few laboratory sessions you will be introduced to some basic rules, equipment, and techniques and some situations where you use them. These include laboratory safety rules, *Le Système International d'Unités* (SI Units), the Bunsen burner, and the analytical balance. Additional laboratory techniques are illustrated under Laboratory Techniques pages 11–34; others are introduced as the need arises.

DRY LAB PROCEDURE

Procedure Overview: Laboratory procedures are introduced. A familiarity with laboratory apparatus, the policies regarding laboratory safety, the procedures for presenting laboratory data, and an encounter with the SI units are emphasized.

A. Laboratory Check-in

At the beginning of the first laboratory period you are assigned a lab station containing laboratory equipment. Place the laboratory equipment on the laboratory bench and, with the check-in list provided on page 9, check off each piece of chemical apparatus as you return it to the drawer. If you are unsure of a name for the apparatus, refer to the list of chemical "kitchenware" on pages 8–9. Ask your instructor about any items on the check-in list that are not at your lab station.

A good scientist is always neat and well organized; keep your equipment clean and arranged in an orderly manner so that it is ready for immediate use. Have at your lab station dishwashing soap (or detergent) for cleaning glassware and paper towels or a rubber lab mat on which to set the clean glassware for drying.

Obtain your laboratory instructor's approval for the completion of the check-in procedure. Refer to Part A of the Report Sheet.

B. Laboratory Safety and Guidelines

Your laboratory instructor will discuss laboratory safety and other basic laboratory procedures with you. Remember, however, that your laboratory instructor cannot practice laboratory safety for you or for those who work with you—it is your responsibility to *play it safe!*

On the inside front cover of this manual is space to list the location of some important safety equipment and important information for reference in the laboratory. Fill this out. Obtain your laboratory instructor's approval for its completion.

Read and study the Laboratory Safety and Guidelines section on pages 1–4. Complete any other laboratory safety sessions or requirements that are requested by your laboratory instructor. Answer the laboratory safety questions on the Report Sheet.

C. Laboratory Data

Each of the experiments in this manual will require you to observe, record and report data, perform calculations on the data, and then analyze and interpret the data. Scientists are careful in the procedures that are used for handling data so that the reliability and credibility of the experimental data are upheld.

It is important that these procedures are followed at the outset of your laboratory experience. Read and study the Laboratory Data section on pages 5–8. Answer the laboratory data questions on the Report Sheet.

D. *Le Système International d'Unités* (SI Units)

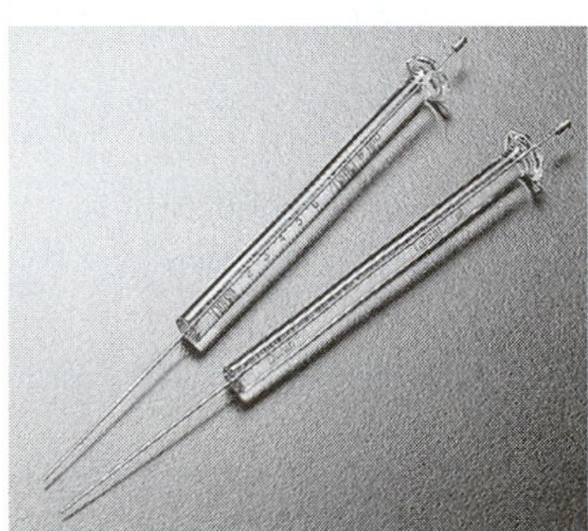

A microliter syringe.

The SI, a modern version of the metric system, provides a logical and interconnected framework for all basic measurements. The SI, and some of its slight modifications, is used throughout the world by scientists and engineers as the international system for scientific measurements and, in most countries, for everyday measurements. For laboratory measurements, the SI base unit of mass is the **kilogram (kg)** [chemists are most familiar with the **gram (g)**, where 10^3 g = 1 kg], the SI base unit for length is the **meter (m)** [chemists are most familiar with several subdivisions of the meter], and the SI derived unit for volume is the **cubic meter (m^3)** [chemists are most familiar with the **liter (L)**, where 1 L = 10^{-3} m^3 = 1 dm^3]. Subdivisions and multiples of each unit are related to these units by a power of ten. The prefixes used to denote these subdivisions and multiples are shown in Table D1.1. Memorize these prefixes and their meanings.

Table D1.1 Prefixes in *Le Système International d'Unités*

Prefix	Abbreviation	Meaning (power of ten)	Example Using "grams"
femto-	f	10^{-15}	fg = 10^{-15} g
pico-	p	10^{-12}	pg = 10^{-12} g
nano-	n	10^{-9}	ng = 10^{-9} g
micro-	μ	10^{-6}	μg = 10^{-6} g
milli-	m	10^{-3}	mg = 10^{-3} g
centi-	c	10^{-2}	cg = 10^{-2} g
deci-	d	10^{-1}	dg = 10^{-1} g
kilo-	k	10^{3}	kg = 10^{3} g
mega-	M	10^{6}	Mg = 10^{6} g
giga-	G	10^{9}	Gg = 10^{9} g

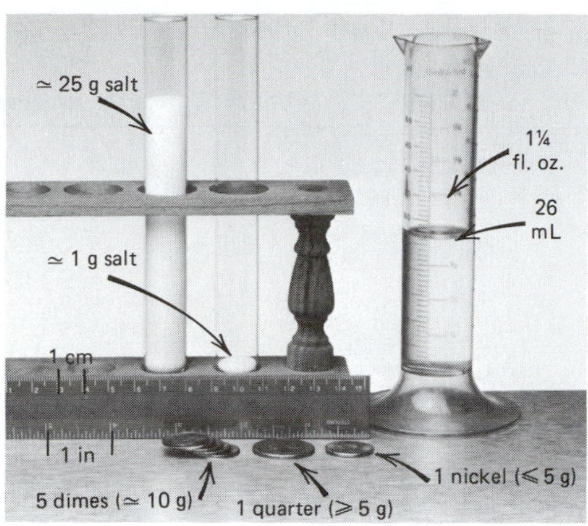

Figure D1.1 Comparisons of SI and English measurements.

In Table D1.1, *the prefix indicates the power of ten.* For example 4.3 *milli*grams means 4.3×10^{-3} grams; *milli* has the same meaning as "$\times\ 10^{-3}$." Figure D1.1 shows representative SI units for mass, length, and volume.

Conversions of measurements within SI are quite simple if the definitions for the prefixes are known and unit conversion factors for problem solving are used. To illustrate the use of unit conversion factors, consider the following example.

Example D1.1 Convert 4.3 milligrams to micrograms.
Solution. From Table D1.1, note that milli- and 10^{-3} are equivalent and that micro- and 10^{-6} are equivalent. Considering mass, it also means that mg $= 10^{-3}$ g *and* μg 10^{-6} g. This produces two equivalent unit conversion factors for each equality:

$$\frac{10^{-3}\text{ g}}{\text{mg}}, \quad \frac{\text{mg}}{10^{-3}\text{ g}} \quad \text{and} \quad \frac{10^{-6}\text{ g}}{\mu\text{g}}, \quad \frac{\mu\text{g}}{10^{-6}\text{g}}$$

Beginning our solution to the problem with the measured quantity (i.e., 4.3 mg), we need to convert mg to g and g to μg. We multiply by the appropriate conversion factors to obtain proper unit cancellation:

$$4.3\ \cancel{\text{mg}} \times \frac{10^{-3}\ \cancel{\text{g}}}{\cancel{\text{mg}}} \times \frac{\mu\text{g}}{10^{-6}\ \cancel{\text{g}}} = 4.3 \times 10^{3}\ \mu\text{g}$$

mg cancel g cancel

$$\text{mg} \rightarrow \text{g} \rightarrow \mu\text{g} = \mu\text{g}$$

The conversion factors in Example D1.1 have no effect on the magnitude of the mass measurement (the conversion factor = 1), only the units by which it is expressed.

The SI is compared with the English system in Table D1.2. SI units of measurement that chemists commonly use in the laboratory are listed in brackets. Appendix A has a more comprehensive table of conversion factors. Conversions between the SI and the English system are quite valuable, especially to Americans because international science and trade communications are in SI or metric units.

Appendix A

Example D1.2 Using Tables D1.1 and D1.2, determine the volume of 3.00 pints of water in terms of cubic centimeters.
Solution. As 1 cm $= 10^{-2}$ m, then 1 cm$^3 = (10^{-2}\text{ m})^3 = (10^{-2})^3\text{ m}^3$.

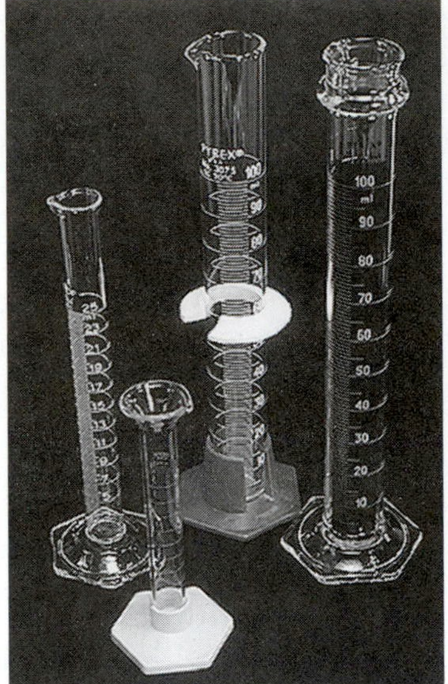

Graduated cylinders of different volumes.

Table D1.2 Comparison of *Le Système International d'Unités* and English System of Measurements

Physical Quantity	SI Unit	Conversion Factor
Length	meter (m)	1 km = 0.6214 mi 1 m = 39.37 in. 1 in. = 0.0254 m = 2.54 cm
Volume	cubic meter (m^3) [liter (L)][a]	1 L = 10^{-3} m^3 = 1 dm^3 = 10^3 mL 1 L = 1.057 qt 1 oz (fluid) 29.57 mL
Mass	kilogram (kg) [gram (g)]	1 lb = 453.6 g 1 kg = 2.205 lb
Pressure	pascal (Pa) [atmosphere (atm)]	1 Pa = 1 N/m^2 1 atm = 101.325 kPa = 760 torr 1 atm = 14.70 lb/in^2 (psi)
Temperature	kelvin (K) [degrees Celsius (°C)]	K = 273 + °C $°C = \frac{°F - 32}{1.8}$
Energy	joule (J)	1 cal = 4.184 J 1 Btu = 1054 J

[a]The SI units enclosed in brackets are commonly used in chemical measurements and calculations.

From Table D1.2, we need conversion factors for pints → quarts, quarts → liters, liters → m^3, and finally m^3 → cm^3. Starting with 3.00 pt (our measured value in the problem), we have:

$$3.00\ \cancel{pt} \times \frac{1\ \cancel{qt}}{2\cancel{pt}} \times \frac{1\ \cancel{L}}{1.057\ \cancel{qt}} \times \frac{10^{-3}\ \cancel{m^3}}{1\ \cancel{L}} \times \frac{cm^3}{(10^{-2})^3\ \cancel{m^3}} = 1.42 \times 10^3\ cm^3$$

pt cancel qt cancel L cancel m^3 cancel

pt → qt → L → m^3 → cm^3 = cm^3

Note again that the conversion factors in Example D1.2 do not change the magnitude of the measurement, only its form of expression.

The Report Sheet, Part D, further acquaints you with conversions within the SI and between SI and the English system. Ask your laboratory instructor which of the questions from Part D you are to complete on the Report Sheet.

Dry Lab 1 *Report Sheet*

The Laboratory and SI

Date __________ Lab Sec. ______ Name ____ Desk No. __________

A. Laboratory Check-in

Instructor's approval ____

B. Laboratory Safety and Guidelines

Instructor's approval for completion of inside front cover ____

Read the Laboratory Safety and Guidelines section on pages 1–4 and answer the following as **true** or **false**.

_________ **1.** Prescription glasses, which are required by law to be "safety glasses," can be worn in place of safety goggles in the laboratory.

_________ **2.** Sleeveless blouses and tank tops are *not* appropriate attire for the laboratory.

_________ **3.** Only shoes that shed liquids are permitted in the laboratory.

_________ **4.** "I just finished my tennis class. I can wear my tennis shorts to lab just this one time, right?"

_________ **5.** Your laboratory has an eye wash fountain.

_________ **6.** "Oops! I broke a beaker containing deionized water in the sink." An accident as simple as that does not need to be reported to the laboratory instructor.

_________ **7.** A beaker containing an acidic solution has broken on the bench top and spilled on your clothes from the waist down and it burns. Ouch! You should immediately proceed to the safety shower and flood the affected area.

_________ **8.** You touched a piece of hot glass and burned your fingers. Immediately go to the medicine cabinet to apply a salve.

_________ **9.** It is good laboratory protocol to inform other students when they are not practicing good laboratory safety procedures. If they continue to not follow the safety procedures, you should "rat" on them . . . tell the laboratory instructor.

_________ **10.** You work more efficiently with the accompaniment of your favorite CD. Therefore, your music played aloud or with your MP3 player in the laboratory is an acceptable laboratory procedure.

_________ **11.** Your friend is a senior chemistry major and thoroughly understands the difficult experiment that you are performing. Therefore, it is advisable (even recommended) that you invite him/her into the laboratory for direct assistance.

_________ **12.** Discuss any interpretation of your laboratory data only with your laboratory instructor, not with your classmates.

Write a short response for the following questions.

1. What does the phrase "neck to knee to wrist" mean with regard to laboratory safety?

2. The first action after an accident occurs is:

3. You want to try a variation of the Experimental Procedure because of your chemical curiosity. What is the proper procedure for performing the experiment?

4. A chemical spill has occurred. What should be your first and second action in treating the chemical spill?

5. Describe how you will be dressed when you are about to begin an experiment in the laboratory.

C. Laboratory Data

Read the Laboratory Data section on pages 5–8 and answer the following as **true** or **false**.

_________ 1. "Quick data," such as that of a mass measurement on a balance located at the far side of the laboratory, can be recorded on a paper scrap and then transferred to the Report Sheet at your lab station.

_________ 2. Data that has been mistakenly recorded on the Report Sheet can be erased and replaced with the correct data. This is to maintain a neat Report Sheet.

_________ 3. All data should be recorded in permanent ink!

_________ 4. The laboratory equipment and instrumentation determine the number of significant figures used to record quantitative data.

_________ 5. Zeros recorded in a measurement are *never* significant figures.

_________ 6. Your laboratory instructor requires you to use a laboratory notebook for recording your data. (If *true*, read Laboratory Data, Part D very closely; additional questions may be asked.)

D. *Le Système International d'Unités* (SI Units)

Circle the questions that have been assigned.

1. Name the SI unit of measurement for:

a. length _________________________ d. temperature _________________________

b. volume _________________________ e. pressure _________________________

c. mass _________________________ f. energy _________________________

2. Complete the following table.

	SI Expression	Power of Ten Expression		SI Expression	Power of Ten Expression
Example	1.2 mg	1.2×10^{-3} g			
a.	3.3 pg	__________	d.	__________	6.72×10^{6} J
b.	__________	7.6×10^{-6} L	e.	2.16 kilowatts	__________
c.	__________	4.3×10^{-3} ampere	f.	__________	1.99×10^{-12} g

3. Convert each of the following using the definitions in Table D1.1 and unit conversion factors. Show the cancellation of units.

a. 2.63 μm $\times$ $\dfrac{\text{\_\_\_\_\_\_\_\_\_\_ m}}{\mu\text{m}}$ $\times$ $\dfrac{\text{\_\_\_\_\_\_\_\_\_\_ pm}}{\text{m}}$ = __________________pm

b. 250 mL $\times$ __________________ $\times$ __________________ = __________________cL

c. 9.04 mg $\times$ __________________ $\times$ __________________ = __________________dg

4. The current US penny has a diameter of 19 mm and a mass of 2.50 g. Convert these measurements to inches and ounces respectively.

5. Determine the volume (in mL) of 1.0 teaspoon. 1 tablespoon = 3 teaspoons; 1 tablespoon = ½ fluid ounce

6. **a.** Express the "inches of rain" received to "centimeters of rain."

***b.** If the indicated amount of rain fell on a football field (160 ft × 300 ft), how many liters and gallons of rain would fall?

7. Measure the inside diameter and length of a test tube in centimeters:

diameter = _______________ length = _______________

Using the equation $V = \pi r^2 l$, calculate the volume of the test tube.

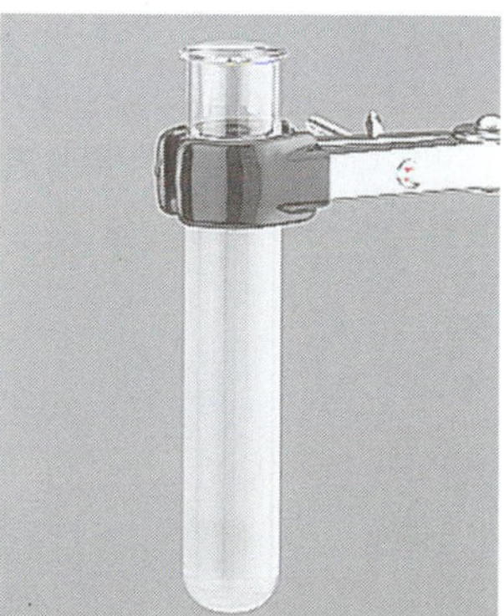

8. The standard width between rails on North American and most European railroads is 4 ft 8 in. Calculate this distance in meters and centimeters.

9. The heat required to raise the temperature of a large cup of water (for coffee) from room temperature to boiling is approximately 100 kJ. Express this quantity of heat in kilocalories and British thermal units. Show your calculation

10. An aspirin tablet has a mass of about 325 mg (5 grains). Calculate the total mass of aspirin tablets in a 250-tablet bottle in grams and ounces.

11. A concrete pile from a waterfront pier was pulled from the harbor water at Port Hueneme, California. Its dimensions were 14 inches × 14 inches × 15 feet (1 meter = 3.28 feet).

a. What is the surface area, expressed in square feet, of a single face of the pile? Exclude the ends of the pile.

b. Determine the total surface area, expressed in square meters, of the pile, including the ends of the pile.

c. How many cubic meters of concrete were used to make the pile?

Experiment 1

Basic Laboratory Operations

A properly adjusted Bunsen flame burns with a blue, nonluminous flame.

OBJECTIVES

- To light and properly adjust the flame of a Bunsen burner
- To develop the skill for properly operating a balance
- To develop the technique of using a pipet
- To determine the density of an unknown substance

TECHNIQUES

The following techniques are used in the Experimental Procedure

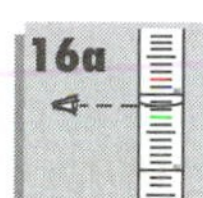

INTRODUCTION

You will use a number of techniques repeatedly throughout your laboratory experience. Seventeen principal techniques are fully described under Laboratory Techniques in this manual. Become familiar with each of these techniques before you are required to use them in an experiment.

In this experiment you will learn several common techniques that are used repeatedly throughout this laboratory manual—you will learn to light and adjust a Bunsen burner, to use a laboratory balance, and to use a pipet. With the skills developed in using a balance and pipet, you will determine the density of a metal and a liquid.

Bunsen Burner

Laboratory burners come in many shapes and sizes, but all accomplish one main purpose: a combustible gas–air mixture yields a hot, efficient flame. Because Robert Bunsen (1811–1899) was the first to design and perfect this burner, his name is given to most burners of this type used in the general chemistry laboratory (Figure 1.1).

The combustible gas used to supply the fuel for the Bunsen burner in most laboratories is natural gas. Natural gas is a mixture of gaseous **hydrocarbons,** but primarily the hydrocarbon methane, CH_4. If sufficient oxygen is supplied, methane burns with a blue, **nonluminous** flame, producing carbon dioxide and water as combustion products.

Hydrocarbon: a molecule consisting of only the elements carbon and hydrogen

Nonluminous: nonglowing or nonilluminating

$$CH_4(g) + 2O_2(g) \rightarrow CO_2(g) + 2H_2O(g)$$

With an insufficient supply of oxygen, small carbon particles are produced which, when heated to **incandescence,** produce a yellow, luminous flame. The combustion products may, in addition to carbon dioxide and water, include carbon monoxide.

Incandescence: glowing with intense heat

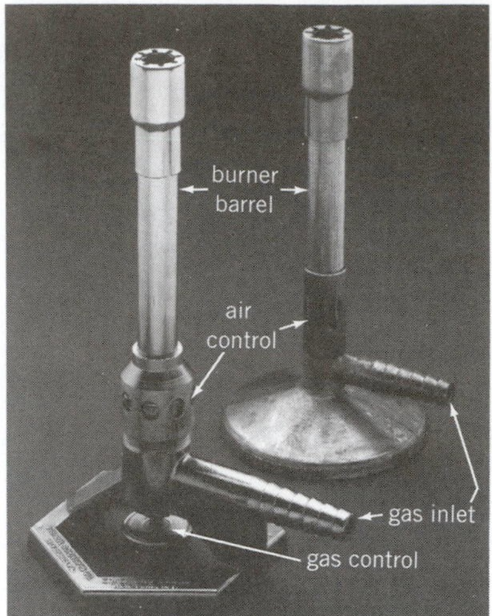

Figure 1.1 Bunsen-type burners.

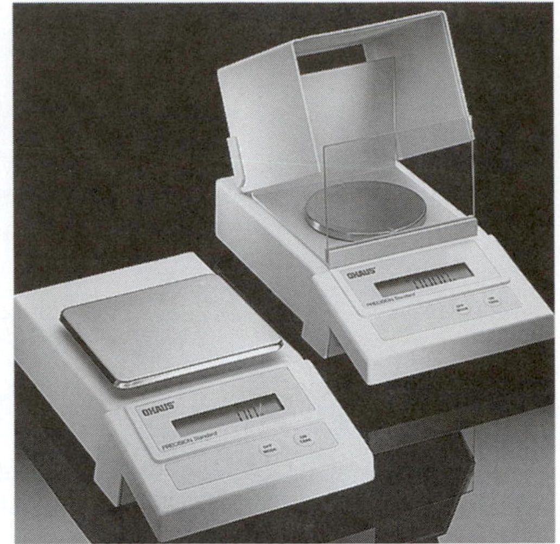

Figure 1.2 Two balances with different sensitivities.

Laboratory Balances

The laboratory balance is perhaps the most common and most often used piece of laboratory apparatus. Balances, *not* scales, are of different makes, models, and sensitivities (Figure 1.2). The selection of the appropriate balance for a mass measurement depends upon the degree of precision required for the analysis. The triple-beam and top-loading balances are the most common. Read Technique 6 under Laboratory Techniques for instructions on the proper operation of the laboratory balance.

Density

Intensive property: property independent of sample size

$$\text{density} = \frac{\text{mass}}{\text{volume}}$$

Each pure substance exhibits its own set of **intensive properties.** One such property is **density,** the mass of a substance per unit volume. In the English system the density of water at 4°C is 8.34 lb/gal or 62.2 lb/ft^3, whereas in SI the density of water at 4°C is 1.00 g/cm^3 or 1.00 g/mL. By measuring the mass and volume of a substance, its density can be calculated.

In this experiment, the density of a water-insoluble solid and the density of an unknown liquid are determined. The data for the mass and the volume of water displaced (see Figure 1.6) are used to calculate the density of the water-insoluble solid; the density of the unknown liquid is calculated from mass and volume measurements of the liquid.

Chemists conventionally express the units for density in the SI as g/cm^3 for solids, g/mL for liquids, and g/L for gases at specified temperatures and pressures.

Experimental Procedure

Procedure Overview: A Bunsen flame is ignited, adjusted, and analyzed. Various laboratory balances are operated and used. Mass and volume data are collected and used to determine the density of a solid and of a liquid.

Perform the experiment with a partner. At each circled superscript (1–7) in the procedure, *stop*, and record your observation on the Report Sheet. Discuss your observations with your lab partner and your instructor.

A. Bunsen Burner

1. **Lighting the Burner.** Properly light a burner using the following sequence of steps:
   a. Attach the tubing from the burner to the gas outlet on the lab bench. Close the gas control valve on the burner (see Figure 1.1) and fully open the gas valve at the outlet.
   b. Close the air holes at the base of the burner and slightly open the gas control valve.

c. Bring a lighted match or striker up the outside of the burner barrel until the escaping gas at the top ignites.

d. After the gas ignites, adjust the gas control valve until the flame is pale blue and has two or more distinct cones.

e. Slowly open the air control valve until you hear a slight buzzing. This sound is characteristic of the hottest flame from the burner. Too much air may blow the flame out. When the *best* adjustment is reached, three distinct cones are visible (Figure 1.3).

f. *If the flame goes out,* immediately close the gas valve at the outlet and repeat the procedure for lighting the burner.①

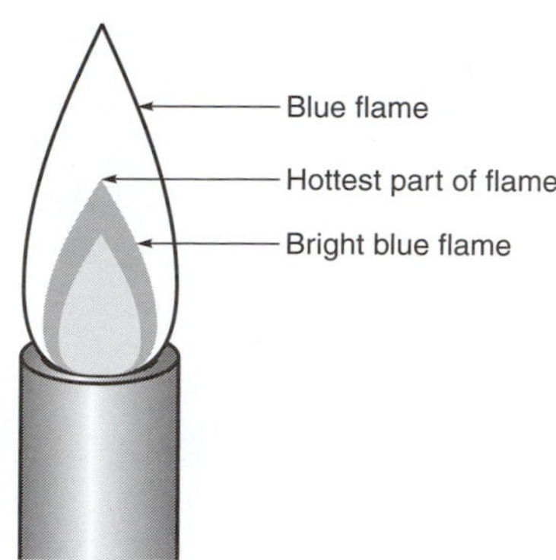

Figure 1.3 Flame of a properly adjusted Bunsen burner.

2. **Observing Flame Temperatures Using a Wire Gauze.** Temperatures within the second (inner) cone of a nonluminous blue flame approach 1500°C.

   a. Using crucible tongs (or forceps), hold a wire gauze parallel to the burner barrel about 1 cm above the burner top (Figure 1.4). Observe the relative heat zones of the flame. Sketch a diagram of your observations on the Report Sheet.

   b. *Close* the air control valve and repeat the observation with a luminous flame.

3. **Observing Flame Temperatures Using the Melting Points of Metals.** a. Adjust the burner to a nonluminous flame. Use crucible tongs to hold 2-cm strips of copper wire (melting point 1083°C), iron wire (melting point 1535°C), and aluminum (melting point 660°C) in the various regions of the flame.

   b. On the Report Sheet record the estimated temperature of the flame in the regions designated in Figure 1.5.② Extinguish the flame when it is not being used by turning off the gas valve at the outlet on the lab bench.

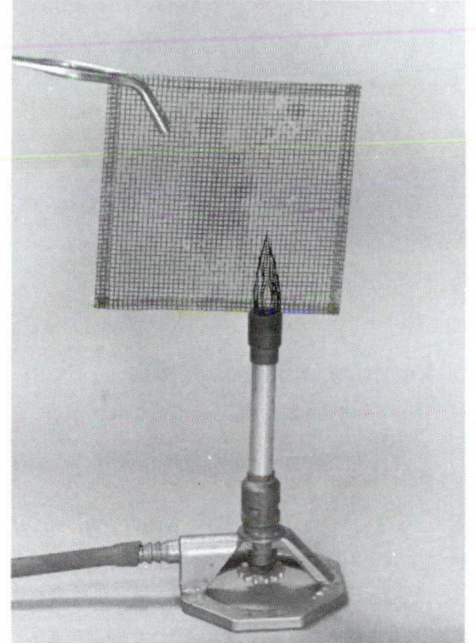

Figure 1.4 Hold the wire gauze parallel to the burner barrel.

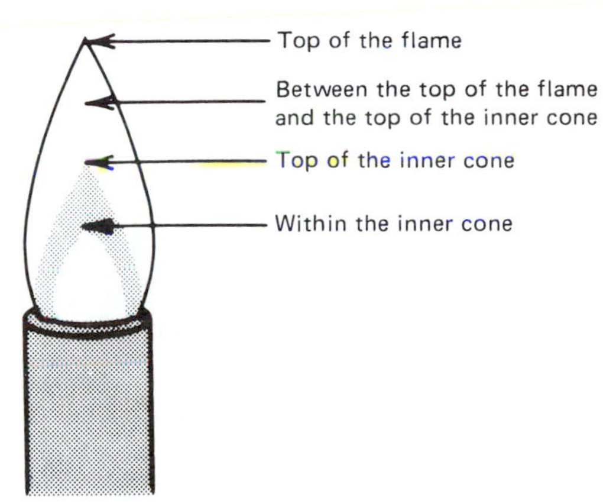

Figure 1.5 Regions of the flame for temperature measurements.

B. Laboratory Balances

1. **Practice Using the Balances.** Refer to the Report Sheet. As suggested, measure the mass of several objects. Use the top-loading balance only after the instructor explains its operation. Be sure to record the mass of the objects according to the sensitivity of the balance.③ Refer to Technique 6.

C. Density

Ask your instructor which balance you are to use to determine the densities of your unknowns. Write the balance number on the Report Sheet.④

1. **Water-Insoluble Solid.** a. Obtain an unknown solid and record its number.⑤ Using the assigned balance tare the mass of a piece of weighing paper, place the

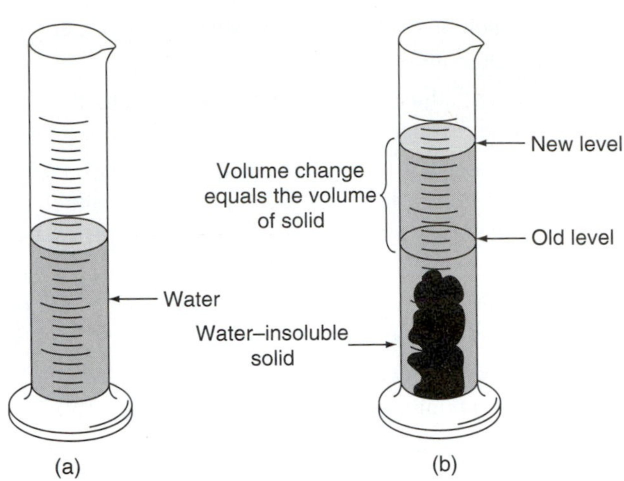

Figure 1.6 Apparatus for measuring the density of a water-insoluble solid.

solid on the weighing paper, and measure its mass. Record the mass to the sensitivity allowed by the balance.

b. Half-fill a 10-mL graduated cylinder with water and record its volume (Figure 1.6a). Refer to Technique 16A for reading and recording a volume.

c. Gently slide the known mass of solid into the graduated cylinder held at a 45° angle. Roll the solid around in the cylinder, removing any air bubbles that are trapped or that adhere to the solid. Record the new water level (Figure 1.6b). The volume of the solid is the difference between the two water levels. Repeat this procedure with a second "dry" solid sample for Trial 2.

Disposal: Check with your instructor for the procedure of properly returning the solid sample.

Refer to Appendix B for calculating average values.

2. **Liquid, Water.** a. Clean and dry your smallest laboratory beaker. Using your assigned balance, tare its mass. Pipet 5 mL of water into the beaker.

   b. Measure the mass of the water. Calculate the density of water from the available data. Repeat the density determination for Trial 2.

   c. Collect and record the density value of water at room temperature from five additional laboratory measurements from classmates.⑥ Calculate the average density of water at room temperature.

Flammable: capable of igniting in air (generally initiated with a flame or spark)

3. **Liquid, Unknown.** a. Dry the beaker and pipet. Ask the instructor for a liquid unknown and record its number.⑦ (**Caution:** *The unknown liquid may be* ***flammable.*** *Do not inhale the fumes of the liquid; extinguish all flames.*)

   b. Rinse the pipet with two 1-mL quantities of the unknown liquid and discard. Repeat the measurements of Part C.2, substituting the unknown liquid for the water. Repeat this experiment for Trial 2. Calculate the average density of the liquid.

Disposal: Check with your instructor. Dispose of the unknown liquid and the rinses in the "Waste Liquids" container.

Experiment 1 *Prelaboratory Assignment*

Basic Laboratory Operations

Date __________ Lab Sec. ______ Name ____ Desk No. _________

1. a. What is the dominant color of a nonluminous flame from a Bunsen burner? Explain.

   b. Is the temperature of a luminous flame greater or less than that of a nonluminous flame? Explain.

   c. Why does a luminous flame appear yellow?

2. Experimental Procedure, Part C.1. What is the meaning of the phrase, "tare the mass of a piece of weighing paper?"

3. A fire in the "pits" at the Indianapolis 500 Motor Speedway is especially dangerous because the flame from the fuel used in the race cars is nearly colorless and nonluminous, unlike that of a gasoline fire. How might the fuel used in the Indianapolis 500 race cars differ from that of gasoline?

4. Refer to Technique 16B.
   a. Remove the drop suspended from a pipet tip by . . .

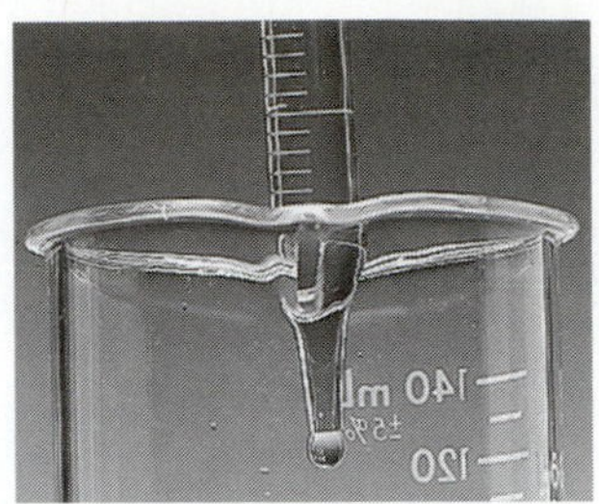

   b. The finger used to control the delivery of liquid from a pipet is the . . .

   c. A pipet is filled with the aid of a . . .

   d. Most pipets are calibrated as "TD 20°C". Define "TD" and what is its meaning regarding the volume of liquid a pipet delivers?

5. Experimental Procedure, Part C.1. The density of diamond is 3.51 g/cm^3 and the density of lead is 11.3 g/cm^3. If equal masses of diamond and lead are transferred to equal volumes of water in separate graduated cylinders, which graduated cylinder would have the greatest volume change? Explain.

6. Experimental Procedure, Part C.3. The mass of a beaker is 5.333 g. After 5.00 mL of hexane, C_6H_{14}, is pipetted into the beaker, the combined mass of the beaker and the hexane sample is 8.613 g. From the data, what is the measured density of hexane?

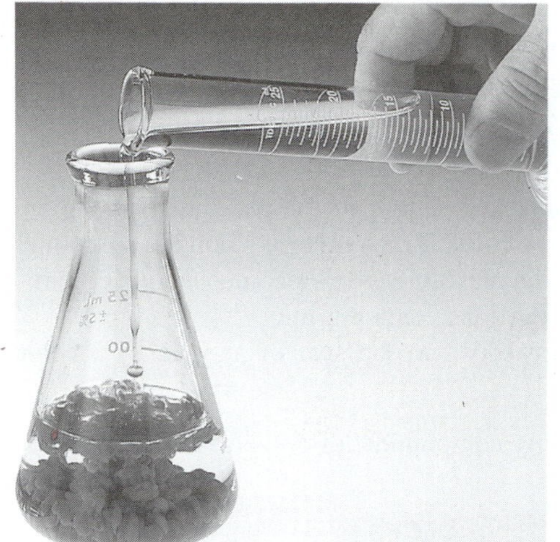

Experiment 2

Identification of a Compound: Chemical Properties

A potassium chromate solution added to a silver nitrate solution results in the formation of insoluble silver chromate.

Objectives

- To identify a compound on the basis of its **chemical properties**
- To design a systematic procedure for determining the presence of a particular compound in aqueous solution

Chemical property: characteristic of a substance that is dependent on its chemical environment

Techniques

The following techniques are used in the Experimental Procedure

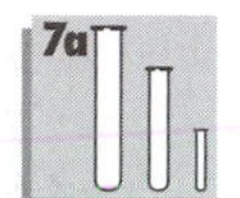

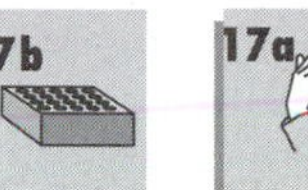

Introduction

Chemists, and scientists in general, develop and design experiments in an attempt to understand, explain, and predict various chemical phenomena. Carefully controlled (laboratory) conditions are needed to minimize the many parameters that affect the observations. Chemists organize and categorize their data, and then systematically analyze the data to reach some conclusion; often, the conclusion may be to carefully plan more experiments!

It is presumptuous to believe that a chemist must know the result of an experiment before it is ever attempted; most often an experiment is designed to determine the presence or absence of a **substance** or to determine or measure a parameter. A goal of the environmental or synthesis research chemist is, for example, to separate the substances of a reaction mixture (either one generated in the laboratory or one found in nature) and then identify each substance through a systematic, or sometimes even a **trial-and-error,** study of their chemical and physical properties. As you will experience later, Experiments 34–37 are designed to identify a specific ion (by taking advantage of its unique chemical properties) in a mixture of ions through a systematic sequence of analyses.

Substance: a pure element or compound having a unique set of chemical and physical properties

Trial-and-error study: a method that is often used to seek a pattern in the accumulated data

On occasion an experiment reveals an observation and data that are uncharacteristic of any other set of properties for a known substance, in which case either a new compound has been synthesized/discovered or experimental errors and/or interpretations have infiltrated the study.

In this experiment, you will observe chemical reactions that are characteristic of various compounds under controlled conditions. After collecting and organizing your data, you will be given an unknown compound, a compound that you have previously investigated. The interpretations of the collected data will assist you in identifying your compound.

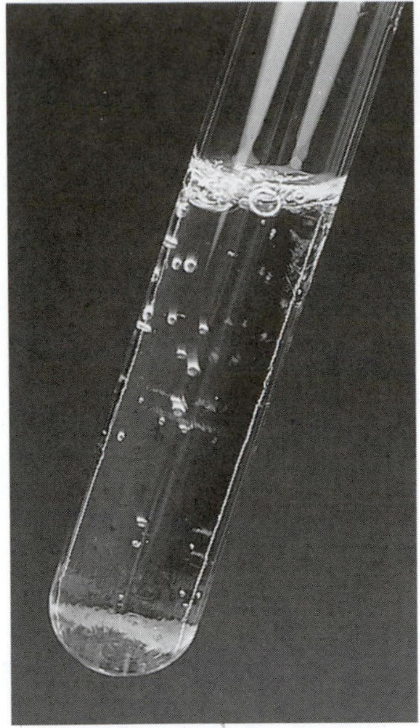

Figure 2.1 A reaction mixture of $NaHCO_3(aq)$ and $HCl(aq)$ produces CO_2 gas.

What observations will you be looking for? Chemical changes are generally accompanied by one or more of the following:

- A *gas* is evolved. This evolution may be quite rapid or it may be a "fizzing" sound (Figure 2.1).
- A *precipitate* appears (or disappears). The nature of the precipitate is important; it may be crystalline, it may have color, it may merely cloud the solution.
- *Heat* may be evolved or absorbed. The reaction vessel becomes warm if the reaction is exothermic or cools if the reaction is endothermic.
- A *color change* occurs. A substance added to the system may cause a color change.

The chemical properties of the following compounds, dissolved in water, are investigated in Part A of this experiment:

Sodium chloride	$NaCl(aq)$
Sodium carbonate	$Na_2CO_3(aq)$
Magnesium sulfate	$MgSO_4(aq)$
Ammonium chloride	$NH_4Cl(aq)$
Water	$H_2O(l)$

The following test **reagents** are used to identify and characterize these compounds:

Silver nitrate	$AgNO_3(aq)$
Sodium hydroxide	$NaOH(aq)$
Hydrochloric acid	$HCl(aq)$

Reagent: a solid chemical or a solution having a known concentration of solute

In Part B of this experiment the chemical properties of five compounds in aqueous solutions, labeled 1 through 5, are investigated with three reagents, labeled A, B, and C. All possible chemical tests will be performed with these eight solutions. An unknown will then be issued and matched with one of the solutions, labeled 1 through 5.

Experimental Procedure

Procedure Overview: In Part A a series of tests for the chemical properties of known compounds in aqueous solutions are conducted. A similar series of tests are conducted on an unknown set of compounds in Part B. In each case, an unknown compound is identified on the basis of the chemical properties observed.

You should discuss and interpret your observations on the known chemical tests with a partner, but each of you should analyze your own unknown compound. At each circled superscript (1–7) in the procedure, *stop*, and record your observation on the Report Sheet.

To organize your work, you will conduct a test on each known compound in the five aqueous solutions and the unknown compound with a single test reagent. It is therefore extremely important that you are organized and that you describe your observations as completely and clearly as possible. The Report Sheet provides a "reaction matrix" for you to describe your observations; because the space is limited, you may want to devise a code, e.g.,

- p—precipitate + color
- c—cloudy + color
- nr—no reaction
- g—gas, no odor
- go—gas, odor

A. Chemical Properties of Known Compounds

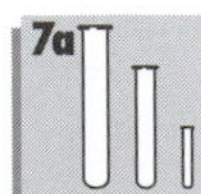

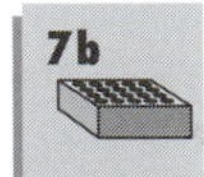

1. **Observations with Silver Nitrate Test Reagent.** a. Use a permanent marker to label five small, clean test tubes (Figure 2.2a) or set up a clean 24-well plate (Figure 2.2b). Ask your instructor which setup you should use. Place 5–10 drops of each of the five "known" solutions into the labeled test tubes (or wells A1–A5).

   b. Use a dropper pipet (or a dropper bottle) for the delivery of the silver nitrate solution. (**Caution:** *$AgNO_3$ forms black stains on the skin. The stain, caused by*

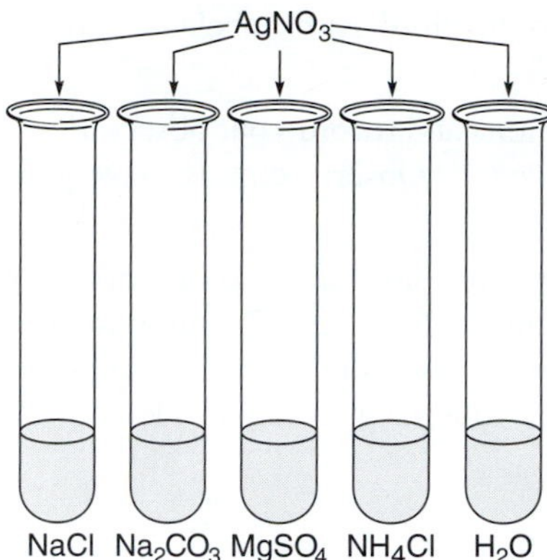

Figure 2.2a Arrangement of test tubes for testing with the silver nitrate reagent.

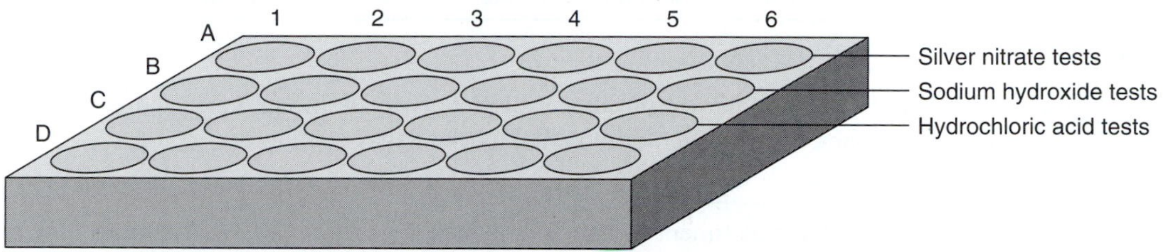

Figure 2.2b Arrangement of test solutions in the 24-well plate for testing salts.

silver metal, causes no harm.) If after adding several drops you observe a chemical change, add 5–10 drops to see if there are additional changes. Record your observations in the reaction matrix on the Report Sheet.① Save your test solutions. Write the formulas for each of the precipitates that formed.

Appendix G

2. **Observations with Sodium Hydroxide Test Reagent.** a. Use a permanent marker to label five additional small, clean test tubes (Figure 2.3). Place 5–10 drops of each of the five "known" solutions into this second set of labeled test tubes (or wells B1–B5).

   b. To each of these solutions slowly add 5–10 drops of the sodium hydroxide solution; make observations as you add the solution. What observations are made? Check to see if a gas evolves in any of the tests. Check for odor. What is the nature of any precipitates that form? Observe *closely*.② Save your test solutions for reference in Part A.4. Write the formulas for each of the precipitates that formed.

Appendix G

3. **Observations with Hydrochloric Acid Test Reagent.** a. Use a permanent marker to label five additional small, clean test tubes (Figure 2.4). Place 5–10

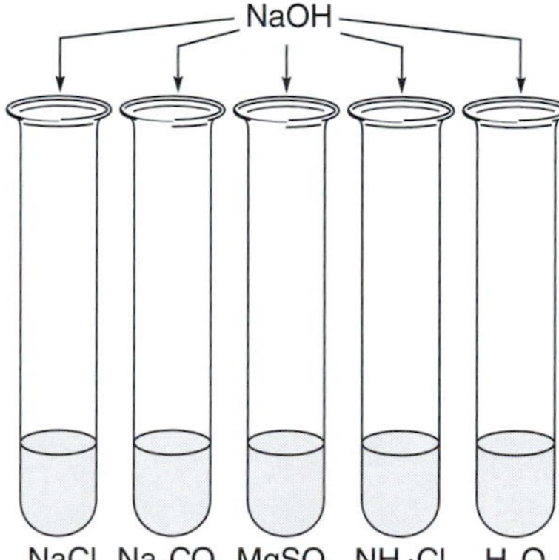

Figure 2.3 Arrangement of test tubes for testing with the sodium hydroxide reagent.

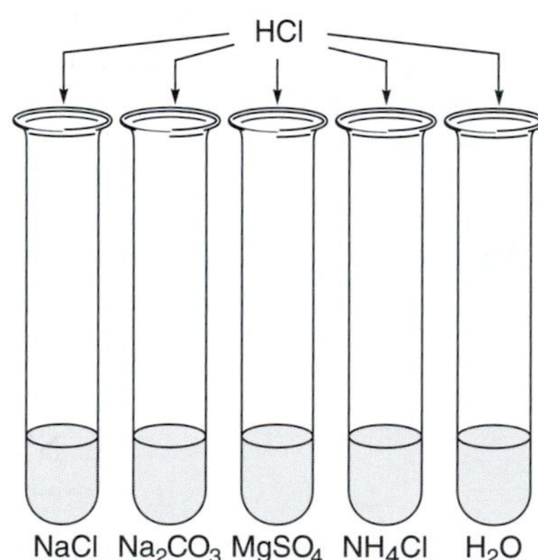

Figure 2.4 Arrangement of test tubes for testing with the hydrochloric acid reagent.

drops of each of the five "known" solutions into this third set of labeled test tubes (or wells C1–C5).

b. Add the hydrochloric test reagent to the solutions and record your observations. Check to see if any gas is evolved. Check for odor.③ Observe closely. Save your test solutions for reference in Part A.4.

4. **Identification of Unknown.** Obtain an unknown for Part A from your laboratory instructor. Repeat the three tests with the reagents in Parts A.1, 2, and 3 on your unknown. On the basis of the data from the "known" solutions (collected and summarized in the reaction matrix) and that of your unknown solution, identify the compound in your unknown solution.④

Disposal: Discard the test solutions in the "Waste Salts" container.

CLEANUP: Rinse the test tubes or well plate twice with tap water and twice with deionized water. Discard each rinse in the "Waste Salts" container.

B. Chemical Properties of Unknown Compounds

The design of the experiment in Part B is similar to that of Part A. Therefore, fifteen test tubes or a 24-well plate is necessary.

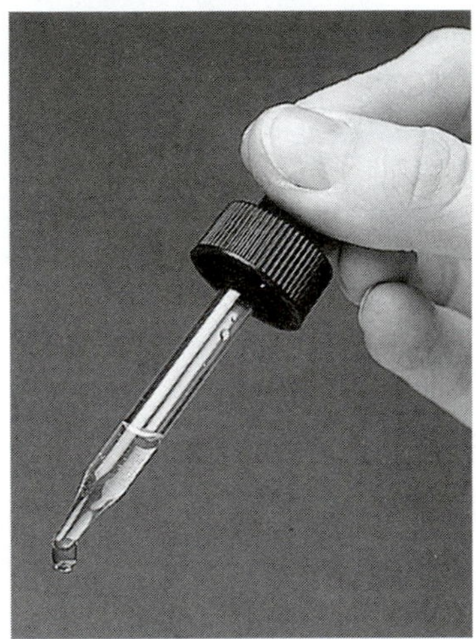

A dropper pipet.

1. **Preparation of Solutions.** On the reagent shelf are five solutions, labeled 1 through 5, each containing a different compound. Use small clean test tubes or the well plate as your testing laboratory. About 1 mL of each test solution is necessary for analysis.
2. **Preparation of Reagents.** Also on the reagent shelf are three reagents labeled A, B, and C. Use a dropper pipet (or dropper bottle) or a Beral pipet to deliver Reagents A through C to the solutions.
3. **Testing the Solutions.** Test each of the five solutions with drops (and then excess drops) of Reagent A. If, after adding several drops, you observe a chemical change, add 5–10 drops more to see if there are additional changes.⑤ Observe closely and describe any evidence of chemical change; record your observations. With a fresh set of solutions 1–5 in clean test tubes (or wells), test each with Reagent B.⑥ Repeat with Reagent C.⑦
4. **Identification of Unknown.** An unknown solution will be issued that is one of the five solutions from Part B.1. On the basis of the data in your reaction matrix and the data you have collected, identify which of the five solutions is your unknown.

Disposal: Discard the test solutions in the "Waste Salts" container.

CLEANUP: Rinse the test tubes or well plate twice with tap water and twice with deionized water. Discard each rinse in the "Waste Salts" container.

Experiment 2 *Prelaboratory Assignment*

Identification of a Compound: Chemical Properties

Date __________ Lab Sec. ______ Name ____ Desk No. __________

1. Experimental Procedure, Part A. What is the size and volume of a "small, clean test tube?" See Technique 7A.

2. a. Depending upon the tip of a dropper pipet, there are approximately 25 drops per milliliter of water. What is the approximate volume (in mL) of one drop of an aqueous solution?

   b. A micropipet delivers 153 drops of alcohol for each milliliter. Calculate the volume (in mL) of alcohol in each drop.

3. Write a balanced equation for the following observed reactions:
   a. Aqueous solutions of sodium sulfate, Na_2SO_4, and barium chloride, $BaCl_2$, are mixed. A white precipitate of barium sulfate forms.

   b. Aqueous solutions of sodium hydroxide, NaOH, and sulfuric acid, H_2SO_4, are mixed. The neutralization product of water is formed with the evolution of heat.

   c. Aqueous solutions of copper(II) nitrate, $Cu(NO_3)_2$, and sodium carbonate, Na_2CO_3, are mixed. A blue precipitate of copper(II) carbonate forms.

4. Part A. The substances, NaCl, Na_2CO_3, $MgSO_4$, and NH_4Cl used for test solutions, are all soluble ionic compounds. For each substance indicate the ions present in its respective test solution.

5. Three colorless solutions in test tubes, with no labels, are in a test tube rack on the laboratory bench. Lying beside the test tubes are three labels: potassium iodide, KI, silver nitrate, $AgNO_3$, and sodium sulfide, Na_2S. You are to place the labels on the test tubes using only the three solutions present. Here are your tests:

   - A portion of Test Tube 1 added to a portion of Test Tube 3 produces a yellow, silver iodide precipitate.
   - A portion of Test Tube 1 added to a portion of Test Tube 2 produces a black, silver sulfide precipitate.

   Your conclusions are:

   Test Tube 1 ______________________________

   Test Tube 2 ______________________________

   Test Tube 3 ______________________________

Experiment 2 *Report Sheet*

Identification of a Compound: Chemical Properties

Date __________ Lab Sec. ______ Name ______________________________ Desk No. __________

A. Chemical Properties of Known Compounds

Test	$NaCl(aq)$	$Na_2CO_3(aq)$	$MgSO_4(aq)$	$NH_4Cl(aq)$	$H_2O(l)$	Unknown
① $AgNO_3(aq)$	________	________	________	________	________	________
② $NaOH(aq)$	________	________	________	________	________	________
③ $HCl(aq)$	________	________	________	________	________	________

Sample No. of unknown for Part A ____________________

④ Compound in unknown solution ____________________

Write formulas for the precipitates that formed in Part A. (See Appendix G)

B. Chemical Properties of Unknown Compounds

Solution No.	1	2	3	4	5	Unknown
⑤ Reagent A	________	________	________	________	________	________
⑥ Reagent B	________	________	________	________	________	________
⑦ Reagent C	________	________	________	________	________	________

Sample No. of unknown for Part B ____________________

Compound of unknown is the same as Solution No. ____________________

Laboratory Questions

Circle the questions that have been assigned.

1. Part A. a. Identify the silver salts that form precipitates. Write the formulas of the precipitates.
   b. Identify the hydroxide salts that form precipitates. Write the formulas of the precipitates.

2. Three colorless solutions in test tubes, with no labels, are in a test tube rack on the laboratory bench. Lying beside the test tubes are three labels: silver nitrate, $AgNO_3$, hydrochloric acid, HCl, and sodium carbonate, Na_2CO_3. You are to place the labels on the test tubes using only the three solutions present. Here is your analysis procedure:

   - A portion of Test Tube 1 added to a portion of Test Tube 2 produces carbon dioxide gas, CO_2.
   - A portion of Test Tube 2 added to a portion of Test Tube 3 produces a white, silver carbonate precipitate.

   a. On the basis of your observations how would you label the three test tubes?
   b. What would you expect to happen if a portion of Test Tube 1 is added to a portion of Test Tube 3?

3. For a solution of the cations Ag^+, Ba^{2+}, Mg^{2+}, or Cu^{2+}, the following experimental observations were collected:

	$NH_3(aq)$	$HCl(aq)$	$H_2SO_4(aq)$
Ag^+	No change	White ppt[a]	No change
Ba^{2+}	No change	No change	White ppt
Mg^{2+}	White ppt	No change	No change
Cu^{2+}	Blue ppt/deep blue soln with excess	No change	No change

[a]Example: when an aqueous solution of hydrochloric acid is added to a solution containing Ag^+, a white precipitate (ppt) forms.

   a. Identify a reagent that distinguishes the chemical properties of Ag^+ and Mg^{2+}. Explain.
   b. Identify a reagent that distinguishes the chemical properties of HCl and H_2SO_4. Explain.
   c. Identify a reagent that distinguishes the chemical properties of Ba^{2+} and Cu^{2+}. Explain.
   *d. Identify a reagent that distinguishes the chemical properties of Cu^{2+} and Mg^{2+}. Explain.

4. Three colorless solutions in test tubes, with no labels, are in a test tube rack on the laboratory bench. Lying beside the tests tubes are three labels: 0.10 *M* Na_2CO_3, 0.10 *M* HCl, and 0.10 *M* KOH. You are to place the labels on the test tubes using only the three solutions present. Here are your tests:

   - A few drops of the solution from test tube #1 added to a similar volume of the solution in test tube #2 produces no visible reaction but the solution becomes warm.
   - A few drops of the solution from test tube #1 added to a similar volume of the solution in test tube #3 produces carbon dioxide gas.

   Identify the labels for test tube #1, #2, and #3.

Experiment 3

Water Analysis: Solids

"Clear" water from streams contain small quantities of dissolved and suspended solids.

Objective

- To determine the total, dissolved, and suspended solids in a water sample

Techniques

The following techniques are used in the Experimental Procedure:

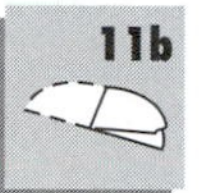

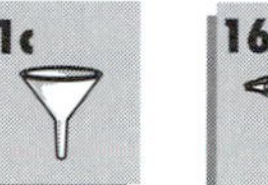

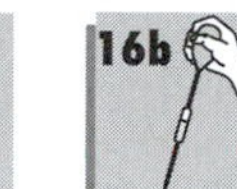

16b

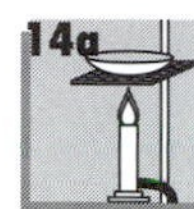

Introduction

Surface water is used as the primary drinking water source for most large municipalities. The water is piped into a water treatment facility where impurities are removed and bacteria are killed before the water is placed into the distribution lines. The contents of the surface water must be known and predictable so that the treatment facility can properly and adequately remove these impurities. Tests are used to determine the contents of the surface water.

Water in the environment has a large number of impurities with an extensive range of concentrations. **Dissolved solids** are water-soluble substances, most often salts although some dissolved solids may come from organic sources. Naturally occurring dissolved salts generally result from the movement of water over or through mineral deposits, such as limestone. These dissolved solids, characteristic of the **watershed,** generally consist of the sodium, calcium, magnesium, and potassium cations and the chloride, sulfate, bicarbonate, carbonate, bromide, and fluoride anions. The dissolved salts are responsible for the "hard" water that exists in some locales. Anthropogenic (human-related) dissolved solids include nitrates from fertilizer runoff and human wastes, phosphates from detergents and fertilizers, and organic compounds from pesticides, sewage runoff, and industrial wastes.

Watershed: the land area from which the natural drainage of water occurs

Salinity, a measure of the total salt content in a water sample, is expressed as the grams of dissolved salts per kilogram of water or as parts per thousand (ppt). The average ocean salinity is 35 ppt whereas fresh water salinity is usually less than 0.5 ppt. Brackish water, where fresh river water meets salty ocean water, varies from 0.5 ppt to 17 ppt. For water samples with low organic levels, the salinity of a water sample approximates that of total dissolved solids (TDS) content.

Suspended solids are very finely divided particles, kept in suspension by the turbulent action of the moving water, are *insoluble* in water but are filterable. Total suspended solids (TSS) is a measure of the **turbidity** and the clarity of the water. Examples of suspended solids include decayed organic matter, sand, salt, and clay.

Suspended solids: solids that exhibit colloidal properties or solids that remain in the water because of turbulence

Total solids are the sum of the dissolved and suspended solids in the water sample. In this experiment the total solids and the dissolved solids are determined directly; the suspended solids are assumed to be the difference, since

$$\text{total solids} = \text{total dissolved solids (TDS)} + \text{total suspended solids (TSS)} \quad (3.1)$$

The U.S. Public Health Service recommends that drinking water not exceed 500 mg total solids/kg water. However, in some localities, the total solids content may range up to 1000 mg/kg of potable water; that's 1 g/L!! An amount over 500 mg/kg water does not mean the water is unfit for drinking; an excess of 500 mg/kg is merely not recommended.

Experimental Procedure

Filtrate: the solution that passes through the filter into a receiving flask.

Turbid sample: a cloudy suspension due to stirred sediment.

Procedure Overview: The total, dissolved, and suspended solids in a water sample are determined in this experiment. The water sample is filtered to remove the suspended solids and the **filtrate** is evaporated to dryness to determine the dissolved solids. The water sample may be from the ocean, a lake, a stream, or from an underground aquifer.

Obtain 100 mL of a water sample from your instructor. Preferably the water sample is high in **turbidity.** Record the sample number and write a short description of the sample on the Report Sheet. With approval, bring your own "environmental" water sample to the laboratory for analysis. Ask your instructor whether evaporating dishes or 250 mL beakers are to be used for the analysis.

A. Total Dissolved Solids (TDS)

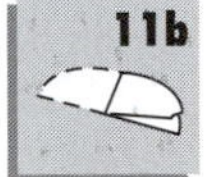

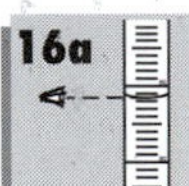

Cool flame: a Bunsen flame of low intensity—a slow rate of natural gas is flowing through the burner barrel.

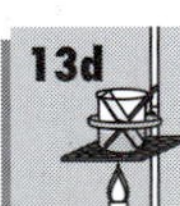

1. **Filter the Water Sample.** Gravity filter about 50 mL of a thoroughly stirred or shaken water sample into a clean, dry 100-mL beaker. While waiting for the filtration to be completed, proceed to Part B.
2. **Evaporate the Filtrate to Dryness.** a. Clean, dry, and tare the mass ($\pm$0.001 g) of an evaporating dish (or 250-mL beaker).
   b. Pipet a 25-mL aliquot (portion) of the filtrate into the evaporating dish (250-mL beaker). Determine the tared mass of the sample.
   c. Place the dish (or beaker) containing the sample on a wire gauze (Figure T.14a or T.13c) and *slowly* heat—*do not boil*—the mixture to dryness. A hot plate may be substituted for a direct, **"cool" flame** (Figure T.13e).
   d. As the mixture nears dryness, cover the evaporating dish (beaker) with a watch glass and reduce the intensity of the flame.[1] If spattering occurs, allow the dish to cool to room temperature, rinse the adhered solids from the watch glass (Figure 3.1) and return the rinse to the dish.
3. **A Final Heating to Dryness.** Again heat slowly, being careful to avoid further spattering. After all of the water has evaporated, maintain a "cool" flame beneath the dish for 3 minutes. Allow the dish to cool to room temperature and determine its final mass. Cool the evaporating dish and sample in a desiccator, if available.

[1]This reduces the spattering of the remaining solid and its subsequent loss in the analysis.

Figure 3.1 Wash the spattered material from the convex side of the watch glass.

B. Total Solids and Total Suspended Solids (TSS)

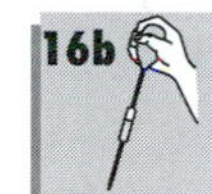

1. **Evaporate an Original Water Sample to Dryness.** a. Clean, dry, and tare the mass (±0.001 g) of a second evaporating dish (or 250-mL beaker).

   b. Thoroughly stir or agitate 100 mL of the original water sample; pipet a 25-mL aliquot of this sample into the evaporating dish (250-mL beaker).[2] Record the tared mass of the water sample.

   c. Evaporate *slowly* the sample to dryness as described in Part A.2. Record the tared mass of the solids remaining in the evaporating dish.

2. **Total Suspended Solids.** Collect the appropriate data to determine the total suspended solids in the water sample.

C. Analysis of Data

Appendix B

1. **Precision of Data?** Compare your TDS and TSS data with three other chemists in your laboratory who have analyzed the *same* water sample. Record their results on the Data Sheet. Calculate the average value for the TSS in the water sample.

[2]If the solution appears to be so turbid that it may plug the pipet tip, use a 25-mL graduated cylinder to measure the water sample as accurately as possible.

D. Chemical Tests[3]

1. **Test for Carbonates and Bicarbonates.** With your spatula loosen small amounts of the dried samples from Part A and Part B and transfer each to *separately* marked 75-mm test tubes. Add several drops of 6 *M* HNO_3 (**Caution:** HNO_3 is corrosive and a severe skin irritant) and observe. What can you conclude from your observation?
2. **Test for Chlorides (Halides).** To each of the test tubes add 10 drops of water, agitate the solution, and add several drops of 0.01 *M* $AgNO_3$ (**Caution:** $AgNO_3$ is a skin irritant) and observe. What can you conclude from your observation?

Appendix G

3. **Test for Sulfates.** With your spatula loosen small amounts of the dried samples from Part A and Part B and transfer each to *separately* marked 75-mm test tubes. Add about 10 drops of water, agitate the solution, and add several drops of 1 *M* $Ba(NO_3)_2$ and observe. What can you conclude from your observation?
4. **Test for Calcium Ion.** With your spatula loosen small amounts of the dried samples from Part A and Part B and transfer each to *separately* marked 75-mm test tubes. Add about 10 drops of water, agitate the solution, and add several drops of 1 *M* $K_2C_2O_4$ and observe. What can you conclude from your observation?

Disposal: Discard the dried salts from Part A and B and the test solutions from Part D in the "Waste Salts" container.

CLEANUP: Rinse the test tubes and evaporating dishes (250-mL beakers) with tap water and twice with deionized water.

[3]For each of the tests in Part D there are other ions that may show a positive test. However the ions being tested are those most common in environmental water samples.

Experiment 3 *Prelaboratory Assignment*

Water Analysis: Solids

Date __________ Lab Sec. ______ Name ______________________________ Desk No. __________

1. List several cations and anions, by formula, that contribute to the salinity of a water sample.

2. Distinguish between and characterize the "total dissolved solids" and "total suspended solids" in a water sample?

3. Experimental Procedure, Part A.2. a. What does the expression, "tare the mass ($\pm$0.001 g) of an evaporating dish" mean? See Technique 6.

   b. What does the expression, "the tared mass of the sample" mean?

4. a. What is an **aliquot** of a sample?

   b. What is the **filtrate** in a gravity filtration procedure?

   c. How full (the maximum level) should a funnel be filled with solution in a filtration procedure?

5. Experimental Procedure, Part A.3. In evaporating a solution, why must the intensity of the heat be reduced as the mixture nears dryness?

6. Experimental Procedure, Part D. What observation is "expected" when
   a. an acid (nitric acid, HNO_3) is added to a solution containing carbonate or bicarbonate ions? See Experiment 2, Experimental Procedure, Part A.3

   b. silver ion is added to a solution containing chloride (or bromide or iodide) ions? See Appendix G.

   c. barium ion is added to a solution containing sulfate ions? See Appendix G.

7. A 25.0-mL aliquot of a well-shaken sample of river water is pipetted into a 27.211-g evaporating dish. After the mixture is evaporated to dryness, the dish and dried sample has a mass of 43.617 g. Determine the total solids in the sample; express total solids in units of g/kg sample (parts per thousand, ppt). Assume the density of the river water to be 1.01 g/mL.

Experiment 3 *Report Sheet*

Water Analysis: Solids

Date __________ Lab Sec. ______ Name ____ Desk No. __________

Sample Number: ____________ Describe the nature of your water sample, i.e., its color, turbidity, etc.

A. Total Dissolved Solids (TDS) *Trial 1*

1. Tared mass of water sample (g) ______________________
2. Tared mass of *dried* sample (g) ______________________
3. Mass of dissolved solids in 25-mL aliquot of filtered sample (g) ______________________
4. Mass of dissolved solids per total mass of sample (g solids/g sample) ______________________
5. Total dissolved solids or salinity (g solids/kg sample, ppt) ______________________

B. Total Solids and Total Suspended Solids (TSS)

1. Tared mass of water sample (g) ______________________
2. Tared mass of *dried* sample (g) ______________________
3. Mass of total solids in 25-mL aliquot of unfiltered sample (g) ______________________
4. Mass of total solids per total mass of sample (g solids/g sample) ______________________
5. Total solids (g solids/kg sample, ppt) ______________________
6. Total suspended solids (g solids/kg sample, ppt), ss_1 ______________________

C. Analysis of Data

	Chemist #1 (you)	Chemist #2	Chemist #3	Chemist #4
TDS (g/kg)				
Total Solids (g/kg)				
TSS (g/kg)	$ss_1 =$	$ss_2 =$	$ss_3 =$	$ss_4 =$

1. Average value of suspended solids from four chemists (x) = __________

D. Chemical Tests.

1. Test for carbonates and bicarbonates. Observation and conclusion.

2. Test for chlorides (halides). Observation and conclusion.

3. Test for sulfates. Observation and conclusion.

4. Test for calcium ion. Observation and conclusion.

Laboratory Questions

Circle the questions that have been assigned.

1. Part A.2. The evaporating dish was not properly cleaned of a volatile material before its tared mass was determined. When the sample is heated to dryness the volatile material is removed. How does this error in technique affect the reported TDS for the water sample? Explain.
2. Part A.2. Some spattering of the sample does occur onto the watchglass near dryness. In a hurry to complete the analysis, the chemist chooses not to return the spattered solids to the original sample and skip the first part of Part A.3. Will the reported TDS for the water sample be too high or too low? Explain.
3. Part A.3. The sample in the evaporating dish is *not* heated to total dryness. How will this error in technique affect the reported value for
   a. TDS? Explain.
   b. TSS? Explain.
4. Part A.3. As the sample cools moisture from the atmosphere condenses on the outside of the evaporating dish before the mass is measured. How does the presence of the condensed moisture affect the reported TDS for the water sample? Explain.
5. Part B.1. The sample in the evaporating dish is *not* heated to total dryness. How will this error in technique affect the reported value for
   a. total solids? Explain.
   b. TSS? Explain.
6. In this Experimental Procedure, two water samples were needed, one for Part A and a second for Part B. How could this experimental procedure have been rewritten so that only a single sample would provide the same data?

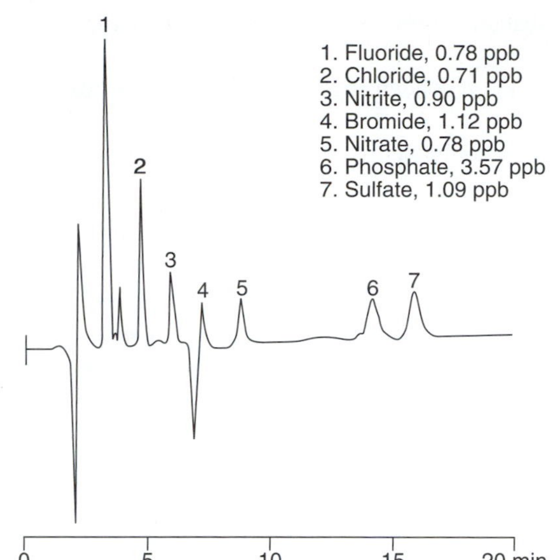

Experiment 4

Paper Chromatography

Trace levels of anions in high purity water can be determined by chromatography.

OBJECTIVES

- To become familiar with a technique for separating a mixture by chromatography
- To separate a mixture of transition metal cations by paper chromatography

TECHNIQUES

The following techniques are used in the Experimental Procedure

INTRODUCTION

Most substances found in nature, and many prepared in the laboratory, are impure; that is, they are a part of a mixture. One goal of chemical research is to devise methods for separating mixtures to remove impurities from the chemical of interest.

A **mixture** is a physical combination of two or more pure substances wherein each substance retains its own chemical identity. For example, each component in a sodium chloride/water mixture possesses the same chemical properties as in the pure state: water consists of H_2O molecules, and sodium chloride is sodium ions, Na^+, and chloride ions, Cl^-.

The method chosen for separating a mixture is based on the differences in the chemical and/or physical properties of the components of the mixture. Some common *physical* methods for separating the components of a mixture include:

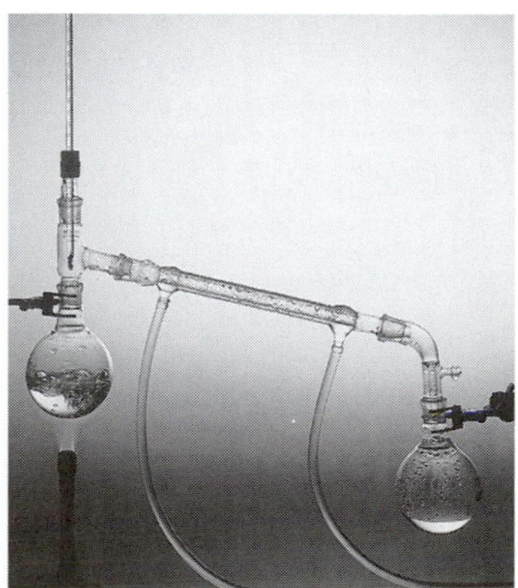
A distillation apparatus.

- **Filtration:** removing a solid substance from a liquid by passing the suspension through a filter (See Techniques 11B–E and Experiment 3 for details.)
- **Distillation:** vaporizing a liquid from a solid (or another liquid) and condensing the vapor (See margin photo.)
- **Crystallization:** forming a crystalline solid by decreasing its solubility by cooling the solution, evaporating the solvent, or adding a solvent in which the substance is less soluble
- **Extraction:** removing a substance from a solid or liquid mixture by adding a solvent in which the substance is more soluble

- **Centrifugation:** removing a substance from a solution using a centrifuge (See Technique 11F for details.)
- **Sublimation:** vaporizing a solid and recondensing its vapor (not all solids sublime, however)
- **Chromatography:** separating the components of a mixture on the basis of their differing adsorptive tendencies on a stationary phase

Mobile phase: the phase (generally liquid) in which the components of the mixture exist

Eluent: the solvent in which the components of the mixture are moved along the stationary phase

Stationary phase: the phase (generally solid) to which the components of the mobile phase are characteristically adsorbed

In the chromatography[1] technique, two phases are required for the separation of the components (or compounds) of a mixture, the mobile phase and the stationary phase. The **mobile phase** consists of the components of the mixture and the solvent—the solvent being called the **eluent** or **eluting solution** (generally a mixture of solvents of differing polarities). The **stationary phase** is an adsorbent that has an intermolecular affinity not only for the solvent,[2] but also for the individual components of the mixture.

As the mobile phase passes over the stationary phase, the chromatogram develops. The different components of the mobile phase have different affinities for the stationary phase. The components with a stronger affinity for the stationary phase move shorter distances while those with a lesser affinity move greater distances along the stationary phase during the time the chromatogram is being developed. Separation, and subsequent identification, of the components of the mixture is thus achieved. The leading edge of the mobile phase in the chromatogram is called the **solvent front.** When the solvent front reaches the edge of the chromatographic paper, the developing of the chromatogram stops.

Of the different types of chromatography including, for example, gas–solid chromatography, liquid–solid chromatography, column chromatography, thin-layer chromatography, and ion chromatography, this experiment uses paper chromatography as a separation technique.

In this experiment, chromatographic paper (similar to filter paper), a paper that consists of polar cellulose molecules, is the stationary phase. The mobile phase consists of one or more of the transition metal cations, Mn^{2+}, Fe^{3+}, Co^{2+}, Ni^{2+}, and Cu^{2+}, dissolved in an acetone–hydrochloric acid eluent.[3] In practice, the chromatographic paper is first "spotted" (and marked with a pencil *only*) with individual solutions for *each* of the cations and then for a solution with a mixture of cations. The paper is then dried. The spotted chromatographic paper is then placed in contact with the eluent to form the mobile phase. By capillary action the transition metal ions are transported along the paper. Each transition metal has its own (unique) adsorptive affinity for the polar, cellulose chromatographic paper; some are more strongly adsorbed than others. Also, each ion has its own solubility in the eluting solution. As a result of these two factors, some transition metal ions move further along the chromatographic paper than others to form **bands** at some distance from the origin, therefore indicating that the ions are separating.[4]

Band: the identifying position of the component on the chromatography paper

For a given eluting solution, stationary phase, temperature, and so on, each ion is characterized by its own R_f (ratio of fronts) factor:

Eluent: eluting solution

$$R_{f,\,ion} = \frac{\text{distance from origin to final position of ion}}{\text{distance from origin to eluent front}} = \frac{D_{ion}}{D_{solvent}} \qquad (4.1)$$

---

[1]*Chromatography* means "the graphing of colors" where historically this technique was used to separate the various colored compounds from naturally occurring colored substances (e.g., dyes, plant pigments).

[2]This attraction is due to the similar type of intermolecular forces between the adsorbent (or the stationary phase) and the components of the mobile phase (e.g., dipole–dipole, ion–dipole, dipole–hydrogen bonding).

[3]The eluting solution (acetone–hydrochloric acid) actually converts the transition metal ions to their chloro ions, and it is in this form that they are separated along the paper.

[4]Consider, for example, a group of students moving through a cafeteria food line: the stationary phase is the food line; the mobile phase consists of students. You are well aware that some students go through the food line quickly because they are not attracted to the food items. Others take forever to pass through the food line because they must stop to consider each item; they thus have a longer residence time in the food line.

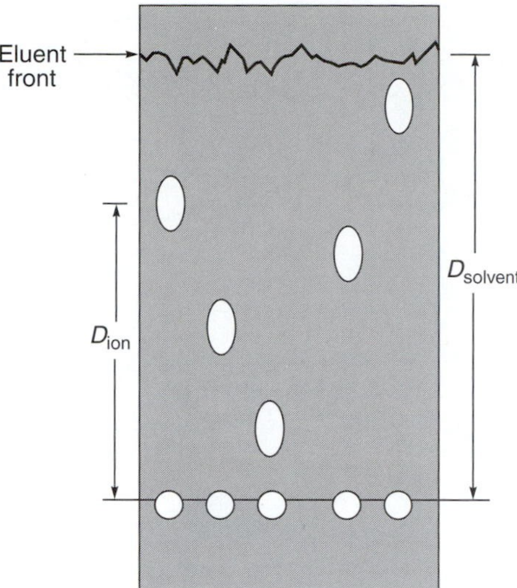

Figure 4.1 Determination of the R_f value for a transition metal ion.

Chromatogram: the "picture" of the separated components on the chromatography paper

The origin of the **chromatogram** is defined as the point where the ion is "spotted" on the paper. The eluent front is defined as the most advanced point of movement of the mobile phase along the paper from the origin (Figure 4.1).

The next step is to identify the position of these bands. Some transition metal ions are already colored; for others, a characteristic reagent for each metal ion is used to enhance the metal ion's appearance and location on the stationary phase.

In this experiment you will prepare a chromatogram for each transition metal ion, locate its characteristic band, and determine its R_f value. A chromatogram of a test solution(s) will also be analyzed to identify the transition metal ions present in the solution.

Experimental Procedure

Procedure Overview: Chromatographic paper is used to separate an aqueous mixture of transition metal cations. An eluent solution is used to move the ions along the paper; the relative solubility of the cations in the solution versus the relative adsorptivity of the cations for the paper results in their separation on the paper. An enhancement reagent is used to intensify the presence of the metal cation band on the paper.

Obtain about 2 mL of three unknown cations mixtures from your instructor in carefully marked 75-mm test tubes. Record the number of each unknown on the Report Sheet.

A. Preparation of Chromatography Apparatus

1. **Developing Chamber.** Obtain a 600-mL beaker and enough plastic wrap (e.g., Saran Wrap®) for a cover (Figure 4.2). Prepare 10 mL of an eluting solution that consists of 9 mL of acetone and 1 mL of 6 *M* HCl. (**Caution:** *Acetone is flammable; extinguish all flames; HCl is corrosive.*) Pour this eluent into the middle of the beaker using a stirring rod; be careful not to wet the sides of the beaker. The depth of the eluent in the beaker should be 0.75–1.25 cm but *less than* 1.5 cm. Cover the beaker with the plastic wrap for about 10 minutes.[5] Hereafter, this apparatus is called the **developing chamber.**

   Figure 4.3 shows a commercial developing chamber.
2. **Ammonia Chamber.** The ammonia chamber may already be assembled in the hood. *Ask your instructor.* Obtain a dry, 1000-mL beaker and place it in the fume hood. Pour 5 mL of conc NH_3 into a 30-mL beaker and position it at the center of the 1000-mL beaker (Figure 4.4). (**Caution:** *Do not inhale the ammonia fumes.*)

Figure 4.2 Use plastic wrap to cover the developing chamber.

[5] The "atmosphere" in the beaker should become saturated with the vapor of the eluent.

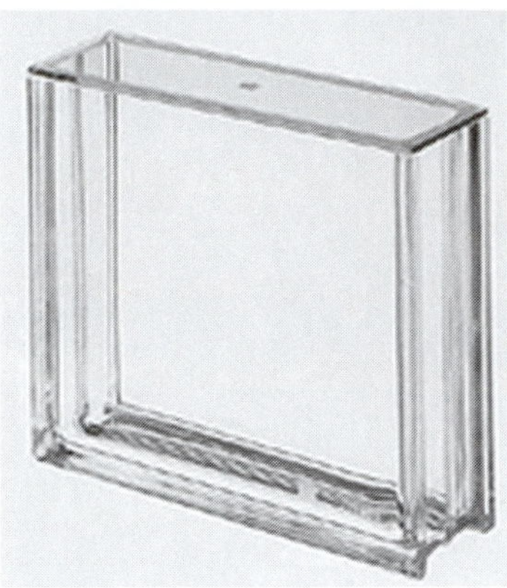

Figure 4.3 A commercial developing chamber.

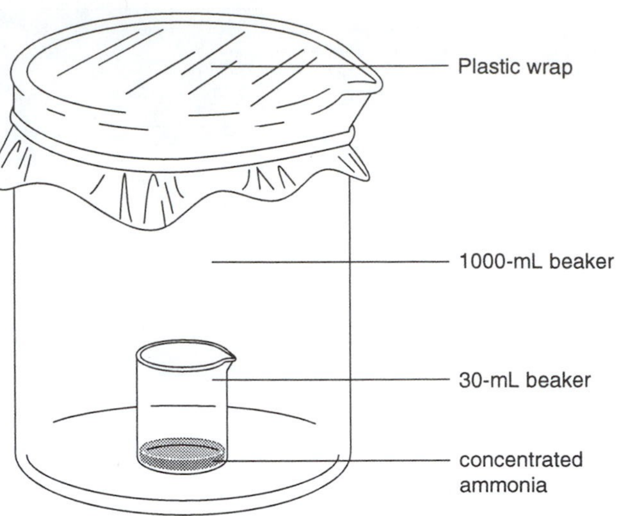

Figure 4.4 Apparatus for the ammonia chamber.

Cover the top of the 1000-mL beaker with plastic wrap. This apparatus will be used in Part C.3.

3. **Capillary Tube.** Obtain eight capillary tubes from the stockroom. When a capillary tube touches a solution, the solution should be drawn into the tip. When the tip is then touched to a piece of filter paper it should deliver a microdrop of solution. Try "spotting" a piece of filter paper with a known solution or water until the diameter of the drop is only 2–4 mm.
4. **Stationary Phase.** Obtain one piece of chromatographic paper (approximately 10 × 20 cm). Handle the paper only along its top 20-cm edge (by your designation) and lay it flat on a clean piece of paper, *not* directly on the lab bench. Draw a *pencil* line 1.5 cm from the *bottom* 20-cm edge of the paper (Figure 4.5). Starting 2 cm from the 10-cm edge and along the 1.5-cm line, make eight X's with a 2-cm separation. Use a pencil to label each X *below* the 1.5-cm line with the five cations being investigated and unknowns U1, U2, and U3.

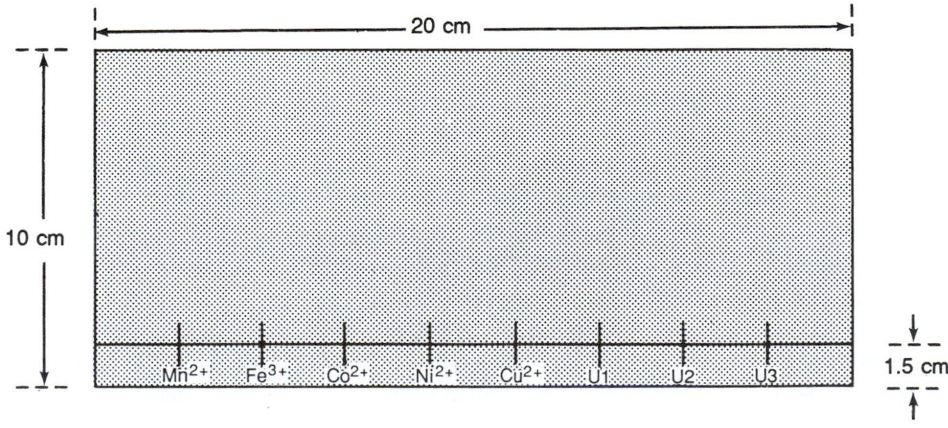

Figure 4.5 Labeling the stationary phase (the chromatographic paper) for the chromatogram.

B. Preparation of the Chromatogram

1. **Spot the Stationary Phase with the Knowns and Unknown(s).** Using the capillary tubes (remember you'll need *eight* of them, one for each solution) "spot" the chromatographic paper at the marked X's with the five known solutions containing the cations and the three unknown solutions. The microdrop should be 2–4 mm in diameter. Allow the "spots" to dry; a heat lamp or hair dryer may be used to hasten the drying—do *not* touch the paper along this bottom edge.

   Repeat the spotting/drying procedure two more times in order to increase the amount of metal ion at the "spot" on the chromatographic paper. Be sure to dry the sample between applications. Dry the paper with **caution:** The heat lamp or hair dryer is hot!

2. **Prepare the Stationary Phase for Elution.** Form the chromatographic paper into a cylinder and, near the top, attach the ends with tape, a staple, or small paper clip; do *not* allow the two ends of the paper to touch (Figure 4.6). Be sure the spots are dry.
3. **Develop the Chromatogram.** Place the paper cylinder into the developing chamber (Part A.1). The entire "bottom" of the cylindrical chromatographic paper must sit on the bottom of the developing chamber. Do *not* allow the paper to touch the wall. Make certain that the eluent is *below* the 1.5-cm line. Replace the plastic wrap and wait until the eluent front has moved to within 1.5 cm of the top of the chromatographic paper. Do not disturb the developing chamber once the paper has been placed inside.

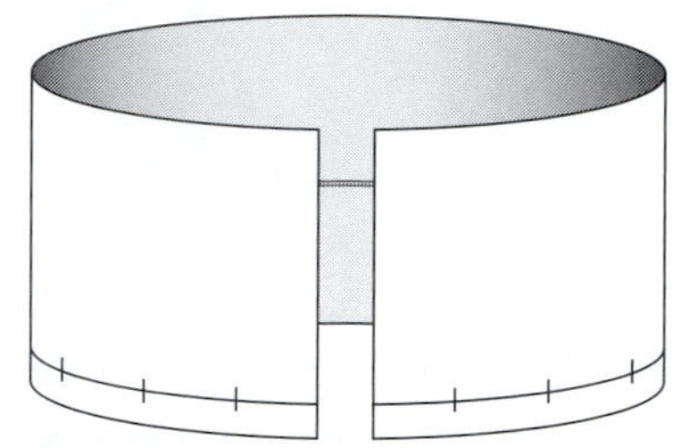

Figure 4.6 Formation of the stationary phase for placement in the developing chamber.

C. Analysis of the Chromatogram

1. **Detection of Bands.** Remove the paper from the developing chamber and *quickly* mark (with a pencil) the position of the eluent front.[6] Allow the chromatogram to dry. While the chromatogram is drying, cover the developing chamber with the plastic wrap. Analyze the paper and circle any colored bands, those from the solutions containing the known cations and those from the unknown solutions.[7]
2. **Enhancement of the Chromatogram.** To enhance the appearance and locations of the bands, move the chromatogram to the fume hood. Position the paper in the ammonia chamber and cover the 1000-mL beaker with the plastic wrap.

   After the deep blue color of Cu^{2+} is evident, remove the chromatogram and circle any new transition metal ion bands that appear. See Table 4.1. Mark the *center* of each band with a pencil. Allow the chromatogram to dry.

Table 4.1 Spot Solutions That Enhance the Band Positions of the Various Cations

	Cation				
	Mn^{2+}	Fe^{3+}	Co^{2+}	Ni^{2+}	Cu^{2+}
NH_3 test	Tan	Red-brown	Pink (brown)	Light blue	Blue
Spot solution	0.1 *m* NaB iO_3 (acidic) (purple)	0.2 *M* KSCN (blood red)	Satd KSCN in acetone (blue-green)	0.1 *M* NaHDMG (brick-red)	0.2 *M* $K_4[Fe(CN)_6]$ (red)

3. **Band Enhancement** (Optional, seek advice from your instructor). The "exact" band positions for the known cations and those of the mixture may be better defined using a second "spot solution." As necessary, use a capillary tube to spot the center of each band (from the known and unknown test solutions) with the corresponding spot solution identified in Table 4.1. This technique more vividly locates the position of the band. Remember, the unknown test solution(s) may have more than one cation.

[6]Do this quickly because the eluent evaporates.
[7]For the unknown solutions, there will most likely be more than one brand.

4. **Analysis of Your Chromatogram.** Where is the center of each band? Mark the center of each band with a pencil. What is the color of each transition metal ion on the chromatogram?

   Calculate the R_f values for each transition metal ion and tabulate them in order of increasing R_f value.

5. **Composition of the Unknown(s).** Look closely at the bands for your unknown(s). What is the R_f value for each band in each unknown? Which ion(s) is(are) present in your unknown(s)? Record your conclusions on the Report Sheet. Submit your chromatogram to your instructor for approval.

Disposal: Dispose of the eluting solution in the "Waste Organics" container. Allow the conc NH_3 in the ammonia chamber to evaporate in the hood.

Attach or Sketch the Chromatogram That Was Developed in the Experiment

Experiment 4 *Prelaboratory Assignment*

Paper Chromatography

Date __________ Lab Sec. ______ Name ______________________________ Desk No. __________

1. a. Define the mobile phase in chromatography.

   b. What is the chemical composition of the mobile phase in this experiment?

2. a. Define the stationary phase in chromatography.

   b. What is the stationary phase in this experiment?

3. a. Define the eluent in chromatography.

   b. What is the eluent (eluting solution) in this experiment?

4. a. Define the R_f value for a cation.

   b. How is it measured in this experiment?

5. A representation of an eluted and developed paper chromatogram of a mixture of dyes is shown at right.

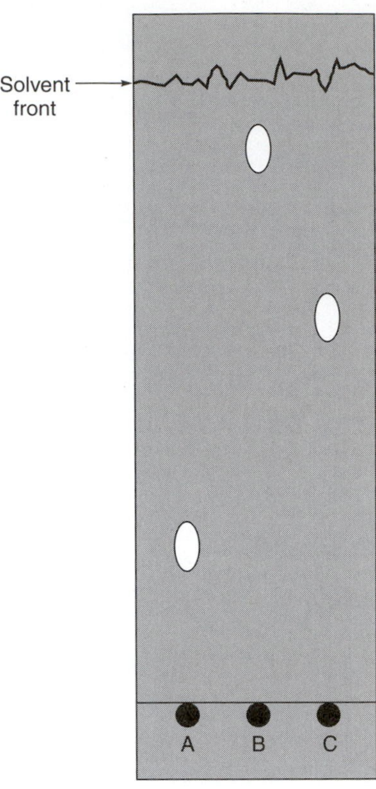

Use a ruler to determine the R_f value for each dye in the mixture.

6. A student developed a chromatogram and found that the eluent front traveled 69 mm and the Zn^{2+} cation traveled 24 mm. In the development of a chromatogram of a mixture of cations, the eluent front traveled 52 mm. If Zn^{2+} is a cation of the unknown mixture, where will its band appear in the chromatogram?

7. The opening photo to this experiment is a chromatogram obtained from an ion chromatograph instrument. The chromatogram shows the presence of a number of anions and their concentrations in a sample of high purity water. The eluent is 0.0007 *M* $NaHCO_3$ and 0.0012 *M* Na_2CO_3. The stationary phase is called Allsep Anion. In this instrument there is an injection port, a 150 × 4.6-mm column (metal tubing through which the mobile phase passes), and a detector that measures the conductivity of the solution. Describe what happens to the sample (i.e., its journey, if you like) as it flows through the chromatograph from the time of its injection until the time of its detection.

Experiment 4 *Report Sheet*

Paper Chromatography

Date __________ Lab Sec. ______ Name ____ Desk No. __________

C. Analysis of the Chromatogram

Distance of solvent front from the origin: __________ mm

	Color (original)	Color (with NH_3)	Color (with spot solution)	Distance (mm) Traveled	R_f
Mn^{2+}	____________	____________	____________	____________	____________
Fe^{3+}	____________	____________	____________	____________	____________
Co^{2+}	____________	____________	____________	____________	____________
Ni^{2+}	____________	____________	____________	____________	____________
Cu^{2+}	____________	____________	____________	____________	____________

Instructor's approval of chromatogram ____

Show your calculations for R_f.

Unknown number(s) U1=________ U2=________ U3=________

	Band 1		Band 2		Band 3		Band 4	
Unknown	Distance (mm) Traveled	R_f	Distance (mm) Traveled	R_f	Distance (mm) Traveled	R_f	Distance (mm) Traveled	R_f
U1	________	_____	________	_____	________	_____	________	_____
U2	________	_____	________	_____	________	_____	________	_____
U3	________	_____	________	_____	________	_____	________	_____

Show your calculations for R_f.

Cations present in U1 ________ ________ ________ ________

Cations present in U2 ________ ________ ________ ________

Cations present in U3 ________ ________ ________ ________

Laboratory Questions

Circle the questions that have been assigned.

1. Part A.1 and Part B.3.
   a. Why is it important to keep the developing chamber covered with plastic wrap during the development of the chromatogram?
   b. The developing chamber is *not* covered with plastic wrap during the development of the chromatogram. How does this technique error affect the D_{ion} for a cation? Explain.
2. Part A.4. Why was a pencil used to mark the chromatogram and not a ballpoint or ink pen?
3. Part B.1. Explain why the cation samples are repeatedly "spotted" and dried on the chromatographic paper.
4. Part B.3. The eluent is to be below the 1.5-cm line on the chromatographic paper. Describe the expected observation if the eluent were above the 1.5-cm line.
5. Part C.2. Explain why the *center* of the band is used to calculate the R_f value for a cation rather than the leading edge of the band.
6. Part C.2. The ammonia chamber is to be covered during the enhancement of the chromatogram. What is the consequence of the ammonia chamber being left uncovered?
7. Suppose two cations have the same R_f value. How might you resolve their presence in a mixture using paper chromatography?

Experiment 5

Chemistry of Copper

The reaction of copper with nitric acid is spontaneous, producing nitrogen dioxide gas and copper(II) ion.

OBJECTIVES

- To observe the chemical properties of copper in a series of chemical reactions
- To use several separation and recovery techniques to isolate the copper compounds from solution
- To determine percent recovery of starting material through a cycle of reactions.

TECHNIQUES

The following techniques are used in the Experimental Procedure

INTRODUCTION

Copper is an element that is chemically combined into a variety of compounds in nature, but most commonly in the form of a sulfide, as in chalcocite, Cu_2S, and chalcopyrite, $FeCuS_2$. Copper metal is an excellent conductor of heat and electricity and is an **alloying element** in bronze and brass. Copper is a soft metal with a characteristic bright orange-brown color, which we often call "copper color" (Figure 5.1). Copper is relatively inert chemically; it does not readily air oxidize (react with oxygen in air) and is not attacked by simple inorganic acids such as sulfuric and hydrochloric acids. Copper metal that does oxidize in air is called **patina.**

Alloying element: an element of low percent composition in a mixture of metals, the result of which produces an alloy with unique, desirable properties

Copper(II) ion forms a number of very colorful compounds; most often, these compounds are blue or blue-green, although other colors are found depending on the copper(II) compound.

We will observe several chemical and physical properties of copper through a sequence of reactions that produce a number of colorful compounds.

Starting with metallic copper at the top, the sequence of products formed is shown on the diagram:

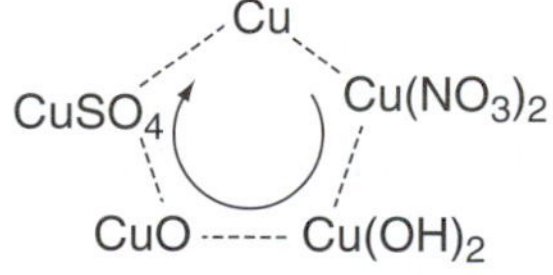

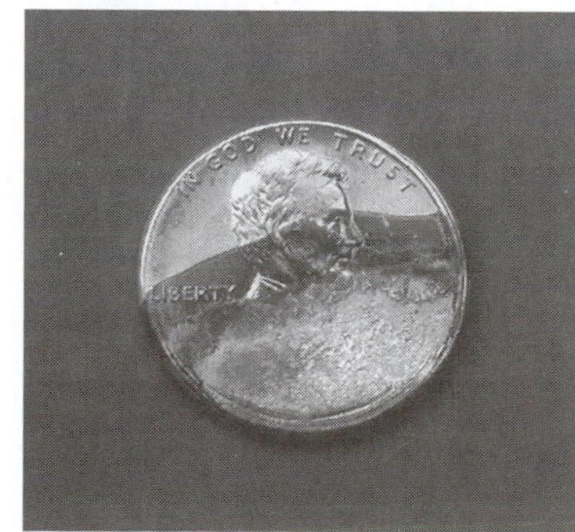

Figure 5.1 The penny is made of zinc (bottom) with a thin copper coating (top).

Dissolution of Copper Metal

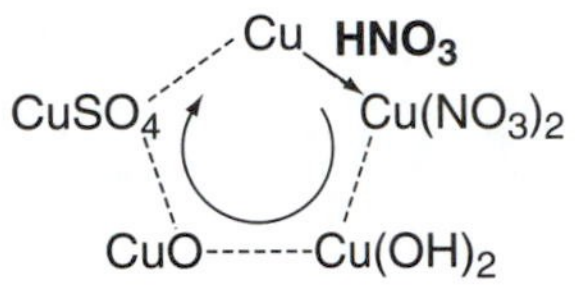

Cu HNO$_3$
CuSO$_4$ Cu(NO$_3$)$_2$
CuO Cu(OH)$_2$

Copper reacts readily with substances called **oxidizing agents** (substances that remove electrons from other substances). In this experiment aqueous nitric acid, HNO_3, is the substance that oxidizes copper metal to the copper(II) ion (opening photo).

$$Cu(s) + 4\ HNO_3(aq) \rightarrow Cu(NO_3)_2(aq) + 2\ NO_2(g) + 2\ H_2O(l) \qquad (5.1)$$

As products of this reaction, copper(II) nitrate, $Cu(NO_3)_2$, is a water-soluble salt that produces a blue solution, and nitrogen dioxide, NO_2, is a dense, toxic, red-brown gas. The solution remains acidic because an excess of nitric acid is used for the reaction.

Precipitation of Copper(II) Hydroxide from Solution

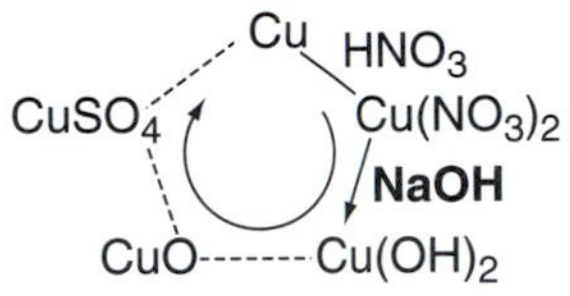

Cu HNO$_3$
CuSO$_4$ Cu(NO$_3$)$_2$
NaOH
CuO Cu(OH)$_2$

Appendix G

For Part B of the Experimental Procedure, the solution containing the soluble $Cu(NO_3)_2$ is treated with sodium hydroxide, NaOH, a base. Copper(II) hydroxide, $Cu(OH)_2$, a light blue solid, precipitates from the solution.

$$Cu(NO_3)_2(aq) + 2\ NaOH(aq) \rightarrow Cu(OH)_2(s) + 2\ NaNO_3(aq) \qquad (5.2)$$

Sodium nitrate, $NaNO_3$, is a colorless salt that remains dissolved in solution.

Conversion of Copper(II) Hydroxide to a Second Insoluble Salt

Heat applied to solid copper(II) hydroxide causes black, insoluble copper(II) oxide, CuO, to form.

$$Cu(OH)_2(s) \xrightarrow{\Delta} CuO(s) + H_2O(g) \qquad (5.3)$$

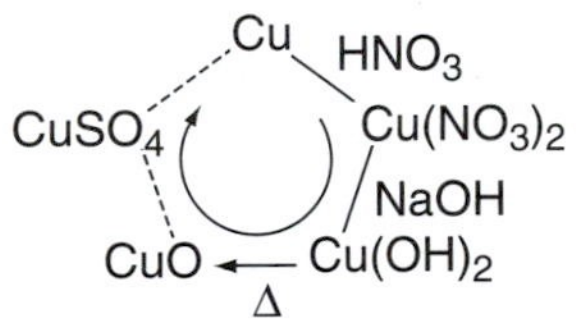

Cu HNO$_3$
CuSO$_4$ Cu(NO$_3$)$_2$
NaOH
CuO Cu(OH)$_2$
Δ

Dissolution of Copper(II) Oxide

Copper(II) oxide dissolves readily with the addition of aqueous sulfuric acid, H_2SO_4, forming a sky blue solution as a result of the formation of the water-soluble salt, copper(II) sulfate, $CuSO_4$.

$$CuO(s) + H_2SO_4(aq) \rightarrow CuSO_4(aq) + H_2O(l) \qquad (5.4)$$

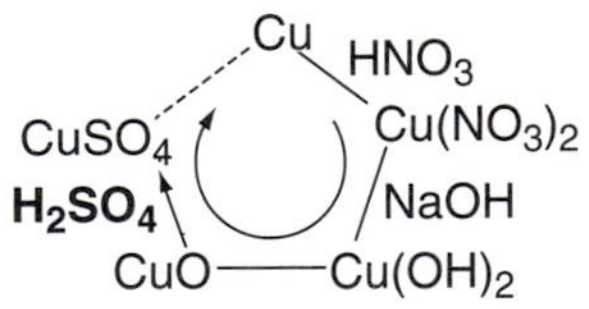

Cu HNO$_3$
CuSO$_4$ Cu(NO$_3$)$_2$
H$_2$SO$_4$ NaOH
CuO Cu(OH)$_2$

Re-formation of Copper Metal

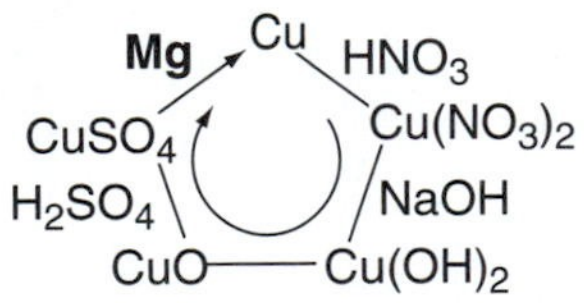

Mg Cu HNO$_3$
CuSO$_4$ Cu(NO$_3$)$_2$
H$_2$SO$_4$ NaOH
CuO Cu(OH)$_2$

Finally, the addition of magnesium metal, Mg, to the copper(II) sulfate solution completes the copper cycle.

In this reaction, magnesium serves as a **reducing agent** (a substance that provides electrons to another substance). Magnesium reduces copper(II) ion from the copper(II) sulfate solution to copper metal and water-soluble magnesium sulfate, $MgSO_4$.

$$CuSO_4(aq) + Mg(s) \rightarrow Cu(s) + MgSO_4(aq) \qquad (5.5)$$

Magnesium metal also reacts with sulfuric acid. Therefore when magnesium metal is added to the acidic copper(II) sulfate solution, a second reaction occurs that produces hydrogen gas, H_2, and additional magnesium sulfate.

$$Mg(s) + H_2SO_4(aq) \rightarrow H_2(g) + MgSO_4(aq) \qquad (5.6)$$

Therefore hydrogen gas bubbles are observed during this reaction step of the cycle. This reaction is also used to remove any excess magnesium metal that remains after the copper metal has been recovered.

Reaction Types That Occur in This Experiment

As you proceed in this course, you will often observe similar types of reactions. An organization of chemical reactions helps to simplify our overall understanding, interpretation, and predictability of chemical behavior. Some of the reaction types that occur in this experiment are listed.

Equation 5.1. A single displacement reaction *and* an oxidation–reduction reaction

Equation 5.2. A double displacement (or metathesis) reaction

Equation 5.3. A decomposition reaction

Equation 5.4. A double displacement (or metathesis) reaction *and* an acid–base reaction

Equation 5.5. A single displacement reaction *and* an oxidation–reduction reaction

Equation 5.6. A single displacement reaction *and* an oxidation–reduction reaction

EXPERIMENTAL PROCEDURE

Procedure Overview: Copper metal is successively treated with nitric acid, sodium hydroxide, heat, sulfuric acid, and magnesium in a cycle of chemical reactions to regenerate the copper metal. The chemical properties of copper are observed in the cycle.

You will need to obtain an instructor approval after each step in the Experimental Procedure. Perform the experiment with a partner. At each circled superscript (1–5) in the procedure, *stop,* and record your observation on the Report Sheet. Discuss your observations with your lab partner and your instructor.

A. Copper Metal to Copper(II) Nitrate

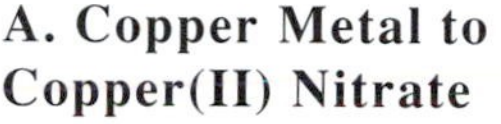

Perform the series of reactions in a test tube that is compatible with your laboratory centrifuge. Consult with your laboratory instructor.

1. **Preparation of the Copper Metal Sample.** Obtain a *less than* 0.02-g sample of Cu wool. Tare the mass (±0.001 g) of the selected test tube.[1] Roll and place the Cu wool into the test tube, measure and record the mass of the copper sample.

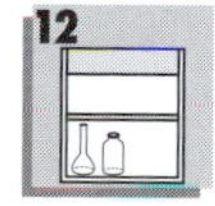

2. **Reaction of the Copper Metal.** Perform this step in the fume hood because of the evolution of toxic $NO_2(g)$. Using a dropper bottle or a dropper pipet, add drops ($\leq$10 drops) of conc HNO_3 to the copper sample until no further evidence of a chemical reaction is observed. Do not add an excess! (**Caution:** *Concentrated* HNO_3 *is very corrosive. Do not allow it to touch the skin. If it does, wash immediately with excess water. Nitric acid will turn your skin yellow, a way to check your laboratory technique!*)

   At this point the Cu metal has completely reacted. (**Caution:** *Do not inhale the evolved nitrogen dioxide gas.*) What is the color of the gas? Add 10 drops of deionized water. Show the resulting solution to your laboratory instructor for approval (1) and save the solution for Part B.

B. Copper(II) Nitrate to Copper(II) Hydroxide

1. **Preparation of Copper(II) Hydroxide.** Agitate or continuously stir with a stirring rod the solution from Part A.2 while slowly adding 10 drops of 6 *M* NaOH. (**Caution:** *Wash with water immediately if the* NaOH *comes into contact with the skin.*) This forms the $Cu(OH)_2$ precipitate. After the first 10 drops are added, add 10 more drops of 6 *M* NaOH. Using a wash bottle and deionized water, rinse the stirring rod, allowing the rinse water to go into the test tube. Centrifuge the solution for 30 seconds (ask your instructor for instructions in operating the centrifuge).

[1] Consult with your instructor.

Supernatant: the clear solution in the test tube

2. **A Complete Precipitation.** Test for a complete precipitation of $Cu(OH)_2$ by adding 2–3 more drops of 6 *M* NaOH to the **supernatant.** If additional precipitate forms, add 4–5 more drops and again centrifuge. Repeat the test until no further formation of the $Cu(OH)_2$ occurs. The solution should appear colorless and the precipitate should be light blue. Obtain your laboratory instructor's approval② and save for Part C.

C. Copper(II) Hydroxide to Copper(II) Oxide

Cool flame: an adjusted Bunsen flame having a low supply of fuel

1. **Heating to Dryness.** Decant (pour off) and discard the supernatant from the test tube. *Carefully* (***very carefully!***) and slowly heat the test tube with a **cool flame**[2] until the $Cu(OH)_2$ precipitate changes color. **Read footnote 2!** You need *not* heat the contents to dryness. Avoid ejection (and projection) of your copper compound by *not* holding the test tube over the direct flame for a prolonged period of time. If the contents of the test tube are ejected, you will need to restart the Experimental Procedure at Part A. Obtain your instructor's approval③ and save for Part D.

D. Copper(II) Oxide to Copper(II) Sulfate

1. **Dissolution of Copper(II) Oxide.** To the solid CuO in the test tube from Part C, add drops ($\leq$20 drops, 1 mL) of 6 *M* H_2SO_4 with agitation until the CuO dissolves. (**Caution:** *Do not let sulfuric acid touch the skin!*) (Slight heating *may* be necessary, but be careful not to eject the contents!) The solution's sky blue appearance is evidence of the presence of soluble $CuSO_4$. Obtain your instructor's approval④ and save for Part E.

E. Copper(II) Sulfate to Copper Metal

1. **Formation of Copper Metal.** Using fine steel wool, polish about 5–7 cm of Mg ribbon. Cut or tear the ribbon into a 1-cm lengths. Dilute the solution from Part D with deionized water until the test tube is half-full. Add a Mg strip 1-cm, to the solution. When the Mg strip has reacted (disappeared), add a second Mg strip and so on until the "blue" has disappeared from the solution. Describe what is happening. What is the coating on the magnesium ribbon? What is the gas?⑤

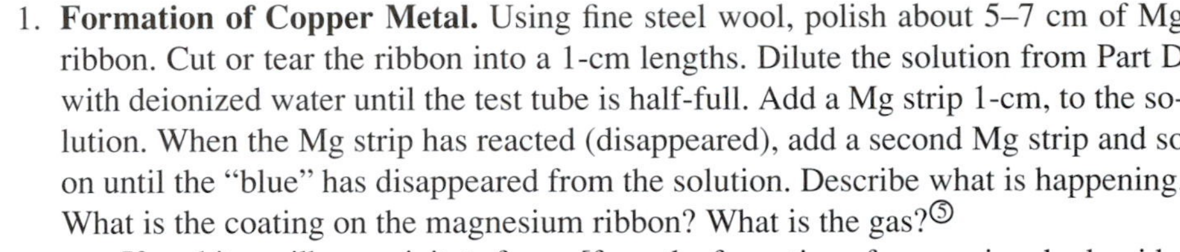

   If a white, milky precipitate forms [from the formation of magnesium hydroxide, $Mg(OH)_2$], add several drops of 6 *M* H_2SO_4. (**Caution:** *Avoid skin contact.*) Break up the red-brown Cu coating on the Mg ribbon with a stirring rod. *After* breaking up the Cu metal *and* after adding several pieces of Mg ribbon, centrifuge the mixture.

2. **Washing.** Add drops of 6 *M* H_2SO_4 to dissolve any excess Mg ribbon. (**Caution:** *Avoid skin contact.*) Do this by breaking up the Cu metal with a stirring rod to expose the Mg ribbon, coated with Cu metal, to the H_2SO_4 solution. Centrifuge for 30 seconds, decant, and discard the supernatant. Wash the red-brown Cu metal with three 1-mL portions of deionized water.[3] Rinse the stirring rod in the test tube. Centrifuge, decant, and discard each washing.

3. **Determination of the Mass of Recovered Copper.** *Very gently* dry the Cu in the test tube over a *cool* flame (***see footnote 2***). Allow the tube and contents to cool and determine the mass ($\pm$0.001 g). Repeat the heating procedure until a reproducibility in mass of $\pm$1% is obtained. Record the mass of Cu recovered in the experiment.

Disposal: All solutions used in the procedure can be disposed of in the "Waste Salts" container. Dispose of the copper metal in the "Waste Solids" container. Check with your instructor.

CLEANUP: Rinse all glassware twice with tap water and twice with deionized water. Discard all rinses in the sink.

---

[2]From Technique 13A, "If you can feel the heat of the flame with the hand holding the test tube clamp, the flame is too hot!"

[3]Wash the Cu metal by adding water, stirring the mixture with a stirring rod, and allowing the mixture to settle.

Experiment 5 *Prelaboratory Assignment*

Chemistry of Copper

Date __________ Lab Sec. ______ Name ____ Desk No. __________

1. Explain why a "new" copper penny does not maintain its luster after many years in circulation.

2. What is/are the color(s) of most copper compounds?

3. Copper forms many different compounds in this experiment. Identify the reagent that
   a. changes copper metal to a copper ion that is blue is solution.

   b. changes a solid black compound of copper to a sky-blue solution of copper ion.

   c. changes a sky-blue solution of copper ion to copper metal.

4. Experimental Procedure, Part C.1 and Part E.3 are critical safety steps (for the same reason) in the experiment. Why?

5. A number of **"Caution"** chemicals and solutions are used in this experiment. Refer to the Experimental Procedure and the corresponding sections to identify the specific chemical or solution that must be handled with care.

Experimental Procedure	Precautionary Chemical/Solution
Part A.2	
Part A.2	
Part B.1	
Part D.1	
Part E.1	

6. Refer to Technique 11F.
   a. What function does a centrifuge perform?

   b. Describe the technique for "balancing a centrifuge" when centrifuging a sample.

7. A 0.0217-g sample of copper metal is recycled through the series of reactions in this experiment. In Part E.3, a 0.0183-g sample of copper is isolated. What is the percent recovery of the copper metal?

*8. What volume, in drops, of 16 *M* HNO_3 is required to react with 0.0191 g of Cu metal? See Equation 5.1. Assume 20 drops per milliliter. The expression "16 *M* HNO_3" indicates that 16 mol HNO_3 is present in each liter of the HNO_3 solution.

Experiment 5 *Report Sheet*

Chemistry of Copper

Date ________ Lab Sec. ______ Name ______________________________ Desk No. ________

Data for Copper Sample

1. Tared mass of Cu sample (g) ______________

Synthesis of	Observation	Instructor Approval	Balanced Equation
A.① $Cu(NO_3)_2$	______________	______	______________
B.② $Cu(OH)_2$	______________	______	______________
C.③ CuO	______________	______	______________
D.④ $CuSO_4$	______________	______	______________

E.⑤ Copper(II) sulfate to copper metal. Write a full description of the reactions that occurred. Include a balanced equation in your discussion.

2. Tared mass of Cu recovered 1st mass (g) ________________

2nd mass (g) ________________

3rd mass (g) ________________

3. Mass of Cu recovered (g) ________________

4. Percent recovery, $\frac{\text{mass Cu (recovered)}}{\text{mass Cu (original)}} \times 100$ ________________

5. Discuss the variance of your percent recovery from the theoretical recovery.

Laboratory Questions

Circle the questions that have been assigned.

1. Part A.2 What is the formula *and* the color of the gas that is evolved?
2. Part B.1. When the NaOH solution is added, $Cu(OH)_2$ does not precipitate immediately. What else present in the reaction mixture from Part A reacts with the NaOH before the copper ion? Explain.
3. Part C.1. The sample in Part B was *not* centrifuged. Why? Perhaps the student chemist had to be across campus for another appointment. Because of the student's "other priorities" how will this
   a. affect the experimental procedure for Part C? Explain.
   b. affect the percent recovery of copper in the experiment? Explain.
4. Part D.1. All of the CuO does *not* react with the sulfuric acid. Will the reported percent recovery of copper in the experiment be too high or too low? Explain.
5. Part E. Sulfuric acid has a dual role in the chemistry. What are its two roles in the recovery of the copper metal?
6. Part E.2. The 6 *M* H_2SO_4 is not available but 6 *M* HNO_3 is available. What changes will be observed if the nitric acid is used in place of the sulfuric acid? Explain.
7. Part E.2. The copper metal is not broken up and thereby occludes the magnesium metal, but the experimental procedure is continued. How does this affect the reported percent recovery of the copper metal? Explain.

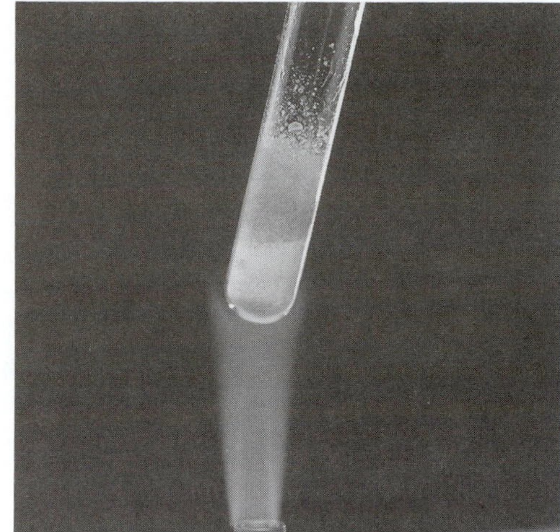

Experiment 6

Percent Water in a Hydrated Salt

Heat readily removes the hydrated water molecules (top of sample in test tube) from copper(II) sulfate pentahydrate forming anhydrous copper(II) sulfate (bottom).

OBJECTIVES

- To determine the percent by mass of water in a hydrated salt
- To establish the formula of a hydrated salt

TECHNIQUES

The following techniques are used in the Experimental Procedure

INTRODUCTION

Hydrate: water molecules are chemically bound to the ions of the salt as part of the structure of the compound

Anhydrous: without water

Many salts occurring in nature or purchased from the grocery shelf or from chemical suppliers are **hydrated;** that is, a number of water molecules are chemically bound to the ions of the salt in its crystalline structure. These water molecules are referred to as **waters of crystallization.** The number of moles of water per mole of salt is usually a constant. For example, iron(III) chloride is purchased as $FeCl_3{\bullet}6H_2O$, not as $FeCl_3$, and copper(II) sulfate as $CuSO_4{\bullet}5H_2O$, not as $CuSO_4$. For some salts, the water molecules are so loosely bound to the ions that heat removes them to form the **anhydrous** salt. Hydrated salts that spontaneously (without heat) lose water molecules to the atmosphere are **efflorescent,** whereas salts that readily absorb water are **deliquescent.**

For Epsom salt (magnesium sulfate heptahydrate, Figure 6.1), the anhydrous salt, $MgSO_4$, forms with gentle heating.

$$MgSO_4{\cdot}7H_2O(s) \xrightarrow{\Delta(>200°C)} MgSO_4(s) + 7\ H_2O(g) \tag{6.1}$$

In other salts such as $FeCl_3{\bullet}6H_2O$, the water molecules are so strongly bound to the salt that anhydrous $FeCl_3$ cannot form regardless of the intensity of the heat.

In Epsom salt 7 moles of water, or 126.1 g of H_2O, are bound to each mole of magnesium sulfate, or 120.4 g of $MgSO_4$. The percent by mass of water in the salt is

$$\frac{126.1\text{ g }H_2O}{(126.1\text{ g} + 120.4\text{ g})\text{ salt}} \times 100 = 51.16\%\ H_2O \tag{6.2}$$

A **gravimetric analysis** is an analytical method that relies almost exclusively on mass measurements for the analysis. Generally the substance being analyzed must have a mass large enough to be measured easily with the balances that are available in the laboratory.

This experiment uses the gravimetric analysis method to determine the percent by mass of water in a hydrated salt. The mass of a hydrated salt is measured, the sample is heated to drive off the hydrated water molecules (the waters of crystallization), and then the mass of remaining sample is measured again. Cycles of heating and measuring of the sample's mass are continued until reproducibility of the mass measurements is attained.

Figure 6.1 Epsom salt, $MgSO_4 \cdot 7H_2O$.

Experimental Procedure

Procedure Overview: The mass of a hydrated salt is measured before and after it is heated to a high temperature. This mass difference and the mass of the anhydrous salt are the data needed to calculate the percent water in the original hydrated salt.

You are to complete *at least* two trials in this experiment. Obtain a hydrated salt from your instructor. Record the unknown number of your hydrated salt on the Report Sheet.

A. Self-Protection

Fired: heated to a very high temperature to volatilize impurities

1. **Prepare a Clean Crucible.** Obtain a clean crucible and lid. Check the crucible for stress fractures or fissures, which are common in often-used crucibles. If none are found, support the crucible and lid on a clay triangle and heat with an intense flame for 5 minutes (see Technique 15C, Figure T.15c). Allow them to cool slowly. (**Caution:** *Do not set them on the lab bench for fear of contamination.*)[1,2]

   Caution: hot and cool crucibles look the same—do not touch! Determine the mass (±0.001 g) of the **fired**, *cool* crucible and lid and record. Handle the crucible and lid with the crucible tongs for the remainder of the experiment (Figure 6.2); do *not* use your fingers—oil from the fingers can contaminate the surface of the crucible and lid.

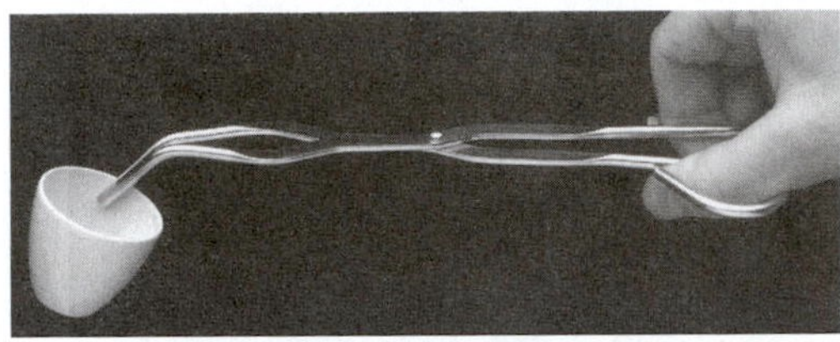

Figure 6.2 Handle the crucible with tongs after heating.

2. **Determine the Mass of Sample.** Add no more than 3 g of your hydrated salt to the crucible and measure the combined mass (±0.001 g) of the crucible, lid, and hydrated salt. Record the mass of the hydrated salt.
3. **Adjust the Crucible Lid.** Return the crucible (use crucible tongs only) with the sample to the clay triangle; set the lid just off the lip of the crucible to allow the evolved water molecules to escape on heating (Figure T.15e).

B. Thermal Decomposition of the Sample

1. **Heat the Sample.** Initially heat the sample slowly and then gradually intensify the heat. Do *not* allow the crucible to become red hot. This may cause the anhydrous salt to decompose as well. Maintain the high temperature on the sample for 10 minutes.

   Cover the crucible with the lid; allow them to cool to room temperature (see footnote 1). Determine the combined mass of the crucible, lid, and anhydrous salt on the same balance that was used for earlier measurements.

2. **Have You Removed All of the Water?** Reheat the sample for 2 minutes, but do *not* intensify the flame—avoid the decomposition of the salt. Again measure the combined mass. If this second mass measurement of the anhydrous salt disagrees by greater than ±0.010 g from that in Part B.1, repeat Part B.2.
3. **Repeat with a New Sample.** Repeat the experiment two more times with an original hydrated salt sample.

Disposal: Dispose of all waste anhydrous salt in the "Waste Solids" container.

CLEANUP: Rinse the crucible with 2–3 milliliters of 1 *M* HCl and discard in the "Waste Acids" container. Then rinse several times with tap water and finally deionized water. Each water rinse can be discarded in the sink.

[1] Place the crucible and lid in a desiccator (if available) for cooling.

[2] If the crucible remains dirty after heating, move the apparatus to the fume hood, add 1–2 mL of 6 *M* HNO_3, and gently evaporate to dryness. (**Caution:** *Avoid skin contact, flush immediately with water.*)

Percent Water in a Hydrated Salt

Date __________ Lab Sec. ______ Name ____ Desk No. __________

1. Calcium chloride, a deliquescent salt, is used as a desiccant in laboratory desiccators (Technique 15B). Explain.

2. Experimental Procedure, Part A.1. What is the purpose of firing the crucible?

3. A 1.803-g sample of gypsum, a hydrated salt of calcium sulfate, $CaSO_4$, is heated in a crucible until a constant mass is reached. The mass of the anhydrous $CaSO_4$ salt is 1.426 g.

   a. Calculate the percent by mass of water in the hydrated calcium sulfate salt.

   b. Calculate the moles of water removed and the moles of anhydrous $CaSO_4$ remaining in the crucible.

   c. What is the formula of the hydrated calcium sulfate; that is, what is the whole-number mole ratio of calcium sulfate to water?

4. The gravimetric analysis of this experiment is meant to be quantitative; therefore, all precautions should be made to minimize errors in the analysis. Review Technique 15C.
   a. The crucible and lid are handled exclusively with crucible tongs in the experiment. How does this technique maintain the integrity of the analysis?

   b. Mass measurements of the crucible, lid, and sample are performed only at room temperature. Why is this technique necessary for a gravimetric analysis?

   c. Why is the position of the crucible lid critical to the dehydration of the salt during the heating process? Explain.

5. The following data were collected from the gravimetric analysis of a hydrated salt:

Mass of crucible and lid (g)	18.733
Mass of crucible, lid, and hydrated salt (g)	21.171
Mass of crucible, lid, and anhydrous salt (g)	20.122

   Determine the percent water in the hydrated salt.

6. a. What is the percent by mass of water in copper(II) sulfate pentahydrate, $CuSO_4{\cdot}5H_2O$? See opening photo.

   b. What mass due to waters of crystallization is present in a 50.0-g sample of $CuSO_4{\cdot}5H_2O$?

Experiment 6 *Report Sheet*

Percent Water in a Hydrated Salt

Date __________ Lab Sec. ______ Name ______________________________ Desk No. __________

Unknown No. ______________________

	*Trial 1**	*Trial 2*	*Trial 3*
1. Mass of crucible and lid (g)			
2. Mass of crucible, lid, and hydrated salt (g)			
3. Mass of crucible, lid, and anhydrous salt			
1st mass measurement (g)			
2nd mass measurement (g)			
3rd mass measurement (g)			

Calculations

1. Mass of hydrated salt (g)			
2. Mass of anhydrous salt (g)			
3. Mass of water lost (g)			
4. Percent by mass of volatile water in hydrated salt			

5. Average percent H_2O in hydrated salt ______________________

*Calculations for Trial 1. Show your work.

Laboratory Questions

Circle the questions that have been assigned.

1. Part A.1. During the cooling of the fired crucible, water vapor condensed on the crucible wall before its mass measurement. The condensation did not occur following thermal decomposition of the hydrated salt in Part B. Will the reported percent water in the hydrated salt be reported too high or too low? Explain.
2. Part A.1. The fired crucible is handled with (oily) fingers before its mass measurement. Subsequently in Part B.1, the oil from the fingers is burned off. How does this technique error affect the reported percent water in the hydrated salt? Explain.
3. Part A.1. The crucible is handled with (oily) fingers after its mass measurement but before the ~3 g sample of the hydrated salt is measured (Part A.2). Subsequently in Part B.1, the oil from the fingers is burned off. How does this technique error affect the reported percent water in the hydrated salt? Explain.
4. Part A.1. Suppose the original sample is unknowingly contaminated with a second anhydrous salt. Will the reported water to salt mole ratio be too high, too low, or unaffected by its presence? Explain.

*5. After heating the crucible in Part A.1, the crucible is set on the lab bench where it is contaminated with the cleaning oil used to clean the lab bench but before its mass is measured. The analysis continues through Part B.1 where the mass of the anhydrous salt is determined. While heating, the cleaning oil is burned off the bottom of the crucible. Describe the error that has occurred; that is, is the mass of the anhydrous salt remaining in the crucible reported as being too high or too low? Explain.

6. Part B.1. The hydrated salt is overheated and the anhydrous salt thermally decomposes, one product being a gas. Will the reported percent water in the hydrated salt be reported too high, too low, or be unaffected? Explain.
7. Part B.1. Some of the hydrated salt spatters out of the crucible because of a too rapid heating process. How does this technique error affect the reported percent water in the hydrated salt? Explain.

Experiment 7

Empirical Formulas

Crucibles are "fired" at high temperatures to volatilize impurities.

OBJECTIVE

- To determine the empirical formulas of three compounds, two by a combination reaction and the other by a decomposition reaction

TECHNIQUES

The following techniques are used in the Experimental Procedure

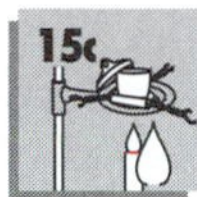

INTRODUCTION

The **empirical formula** of a compound is the simplest whole-number ratio of moles of elements in the compound. The experimental determination of the empirical formula of a compound from its elements requires three steps:

- determine the mass of each element in the sample
- calculate the number of moles of each element in the sample
- express the ratio of the moles of each element as small whole numbers

For example, an analysis of a sample of table salt shows that 2.75 g of sodium and 4.25 g of chlorine are present. The numbers of moles of these elements are

$$2.75 \text{ g} \times \frac{\text{mol Na}}{22.99 \text{ gNa}} = 0.120 \text{ mol Na}, \qquad 4.25 \text{ g} \times \frac{\text{mol Cl}}{35.45 \text{ g Cl}} = 0.120 \text{ mol Cl}$$

The mole ratio of sodium to chlorine is 0.120 to 0.120. As the empirical formula *must* be expressed in a ratio of small whole numbers, the whole-number ratio is 1 to 1 and the empirical formula of sodium chloride is Na_1Cl_1 or simply NaCl.

The empirical formula also provides a mass ratio of the elements in the compound. The formula NaCl states that 22.99 g (1 mol) of sodium combines with 35.45 g (1 mol) of chlorine to form 58.44 g (1 mol) of sodium chloride.

The mass percentages of sodium and chlorine in sodium chloride are:

$$\frac{22.99}{22.99 + 35.45} \times 100 = 39.34\% \text{ Na} \qquad \frac{35.45}{22.99 + 35.45} \times 100 = 60.66\% \text{ Cl}$$

The mass percent of sodium and chlorine in sodium chloride is always the same (constant and definite), a statement of the **law of definite proportions.**

We can determine the empirical formula of a compound from either a combination reaction or a decomposition reaction. In the **combination reaction,** a known mass of one reactant and the mass of the product are measured. An example of a combination

Combination reaction: two elements combine to form a compound

reaction is the reaction of iron with chlorine: the initial mass of the iron and the final mass of the product, iron(III) chloride, are determined. From the difference between the masses, the mass of chlorine that reacts and, subsequently, the moles of iron and chlorine that react to form the product are calculated. This mole ratio of iron to chlorine yields the empirical formula of iron(III) chloride.

Decomposition reaction: a compound decomposes into two or more elements or simpler compounds

In the **decomposition reaction,** the initial mass of the compound used for the analysis and the final mass of at least one of the products are measured. An example would be the decomposition of mercury(II) oxide to mercury metal and oxygen gas: the initial mass of mercury(II) oxide and the final mass of the mercury metal are determined. The difference between the measured masses is the mass of oxygen in the mercury(II) oxide. The moles of mercury and oxygen in the original compound are then calculated to provide a whole-number mole ratio of mercury to oxygen.

In Part B of this experiment a combination reaction of magnesium and oxygen is used to determine the empirical formula of magnesium oxide. The initial mass of the magnesium and the mass of the product are measured.

In Part C of this experiment a decomposition reaction of a pure compound into calcium oxide, CaO, and carbon dioxide, CO_2, is analyzed. The masses of the compound and the calcium oxide are measured. Further analysis of the data provides the empirical formula of the compound.

In Part D of this experiment a combination reaction of tin and oxygen is used to determine the empirical formula of a tin oxide. Since tin forms more than one oxide, a difference in laboratory technique and persistence may lead to different determinations of the reported empirical formula. The initial measured mass of tin is not reacted directly with the oxygen of the air, but rather with nitric acid to produce the oxide. Because the mass of the final product consists only of tin and oxygen its empirical formula is calculated.

Experimental Procedure

Procedure Overview: A crucible of constant mass is used to thermally form (Part B) or decompose (Part C) a compound of fixed composition or form a compound of variable composition (Part D). Mass measurements before and after the heating procedure are used to calculate the empirical formula of the compound.

Ask your instructor which part(s) of the experiment you are to complete. If you are to perform more than one part, you will need to organize some of your data on a separate sheet of paper for the Report Sheet.

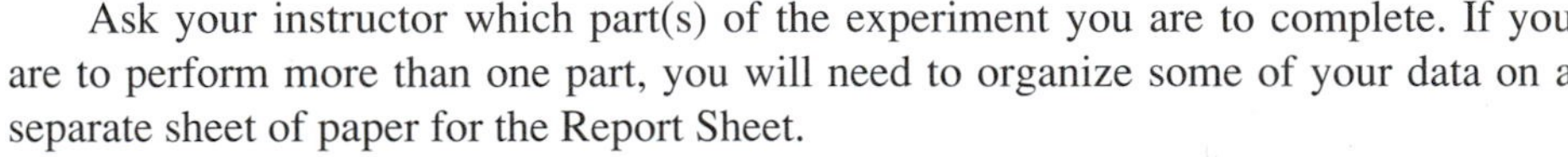

You are to complete at least two trials for each part of the experiment that you are assigned.

A. Preparation of a Crucible

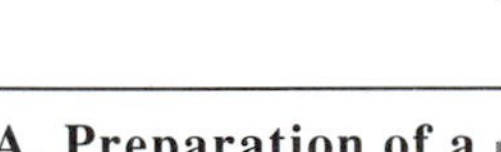

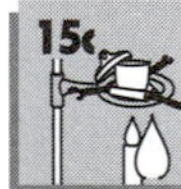

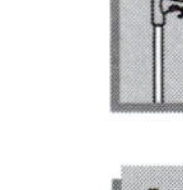

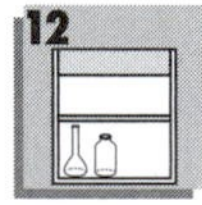

1. **Prepare a Clean Crucible.** Obtain a clean crucible and lid. Often-used crucibles tend to form stress fractures or fissures. Check the crucible for flaws; if any are found obtain a second crucible. Support the crucible and lid on a clay triangle and heat with an intense flame for 5 minutes. Allow them to cool to room temperature.[1] If the crucible remains dirty after heating, *upon the advice of your instructor,* move the apparatus to the fume hood, add 1–2 mL of 6 *M* HNO_3, and gently evaporate to dryness. (**Caution:** *Avoid skin contact, flush immediately with water.*)

   Tare the mass of the **fired,** *cool* crucible and lid. Use only crucible tongs to handle the crucible and lid for the remainder of the experiment; do *not* use your fingers (**Caution:** *hot and cold crucibles look the same—do not touch!*).

Fired: heated to a very high temperature to volatilize impurities

B. Combination Reaction of Magnesium and Oxygen

1. **Prepare the Sample.** Polish (with steel wool) 0.15–20 g of magnesium ribbon; curl the ribbon to lie in the crucible. Record the tared mass (±0.001 g) of the magnesium sample in the crucible with the lid.

[1] Cool the crucible and lid to room temperature (and perform all other cooling processes in the Experimental Procedure) in a desiccator if one is available. When cool, remove the crucible and lid from the desiccator with crucible tongs and measure the mass of the crucible and lid.

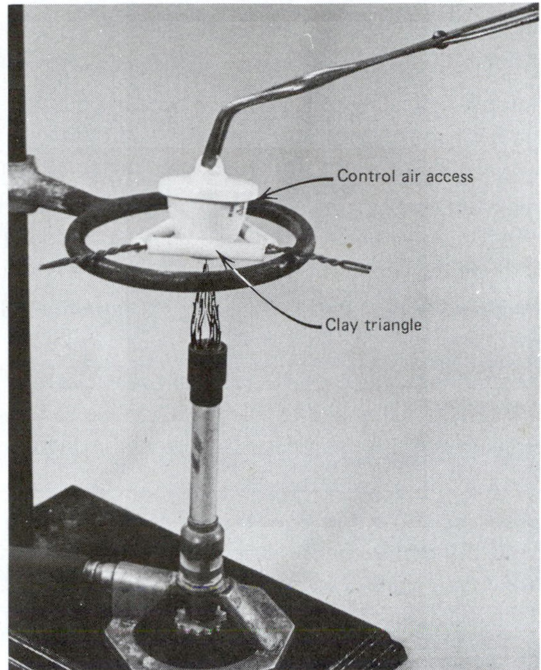

Figure 7.1 Controlling access of air to the magnesium ribbon.

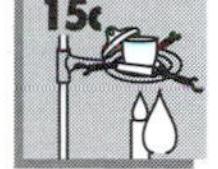

2. **Heat the Sample in Air.** Place the crucible containing the Mg ribbon and lid on the clay triangle. Heat *slowly,* occasionally lifting the lid to allow air to reach the Mg ribbon (Figure 7.1).

   If too much air comes in contact with the Mg ribbon, rapid oxidation of the Mg occurs and it burns brightly. (**Caution:** *You do not want this to happen. If it does, do not watch the burn; it may cause temporary blindness!*) Immediately return the lid to the crucible, allow the apparatus to cool, and return to Part A to repeat the experiment.

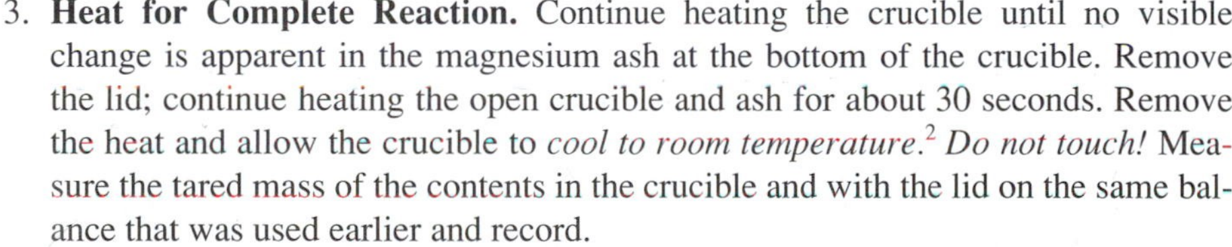

3. **Heat for Complete Reaction.** Continue heating the crucible until no visible change is apparent in the magnesium ash at the bottom of the crucible. Remove the lid; continue heating the open crucible and ash for about 30 seconds. Remove the heat and allow the crucible to *cool to room temperature.*[2] *Do not touch!* Measure the tared mass of the contents in the crucible and with the lid on the same balance that was used earlier and record.
4. **Test for Complete Reaction.** Add a few drops of water to decompose any magnesium nitride[3] that may have formed during combustion. Reheat the sample for 1 minute, but do not intensify the flame. Allow the apparatus to cool and again measure the tared mass of the contents. If this second mass is greater than $\pm 1\%$ from that recorded in Part B.3, repeat Part B.4.
5. **Calculations.** Determine the mole ratio of magnesium to oxygen, and thus the formula, of the pure compound.

Disposal and Cleanup: Wash the cool crucible with a dilute solution of hydrochloric acid and discard in the "Waste Acids" container. Rinse twice with tap water and twice with deionized water.

[2]Place the crucible and lid in a desiccator (if available) for cooling.

[3]At higher temperatures magnesium metal reacts with the nitrogen in the air to form magnesium nitride:

$$3\ Mg(s) + N_2(g) \rightarrow Mg_3N_2(s)$$

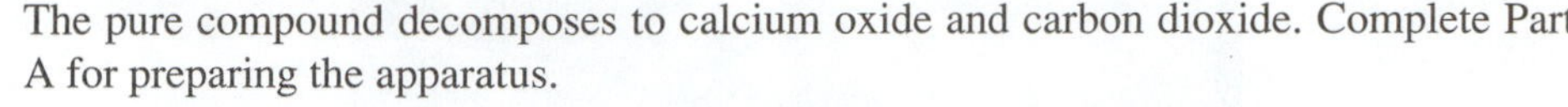

C. Decomposition Reaction of a Pure Compound

The pure compound decomposes to calcium oxide and carbon dioxide. Complete Part A for preparing the apparatus.

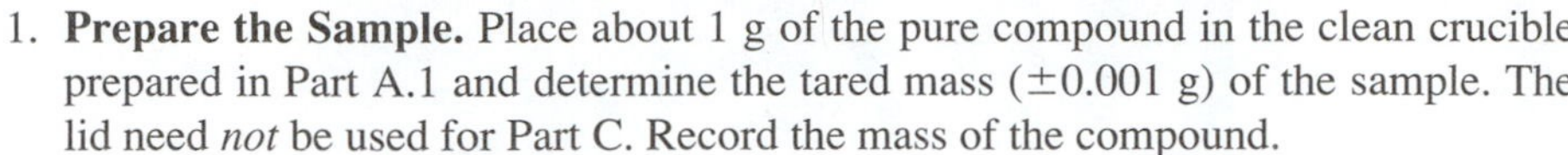

1. **Prepare the Sample.** Place about 1 g of the pure compound in the clean crucible prepared in Part A.1 and determine the tared mass (±0.001 g) of the sample. The lid need *not* be used for Part C. Record the mass of the compound.

2. **Heat the Sample.** Heat the crucible, gradually intensifying the heat.[4] Maintain the intense flame for 20–25 minutes. Allow the sample to cool (in a desiccator if available). Determine the tared mass of the contents. *Do not touch!*
3. **Analyze the Product.** Repeat Part C.2 until ±1% reproducibility of the mass is obtained.

4. **Calculations.** Determine the mole ratio of calcium oxide to carbon dioxide in the pure compound. Considering the formulas of calcium oxide and carbon dioxide to be CaO and CO_2, respectively, what is the empirical formula of the original pure compound?

Disposal and Cleanup: Wash the cool crucible with a dilute solution of hydrochloric acid and discard in the "Waste Acids" container. Rinse twice with tap water and twice with deionized water.

D. Combination Reaction of Tin and Oxygen

1. **Prepare the Sample.** Place about 0.5 g of granulated tin in the clean crucible prepared in Part A.1, cover with the crucible lid, and determine its tared mass (±0.001 g). Record the tared mass of the tin sample. Only use crucible tongs to handle the crucible and lid.

2. **Transfer the Sample to a Fume Hood.** Add drops of 6 *M* HNO_3 (**Caution:** *HNO_3 is very corrosive and a severe skin irritant. If it contacts the skin wash immediately with excess water*) to the tin sample until no further reaction is apparent with the tin (**Caution:** *do not inhale the gaseous vapors!*). Add 4–5 additional drops of the 6 *M* HNO_3. Keep the crucible and sample in the fume hood until no further gaseous vapors from the reaction are visible.

3. **Heat to Dryness.** Upon approval of your laboratory instructor, return the sample to the laboratory bench.
   a. *Initial dryness.* See Figure T.15e. *Slowly,* and with a "**cool flame,**" heat the sample until the solid first appears dry (*avoid any popping or spattering of the sample throughout the heating process*).
   b. *Final dryness.* Break up the solid with a stirring rod and resume heating, now with a more intense flame, until the solid appears a pale yellow.
   c. *Cool.* Allow the crucible, lid, and sample to cool (in a desiccator if available) to room temperature. Determine the tared mass of the tin compound.

Cool flame. If you can feel the heat of the flame with a hand held beside the crucible, the flame it too hot!

4. **Constant Mass.** Repeat Part D.3 until the tared mass of the tin compound has a ±1% reproducibility.
5. **Calculations.** Determine the mole ratio of tin to oxygen in the compound and the empirical formula of the tin oxide.

Disposal and Cleanup: Wash the cool crucible with a dilute solution of hydrochloric acid and discard in the "Waste Acids" container. Rinse twice with tap water and twice with deionized water.

[4]The pure compound decomposes near 825°C. See Experiment 1 for data on the temperatures of a Bunsen burner flame.

Experiment 7 *Prelaboratory Assignment*

Empirical Formulas

Date __________ Lab Sec. ______ Name ______________________________ Desk No. __________

1. Elemental mercury was first discovered when a mercury oxide was decomposed with heat, forming mercury metal and oxygen gas. When a 1.048-g sample of the mercury oxide is heated, 0.971 g of mercury metal remains. *Note:* do not attempt this experiment in the laboratory because of the release of toxic mercury vapor.
   a. What is the mole ratio of mercury to oxygen in the sample?

   b. What is the empirical formula of the mercury oxide?

   c. What is the percent by mass of mercury and oxygen in the sample?

2. A 2.60-g sample of titanium metal chemically combines with chlorine gas to form 10.31 g of a titanium chloride.
   a. What is the empirical formula of the titanium chloride?

   b. What is the percent by mass of titanium and chlorine in the sample?

3. a. Experimental Procedure, Part A. List two reasons for using crucible tongs to handle the crucible and lid after their initial firing.

   b. Why is it best to cool the crucible and lid (and sample) in a desiccator rather than on the laboratory bench?

4. Experiment Procedure, Part D.3 and Technique 13a. Characterize a "cool flame."

5. A chemist (as all scientists) collects and interprets laboratory data to present results. Paraphrase the introduction to "Laboratory Data" at the front of the manual.

Experiment 8

Limiting Reactant

Gravity filtration is used to filter finely divided precipitates.

Objectives

- To determine the limiting reactant in a mixture of two soluble salts
- To determine the **percent composition** of each substance in a salt mixture

Techniques

The following techniques are used in the Experimental Procedure

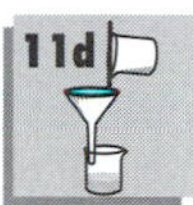

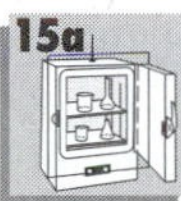

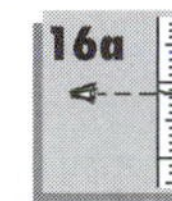

Percent composition: the mass ratio of the components of a mixture or compound to the total mass of the sample

Introduction

Two factors affect the yield of products in a chemical reaction: (1) the amounts of starting materials (reactants) and (2) the **percent yield** of the reaction. Many experimental conditions, for example, temperature and pressure, can be adjusted to increase the yield of a desired product in a chemical reaction, but because chemicals react according to fixed mole ratios (**stoichiometrically**), only a limited amount of product can form from given amounts of starting materials. The reactant determining the amount of product generated in a chemical reaction is called the **limiting reactant** in the chemical system.

Percent yield:

$$\left(\frac{\text{actual yield}}{\text{theoretical yield}}\right) \times 100$$

Stoichiometrically: by a study of a chemical reaction using a balanced equation

To better understand the concept of the limiting reactant, let us look at the reaction under investigation in this experiment, the reaction of sodium phosphate dodecahydrate, $Na_3PO_4{\cdot}12H_2O$, and barium chloride dihydrate, $BaCl_2{\cdot}2H_2O$, in an aqueous system. The molecular form of the equation for the reaction in aqueous solution is

$$2\,Na_3PO_4{\cdot}12H_2O(aq) + 3\,BaCl_2{\cdot}2H_2O(aq) \rightarrow Ba_3(PO_4)_2(s) + 6\,NaCl(aq) + 30\,H_2O(l) \quad (8.1)$$

As the two reactant salts and sodium chloride are soluble in water but barium phosphate is insoluble, the ionic equation for the reaction is

$$6\,Na^+(aq) + 2\,PO_4^{3-}(aq) + 24\,H_2O(l) + 3\,Ba^{2+}(aq) + 6\,Cl^-(aq) + 6\,H_2O(l) \rightarrow Ba_3(PO_4)_2(s) + 6\,Na^+(aq) + 6\,Cl^-(aq) + 30\,H_2O(l) \quad (8.2)$$

Spectator ions: cations or anions that do not participate in any observable or detectable chemical reaction

Net ionic equation: an equation that includes only those ions that participate in the observed chemical reaction

Presenting only the ions that show evidence of a chemical reaction occurring, i.e., the formation of a precipitate (and removing **spectator ions** from the equation), we have the **net ionic equation** for the observed reaction:

$$2\ PO_4^{3-}(aq) + 3\ Ba^{2+}(aq) \rightarrow Ba_3(PO_4)_2(s) \quad (8.3)$$

From the balanced net ionic equation, 2 mol of phosphate ion (from the 2 mol of $Na_3PO_4 \bullet 12H_2O$, molar mass = 380.12 g/mol, or 760.24 g) reacts with 3 mol of barium ion (from 3 mol of $BaCl_2 \bullet 2H_2O$, molar mass = 244.27 g/mol, or 732.81 g) *if* the reaction proceeds to completion. The equation also predicts the formation of 1 mol of $Ba_3(PO_4)_2$ (molar mass = 601.93 g/mol), or 601.93 g.

In Part A of this experiment the solid salts $Na_3PO_4 \bullet 12H_2O$ and $BaCl_2 \bullet 2H_2O$ form a heterogeneous mixture of unknown composition. The mass of the solid mixture is measured and then added to water—insoluble $Ba_3(PO_4)_2$ forms. The $Ba_3(PO_4)_2$ precipitate is collected, via gravity filtration and dried, and its mass is measured.

The percent composition of the salt mixture is determined by first testing for the limiting reactant. In Part B, the limiting reactant for the formation of solid barium phosphate is determined from two precipitation tests of the solution: (1) the solution is tested for an excess of barium ion with a phosphate reagent—observed formation of a precipitate indicates the presence of an excess of barium ion (and a limited amount of phosphate ion) in the salt mixture; (2) the solution is also tested for an excess of phosphate ion with a barium reagent—observed formation of a precipitate indicates the presence of an excess of phosphate ion (and a limited amount of barium ion) in the salt mixture.

Calculations

The calculations for the analysis of the data in this experiment require some attention. The question, after collection of all of the data, becomes, "How do I proceed to determine the percent composition of a salt mixture containing the salts $Na_3PO_4 \bullet 12H_2O$ and $BaCl_2 \bullet 2H_2O$ from the data of a precipitation reaction?" Consider the following scenario: A 0.942-g sample of the salt mixture is added to water and 0.188 g of $Ba_3(PO_4)_2$ precipitate forms. Tests reveal that $BaCl_2 \bullet 2H_2O$ is the limiting reactant. What is the percent composition of the salt mixture?

Using the experimental fact that $BaCl_2 \bullet 2H_2O$ is the limiting reactant, the stoichiometry of the reaction indicates that 1 mol $Ba_3(PO_4)_2$ precipitate requires 3 mol Ba^{2+} (and therefore 3 mol $BaCl_2 \bullet 2H_2O$) to form (Equation 8.2). As 0.188 g $Ba_3(PO_4)_2$ forms, then

$$0.188\text{ g } Ba_3(PO_4)_2 \times \frac{1\text{ ml } Ba_3(PO_4)_2}{601.93\text{ g}} \times \frac{3\text{ mol } Ba^{2+}}{1\text{ mol } Ba_3(PO_4)_2} = 9.37 \times 10^{-4}\text{ mol } Ba^{2+}$$

reacts, and

$$9.37 \times 10^{-4}\text{ mol } Ba^{2+} \times \frac{1\text{ mol } BaCl_2 \cdot 2H_2O}{1\text{ mol } Ba^{2}} \times \frac{244.27\text{ g } BaCl_2 \cdot 2H_2O}{1\text{ mol } BaCl_2 \cdot 2H_2O} = 0.229\text{ g}$$

$BaCl_2 \bullet 2H_2O$ is a part of the original salt mixture.

The mass of the $Na_3PO_4 \bullet 12H_2O$ in the salt mixture must be the difference between the total mass of the original salt sample and the mass of the $BaCl_2 \bullet 2H_2O$, or (0.942 g − 0.229 g =) 0.713 g.

The percent $BaCl_2 \bullet 2H_2O$ in the salt mixture is

$$\frac{0.229\text{ g}}{0.942\text{ g}} \times 100 = 24.3\%$$

The percent $Na_3PO_4 \bullet 12H_2O$ in the salt mixture is

$$\frac{0.713\text{ g}}{0.942\text{ g}} \times 100 = 75.7\%$$

EXPERIMENTAL PROCEDURE

Supernatant: the clear solution that exists after the precipitate has settled

Procedure Overview: In Part A a measured mass of a solid $Na_3PO_4 \cdot 12H_2O/BaCl_2 \cdot 2H_2O$ salt mixture of unknown composition is added to water. The precipitate that forms is digested, filtered, and dried, and its mass measured. Observations from tests on the **supernatant** solution in Part B determine which salt in the mixture is the limiting reactant. An analysis of the data provides the determination of the percent composition of the salt mixture.

Two trials are recommended for this experiment. To hasten the analyses, measure the mass of duplicate unknown solid salt mixtures and simultaneously follow the procedure for each. Label the beakers accordingly for Trial 1 and Trial 2 to avoid the intermixing of samples and solutions.

Obtain about 2–3 g of an unknown $Na_3PO_4 \cdot 12H_2O/BaCl_2 \cdot 2H_2O$ salt mixture.

A. Precipitation of $Ba_3(PO_4)_2$ from the Salt Mixture

1. **Prepare the Salt Mixture.** a. Measure (±0.001 g) on a weighing paper or dish about 1 g of the salt mixture. Record this tared mass for Trial 1 on the Report Sheet. Repeat for Trial 2.
   b. Transfer the salt mixture to a labeled 250-mL beaker and add ~150 mL of deionized water. Stir the mixture with a stirring rod for 2–3 minutes and then allow the precipitate to settle. Leave the stirring rod in the beaker.
2. **Digest the Precipitate.** a. Cover the beaker with a watchglass and warm the solution (80–90°C) over a steam bath or with a **cool flame** (***don't boil!***) for 15 minutes[1] (Figure 8.1). While the precipitate is being kept warm, proceed to Part A.3. Periodically check on the progress of the heating.
   b. After 15 minutes, remove the heat and allow the precipitate to settle; the solution does *not* need to cool to room temperature.
   c. While the precipitate is settling, heat (70–80°C) about 30 mL of deionized water for use as wash water in Part A.5.

Cool flame: a nonluminous flame with a limited amount of gas being burned

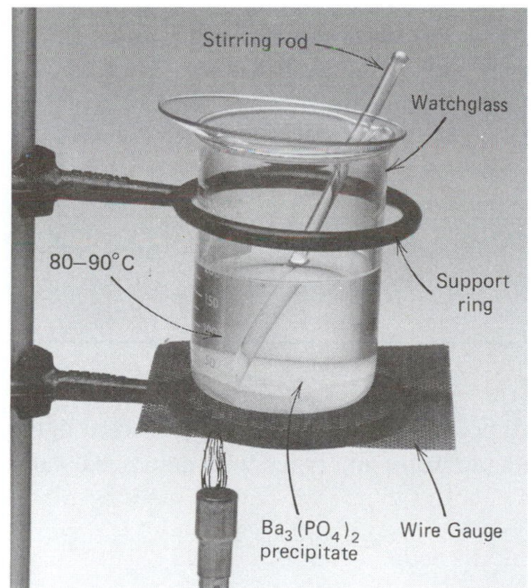

Figure 8.1 Warming and digesting the precipitate.

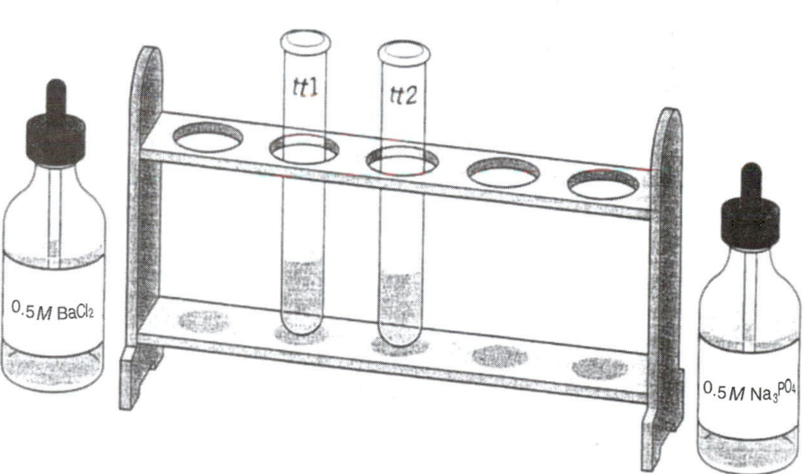

Figure 8.2 Testing for the excess (and the limiting) reactant.

[1]This is called *digesting the precipitate.* This heating procedure causes the formation of larger $Ba_3(PO_4)_2$ particles for the purpose of a better separation during the filtering procedure.

3. **Set Up a Gravity (or Vacuum[2]) Filtering Apparatus.** Place your initials (in pencil) on a piece of Whatman No. 42 or Fisher*brand* Q2 filter paper,[3] fold, and tear off its corner. Determine its mass (±0.001 g). Seal the filter paper into the filter funnel with a small amount of deionized water. Discard the deionized water from the receiving flask. Have your instructor inspect your apparatus before continuing (see opening photo). Return to Part A.2b.

4. **Withdraw and Save Supernatant.** Once the precipitate has settled and the supernatant has cleared in Part A.2b, use a dropping pipet to withdraw enough supernatant to half-fill two 75-mm test tubes, labeled "1" and "2." Save for Part B.

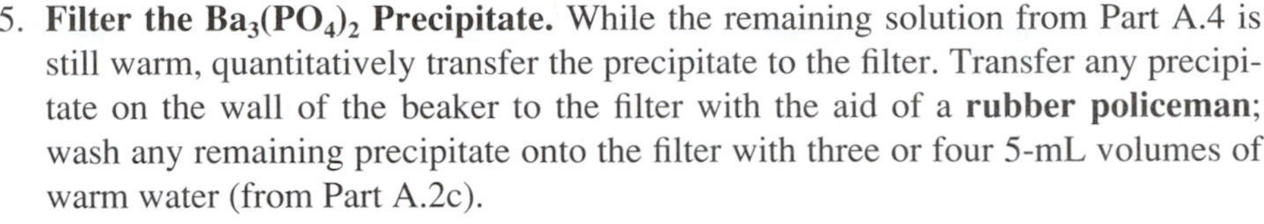

Rubber policeman: a spatula-like rubber tip attached to a stirring rod

5. **Filter the $Ba_3(PO_4)_2$ Precipitate.** While the remaining solution from Part A.4 is still warm, quantitatively transfer the precipitate to the filter. Transfer any precipitate on the wall of the beaker to the filter with the aid of a **rubber policeman**; wash any remaining precipitate onto the filter with three or four 5-mL volumes of warm water (from Part A.2c).

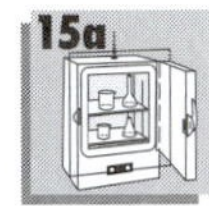

6. **Dry and Measure the Amount of $Ba_3(PO_4)_2$ Precipitate.** Remove the filter paper and precipitate from the filter funnel. Air-dry the precipitate on the filter paper until the next laboratory period or dry in a 110°C constant temperature drying oven for 30–40 minutes or overnight.[4] Determine the combined mass (±0.001 g) of the precipitate and filter paper. Record.

B. Determination of the Limiting Reactant

From the following two tests (Figure 8.2) you can determine the limiting reactant in the original salt mixture. Some cloudiness may appear in both tests, but one will show a definite formation of precipitate.

1. **Clarify the Supernatant.** Centrifuge the two collected supernatant samples from Part A.4.
2. **Test for *Excess* PO_4^{3-}.** Add 2 drops of the test reagent 0.5 *M* $BaCl_2$ to the supernatant liquid in test tube 1. If a precipitate forms, the PO_4^{3-} is *in excess* and Ba^{2+} is the limiting reactant in the original salt mixture.
3. **Test for *Excess* Ba^{2+}.** Add 2 drops of the test reagent 0.5 *M* Na_3PO_4 to the supernatant liquid in test tube 2. If a precipitate forms, the Ba^{2+} is *in excess* and PO_4^{3-} is the limiting reactant in the original salt mixture. An obvious formation of precipitate should appear in only one of the tests.

Disposal: Dispose of the barium phosphate, including the filter paper, in the "Waste Solids" container. Dispose of the waste solutions in the "Waste Liquids" container.

CLEANUP: Rinse each beaker with small portions of warm water and discard in the "Waste Liquids" container. Rinse twice with tap water and twice with deionized water and discard in the sink.

---

[2]A vacuum filtering apparatus (Technique 11E) can also be used; the filtering procedure will be more rapid, but because of the small crystal size, more precipitate will pass through the filter paper.

[3]Whatman No. 42 and Fisher*brand* Q2 filter papers are both fine-porosity filter papers; a fine-porosity filter paper is required to filter the finely divided $Ba_3(PO_4)_2$ precipitate.

[4]The drying time can be reduced by placing the precipitate and filter paper into a microwave oven.

Experiment 8 *Prelaboratory Assignment*

Limiting Reactant

Date __________ Lab Sec. ______ Name ____ Desk No. __________

1. Experimental Procedure, Part A.2. What is the procedure and purpose of "digesting the precipitate?"

2. Barium phosphate is a very finely divided precipitate. Identify two special steps in the procedure that are incorporated into the experiment that minimize the loss of barium phosphate in its analysis.

3. The $Ba_3(PO_4)_2$ (molar mass = 601.93 g/mol) precipitate that formed from a salt mixture has a mass of 0.417 g. Experimental tests revealed that $Na_3PO_4 \cdot 12H_2O$ (molar mass = 380.12 g/mol) was the limiting reactant in the formation of the precipitate and that $BaCl_2 \cdot 2H_2O$ was the excess reactant in the salt mixture. Determine the mass of $Na_3PO_4 \cdot 12H_2O$ in the salt mixture.

4. A 1.276-g mixture of the solid salts Na_2SO_4 and $Pb(NO_3)_2$ forms an aqueous solution with the precipitation of $PbSO_4$. The precipitate was filtered and dried, and its mass was determined to be 0.717 g. The limiting reactant was determined to be $Pb(NO_3)_2$.

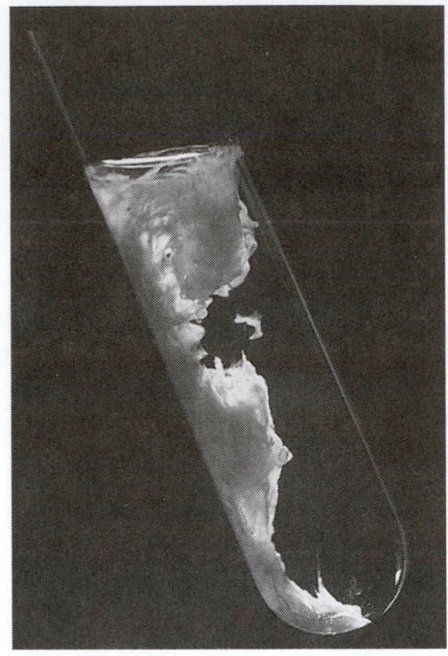

a. Write the molecular form of the equation for the reaction.

b. Write the net ionic equation for the reaction.

c. How many moles and grams of $Pb(NO_3)_2$ are in the reaction mixture?

d. What is the percent by mass of each salt in the mixture?

Experiment 8 *Report Sheet*

Limiting Reactant

Date __________ Lab Sec. ______ Name ____ Desk No. __________

A. Precipitation of $Ba_3(PO_4)_2$ from the Salt Mixture

Unknown number ______________	*Trial 1*	*Trial 2*
1. Tared mass of salt mixture (g)	______	______
2. Mass of filter paper (g)	______	______
3. Mass of filter paper and $Ba_3(PO_4)_2$ (g)	______	______
4. Mass of $Ba_3(PO_4)_2$ (g)	______	______

B. Determination of Limiting Reactant

1. Limiting reactant in salt mixture (write complete formula) ______________

2. Excess reactant in salt mixture (write complete formula) ______________

Data Analysis

1. Moles of $Ba_3(PO_4)_2$ precipitated (mol)	______	______
2. Moles of limiting reactant in salt mixture (mol) • formula of limiting hydrate ______________	______	______
3. Mass of limiting reactant in salt mixture (g) • formula of limiting hydrate ______________	______	______
4. Mass of excess reactant in salt mixture (g) • formula of excess hydrate ______________	______	______
5. Percent limiting reactant in salt mixture (%) • formula of limiting hydrate ______________	______	______
6. Percent excess reactant in salt mixture (%) • formula of excess hydrate ______________	______	______

Show all calculations for Trial 1.

Laboratory Questions

Circle the questions that have been assigned.

1. Part A.2. If the step for digesting the precipitate were omitted, what would be the probable consequence of reporting the "percent limiting reactant" in the salt mixture? Explain.
2. Part A.3. A couple of drops of water were accidentally placed on the properly folded filter paper before its mass was measured. However, in Part A.6, the $Ba_3(PO_4)_2$ precipitate and the filter paper were dry. How does this sloppy technique affect the reported mass of the limiting reactant in the original salt mixture? Explain.
3. Part A.5. Because of the porosity of the filter paper and the finely divided precipitate, some of the $Ba_3(PO_4)_2$ precipitate passes through the filter paper. Will the reported percent of the limiting reactant in the original salt mixture be reported too high or too low? Explain.
4. Part A.5. Excessive quantities of wash water are added to the $Ba_3(PO_4)_2$ precipitate. How does this affect the mass of $Ba_3(PO_4)_2$ precipitate reported in Part A.6?
5. Part A.6. The $Ba_3(PO_4)_2$ precipitate is not completely dry when its "dried" mass is determined. Will the reported mass of the limiting reactant in the original salt mixture be reported too high or too low? Explain.
6. Part A.6. Because of the slight solubility, 0.519 mg/L, of $Ba_3(PO_4)_2$, all of the Ba^{2+} and PO_4^{3-} ions do *not* precipitate from solution.
   a. How many milligrams and moles of $Ba_3(PO_4)_2$ dissolve in 150 mL of solution?
   b. How does this loss affect the reported "percent excess reactant" in the original salt mixture? Explain.

Experiment 9

A Volumetric Analysis

A titrimetric analysis requires the careful addition of titrant.

Objectives

- To prepare and standardize a sodium hydroxide solution
- To determine the molar concentration of a strong acid

Techniques

The following techniques are used in the Experimental Procedure

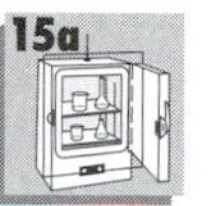

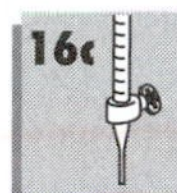

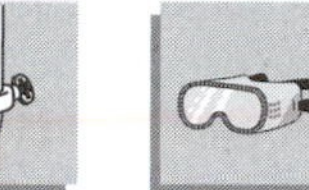

Introduction

A chemical analysis that is performed primarily with the aid of volumetric glassware (e.g., pipets, burets, volumetric flasks) is called **volumetric analysis.** For a volumetric analysis procedure, a known quantity or a carefully measured amount of one substance reacts with a to-be-determined amount of another substance with the reaction occurring in aqueous solution. The volumes of all solutions are carefully measured with volumetric glassware.

The known amount of the substance for an analysis is generally measured and available in two ways:

1. As a **primary standard:** A precise mass (and thus, moles) of a solid substance is measured on a balance, dissolved in water, and then reacted with the substance being analyzed.
2. As a **standard solution:** A measured number of moles of substance is present in a measured volume of solution—a solution of known concentration, generally expressed as the molar concentration (or molarity) of the substance. A measured volume of the standard solution then reacts with the substance being analyzed.

Primary standard: a substance that has a known high degree of purity, a relatively large molar mass, is nonhygroscopic, and reacts in a predictable way

Standard solution: a solution having a very well known concentration of a solute

The reaction of the known substance with the substance to be analyzed, occurring in aqueous solution, is conducted by a titration procedure.

The titration procedure requires a buret to dispense a liquid, called the **titrant,** into a flask containing the **analyte** (Figure 9.1*a*). The titrant may be a solution of known or unknown concentration. The analyte may be a solution whose volume is measured with a pipet or it may be a dissolved solid with a very accurately measured mass. For the acid–base titration studied in this experiment, the titrant is a sodium hydroxide solution and the analyte is an acid.

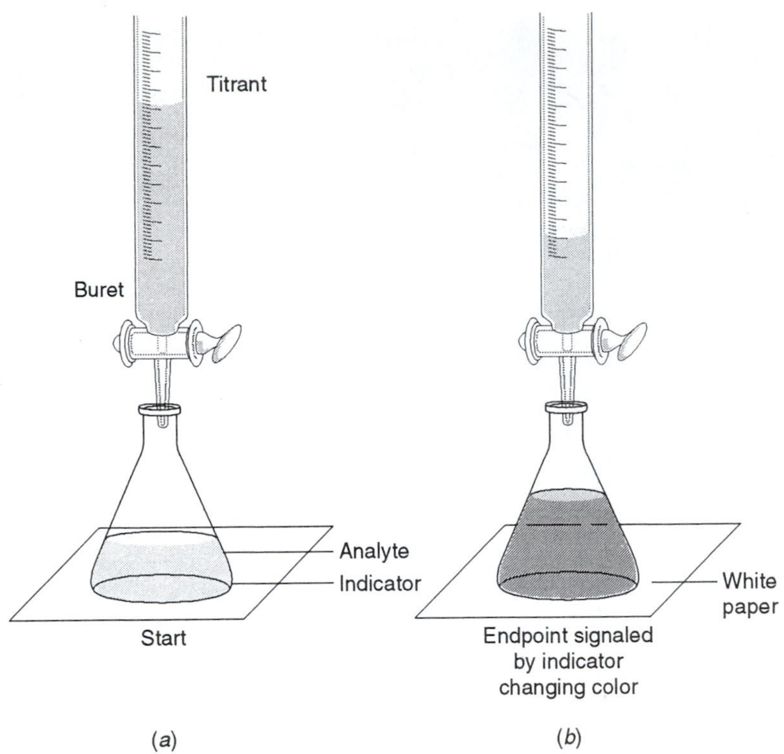

Figure 9.1 (a) Titrant in the buret is dispensed into the analyte until (b) the indicator changes color at its endpoint.

Stoichiometric amounts: amounts corresponding to the mole ratio of the balanced equation

Acid–base indicator: a substance having an acidic structure with a different color than its basic structure

pH: the negative logarithm of the molar concentration of H_3O^+, pH = $-\log[H_3O^+]$

A reaction is complete when **stoichiometric amounts** of the reacting substances are combined. In a titration this is the **stoichiometric point.**[1] In this experiment the stoichiometric point for the acid–base titration is detected using a phenolphthalein **indicator.** Phenolphthalein is colorless in an acidic solution but red (or pink) in a basic solution. The point in the titration at which the phenolphthalein changes color is called the **endpoint** of the indicator (Figure 9.1*b*). Indicators are selected so that the stoichiometric point in the titration coincides (at approximately the same **pH**) with the endpoint of the indicator.

Standardization of a Sodium Hydroxide Solution

Hygroscopic: able to absorb water vapor readily

Solid sodium hydroxide is very **hygroscopic;** therefore its mass cannot be measured accurately to prepare a solution with a well-known molar concentration (a primary standard solution). To prepare a NaOH solution with an exact molar concentration, it must be standardized with an acid that *is* a primary standard.

In Part A of this experiment, *dry* potassium hydrogen phthalate, $KHC_8H_4O_4$, is used as the primary acid standard for determining the molar concentration of a sodium hydroxide solution. Potassium hydrogen phthalate is a white, crystalline, acidic solid. It has the properties of a primary standard because of its high purity, relatively high molar mass, and because it is only *very slightly* hygroscopic. The amount of $KHC_8H_4O_4$ used for the analysis is calculated from its measured mass and molar mass (204.44 g/mol):

COOH
COO⁻K⁺
potassium hydrogen phthlate

$$\text{mass (g) } KHC_8H_4O_4 \times \frac{\text{mol } KHC_8H_4O_4}{204.44 \text{ g } KHC_8H_4O_4} = \text{mol } KHC_8H_4O_4 \qquad (9.1)$$

From the balanced equation for the reaction, one mole of $KHC_8H_4O_4$ reacts with one mole of NaOH according to the net ionic equation:

$$HC_8H_4O_4^-(aq) + OH^-(aq) \rightarrow H_2O(l) + C_8H_4O_4^{2-}(aq) \qquad (9.2)$$

[1]The stoichiometric point is also called the **equivalence point**, indicating the point at which stoichiometrically equivalent quantities of the reacting substances are combined.

In the experimental procedure an accurately measured mass of dry potassium hydrogen phthalate is dissolved in deionized water. A prepared NaOH solution is then dispensed from a buret into the $KHC_8H_4O_4$ solution until the stoichiometric point is reached, signaled by the colorless to pink change of the phenolphthalein indicator. At this point the dispensed volume of NaOH is noted and recorded.

The molar concentration of the NaOH solution is calculated using Equation 9.2 and

$$\text{molar concentration } (M) \text{ of NaOH (mol/L)} = \frac{\text{mol NaOH}}{\text{L of NaOH solution}} \tag{9.3}$$

Once the molar concentration of the sodium hydroxide is calculated, the solution is said to be "standardized" and the sodium hydroxide solution is called a **secondary standard** solution.

Molar Concentration of an Acid Solution

In Part B, an unknown molar concentration of an acid solution is determined. The standardized NaOH solution is used to titrate an accurately measured volume of the acid to the stoichiometric point. By knowing the volume and molar concentration of the NaOH, the number of moles of NaOH used for the analysis is

$$\text{volume (L)} \times \text{molar concentration (mol/L)} = \text{mol NaOH} \tag{9.4}$$

From the stoichiometry of the reaction (your instructor will inform you of the acid type, HA or H_2A), the moles of acid neutralized in the reaction can be calculated.

From the moles of the acid that react and its measured volume, the molar concentration of the acid is calculated:

$$\text{molar concentration of the acid (mol/L)} = \frac{\text{mol acid}}{\text{volume of acid (L)}} \tag{9.5}$$

Experimental Procedure

Procedure Overview: A NaOH solution is prepared with an approximate concentration. A more accurate molar concentration of the NaOH solution (as the titrant) is determined using dry potassium hydrogen phthalate as a primary standard. The NaOH solution, now a secondary standard solution, is then used to determine the molar concentration of an acid solution.

A. The Standardization of a Sodium Hydroxide Solution

You are to complete at least three "good" trials (±1% reproducibility) in standardizing the NaOH solution. Prepare three clean 125-mL or 250-mL Erlenmeyer flasks for the titration.

You will need to use approximately one liter of boiled, deionized water for this experiment. Start preparing that first.

1. **Prepare the Stock NaOH Solution.**[2] One week before the scheduled laboratory period, dissolve about 4 g of NaOH (pellets or flakes) (**Caution:** *NaOH is very corrosive—do not allow skin contact. Wash hands thoroughly with water.*) in 5 mL of deionized water in a 150-mm rubber-stoppered test tube. Thoroughly mix and allow the solution to stand for the precipitation of sodium carbonate, Na_2CO_3.[3]

---

[2]Check with your laboratory instructor to see if the NaOH solution is prepared for Part A.1 (and/or Part A.3) and to see if the $KHC_8H_4O_4$ is dried for Part A.2.

[3]Carbon dioxide, CO_2, from the atmosphere is an **acidic anhydride** (meaning that when CO_2 dissolves in water, it forms an acidic solution). The acid CO_2 reacts with the base NaOH to form the less soluble salt, Na_2CO_3.

$$CO_2(g) + 2\ NaOH(aq) \rightarrow Na_2CO_3(s) + H_2O(l)$$

2. **Dry the Primary Standard Acid.** Dry 2–3 g of $KHC_8H_4O_4$ at 110°C for several hours in a constant temperature drying oven. Cool the sample in a desiccator.

3. **Prepare the Diluted NaOH Solution.** Decant about 4 mL of the NaOH solution prepared in Part A.1 into a 500-mL polyethylene bottle (Figure 9.2). (**Caution:** *Concentrated NaOH solution is extremely corrosive and will cause severe skin removal!*) Dilute to 500 mL with previously boiled,[4] deionized water. Cap the polyethylene bottle to prevent the absorption of CO_2. Swirl the solution and label the bottle.

   Calculate an *approximate* molar concentration of your diluted NaOH solution.

4. **Prepare the Primary Standard Acid.** a. Calculate the mass of $KHC_8H_4O_4$ that will require about 15–20 mL of your diluted NaOH solution to reach the stoichiometric point. Show the calculations on the Report Sheet.

   b. Measure this mass (±0.001 g) of $KHC_8H_4O_4$ on weighing paper (Figure 9.3) and transfer it to a labeled Erlenmeyer flask. Similarly, prepare all three samples while you are occupying the balance. Dissolve the $KHC_8H_4O_4$ in about 50 mL of previously boiled, deionized water and add 2 drops of phenolphthalein.

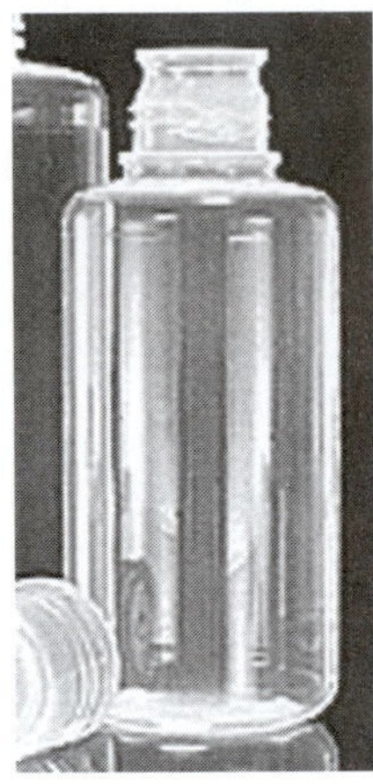

Figure 9.2 A 500-mL polyethylene bottle for the NaOH solution.

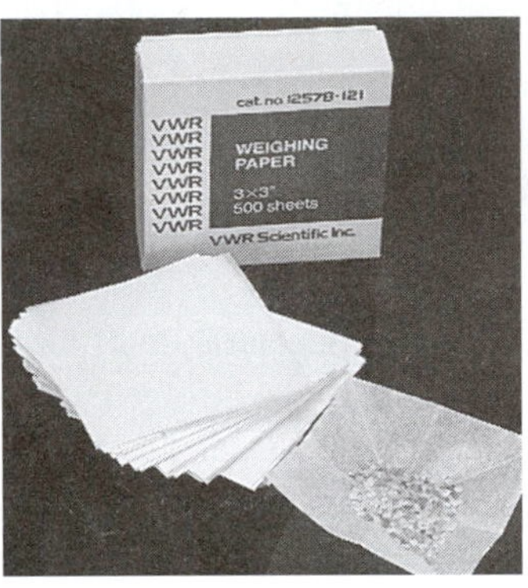

Figure 9.3 Weighing paper for the $KHC_8H_4O_4$ measurements.

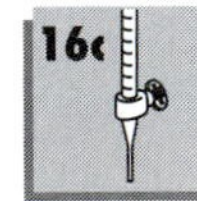

5. **Prepare a Clean Buret.** Wash a 50-mL buret and a funnel thoroughly with soap and water using a long buret brush. Flush the buret with tap water and rinse several times with deionized water. Rinse the buret with three 5-mL portions of the diluted NaOH solution, making certain that the solution wets the entire inner surface. Drain each rinse through the buret tip. Discard each rinse in the "Waste Bases" container. Have the instructor approve your buret and titration setup before continuing.

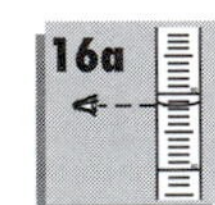

6. **Fill the Buret.** Using a clean funnel, fill the buret with the NaOH solution.[5] After 10–15 seconds, read the volume by viewing the bottom of the meniscus with the aid of a black line drawn on a white card (the buret can be removed from the stand or moved up or down in the buret clamp to make this reading; you need not stand on a lab stool to read the meniscus). Record this initial volume according to the guideline in Technique 16A.2, "using all certain digits (from the labeled calibration marks on the glassware) *plus* one uncertain digit (the last digit which is the best estimate between the calibration marks)." Place a sheet of white paper beneath the Erlenmeyer flask.

---

[4] Boiling the water removes traces of CO_2 that would react with the sodium hydroxide in solution.
[5] Be certain all air bubbles are removed from the buret tip.

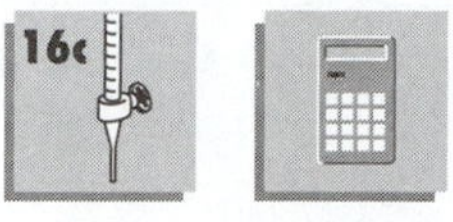

7. **Titrate the Primary Standard Acid.** Slowly add the NaOH titrant to the first acid sample prepared in Part A.4. Swirl the flask (with the proper hand[6]) after each addition. Initially, add the NaOH solution in 1- to 2-mL increments. As the stoichiometric point nears, the color fade of the indicator occurs more slowly. Occasionally rinse the wall of the flask with (previously boiled, deionized) water from your wash bottle. Continue addition of the NaOH titrant until the endpoint is reached. *The endpoint in the titration should be within one-half drop of a slight pink color* (see opening photo). The color should persist for 30 seconds. Read (Figure 9.4) and record the final volume of NaOH in the buret.

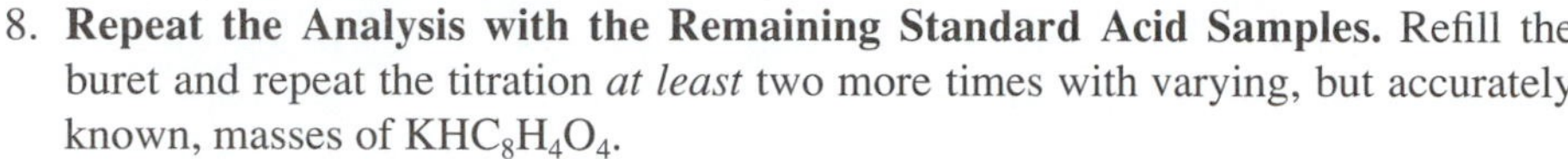

8. **Repeat the Analysis with the Remaining Standard Acid Samples.** Refill the buret and repeat the titration *at least* two more times with varying, but accurately known, masses of $KHC_8H_4O_4$.
9. **Do the Calculations.** Calculate the molar concentration of the diluted NaOH solution. The molar concentrations of the NaOH solution from the three analyses should be within ±1%. Place a corresponding label on the 500-mL polyethylene bottle.

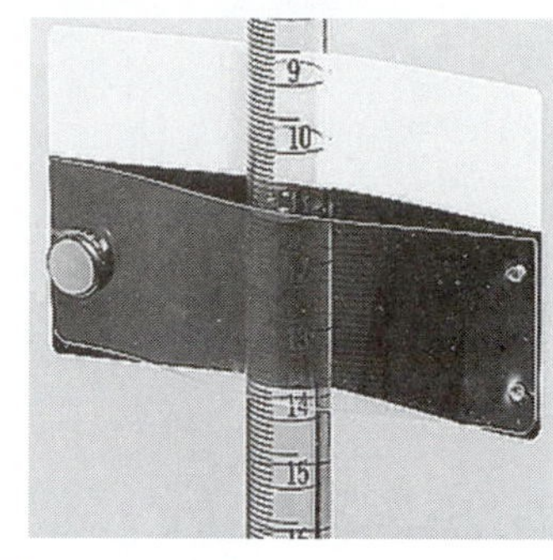

Figure 9.4 Read the volume of titrant with a black background.

> *Disposal:* Dispose of the neutralized solutions in the Erlenmeyer flasks in the "Waste Acids" container.

B. Molar Concentration of an Acid Solution

Three samples of the acid having an unknown concentration are to be analyzed. Ask your instructor for the acid type of your unknown (i.e., HA or H_2A). Prepare three *clean* 125- or 250-mL Erlenmeyer flasks for this determination.

1. **Prepare the Acid Samples of Unknown Concentration.** In an Erlenmeyer flask, pipet 25.00 mL of the acid solution. Add 2 drops of phenolphthalein.

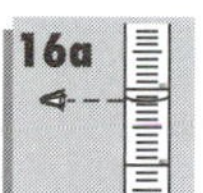

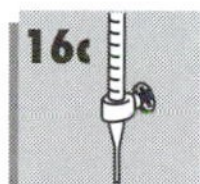

2. **Fill the Buret and Titrate.** Refill the buret with the (now) standardized NaOH solution and, after 10–15 seconds, read and record the initial volume. Refer to Parts A.6 and A.7. Titrate the acid sample to the phenolphthalein endpoint. After 10–15 seconds, read and record the final volume of titrant.
3. **Repeat.** Similarly titrate the other samples of the acid solution.
4. **Save.** Save your standardized NaOH solution in the *tightly capped* 500-mL polyethylene bottle for Experiments 10, 26, 27, and/or 28. Consult with your instructor.

5. **Calculations.** Calculate the average molar concentration of your acid unknown.

> *Disposal:* Dispose of the neutralized solutions in the "Waste Acids" container. Consult with your instructor.

CLEANUP: Rinse the buret and pipet several times with tap water and discard through the tip into the sink. Rinse twice with deionized water. Similarly clean the Erlenmeyer flasks.

Check and clean the balance area. All solids should be discarded in the "Waste Solid Acids" container.

[6]Check Technique 16C.3 of this laboratory manual.

Notes and Calculations

Experiment 9 *Prelaboratory Assignment*

A Volumetric Analysis

Date __________ Lab Sec. ______ Name ____ Desk No. __________

1. a. Define the analyte in a titration.

   b. Is the indicator generally added to the titrant or the analyte in a titration?

2. a. What is the primary standard used in this experiment? Define a primary standard.

   b. What is the secondary standard used in this experiment? Define secondary standard.

3. Distinguish between a stoichiometric point and an endpoint in an acid–base titration.

4. a. When rinsing a buret after cleaning it with soap and water, should the rinse be dispensed through the buret tip or the top opening of the buret? Explain.

   b. In preparing the buret for titration (Experimental Procedure, Part A.5), the final rinse is with the NaOH titrant rather than with deionized water. Explain.

   c. How is a "half-drop" of titrant dispensed from a buret?

5. In Part A.1, a 4-g mass of NaOH is dissolved in 5 mL of water. In Part A.3, a 4-mL aliquot of this solution is diluted to 500 mL of solution. What is the approximate molar concentration of NaOH in the diluted solution? Enter this information on your Report Sheet.

6. a. A 0.397-g sample of potassium hydrogen phthalate, $KHC_8H_4O_4$ (molar mass = 204.44 g/mol) is dissolved with 50 mL of deionized water in a 125-mL Erlenmeyer flask. The sample is titrated to the phenolphthalein endpoint with 16.22 mL of a sodium hydroxide solution. What is the molar concentration of the NaOH solution?

   b. A 25.00-mL aliquot of a nitric acid solution of unknown concentration is pipetted into a 125-mL Erlenmeyer flask and 2 drops of phenolphthalein are added. The *above* sodium hydroxide solution (the titrant) is used to titrate the nitric acid solution (the analyte). If 12.75 mL of the titrant is dispensed from a buret in causing a color change of the phenolphthalein, what is the molar concentration of the nitric acid solution?

Experiment 9 *Report Sheet*

A Volumetric Analysis

Date __________ Lab Sec. ______ Name ______________________________ Desk No. __________

Maintain at least three significant figures when recording data and performing calculations.

A. Standardization of a Sodium Hydroxide Solution

Approximate molar concentration of diluted NaOH solution (Part A.3). Show calculations.

Approximate mass of $KHC_8H_4O_4$ for the standardization of the NaOH solution (Part A.4). Show calculations.

	Trial 1	*Trial 2*	*Trial 3*
1. Tared mass of $KHC_8H_4O_4$ (g)	________	________	________
2. Molar mass of $KHC_8H_4O_4$		204.44 g/mol	
3. Moles of $KHC_8H_4O_4$ (mol)	________	________	________
Titration apparatus approval		________	
4. Buret reading of NaOH, *initial* (mL)	________	________	________
5. Buret reading of NaOH, *final* (mL)	________	________	________
6. Volume of NaOH dispensed (mL)	________	________	________
7. Molar concentration of NaOH (mol/L)	________	________	________
8. Average molar concentration of NaOH (mol/L)		________	

*Show calculation for Trial 1.

B. Molar Concentration of an Acid Solution

Acid type: __________ Unknown No. __________

Balanced equation for neutralization of acid with NaOH.

	Sample 1	Sample 2	Sample 3
1. Volume of acid solution (mL)	25.0	25.0	25.0
2. Buret reading of NaOH, *initial* (mL)	______	______	______
3. Buret reading of NaOH, *final* (mL)	______	______	______
4. Volume of NaOH dispensed (mL)	______	______	______
5. Molar concentration of NaOH (mol/L), Part A		______	
6. Moles of NaOH dispensed (mol)	______	______	______
7. Molar concentration of acid solution (mol/L)	______	______	______
8. Average molar concentration of acid solution (mol/L)		______	

*Show calculations for Sample 1.

Laboratory Questions

Circle the questions that have been assigned.

1. Part A.2. Pure potassium hydrogen phthalate is used for the standardization of the sodium hydroxide solution. Suppose that the potassium hydrogen phthalate is *not* completely dry. Will the reported molar concentration of the sodium hydroxide solution be too high, too low, or unaffected because of the moistness of the potassium hydrogen phthalate? Explain.
2. Part A.4. The potassium hydrogen phthalate primary standard is dissolved in 25 mL instead of the suggested 50 mL of deionized water. How will this volume change affect the calculated molar concentration of the NaOH solution? Explain.
3. Part A.4. Phenolphthalein is a weak organic acid, being colorless in an acidic solution and pink (because of the color of its conjugate base) in a basic solution. The Experimental Procedure suggests the addition of 2 drops of phenolphthalein for the standardization of the sodium hydroxide solution. Explain why the analysis will be less accurate with the addition of a larger amount, e.g., 20 drops, of phenolphthalein.
4. Part A.7. A drop of the NaOH titrant adheres to the side of the buret (because of a dirty buret) between the initial and final readings for the titration. How does this technique error affect the reported molar concentration of the NaOH solution? Explain.
5. Part B.2. The wall of the Erlenmeyer flask is occasionally rinsed with water from the wash bottle (see Part A.7) during the analysis of the acid solution. How does this affect the reported molar concentration of the acid solution? Explain.
6. Part B.2. An air bubble initially entrapped in the buret tip is passed from the buret during the titration. Will the reported molar concentration of the acid be reported too high, too low, or unchanged as a result? Explain.
7. Part B.4. The polyethylene bottle containing the standardized NaOH solution is *not* tightly capped for subsequent experiments. How does this affect the molar concentration of the NaOH solution? Explain.

Experiment 10

Vinegar Analysis

Vinegar is a 4–5% (by mass) solution in acetic acid.

Objective

- To determine the percent by mass of acetic acid in vinegar

Techniques

The following techniques are used in the Experimental Procedure

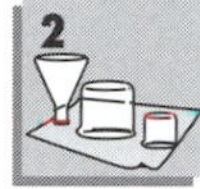

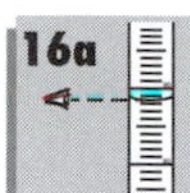

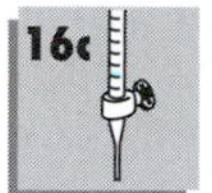

Introduction

$CH_3-C(=O)-OH$

acetic acid

Household vinegar is a 4–5% (by mass) acetic acid, CH_3COOH, solution (4% is the minimum federal standard). Generally, caramel flavoring and coloring are also added to make the product aesthetically more appealing. This experiment compares the acetic acid concentrations in at least two vinegars.

A volumetric analysis using the titration technique (Technique 16C) is the method used for the determination of the percent by mass of acetic acid in vinegar. A measured mass of vinegar is titrated to the phenolphthalein endpoint with a measured volume of a standardized sodium hydroxide solution. As the volume and molar concentration of the standardized NaOH solution are known, the moles of NaOH used for the analysis are also known.

The moles of CH_3COOH are calculated from the balanced equation:

$$CH_3COOH(aq) + NaOH(aq) \rightarrow NaCH_3CO_2(aq) + H_2O(l) \qquad (10.1)$$

The mass of CH_3COOH in the vinegar is calculated from the measured moles of CH_3COOH neutralized in the reaction and its molar mass, 60.05 g/mol:

$$\text{mass (g) of } CH_3COOH = \text{mol } CH_3COOH \times \frac{60.05 \text{ g } CH_3COOH}{\text{mol } CH_3COOH} \qquad (10.2)$$

Finally, the percent by mass of CH_3COOH in vinegar is calculated:

$$\% \text{ by mass of } CH_3COOH = \frac{\text{mass (g) of } CH_3COOH}{\text{mass (g) of vinegar}} \times 100 \qquad (10.3)$$

Experimental Procedure

Procedure Overview: Samples of two vinegars are analyzed for the amount of acetic acid in the sample. A titration setup is used for the analysis, using a standardized NaOH solution as the titrant and phenolphthalein as the indicator. Stoichiometry calculations determine the percent of acetic acid in the vinegars.

Two samples of two vinegars are to be analyzed. At the beginning of the laboratory period, obtain 10 mL of each vinegar in separate 10-mL graduated cylinders.

A standardized NaOH solution was prepared in Experiment 9. If that solution was saved, it is to be used for this experiment. If the solution was not saved, you either must again prepare the solution (Experiment 9, Part A) or obtain about 150 mL of a standardized NaOH solution prepared by stockroom personnel. Record the *exact* molar concentration of the NaOH solution on the Report Sheet. Your instructor will advise you.

A. Preparation of Vinegar Sample

Clean at least two 125- or 250-mL Erlenmeyer flasks. When dry, label each flask and measure its mass (±0.01 g).

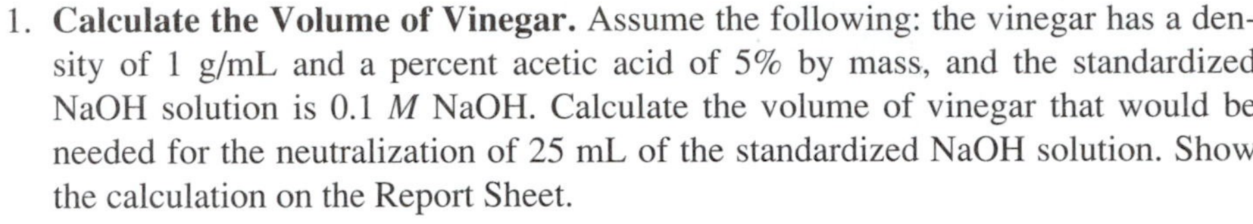

1. **Calculate the Volume of Vinegar.** Assume the following: the vinegar has a density of 1 g/mL and a percent acetic acid of 5% by mass, and the standardized NaOH solution is 0.1 *M* NaOH. Calculate the volume of vinegar that would be needed for the neutralization of 25 mL of the standardized NaOH solution. Show the calculation on the Report Sheet.

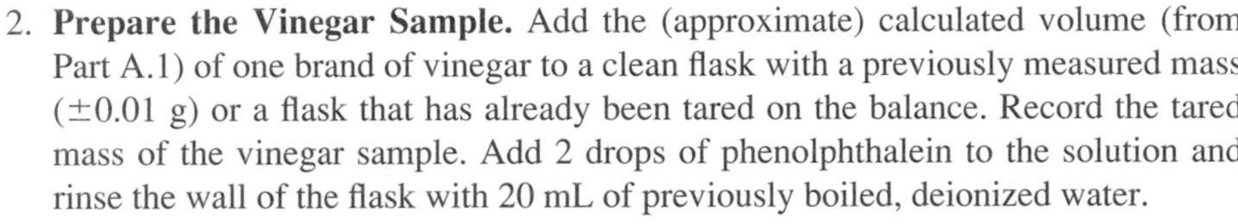

2. **Prepare the Vinegar Sample.** Add the (approximate) calculated volume (from Part A.1) of one brand of vinegar to a clean flask with a previously measured mass (±0.01 g) or a flask that has already been tared on the balance. Record the tared mass of the vinegar sample. Add 2 drops of phenolphthalein to the solution and rinse the wall of the flask with 20 mL of previously boiled, deionized water.

3. **Prepare the Buret and Titration Setup.** Rinse a clean 50-mL buret with the standardized NaOH solution, making certain no drops cling to the inside wall. Fill the buret with the standardized NaOH solution, eliminate all air bubbles in the buret tip, and, after 10–15 seconds, read and record the initial volume, "using all certain digits (from the labeled calibration marks on the buret) *plus* one uncertain digit (the last digit which is the best estimate between the calibration marks)." Place a sheet of white paper beneath the flask containing the vinegar sample.

B. Analysis of Vinegar Sample

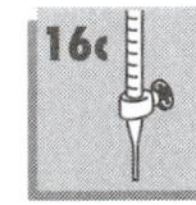

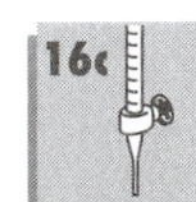

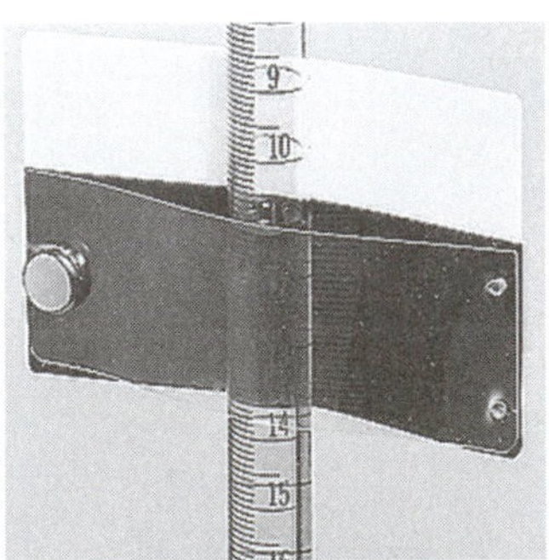

Figure 10.1 Read the volume of titrant with a black background.

1. **Titrate the Vinegar Sample.** Slowly add the NaOH solution from the buret to the acid, swirling the flask (with the proper hand[1]) after each addition. Occasionally, rinse the wall of the flask with (previously boiled, deionized) water from your wash bottle. Continue addition of the NaOH titrant until the endpoint is reached.[2] The endpoint in the titration should be within one-half drop of a slight pink color. Be careful *not* to surpass the endpoint. The color should persist for 30 seconds. After 10–15 seconds, read (Figure 10.1) and record the final volume of NaOH titrant in the buret (see Technique 16A.2).
2. **Repeat with the Same Vinegar.** Refill the buret and repeat the titration *at least* once more with another sample of the same vinegar.
3. **Repeat with Another Vinegar.** Perform at least two analyses of a different vinegar and determine its acetic acid content.
4. **Calculations.** Determine the average percent by mass of acetic acid in the two vinegars.

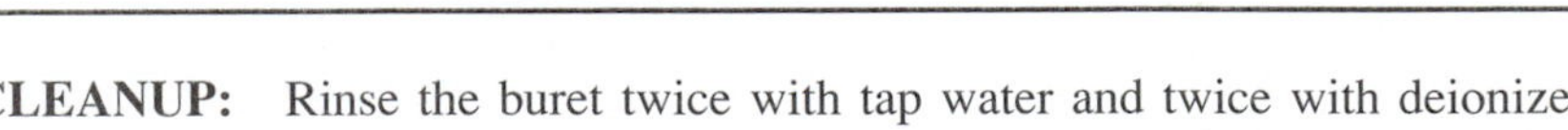

Disposal: All test solutions and the NaOH solution in the buret can be discarded in the "Waste Bases" container.

CLEANUP: Rinse the buret twice with tap water and twice with deionized water, discarding each rinse through the buret tip into the sink. Similarly, rinse the flasks.

[1]Check Technique 16C.3 of this laboratory manual.

[2]The endpoint (and the stoichiometric point) is near when the color fade of the phenolphthalein indicator occurs more slowly with each successive addition of smaller volumes of NaOH solution to the vinegar.

Experiment 10 *Prelaboratory Assignment*

Vinegar Analysis

Date ________ Lab Sec. ______ Name ______________________________ Desk No. ________

1. Assuming the density of a 5% acetic acid solution is 1.0 g/mL, determine the volume of the acetic acid solution necessary to neutralize 25.0 mL of 0.10 *M* NaOH. Also record this calculation on your Report Sheet.

2. A 30.84-mL volume of 0.128 *M* NaOH is required to reach the phenolphthalein endpoint in the titration of a 5.441-g sample of vinegar. Calculate the percent acetic acid in the vinegar.

3. A chemist should wait 10–15 seconds after dispensing a volume of titrant before a reading is made. Explain why the "wait" is good laboratory technique.

   b. The color change at the endpoint should persist for 30 seconds. Explain why the time lapse is a good titration technique.

4. For the titration of an acid analyte with a NaOH titrant, a phenolphthalein endpoint will change from colorless to pink. If the "pink" resulted from perhaps only a half-drop (or less) of NaOH titrant and the Erlenmeyer flask is set aside on the laboratory bench, the pink color may disappear after an hour or so. Explain.

5. Explain why it is quantitatively *not* acceptable to titrate each of the vinegar samples with the NaOH titrant to the same *dark red* endpoint.

6. Lemon juice has a pH of about 2.5. *Assuming* that the acidity of lemon juice is due solely to citric acid, that citric acid is a monoprotic acid, and that the density of lemon juice is 1.0 g/mL, then the citric acid concentration calculates to 0.5% by mass. Estimate the volume of 0.0100 *M* NaOH required to neutralize a 3.00-g sample of lemon juice. The molar mass of citric acid is 190.12 g/mol.

7. Oxalic acid, $H_2C_2O_4$•$2H_2O$ (molar mass = 126.07 g/mol) is often used as a primary standard for the standardization of a NaOH solution. If 0.106 g of oxalic acid dihydrate is neutralized by 19.78 mL of a NaOH solution, what is the molar concentration of the NaOH solution? Oxalic acid is a diprotic acid. (Hint: what is the balanced equation?)

Experiment 10 *Report Sheet*

Vinegar Analysis

Date __________ Lab Sec. ______ Name ______________________________ Desk No. __________

Calculation of the approximate volume of the vinegar sample for the analyses (Part A.1).

Brand of Vinegar or Unknown No. ______________ ______________

Data	***Trial 1***	***Trial 2***	***Trial 1***	***Trial 2***
1. Tared mass of vinegar (g)	________	________	________	________
2. Buret reading of NaOH, *initial* (mL)	________	________	________	________
3. Buret reading of NaOH, *final* (mL)	________	________	________	________
4. Volume of NaOH used (mL)	________	________	________	________
5. Molar concentration of NaOH (mol/L)	________________		________________	

Calculations				
1. Amount of NaOH added (mol)	________*	________	________	________
2. Amount of CH_3COOH in vinegar (mol)	________	________	________	________
3. Mass of CH_3COOH in vinegar (g)	________	________	________	________
4. Percent by mass of CH_3COOH in vinegar	________	________	________	________
5. Average percent by mass of CH_3COOH in vinegar	________________		________________	

*Calculations for Trial 1 of the first vinegar sample.

Discuss briefly a comparison of the two vinegars.

Laboratory Questions

Circle the questions that have been assigned.

1. Part A.2. *Previously boiled* deionized water is unavailable. In a hurry to pursue the analysis, deionized water (not boiled) is added. How does this attempt to expedite the analysis affect the reported percent acetic acid in the vinegar? Explain.
2. Part A.2. If *too much* phenolphthalein is added to the vinegar in the flask, will the reported percent acetic acid in vinegar be reported too high, too low, or unchanged? Explain.
3. Part A.3. The buret is filled with the NaOH titrant and the initial volume reading is immediately recorded without waiting the recommended 10–15 seconds. However in Part B.1, the 10–15 second time lapse does occur before the reading is made. How does this technique error affect the reported percent acetic acid in the vinegar? Explain.
4. Part B.1. The endpoint of the titration is overshot! How will this error affect the reported percent acetic acid in the vinegar? Explain.
5. Part B.1. The wall of the flask is periodically rinsed with the previously boiled, deionized water from the wash bottle. How does this titrimetric technique affect the reported percent acetic acid in the vinegar? Explain.
6. Part B.1. A drop of NaOH titrant adheres to the inner wall of the buret (because of a dirty buret!) between the initial and final buret readings for the titration of the vinegar sample. How does this technique error affect the reported percent acetic acid in the vinegar sample? Explain.

Dry Lab 2A

Inorganic Nomenclature I. Oxidation Numbers

Sodium chloride salt crystals are a one-to-one combination of sodium cations and chloride anions. The sodium cation has a charge of 1+ and the chloride anion has a charge of 1−.

OBJECTIVE

- To become familiar with the oxidation numbers of various elements

INTRODUCTION

You have probably noticed that members of virtually all professions have a specialized language. Chemists are no exception. For chemists to communicate internationally, some standardization of the technical language is needed. Historically, common names for many compounds evolved and are still universally understood—for example, water, sugar (sucrose), and ammonia—but with new compounds being synthesized and isolated daily, a "familiar" system is no longer viable.

The Chemical Abstracts Service (CAS) of the American Chemical Society currently has over 22 million inorganic and organic compounds registered in their database with over 4000 new compounds being added *daily*. Not only does each compound have a unique name, according to a systematic method of nomenclature, but also each compound has a unique molecular structure and set of chemical and physical properties. Therefore, it has become absolutely necessary for the scientific community to properly name compounds according to a universally accepted system of nomenclature.

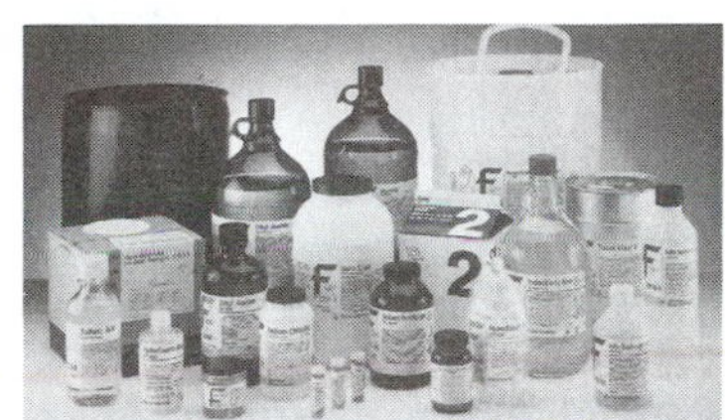

Chemical names may be common names or systematic names.

In the three parts of Dry Lab 2 (A, B, and C) you will learn a few systematic rules, established by the International Union for Pure and Applied Chemistry (IUPAC), for naming and writing formulas for inorganic compounds. You are undoubtedly already familiar with some symbols for the elements and the names for several common compounds. For instance, NaCl is sodium chloride. Continued practice and work in writing formulas and naming compounds will make you even more knowledgeable of the chemist's vocabulary.

Charge and Oxidation Number

Charges are conventionally written "number and charge" (e.g., 3+), but oxidation numbers are written "charge and number" (e.g., −2)

Elements or groups of chemically bonded elements (called **polyatomic** or **molecular groups**) that have **charge** are called **ions**—if the charge is positive they are called **cations,** if negative, they are **anions.** Oftentimes elements combine to form a compound in which the elements may not have an actual charge, but rather an "apparent" charge, called its **oxidation number** (or oxidation state). The oxidation number, the charge an atom would have if the electrons in the bond were assigned to the more electronegative element, may be either positive or negative.

Charges and oxidation numbers are very handy bookkeeping devices for keeping track of what happens to electrons when various elements combine to form compounds. By remembering a relatively few generalizations concerning charges and oxidation numbers, we can write the correct chemical formulas for a large number of compounds. It is *not* necessary to simply memorize disconnected chemical formulas.

As in any system, some rules of accounting must be followed. Some rules for assigning charges and oxidation numbers are as follows:

1. Any element in the free state (not combined with another element) has an oxidation number of zero, regardless of the complexity of the molecule in which it occurs. Each atom in Ne, O_2, P_4, and S_8 has a charge of 0.
2. Monoatomic ions have a charge equal to the charge of the ion. The ions Ca^{2+}, Fe^{3+}, and Cl^- have charges of 2+, 3+, and 1− respectively.
3. Oxygen in compounds has an assigned oxidation number of −2 (except for a −1 in peroxides, e.g., H_2O_2, and +2 in OF_2). The oxidation number of oxygen is −2 in FeO, Fe_2O_3, $KMnO_4$, and KIO_3.
4. Hydrogen in compounds has an oxidation number of +1 (except for a charge of 1− in metal hydrides, e.g., NaH). Its oxidation number is +1 in HCl, $NaHCO_3$, and NH_3.
5. Some elements exhibit only one common charge or oxidation number in certain types of compounds:
   a. Group 1A elements always have a charge of 1+ in compounds.
   b. Group 2A elements always have a charge of 2+ in compounds.
   c. Boron and aluminum always possess an oxidation number of +3 in compounds.
   d. In **binary compounds** *with metals,* the nonmetallic elements of Group 6A generally exhibit a charge of 2−.
   e. In binary compounds *with metals,* the elements of Group 7A have a charge of 1−.

Binary compounds: compounds consisting of only two elements

Polyatomic ions are also called molecular ions

6. **Polyatomic ions** have a charge equal to the sum of the oxidation numbers of the elements of the polyatomic group. The polyatomic ions SO_4^{2-}, NO_3^-, and PO_4^{3-}, have charges of 2−, 1−, and 3−, respectively.
7. In assigning oxidation numbers to elements in a compound, the element closest to fluorine (the most electronegative element) in the periodic table is always assigned the negative oxidation number. In the compound, P_4O_{10}, oxygen has the negative oxidation number of −2.
8. a. For compounds the sum of the charges and oxidation numbers of all atoms in the compound must equal zero.

   Example D2A.1 For Na_2S, the sum of the charges and oxidation numbers equals zero.

   2 Na atoms, 1+ for each (Rule 5a) = 2+

   1 S atom, 2− for each (Rule 5d) = 2−

   Sum of charges (2+) + (2−) = 0

   b. For polyatomic ions the sum of the oxidation numbers of the elements must equal the charge of the ion.

   Example D2A.2 For CO_3^{2-}, the sum of the oxidation numbers of the elements equals a charge of 2−.

   3 O atoms, −2 for each (Rule 3) = −6

   1 C atom which must be +4 = +4

   so that (−6) + (+4) = 2−, the charge of the CO_3^{2-} ion
9. Some chemical elements show more than one charge or oxidation number, depending on the compound. The preceding rules may be used to determine their values. Consider the compounds $FeCl_2$ and $FeCl_3$. Since the chlorine atom has a

charge of 1− when combined with a metal (Rule 5e), the charges of iron are 2+ in $FeCl_2$ and 3+ in $FeCl_3$.

An extensive listing of charges for monoatomic ions and polyatomic ions are presented in Dry Lab 2C, Table D2C.1.

Writing Formulas from Charges and Oxidation Numbers

The formulas of compounds can be written from a knowledge of the charges and oxidation numbers of the elements that make up the compound. The sum of the charges and oxidation numbers for a compound must equal zero.

Example D2A.3 Write the formula of the compound formed from Ca^{2+} ions and Cl^- ions.

As the sum of the charges must equal zero, two Cl^- are needed for each Ca^{2+}; therefore the formula of the compound is $CaCl_2$.

DRY LAB PROCEDURE

Procedure Overview: The charge or oxidation number of an element in a selection of compounds and ions is determined by application of the rules in the Introduction. Additionally, a knowledge of charges of elements and/or polyatomic groups will allow you to write the correct formulas for compounds.

Your instructor will indicate the questions you are to complete. Answer them on a separate piece of paper. Be sure to indicate the date, your lab section, and your desk number on your answer sheet.

1. Indicate the oxidation number of carbon and sulfur in the following compounds.

a. CO	d. $Na_2C_2O_4$	g. SO_2	j. Na_2SO_3	m. SCl_2
b. CO_2	e. CH_4	h. SO_3	k. $Na_2S_2O_3$	n. Na_2S_2
c. Na_2CO_3	f. H_2CO	i. Na_2SO_4	l. $Na_2S_4O_6$	o. $SOCl_2$

2. Indicate the oxidation number of phosphorus, iodine, nitrogen, tellurium, and silicon in the following ions.

a. PO_4^{3-}	d. $P_3O_{10}^{5-}$	g. IO^-	j. NO_2^-	m. $N_2O_2^{2-}$
b. PO_3^{3-}	e. IO_3^-	h. NH_4^+	k. NO^+	n. TeO_4^{2-}
c. HPO_4^{2-}	f. IO_2^-	i. NO_3^-	l. NO_2^+	o. SiO_3^{2-}

3. Indicate the oxidation number of the metallic element(s) in the following compounds.

a. Fe_2O_3	g. CrO_3	m. MnO_2
b. FeO	h. K_2CrO_4	n. PbO_2
c. CoS	i. $K_2Cr_2O_7$	o. Pb_3O_4
d. $CoSO_4$	j. $KCrO_2$	p. ZrI_4
e. K_3CoCl_6	k. K_2MnO_4	q. U_3O_8
f. $CrCl_3$	l. Mn_2O_7	r. UO_2Cl

4. Write formulas of the compounds formed from the following elements (or polyatomic groups of elements) with the indicated charges.

a. Na^+ and SO_4^{2-}	g. Zn^{2+} and CO_3^{2-}	m. Ca^{2+} and $P_3O_{10}^{5-}$
b. Ca^{2+} and SO_4^{2-}	h. Ag^+ and CO_3^{2-}	n. Ca^{2+} and PO_4^{3-}
c. NH_4^+ and SO_4^{2-}	i. Zn^{2+} and Br^-	o. Ca^{2+} and HPO_4^{2-}
d. Li^+ and N^{3-}	j. Co^{3+} and Br^-	p. Os^{8+} and O^{2-}
e. Ca^{2+} and N^{3-}	k. Zr^{4+} and Br^-	q. Ca^{2+} and H^-
f. Cr^{3+} and CO_3^{2-}	l. Ca^{2+} and P^{3-}	r. Al^{3+} and SO_4^{2-}

Dry Lab 2B

Inorganic Nomenclature II. Binary Compounds

Calcium fluoride, magnesium oxide, and sodium chloride (left to right) are binary compounds of a metal and a nonmetal, called **salts.**

OBJECTIVE

- To name and write formulas for the binary compounds of a metal and a nonmetal, of two nonmetals (or metalloid and nonmetal), and of acids.

INTRODUCTION

The oxidation number of an element helps us to write the formula of a compound and to characterize the chemical nature of the element in the compound. Such information may tell us of the reactivity of the compound. In this dry lab, the correct writing and naming of binary compounds become the first steps in understanding the systematic nomenclature of a large array of chemical compounds.

Binary compounds are the simplest of compounds consisting of a chemical combination of two elements, *not* necessarily two atoms. Compounds such as NaCl, SO_3, Cr_2O_3, and HCl are all binary compounds. Binary compounds can be categorized into those of a metal cation and a nonmetal anion, of two nonmetals (or a metalloid and a nonmetal), and of acids.

Metal Cation and Nonmetal Anion—A Binary Salt

Compounds consisting of only two elements are named directly from the elements involved. In naming a binary salt, it is customary to name the more metallic (more electropositive) element first. The root of the second element is then named with the suffix *-ide* added to it. NaBr is sodium brom*ide;* Al_2O_3 is aluminum ox*ide.*

Two polyatomic anions, OH^- and CN^-, have names ending in *-ide,* even though their salts are not binary. Thus, KOH is potassium hydrox*ide* and NaCN is sodium cyan*ide*. One polyatomic cation, NH_4^+, is also treated as a single species in the naming of a compound; NH_4Cl is ammonium chlor*ide,* even though NH_4Cl is not a binary compound.

When a metal cation exhibits more than one charge, the nomenclature must reflect that. Multiple charges normally exist for the transition metal cations. Two systems are used for expressing different charges of a metal cation:

1. The "old" system uses a different suffix to reflect the charge of the metal cation; the suffix is usually added to the root of the Latin name for the cation. The *-ous* ending designates the lower of two charges for the metal cation, whereas the *-ic* ending indicates the higher one.
2. The Stock system uses a Roman numeral after the English name to indicate the charge of the metal cation in the compound.

Examples for the copper, iron, and tin salts are

Formula	"Old" System	Stock System
Cu_2O	*cuprous* oxide	copper(I) oxide
CuO	*cupric* oxide	copper(II) oxide
$FeCl_2$	*ferrous* chloride	iron(II) chloride
$FeCl_3$	*ferric* chloride	iron(III) chloride
$SnCl_2$	*stannous* chloride	tin(II) chloride
$SnCl_4$	*stannic* chloride	tin(IV) chloride

"Old" names are still often used in chemical nomenclature.

In naming compounds of a metal known to have a variable charge, the Stock system is preferred in chemistry today. Following are the metal cations that are commonly named using both the "old" and the Stock system.

Fe^{3+}	ferric	iron(III)	Cu^{2+}	cupric	copper(II)
Fe^{2+}	ferrous	iron(II)	Cu^{+}	cuprous	copper(I)
Sn^{4+}	stannic	tin(IV)	Hg^{2+}	mercuric	mercury(II)
Sn^{2+}	stannous	tin(II)	$Hg_2{}^{2+}$	mercurous	mercury(I)
Cr^{3+}	chromic	chromium(III)	Co^{3+}	cobaltic	cobalt(III)
Cr^{2+}	chromous	chromium(II)	Co^{2+}	cobaltous	cobalt(II)
Pb^{4+}	plumbic	lead(IV)	Mn^{3+}	manganic	manganese(III)
Pb^{2+}	plumbous	lead(II)	Mn^{2+}	manganous	manganese(II)

Two Nonmetals (or a Metalloid and a Nonmetal)

Often more than two compounds form from the chemical combination of two nonmetals or a metalloid and a nonmetal. For example, nitrogen and oxygen combine to form the compounds N_2O, NO, NO_2, N_2O_3, N_2O_4, and N_2O_5. It is obvious that these are all nitrogen ox*ide*s, but each must have a unique name.

To distinguish between the nitrogen oxides and, in general, to name compounds formed between two nonmetals or a metalloid and a nonmetal, Greek prefixes are used to designate the number of atoms of each element present in the molecule. The common Greek prefixes follow:

Prefix	Meaning	Prefix	Meaning
mono-*	one	hepta-	seven
di-	two	octa-	eight
tri-	three	nona-	nine
tetra-	four	deca-	ten
pesnta-	five	dodeca-	twelve
hexa-	six		

*mono- is seldom used, as "one" is generally implied.

The "prefix" system is the *preferred* method for naming the binary compounds of two nonmetals and of a metalloid and a nonmetal. The Stock system is occasionally used, but as you can see, this system does not differentiate between the NO_2 and N_2O_4 nitrogen oxides below.

Formula	"Prefix" System	Stock System
N_2O	*di*nitrogen oxide	nitrogen(I) oxide
NO	nitrogen oxide	nitrogen(II) oxide
NO_2	nitrogen *di*oxide	nitrogen(IV) oxide
N_2O_3	*di*nitrogen *tri*oxide	nitrogen(III) oxide
N_2O_4	*di*nitrogen *tetr*oxide	nitrogen(IV) oxide
N_2O_5	*di*nitrogen *pent*oxide	nitrogen(V) oxide

dinitrogen trioxide

dinitrogen tetroxide

dinitrogen pentoxide

Note that on occasion the end vowel of the prefix is omitted from the spelling for clarity in pronunciation.

Hydrates

Prefixes are also used for the naming of **hydrates,** inorganic salts in which water molecules are a part of the crystalline structure of the solid. The "waters of hydration" (also called waters of crystallization) are bound to the ions of the salt and, in most cases, can be removed with the application of heat (see Experiment 6). For example, barium chloride is often purchased as barium chloride *di*hydrate, $BaCl_2•2H_2O$. The formula of iron(III) chloride *hexa*hydrate is $FeCl_3•6H_2O$.

Binary Acids

A binary acid is an *aqueous* solution of a compound formed by hydrogen and a more electronegative nonmetal. To name the acid, the prefix *hydro-* and suffix *-ic* are added to the root name of the nonmetal.

Formula	Name
$HCl(aq)$	*hydro*chlor*ic acid*
$HBr(aq)$	*hydro*brom*ic acid*
$H_2S(aq)$	*hydro*sulfur*ic acid*

Writing Formulas from Names

In writing formulas for compounds it is mandatory that the sum of the charges or oxidation numbers (see Dry Lab 2A) of the elements equals zero. The name of the compound must dictate how the formula is to be written.

Example D2B.1 Write the formula for calcium nitride.

Calcium is in Group 2A and therefore has a charge of 2+: Ca^{2+}. The nitrogen atom is always 3− when combined with a metal: N^{3-}. Therefore, for the sum of the charges to equal zero there must be three Ca^{2+} (a total of 6+) for every two N^{3-} (a total of 6−): the formula is Ca_3N_2.

Dry Lab Procedure

Procedure Overview: Given the formula of the compound, the proper names for a large number of compounds are to be written, and given the name of the compound, the formulas for a large number of compounds are to be written.

Your instructor will assign the exercises you are to complete. Answer them on a separate piece of paper. Be sure to indicate the date, your lab section, and your desk number on your answer sheet. Use the rules that have been described.

1. Name the following binary-type salts of the representative elements.

a. Na_3P	d. CaC_2	g. Ca_3P_2	j. K_2S	m. NH_4Br	p. $AlCl_3$
b. Na_2O	e. CaI_2	h. KCN	k. K_2Te	n. $(NH_4)_2S$	q. Al_2O_3
c. Na_3N	f. CaH_2	i. KOH	l. K_2O_2	o. NH_4CN	r. AlN

2. Name the following salts according to the Old "*-ic, -ous*" system *and* the Stock system.

a. CrS	g. $HgCl_2$	m. CoO
b. Cr_2O_3	h. Hg_2Cl_2	n. $CoBr_3•6H_2O$
c. $CrI_3•6H_2O$	i. HgO	o. SnF_4
d. CuCl	j. Fe_2O_3	p. SnO_2
e. CuI_2	k. FeS	q. Cu_2O
f. $CuBr_2•4H_2O$	l. $FeI_3•6H_2O$	r. $Fe(OH)_3$

3. Name the following binary acids.
   a. HF(aq)
   b. HI(aq)
   c. H_2Se(aq)
   d. HBr(aq)
   e. H_2Te(aq)
   f. HCl(aq)
4. Name the following binary compounds consisting of two nonmetals or a metalloid and a nonmetal.
   a. SO_2
   b. SO_3
   c. S_4N_4
   d. SF_6
   e. SCl_4
   f. NO_2
   g. N_2O_5
   h. N_2S_4
   i. NF_3
   j. $SiCl_4$
   k. SiO_2
   l. $AsCl_3$
   m. AsH_3
   n. AsF_5
   o. HCl
   p. XeF_4
   q. XeF_6
   r. XeO_3
5. Write formulas for the following compounds consisting of a metal cation and a nonmetal anion.
   a. ferrous sulfide
   b. iron(III) hydroxide
   c. ferric oxide
   d. aluminum iodide
   e. copper(I) chloride
   f. cupric cyanide tetrahydrate
   g. manganese(IV) oxide
   h. nickel(III) oxide
   i. chromium(III) oxide
   j. titanium(IV) chloride
   k. cobalt(II) chloride hexahydrate
   l. cobaltous oxide
   m. mercury(I) chloride
   n. mercuric iodide
6. Write formulas for the following compounds consisting of two nonmetals (or a metalloid and a nonmetal).
   a. hydrochloric acid
   b. hydrosulfuric acid
   c. hydroiodic acid
   d. silicon tetrafluoride
   e. arsenic pentafluoride
   f. xenon hexafluoride
   g. iodine pentafluoride
   h. krypton difluoride
   i. tetrasulfur tetranitride
   j. dichlorine heptoxide
   k. phosphorus trihydride (phosphine)
   l. tetraphosphorus decoxide
7. Write formulas for the following compounds.
   a. dichlorine heptaoxide
   b. copper(II) phosphide
   c. iodine pentafluoride
   d. ferrous cyanide
   e. ammonium iodide
   f. platinum(IV) chloride
   g. cuprous cyanide
   h. cobaltous hydroxide
   i. gold(III) cyanide dihydrate
   j. sulfur tetrafluoride
   k. vanadium(V) oxide
   l. lithium hydride
   m. cobaltic oxide hexahydrate
   n. stannic oxide
8. Write the formulas for the compounds shown in the photo.

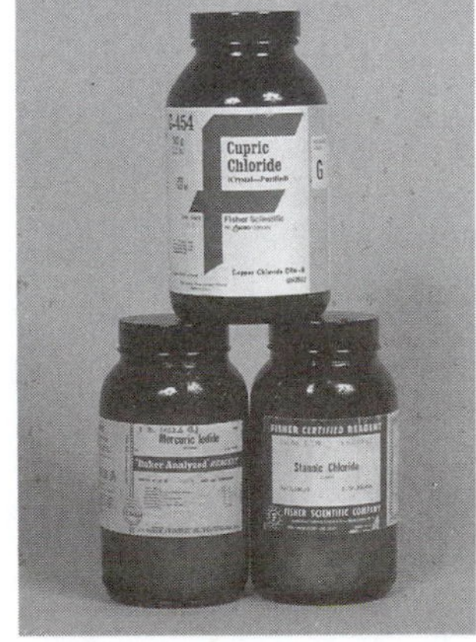

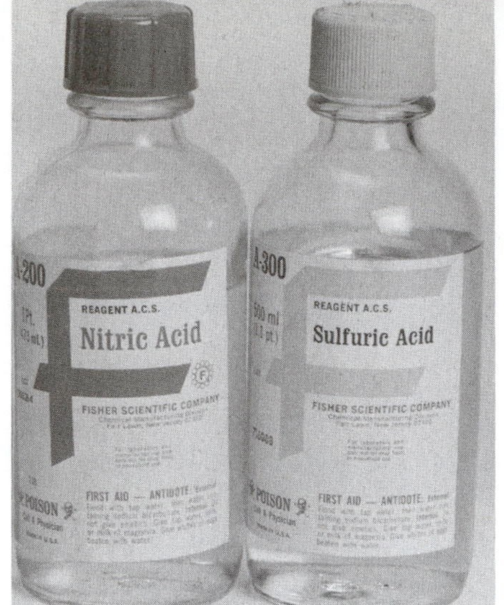

Dry Lab 2C

Inorganic Nomenclature III. Ternary Compounds

Common ternary acids (or oxoacids) in the laboratory are nitric acid and sulfuric acid.

OBJECTIVES

- To name and write formulas for salts and acids containing polyatomic anions
- To name and write formulas for acid salts

INTRODUCTION

In Dry Lab 2B, the naming and the writing of formulas for binary compounds, the simplest of compounds, were introduced.

Ternary compounds are generally considered as having a polyatomic anion containing oxygen. Compounds such as $KMnO_4$, $BaSO_4$, $KClO_3$, and HNO_3 are ternary compounds that have the polyatomic anions MnO_4^-, SO_4^{2-}, ClO_3^-, and NO_3^-, respectively. Ternary compounds can be categorized into salts and acids.

Metal Cation and a Polyatomic Anion Containing Oxygen—A Ternary Salt

Many polyatomic anions consist of one element (usually a nonmetal) and oxygen; the entire grouping of atoms carries a negative charge. The anion is named by using the root name of the element (not the oxygen) with the suffix *-ate*. As examples, SO_4^{2-} is the sulf*ate* ion and NO_3^- is the nitr*ate* ion.

sulfate ion

nitrate ion

phosphate ion

If two polyatomic anions form from an element and oxygen, the polyatomic anion having the element with the higher oxidation number uses the suffix *-ate;* the polyatomic anion having the element with the lower oxidation number uses the suffix *-ite*. As examples,

Ion	Oxidation Number	Name	Ion	Oxidation Number	Name
SO_4^{2-}	S is +6	sulf*ate*	SO_3^{2-}	S is +4	sulf*ite*
NO_3^-	N is +5	nitr*ate*	NO_2^-	N is +3	nitr*ite*
PO_4^{3-}	P is +5	phosph*ate*	PO_3^{3-}	P is +3	phosph*ite*

The formulas and names of common polyatomic anions are listed in Table D2C.1.

Salts that have a polyatomic anion are named in the same manner as the binary salts: the metal cation (along with the *-ic, -ous* or Stock system clarification) is named first followed by the polyatomic anion. Some examples follow:

Salt	*Name*
Na_2SO_4	sodium sulfate
Na_2SO_3	sodium sulfite
$Fe(NO_3)_3$	iron(III) or ferric nitrate
$FeSO_4$	iron(II) or ferrous sulfate

Table D2C.1 Names and Charges of Common Monoatomic and Polyatomic Ions

A. Metallic and Polyatomic Cations

Charge of 1+			
NH_4^+	ammonium	Li^+	lithium
Cu^+	copper(I), cuprous	K^+	potassium
H^+	hydrogen	Ag^+	silver
		Na^+	sodium
Charge of 2+			
Ba^{2+}	barium	Mn^{2+}	manganese(II), manganous
Cd^{2+}	cadmium	Hg^{2+}	mercury(II), mercuric
Ca^{2+}	calcium	Hg_2^{2+}	mercury(I), mercurous
Cr^{2+}	chromium(II), chromous	Ni^{2+}	nickel(II), nickelous
Co^{2+}	cobalt(II), cobaltous	Sr^{2+}	strontium
Cu^{2+}	copper(II), cupric	Sn^{2+}	tin(II), stannous
Fe^{2+}	iron(II), ferrous	UO_2^{2+}	uranyl
Pb^{2+}	lead(II), plumbous	VO^{2+}	vanadyl
Mg^{2+}	magnesium	Zn^{2+}	zinc
Charge of 3+			
Al^{3+}	aluminum	Co^{3+}	cobalt(III), cobaltic
As^{3+}	arsenic(III), arsenious	Fe^{3+}	iron(III), ferric
Cr^{3+}	chromium(III), chromic	Mn^{3+}	manganese(III), manganic
Charge of 4+			
Pb^{4+}	lead(IV), plumbic	Sn^{4+}	tin(IV), stannic
Charge of 5+			
V^{5+}	vanadium(V)	As^{5+}	arsenic(V), arsenic

B. Nonmetallic and Polyatomic Anions

Charge of 1−			
$CH_3CO_2^-$	acetate or	H^-	hydride
$C_2H_3O_2^-$	acetate	ClO^-	hypochlorite
Br^-	bromide	I^-	iodide
ClO_3^-	chlorate	NO_3^-	nitrate
Cl^-	chloride	NO_2^-	nitrite
CN^-	cyanide	ClO_4^-	perchlorate
F^-	fluoride	IO_4^-	periodate
OH^-	hydroxide	MnO_4^-	permanganate
Charge of 2−			
CO_3^{2-}	carbonate	O_2^{2-}	peroxide
CrO_4^{2-}	chromate	SiO_3^{2-}	silicate
$Cr_2O_7^{2-}$	dichromate	SO_4^{2-}	sulfate
MnO_4^{2-}	manganate	SO_3^{2-}	sulfite
O^{2-}	oxide	S^{2-}	sulfide
$C_2O_4^{2-}$	oxalate	$S_2O_3^{2-}$	thiosulfate
Charge of 3−			
N^{3-}	nitride	P^{3-}	phosphide
PO_4^{3-}	phosphate	BO_3^{3-}	borate
PO_3^{3-}	phosphite	AsO_3^{3-}	arsenite
		AsO_4^{3-}	arsenate

If more than two polyatomic anions are formed from a given element and oxygen, the prefixes *per-* and *hypo-* are added to distinguish the additional ions. This appears most often among the polyatomic anions with a halogen as the distinguishing element. The prefixes *per-* and *hypo-* are often used to identify the extremes (the high and low) in oxidation numbers of the element in the anion. Table D2C.2 summarizes the nomenclature for these and all polyatomic anions.

For example, ClO_4^- is *per*chlor*ate* because the oxidation number of Cl is +7, whereas ClO^- is *hypo*chlor*ite* because the oxidation number of Cl is +1. Therefore, $NaClO_4$ is sodium perchlorate and NaClO is sodium hypochlorite. The name of $Co(ClO_3)_2$ is cobalt(II) chlorate because the chlorate ion is ClO_3^- (see Table D2C.1)

Table D2C.2 Nomenclature of Polyatomic Anions and Ternary Acids

Name of Polyatomic Anion	Acid	4A	5A	6A	7A
		Group Number			
		Oxidation Number			
per___ate	per___ic acid	—	—	—	+7
___ate	___ic acid	+4	+5	+6	+5
___ite	___ous acid	—	+3	+4	+3
hypo___ite	hypo___ous acid	—	+1	+2	+1

and the oxidation number of Cl is +5 (see Table D2C.2). The cobalt is Co^{2+} because two chlorate ions are required for the neutral salt.

Ternary Acids or Oxoacids

The ternary acids (also referred to as **oxoacids**) are compounds of hydrogen and a polyatomic anion. In contrast to the binary acids, the naming of the ternary acids does not include any mention of hydrogen. The ternary acids are named by using the root of the element in the polyatomic anion and adding a suffix; if the polyatomic anion ends in *-ate,* the ternary acid is named an *-ic acid.* If the polyatomic anion ends in *-ite,* the ternary acid is named an *-ous acid.* To understand by example, the following are representative names for ternary acids.

sulfuric acid

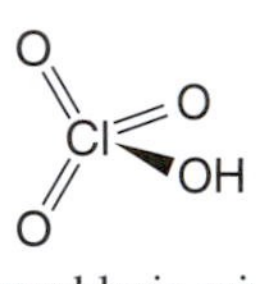
perchloric acid

Ion	Name of Ion	Acid	Name of Acid
SO_4^{2-}	sulf*ate* ion	H_2SO_4	sulfur*ic acid*
SO_3^{2-}	sulf*ite* ion	H_2SO_3	sulfur*ous acid*
CO_3^{2-}	carbon*ate* ion	H_2CO_3	carbon*ic acid*
NO_3^-	nitr*ate* ion	HNO_3	nitr*ic acid*
ClO_4^-	perchlor*ate* ion	$HClO_4$	perchlor*ic acid*
IO^-	hypoiod*ite* ion	HIO	hypoiod*ous acid*

Acid Salts

Acid salts are salts in which a metal cation replaces *less than* all of the hydrogens of an acid having more than one hydrogen (a polyprotic acid). The remaining presence of the hydrogen in the compound is indicated by inserting its name into that of the salt.

Salt	Name
$NaHSO_4$	sodium *hydrogen* sulfate (one Na^+ ion replaces one H^+ ion in H_2SO_4)
$CaHPO_4$	calcium *hydrogen* phosphate (one Ca^{2+} ion replaces two H^+ ions in H_3PO_4)
NaH_2PO_4	sodium *dihydrogen* phosphate (one Na^+ ion replaces one H^+ ion in H_3PO_4)
$NaHCO_3$	sodium *hydrogen* carbonate (one Na^+ ion replaces one H^+ ion in H_2CO_3)
NaHS	sodium *hydrogen* sulfide (one Na^+ ion replaces one H^+ ion in H_2S)

An older system of naming acid salts substitutes the prefix "bi" for a *single* hydrogen before naming the polyatomic anion. For example, $NaHCO_3$ is sodium *bi*carbonate, $NaHSO_4$ is sodium *bi*sulfate, and NaHS is sodium *bi*sulfide.

Dry Lab Procedure

Procedure Overview: Given the formula of the compounds, the proper names for a large number of compounds are to be written, and given the name of the compounds, the formulas for a large number of compounds are to be written.

Your instructor will assign the exercises you are to complete. Answer them on a separate piece of paper. Be sure to indicate the date, your lab section, and your desk number on your answer sheet. Use the rules that have been described.

1. Use Table D2C.2 to name the following polyatomic anions.

a. BrO_3^-	d. $N_2O_2^{2-}$	g. IO_2^-	j. TeO_4^{2-}
b. IO_3^-	e. AsO_2^-	h. SO_3^{2-}	k. SeO_4^{2-}
c. PO_2^{3-}	f. BrO_2^-	i. SiO_3^{2-}	l. NO_2^-

2. Name the following salts of the representative elements.

a. Na_2SO_4
b. K_3AsO_4
c. Li_2CO_3
d. $Ca_3(PO_4)_2$
e. $Ca_3(PO_3)_2$
f. Na_2SiO_3
g. K_2CrO_4
h. $K_2Cr_2O_7$
i. K_2MnO_4
j. $KMnO_4$
k. Li_2SO_3
l. Li_2SO_4
m. $Li_2S_2O_3$
n. $Ba(NO_2)_2$
o. $Ba(NO_3)_2$
p. KCH_3CO_2

3. Name the following salts of the transition and post-transition elements using the Stock system.

a. $Fe(OH)_3$
b. $FePO_4•6H_2O$
c. $FeSO_4•7H_2O$
d. $CuCN$
e. $CuCO_3$
f. $CuSO_4•5H_2O$
g. $Sn(NO_3)_2$
h. $Sn(SO_4)_2$
i. $MnSO_4$
j. $Mn(CH_3CO_2)_2$
k. $Hg_2(NO_3)_2$
l. $Hg(NO_3)_2•H_2O$
m. $CrPO_4$
n. $CrSO_4•6H_2O$
o. $Co_2(CO_3)_3$
p. $CoSO_4•7H_2O$

4. Name the following ternary acids.

a. H_2SO_4
b. H_2SO_3
c. $H_2S_2O_3$
d. H_3PO_4
e. $HMnO_4$
f. H_2CrO_4
g. H_3BO_3
h. HNO_3
i. HNO_2
j. H_2CO_3
k. $H_2C_2O_4$
l. CH_3COOH
m. $HClO_4$
n. $HClO_3$
o. $HClO_2$
p. $HClO$

5. Name the following acid salts. Use the "older" method wherever possible.

a. $NaHCO_3$
b. $Ca(HCO_3)_2$
c. KHC_2O_4
d. NH_4HCO_3
e. $NaHS$
f. $KHSO_3$
g. $NaHSO_4•H_2O$
h. Li_2HPO_4
i. LiH_2PO_4
j. $MgHAsO_4$
k. KH_2AsO_4
l. $KHCrO_4$

6. Write formulas for the following compounds.

a. potassium permanganate
b. potassium manganate
c. calcium carbonate
d. lead(II) carbonate
e. ferric carbonate
f. silver thiosulfate
g. sodium sulfite
h. ferrous sulfate heptahydrate
i. iron(II) oxalate
j. sodium chromate
k. potassium dichromate
l. nickel(II) nitrate hexahydrate
m. chromous nitrite
n. vanadyl nitrate
o. uranyl acetate
p. barium acetate dihydrate
q. sodium silicate
r. calcium hypochlorite
s. potassium chlorate
t. ammonium oxalate
u. sodium borate
v. cuprous iodate

7. Write formulas for the following acids.

a. sulfuric acid
b. thiosulfuric acid
c. sulfurous acid
d. periodic acid
e. iodic acid
f. hypochlorous acid
g. nitrous acid
h. nitric acid
i. phosphorous acid
j. phosphoric acid
k. carbonic acid
l. bromous acid
m. chromic acid
n. permanganic acid
o. manganic acid
p. boric acid
q. oxalic acid
r. silicic acid

8. Write correct formulas for the following hodgepodge of compounds from Dry Labs 2A, 2B, and 2C.
   a. hydrogen fluoride
   b. hydrobromic acid
   c. hypobromous acid
   d. sulfur dioxide
   e. nickel(II) hydroxide
   f. hydrocyanic acid
   g. osmium(VIII) oxide
   h. gold(III) chloride
   i. phosphorus triiodide
   j. cupric iodate
   k. sodium peroxide
   l. potassium oxalate
   m. sodium bisulfite
   n. titanium(IV) chloride
   o. zinc hydride
   p. chromic chloride hexahydrate
   q. aluminum hydroxide
   r. cobalt(II) sulfate hexahydrate
   s. silver cyanide
   t. boric acid
   u. manganous sulfate
   v. silver oxide
   w. silicon dioxide
   x. acetic acid
   y. vanadyl chloride
   z. calcium hydrogen phosphate
9. Write correct formulas for the following hodgepodge of compounds from Dry Labs 2A, 2B, and 2C.
   a. vanadium(V) fluoride
   b. stannic oxide
   c. silicon tetrafluoride
   d. mercuric oxide
   e. lithium hypochlorite
   f. iodine trifluoride
   g. ferrous oxalate
   h. cuprous oxide
   i. copper(I) chloride
   j. calcium hydride
   k. cadmium iodide
   l. barium acetate dihydrate
   m. ammonium sulfide
   n. vanadium(V) oxide
   o. titanium(IV) chloride
   p. scandium(III) nitrate
   q. nickel(II) acetate hexahydrate
   r. mercurous nitrate
   s. lead(II) acetate
   t. ferric phosphate hexahydrate
   u. ferric chromate
   v. dinitrogen tetrasulfide
   w. chromous acetate
   x. calcium nitride
   y. ammonium dichromate
   z. silver acetate
10. Write the correct formulas for the compounds shown in the photo.

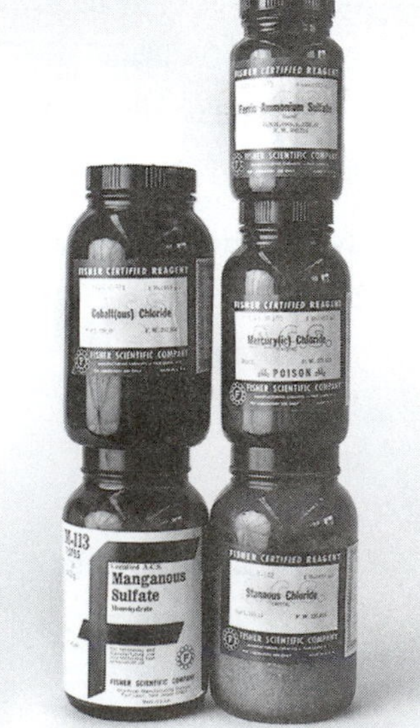

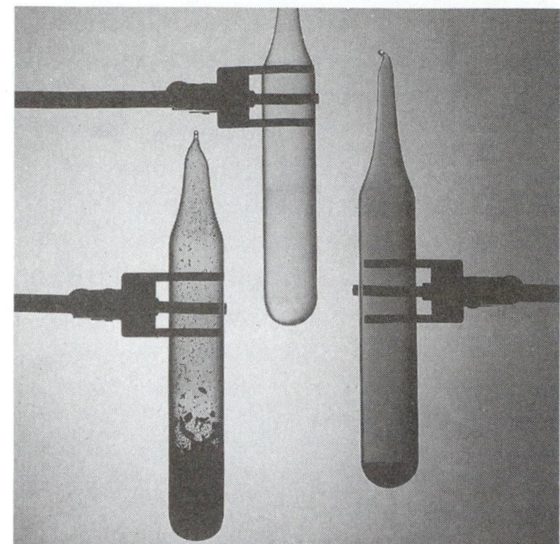

Experiment 11

Periodic Table and Periodic Law

The halogens are, from left to right, solid iodine crystals, chlorine gas, and liquid bromine.

Objectives

- To become more familiar with the periodic table
- To observe and to generalize the trends of various atomic properties within groups and periods of elements
- To observe from experiment the trends of the chemical properties within groups and periods of elements

Techniques

The following techniques are used in the Experimental Procedure

Introduction

Similarities between the chemical and physical properties of the elements were known early in the nineteenth century. In 1817, Johann W. Döbereiner showed in a number of instances that when three elements with similar properties were listed in order of increasing atomic mass, the middle element had properties that were an approximate average of the other two. His groups of "triads" from top to bottom include:

Li	Ca	N	As	S	Cl
Na	Sr	P	Sb	Se	Br
K	Ba	As	Bi	Te	I

Each triad is a part of a family (or group) of elements in the modern periodic table.

In 1866, John Newlands established the "Law of Octaves." Being a lover of music, he theorized that when the elements were arranged in order of increasing atomic mass, every eighth element had similar chemical and physical properties, just like every eighth note on the musical scale has a similar sound! For the lighter elements, this periodicity is surprisingly valid, especially when the noble gases and the transition elements are not considered. In 1866, scientists were unaware of the noble gases.

Musical Scale	Do	Re	Mi	Fa	So	La	Ti
Element Scale	Li	Be	B	C	N	O	F
	Na	Mg	Al	Si	P	S	Cl

It remained until 1869 for two scientists, Dmitri Mendeleev from Russia and Lothar Meyer from Germany, to independently establish a periodic table similar to our present-day arrangement of the elements. Mendeleev showed that with the elements arranged in order of increasing atomic mass, their *chemical* properties recur periodically. When Meyer arranged the elements in order of increasing atomic mass, he found that their *physical*

Dmitri Mendeleev (1834–1907).

properties recur periodically. The two tables, however, were virtually identical. Because he drafted his table earlier in 1869 and because his table included "blanks" for yet-to-be-discovered elements to fit, Mendeleev is considered the "father of the modern periodic table."

Atomic number: the number of protons in the nucleus

In 1913, H. G. J. Moseley's study of the X-ray spectra of the elements refined the periodic table to its current status: **when the elements are arranged in order of increasing atomic *number*, certain chemical and physical properties repeat periodically.** See the inside back cover for a modern version of the periodic table.

In the periodic table each horizontal row of elements is a **period** and each column is a **group** (or **family**). All elements within a group have similar chemical and physical properties. Common terms associated with various sections of the periodic table are

- Representative elements: Group "A" elements
- Transition elements: Groups 3–12
- Inner transition elements: the lanthanide and actinide series
- Metallic elements: elements to the left of the "stairstep line" that runs diagonally from B to At
- Nonmetallic elements: elements to the right of the "stairstep line"
- Metalloids: elements that lie adjacent to the "stairstep line," excluding Al
- Alkali metals: Group 1A elements
- Alkaline earth metals: Group 2A elements
- Chalcogens: Group 6A elements
- Halogens: Group 7A elements
- Noble gases
- Rare earth metals: the lanthanide series of elements
- Coinage metals: Cu, Ag, Au
- Noble metals: Ru, Os, Rh, Ir, Pd, Pt, Ag, Au, Hg

The periodicity of a physical property for a series of elements can be shown by plotting the experimental value of the property versus increasing atomic number. Physical properties studied in this experiment are

- Ionization energy (Figure 11.1): the energy required to remove an electron from a gaseous atom
- Atomic radius (Figure 11.2): the radius of an atom of the element
- Electron affinity (Figure 11.3): the energy released when a neutral gaseous atom accepts an electron
- Melting point (Figure 11.4): the temperature at which the solid and liquid phases of a substance coexist
- Molar volume (Figure 11.5): the volume of a unit cell times Avogadro's number divided by the number of atoms in the unit cell
- Density (Figure 11.6): the mass of a substance per unit volume

Properties of Chlorine

Atomic number	17
Molar mass	35.453 g/mol
Density at 293 K	0.003214 g/cm^3
Molar volume	22.7 cm^3/mol
Melting point	172.22 K
Boiling point	239.2 K
Heat of fusion	3.203 kJ/mol
Heat of vaporization	10.20 kJ/mol
First ionization energy	1251.1 kJ/mol
Second ionization energy	2297.3 kJ/mol
Third ionization energy	3821.8 kJ/mol
Electronegativity	3.16
Electron affinity	349 kJ/mol
Specific heat	0.48 J/g•K
Heat of atomization	121 kJ/mol atoms
Atomic radius	100 pm
Ionic radius (1$^-$ ion)	167 pm
Thermal conductivity	0.01 J/m•s•K

Other physical properties that show trends in groups and periods of elements are listed for chlorine in the table.

Trends in the *chemical* properties of the bold-face elements are also studied in this experiment.

Li	*Be*	*B*	*C*	*N*	*O*	**F**
Na	**Mg**	**Al**	*Si*	*P*	**S**	**Cl**
K	**Ca**	*Ga*	*Ge*	*As*	*Se*	**Br**
Rb	*Sr*	*In*	*Sn*	*Sb*	*Te*	**I**
Cs	**Ba**	*Tl*	*Pb*	*Bi*	*Po*	*As*

In this experiment, the relative acidic and/or basic strength of the hydroxides or oxides in the third period of the periodic table, the relative chemical reactivity of the halogens, and the relative solubility of the hydroxides and sulfates of magnesium, calcium, and barium are observed through a series of qualitative tests. Observe closely the results of each test before generalizing your information.

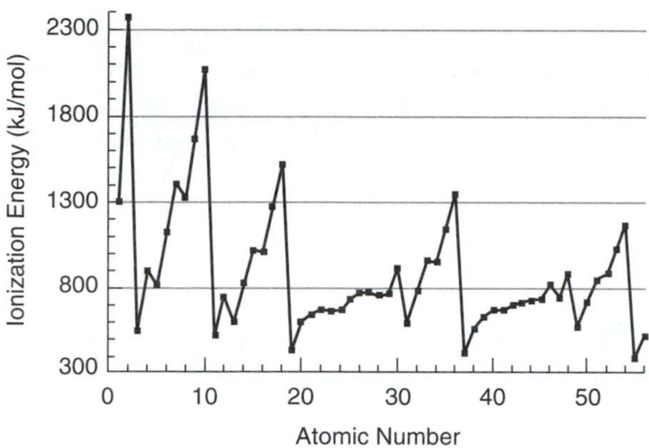

Figure 11.1 Ionization energies (kJ/mol) plotted against atomic number.

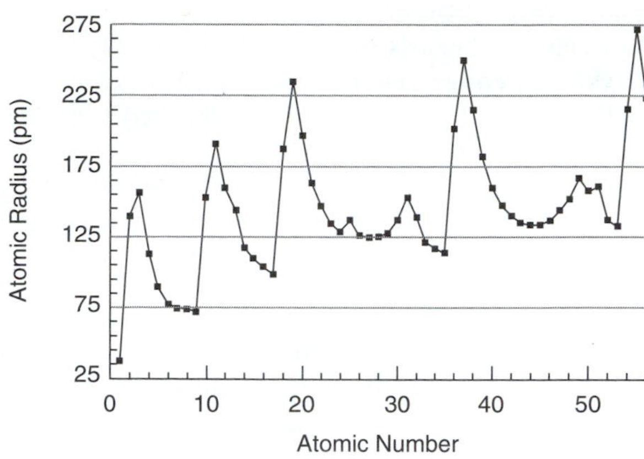

Figure 11.2 Atomic radii (pm) plotted against atomic number.

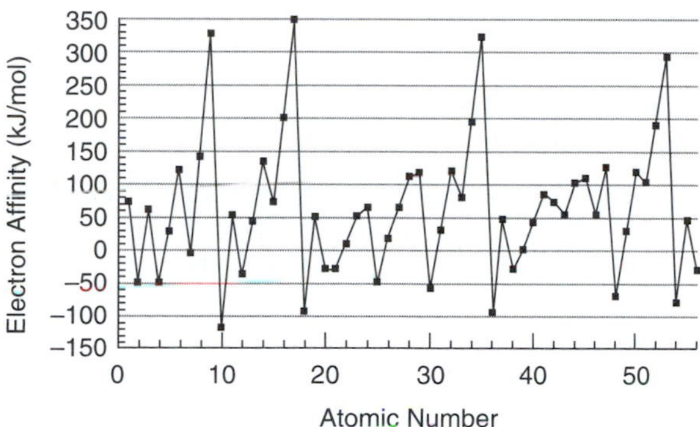

Figure 11.3 Electron affinities (kJ/mol) plotted against atomic number.

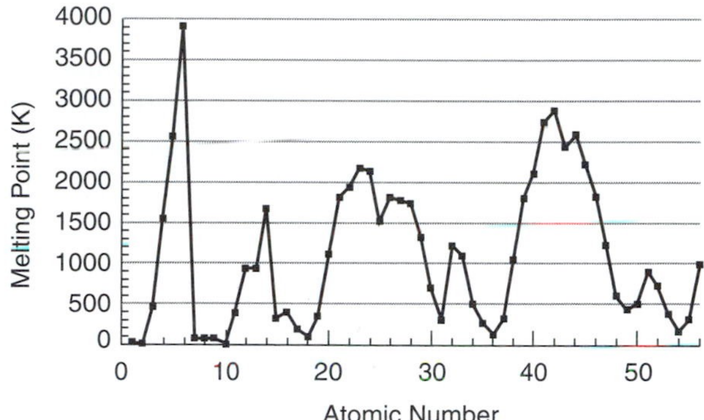

Figure 11.4 Melting points (K) plotted against atomic number.

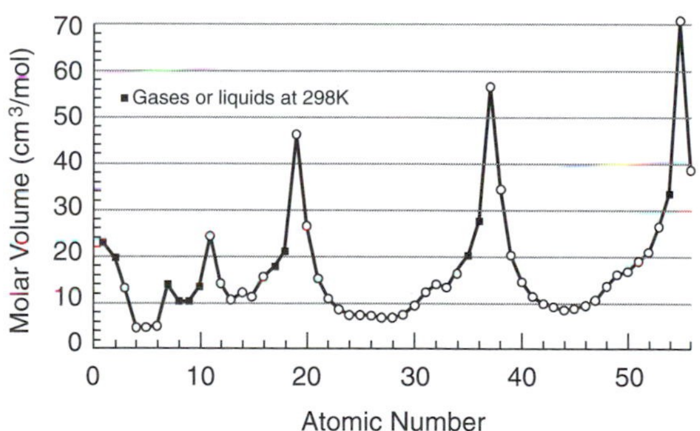

Figure 11.5 Molar volume (cm^3/mol) plotted against atomic number.

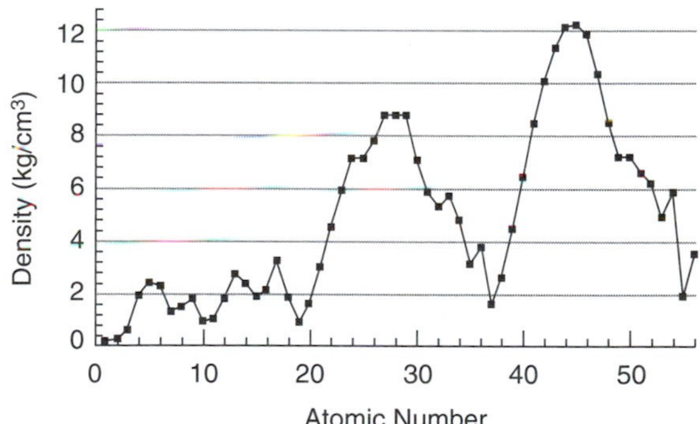

Figure 11.6 Density (kg/m^3) plotted against atomic number.

Experimental Procedure

Procedure Overview: General trends in the physical properties of the elements are observed and studied in Figures 11.1–11.6. Experimental observations of the physical and chemical properties of a number of representative elements are made. Special attention is paid to the chemical properties of the halogens.

Ask your instructor about a working relationship, individuals or partners, for Parts A and B. For Parts C, D, and E perform the experiment with a partner. At each circled superscript [1–18] in the procedure, *stop,* and record your observation on the Report Sheet. Discuss your observations with your lab partner and your instructor.

A. Periodic Trends in Physical Properties (Dry Lab)

Figures 11.1 through 11.6 plot the experimental data of six physical properties of the elements as a function of their atomic number. While actual values cannot be readily obtained from the graphical data, the periodic trends are easily seen. All values requested for the analyses of the graphical data need only be given as your "best possible" estimates from the figure.

The periodic trends for the elements are analyzed through a series of questions on the Report Sheet.

B. Ionization Energies for Aluminum (Dry Lab)

Figure 11.7 is a graphical plot of the ionization energies for the thirteen electrons of aluminum. An interpretation of the experimental data is requested on the Report Sheet.

C. The Appearance of Some Representative Elements

1. **Samples of Elements.** Samples of the third period elements sodium, magnesium, aluminum, silicon, and sulfur are on the reagent table. Note that the Na metal is stored under a nonaqueous liquid to prevent any rapid air oxidation. Polish the Mg and Al metal strips with steel wool for better viewing. Record your observations on the Report Sheet.

   Since some chlorine, bromine, and iodine vapors may escape the test tubes in Parts C.2–4 and D.1–3, *you may want to conduct the experiments in the fume hood.* Consult with your laboratory instructor.

2. **Chlorine.** In a clean, (150-mm) test tube, place 2 mL of a 5% sodium hypochlorite, NaClO, solution (commercial laundry bleach) and 10 drops of cyclohexane. Agitate the mixture (Figure 11.8). Which layer is the cyclohexane layer[1]?

   Add 10 drops of 6 *M* HCl. (**Caution:** 6 *M* HCl *is very corrosive. Wash immediately from the skin.*) Swirl/agitate the mixture (with a stirring rod) so that the HCl mixes with the NaClO solution. Note the color of the chlorine in the cyclohexane layer. Record your observation.① Do not discard—save for Part D.1.

3. **Bromine.** In a second, clean test tube, place 2 mL of 3 *M* KBr solution and 3 drops of cyclohexane. Add 5–10 drops of 8 *M* HNO_3. (**Caution:** HNO_3 *attacks skin tissue; flush the affected area immediately with water.*) Agitate/swirl the mixture so that the 8 *M* HNO_3 mixes with the KBr solution. Note the color of the cyclohexane layer. Record,② but do not discard—save for Part D.2.

4. **Iodine.** Repeat Part C.3 in a third test tube, substituting 3 *M* KI for 3 *M* KBr. Record. Compare the appearance of the three halogens dissolved in the cyclohexane.③ Save for Part D.3.

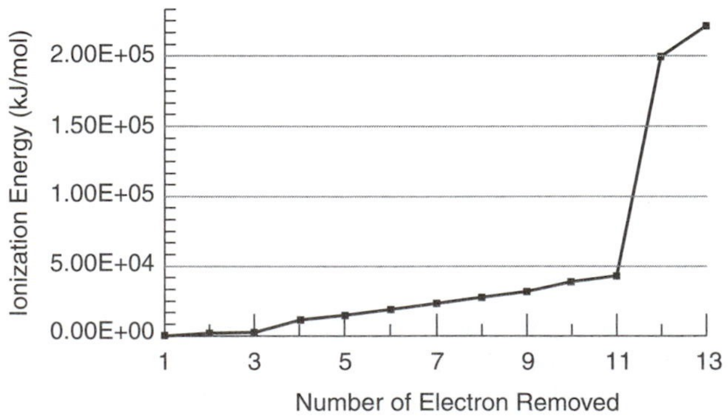

Figure 11.7 The 13 ionization energies for aluminum.

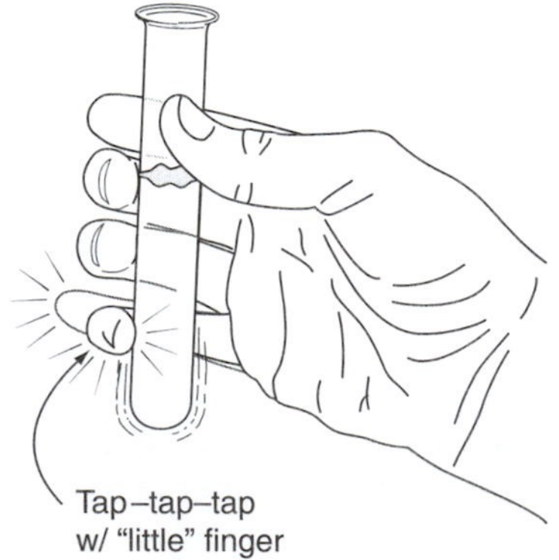

Figure 11.8 Shake the contents of the test tube with the "little" finger.

[1]Mineral oil, or any colorless cooking oil, may be substituted for cyclohexane.

D. The Chemical Properties of the Halogens

For Parts D.1–3, six clean (~75-mm), small test tubes are required. Summarize your observations at the conclusion of Part D.3.

Pinch: a solid mass about the size of a grain of rice

1. **Chlorine and Its Reactions with Bromide and Iodide Ions.** Clean two small test tubes; add a **pinch** (on the end of a spatula) of solid KBr to the *first* test tube and a pinch of KI to the *second*. Use a dropping pipet to withdraw the chlorine/cyclohexane layer from Part C.2 and add an (approximate) equal portion to the two test tubes. Swirl/agitate the solution, observe, and record. Write appropriate net ionic equations.④
2. **Bromine and Its Reactions with Chloride and Iodide Ions.** Add a pinch of solid NaCl to a *third*, small clean test tube and a pinch of KI to the *fourth* test tube. Use a dropping pipet to withdraw the bromine/cyclohexane layer from Part C.3 and add an (approximate) equal portion to the two test tubes. Swirl/agitate the solution, observe and record. Write appropriate net ionic equations.⑤
3. **Iodine and Its Reactions with Chloride and Bromide Ions.** Add a pinch of solid NaCl to a *fifth*, small clean test tube and a pinch of KBr to the *sixth* test tube. Use a dropping pipet to withdraw the iodine/cyclohexane layer from Part C.3 and add an (approximate) equal portion to the two test tubes. Swirl/agitate the solution, observe and record. Write appropriate net ionic equations.⑥

   What can you conclude about the chemical reactivity of the halogens?

> *Disposal:* Dispose of the waste water/halogen mixtures in the "Waste Halogens" container.

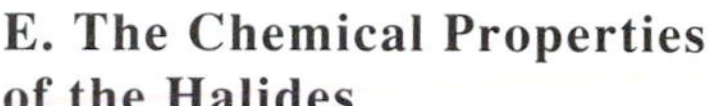

E. The Chemical Properties of the Halides

Twelve clean, small test tubes are required for the chemical reactions observed in Part E. Number each test tube (Figure 11.9).

Appendix G

1. **The Reactions of the Halides with Various Metal Ions.** Label 12 clean, small test tubes and transfer the following to each:
   - Test tubes 1, 2, and 3: a pinch of NaF and 10 drops of water
   - Test tubes 4, 5, and 6: a pinch of NaCl and 10 drops of water
   - Test tubes 7, 8, and 9: a pinch of KBr and 10 drops of water
   - Test tubes 10, 11, and 12: a pinch of KI and 10 drops of water

   a. Slowly add 10 drops of 2 *M* $Ca(NO_3)_2$ to test tubes 1, 4, 7, and 10. Observe closely and over a period of time. Vary the color of the background of the test tubes for observation.⑦

   b. Slowly add 10 drops of 0.1 *M* $AgNO_3$ to test tubes 2, 5, 8, and 11. After about 1 minute, add 10 drops of 3 *M* NH_3.⑧

   c. Add 1 drop of 6 *M* HNO_3 (**Caution!**) and slowly add 10 drops of 0.1 *M* $Fe(NO_3)_3$ to test tubes 3, 6, 9, and 12. Observe closely and over a period of time.⑨

   d. Summarize your observations of the chemical activity for the halides with metal ions.

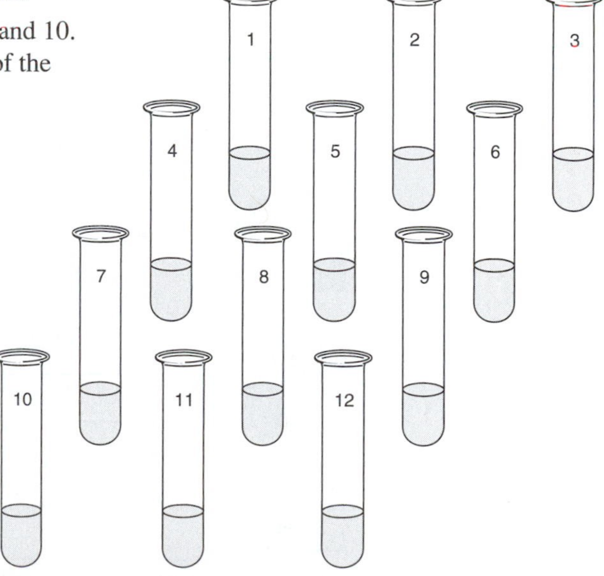

Figure 11.9 Twelve labeled test tubes to test the reactivity of the halides.

Disposal: Dispose of the waste water/halogen mixtures in the "Waste Halogens" container.

CLEANUP: Rinse the test tubes with copious amounts of tap water and twice with deionized water. Discard the rinses in the sink.

F. Chemical Reactivity of Some Representative Elements

1. **Sodium.** *Instructor Demonstration Only.* Wrap a pea-size piece of sodium metal in aluminum foil. Fill a 200-mm Pyrex test tube with water, add 2 drops of phenolphthalein,[2] and invert the test tube in a beaker of water (Figure 11.10). Set the beaker and test tube behind a safety shield. Punch 5–10 holes with a pin in the aluminum foil.

   With a pair of tongs or tweezers, place the wrapped sodium metal in the mouth of the test tube, keeping it under water. What is the evolved gas? Test the gas by holding the mouth of the inverted test tube over a Bunsen flame.⑩ A loud "pop" indicates the presence of hydrogen gas. Account for the appearance of the color change in the solution.⑪

   Six clean, small test tubes are required for Parts F.2–4.

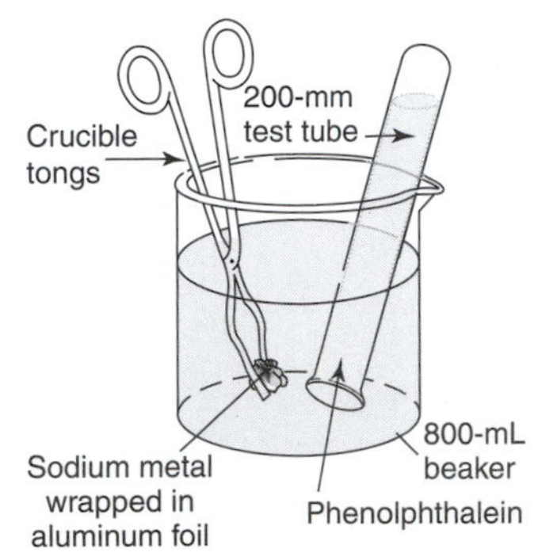

Figure 11.10 Collection of hydrogen gas from the reaction of sodium and water.

2. **Magnesium and Aluminum.** Polish strips of Mg and Al metal; cut 5-mm pieces and place them into separate small test tubes. Add 10 drops of 3 *M* HCl to each test tube. (**Caution:** *Do not allow the* HCl *to touch the skin. Wash the affected area immediately.*) Which metal reacts more rapidly?⑫ What is the gas that is evolved?⑬

   Add (and count) drops of 6 *M* NaOH to each test tube until a precipitate appears. Continue to add NaOH to the test tube containing the aluminum ion until the precipitate dissolves. Add the same number of drops to the test tube containing the magnesium ion. Record your observations.[3] ⑭

   Add drops of 6 *M* HCl until both solutions are again colorless. Observe closely as each drop is added. Record and explain.

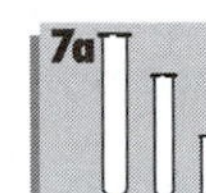

3. **Solubilities of Alkaline-Earth Cations.** a. *Solubility of the hydroxides.* Place 10 drops of 0.1 *M* $MgCl_2$, 0.1 *M* $CaCl_2$, and 0.1 *M* $BaCl_2$ in three, separate, clean test tubes. Count and add drops of 0.050 *M* NaOH until a cloudiness appears in each test tube. Predict the trend in the solubility of the hydroxides of the Group 2A cations.⑮

   b. *Solubility of the sulfates.* Place 10 drops of 0.1 *M* $MgCl_2$, 0.1 *M* $CaCl_2$, and 0.1 *M* $BaCl_2$ in three, separate, clean test tubes. Count and add drops of 0.10 *M* Na_2SO_4 until a cloudiness appears in each test tube. Predict the trend in the solubility of the sulfates of the Group 2A cations.⑯

4. **Sulfurous Acid and Sulfuric Acid.** Because of the possible evolution of a foul-smelling gas, you may want to conduct this part of the experiment in the fume hood. Consult with your instructor.

   Place a "double" pinch of solid sodium sulfite, Na_2SO_3, into a small or medium-sized test tube. Add 5–10 drops of 6 *M* HCl. Test the evolved gas with wet, blue litmus paper. Write a balanced equation for the reaction.⑰

   Test 10 drops of 0.1 *M* H_2SO_4 in a test tube with litmus paper, both by holding moistened, blue litmus paper over the test tube and by a direct test of the solution. Account for any differences or similarities in your observations.⑱

Disposal: Discard the solutions as directed by your instructor.

CLEANUP: Rinse the test tubes twice with tap water and with deionized water. Discard the rinses in the sink.

[2]Phenolphthalein is an acid-base indicator; it is colorless in an acidic solution but pink in a basic solution.
[3]Magnesium ion precipitates as magnesium hydroxide, $Mg(OH)_2$; aluminum ion also precipitates as the hydroxide, $Al(OH)_3$, but redissolves in an excess of OH^- to produce $Al(OH)_4^-$, the aluminate ion.

Experiment 11 *Prelaboratory Assignment*

Periodic Table and Periodic Law

Date __________ Lab Sec. ______ Name ____ Desk No. __________

1. On the blank periodic table, *clearly* locate the following, using your own color code.
   a. Representative elements
   b. Transition elements
   c. Inner transition elements
   d. Chalcogens
   e. Coinage metals
   f. Metalloids
   g. Alkali metals
   h. Alkaline earth metals
   i. Halogens
   j. Noble gases
   k. Noble metals
   l. Lanthanide series
   m. Actinide series

Periodic Table

2. Sketch in the "stairstep" line that separates the metals from the nonmetals on the periodic table.

3. Classify each of the following elements according to the categories of elements identified in Question 1:

   a. Lithium ______________
   b. Plutonium ______________
   c. Argon ______________
   d. Tungsten ______________
   e. Indium ______________
   f. Calcium ______________
   g. Lead ______________
   h. Copper ______________
   i. Iodine ______________
   h. Gadolinium ______________

4. Refer to Figure 11.1. Which of the following has the highest ionization energy?

   a. carbon or oxygen ____________

   b. phosphorus or sulfur ____________ d. magnesium or calcium ____________

   c. magnesium or aluminum ____________ e. xenon or cesium ____________

5. Refer to Figure 11.2. Which of the following has the largest atomic radius?

   a. carbon or oxygen ____________

   b. phosphorus or sulfur ____________ d. magnesium or calcium ____________

   c. magnesium or aluminum ____________ e. xenon or cesium ____________

   Compare your answers for Questions 4 and 5. What correlation can be made?

6. a. Proceeding from left to right across a period of the periodic table the elements become (more, less) metallic.

   b. Proceeding from top to bottom in a group of the periodic table the elements become (more, less) metallic.

7. The chemical reactivity of three halogens are studied in this experiment. Name them and write formulas for their naturally occurring elemental forms.

   a. ____________________ ____________

   b. ____________________ ____________

   c. ____________________ ____________

8. List the name of the scientist who:
   a. studied X-ray spectra of the elements to reveal their periodicity

   b. postulated a periodicity of elements according to physical properties

   c. postulated a periodicity of elements according to chemical properties

9. Why is Mendeleev, rather than Meyer, given the major recognition for the organization of the elements that led to the modern version of the periodic table?

Experiment 11 *Report Sheet*

Periodic Table and Periodic Law

Date __________ Lab Sec. ______ Name ____ Desk No. __________

A. Periodic Trends in Physical Properties (Dry Lab)

Consult with your laboratory instructor as to the procedure and schedule for submitting your responses to the following questions about periodic trends.

1. Figure 11.1, graphical data for the ionization energies of the elements, shows sawtooth trends across the periods of the elements.
   a. Locate the noble gas group of elements. What appears to be the periodic trend in ionization energies "down" the noble gas group (i.e., with increasing atomic number)?
   b. Scanning the graphical data for elements adjacent to and then further removed from the halogens, what *general statement* can summarize the trend in the ionization energies when moving down a group of elements?
   c. What *general statement* can summarize the trend in the ionization energies when moving "across" a period of elements?

2. Figure 11.2, graphical data for the atomic radii of the elements shows generally decreasing trends across a period of elements. The noble gases are an anomaly.
   a. Which group of elements has the largest atomic radii?
   b. Moving down a group of elements (increasing atomic number), what is the general trend for atomic radii?
   c. What *general statement* can summarize the correlation of ionization energies to atomic radii for the elements?

3. Figure 11.3, graphical data for the electron affinities of the elements, shows a number of irregularities, but a general increasing trend in values exists across a period of elements.
   a. Which group of elements has the highest electron affinities?
   b. Is the trend in electron affinities repetitive for Periods 2 and 3? Cite examples.
   c. Is there a correlation of electron affinities to atomic radii for the elements? If so, what is it? Cite examples.

4. Figure 11.4 shows repeated regularities in melting points across the periods of elements.
   a. What is the trend in melting points across a period of elements?
   b. Which element has the highest melting point? Does its congeners (identify them) correspondingly have high melting points?
   c. What is the trend in the melting points across periods of the transition elements?

5. Figure 11.5, graphical data for the molar volume of the elements, is periodic in nature.
   a. Which elements (and in which group) have the largest molar volume?
   b. How does this observation compare to those elements having the largest atomic radii?
   c. How does this observation compare to those elements having the lowest ionization energies?
   d. What is the correlation between the melting points and the molar volumes for the periods of the transition elements? Cite examples.

6. Figure 11.6 shows repeated trends in density for the periods of elements.
   a. What is the general trend in densities for Periods 2 and 3?
   b. What is the trend in the densities across periods of the transition elements?
   c. What *general statement* can summarize the correlation of melting points to densities for the elements?
   d. What is the correlation between the densities and the molar volumes for the periods of the transition elements? Is that correlation also valid for densities and atomic radii? Explain.

B. Ionization Energies for Aluminum (Dry Lab)

Consult with your laboratory instructor as to the procedure and schedule for submitting your responses to the following questions about periodic trends.

1. (Optional) See your laboratory instructor. Figure 11.7 graphs the experimental values for the 13 ionization energies of aluminum.
   a. On the basis of the electron configuration of aluminum, explain why there is a significant increase in the fourth ionization energy for aluminum.
   b. On the basis of the electron configuration of aluminum, explain why there is an even more significant increase in the twelfth ionization energy for aluminum.
   c. Where might you see similar increases in ionization energies for magnesium? For silicon?

C. The Appearance of Some Representative Elements

Element	Physical State (g, l, s)	Physical Appearance and Other Observations	Color
Na			
Mg			
Al			
Si			
S_8			
① Cl_2			
② Br_2			
③ I_2			

D. The Chemical Properties of the Halogens

Observations in the cyclohexane layer

	Cl_2	Br_2	I_2	Net Ionic Equation(s)
④ KCl	XX*			
⑤ KBr		XX		
⑥ KI			XX	

*No reaction expected.

What can you conclude about the relative reactivity of Cl_2, Br_2, and I_2?

E. The Chemical Properties of the Halides

1. The Reactions of the Halides with Various Metal Ions
   Describe the appearance of each mixture in tubes 1–12.

	⑦$Ca(NO_3)_2(aq)$	⑧$AgNO_3(aq)$	⑨$Fe(NO_3)_3(aq)$
$NaF(aq)$	1	2	3
$NaCl(aq)$	4	5	6
$NaBr(aq)$	7	8	9
$NaI(aq)$	10	11	12

From the observed data, answer the following questions.

a. Which of the fluorides are insoluble? ____________________

b. Which of the chlorides are insoluble? ____________________

c. Which of the bromides are insoluble? ____________________

d. Which of the iodides are insoluble? ____________________

How does the addition of ammonia distinguish the chemical reactivity of the halides? Explain.

F. Chemical Reactivity of Some Representative Elements

1. Sodium

   a. ⑩Name the gas evolved in the reaction of Na with water: ______________

   b. ⑪What product is produced from the reaction of Na with water as indicated by the action of phenolphthalein?

   c. Write the chemical equation for the reaction of Na with water:

2. Magnesium and Aluminum

   a. ⑫Which metal reacts more rapidly with HCl? ______________

      ⑬What is the gas that is evolved when Mg and Al react with HCl? ______________

   b. Reaction of metal ion with NaOH

⑭Observations	6 *M* NaOH	Excess 6 *M* NaOH
Mg^{2+} (drops to precipitate)		
Al^{3+} (drops to precipitate)		

   c. Explain the differences in chemical behavior of the magnesium and aluminum hydroxides. Use chemical equations in your discussion.

3. Solubilities of Alkaline-Earth Cations

Observations	⑮0.050 *M* NaOH	⑯0.10 *M* Na_2SO_y
Mg^{2+} (drops to precipitate)		
Ca^{2+} (drops to precipitate)		
Ba^{2+} (drops to precipitate)		

List the hydroxide salts in order of increasing solubility: ______________ < _______________ < _______________

List the sulfate salts in order of increasing solubility: _______________ < _______________ < _______________

What can you conclude about the general trend in solubilities of the Group 2A metal hydroxides and sulfates?

4. Sulfurous Acid and Sulfuric Acid
   a. ⑰What does the litmus test indicate about the chemical stability of sulfurous acid?

   b. What is the gas that is generated? _______________
   c. Write an equation for the decomposition of sulfurous acid.

   d. ⑱What is the color change of the litmus paper held above the sulfuric acid solution? _______________
   e. What chemical property differentiates sulfurous acid from sulfuric acid?

Laboratory Questions

Circle the questions that have been assigned.

1. Part A. a. Which group of elements has the highest ionization energies?
   b. Which group of elements has the lowest ionization energies?
2. Part A. a. Which group of elements has the largest atomic radii?
   b. Which group of elements has the smallest atomic radii?
3. Part A. a. Which group of elements has the highest electron affinity?
   b. Which group of elements has the lowest electron affinity?
4. Part D. Chlorine is used extensively as a disinfectant and bleaching agent. Without regard to adverse effects or costs, would bromine be a more or less effective disinfectant and bleaching agent? Explain.
5. Part D. Is fluorine gas predicted to be more or less reactive than chlorine gas? Explain.
6. Part D. Not much is known of the chemistry of astatine. Would you expect astatine to more or less reactive that iodine? Explain.
7. Part F.2. Is potassium metal or calcium metal predicted to be more reactive in water? Explain.
8. Part F.2. Predict the reactivity of silicon metal relative to that of magnesium and aluminum. Explain.
9. Part F.3. a. Is strontium hydroxide predicted to be more or less soluble that $Ca(OH)_2$? Explain.
   b. Is strontium sulfate predicted to be more or less soluble that $CaSO_4$? Explain.
10. Part F.4. Predict the acid–base properties of other nonmetallic oxides, such as CO_2, NO_2, and P_4O_{10}. Explain.

Dry Lab 3

Atomic and Molecular Structure

Metallic cations heated to high temperatures produce characteristic colors that appear in the starbursts.

OBJECTIVES

- To view and calibrate the visible line spectra
- To identify an element from its visible line spectrum
- To identify a compound from its infrared spectrum
- To predict the three-dimensional structure of molecules and molecular ions

INTRODUCTION

Visible light, as we know it, is responsible for all of the colors of nature—the blue skies, the green trees, the red roses, the orange-red rocks, and the brown deer. Our eyes are able to sense and distinguish the subtleties and the intensities of those colors through a complex naturally designed detection system. The "beauty" of nature is a result of the interaction of sunlight with matter. Since all matter consists of atoms and molecules, then it is obvious that sunlight, in some way, interacts with them to produce nature's colors.

Internally within molecules of compounds there are electrons, vibrating bonds, and rotating atoms, all of which can absorb energy. Since every compound is different, every molecule of a given compound possesses its own "unique" set of electronic, vibrational, and rotational **energy states,** which are said to be **quantized.** When incident electromagnetic (EM) radiation falls on a *molecule,* the radiation absorbed (the absorbed light) is an energy equal to the difference between two energy states, placing the molecule in an "excited state." The remainder of the EM radiation passes through the molecule unaffected.

Energy state: the amount of energy confined within an atom or molecule, which can be changed by the absorption or emission of discrete (quantized) amounts of energy

Quantized: only a definitive amount (of energy)

Atoms of elements interact with EM radiation in much the same way, except there are no bonds to vibrate and atoms to rotate. Only electronic energy states are available for energy absorption.

EM radiation is energy as well as light, and light has wavelengths and frequencies. The relationship between the energy, E, and its **wavelength,** λ, and **frequency,** ν, is expressed by the equation:

Wavelength: the distance between two crests of a wave

Frequency: the number of crests that pass a given point per second

$$E = \frac{hc}{\lambda} = h\nu \qquad \text{(D3.1)}$$

where h is Planck's constant, 6.63×10^{-34} J•s/photon; c equals the velocity of the EM radiation, 3.00×10^{8} m/s; λ is the wavelength of the EM radiation in meters; and ν (pronounced "new") is its frequency in reciprocal seconds (s^{-1}).

EM radiation includes not only the wavelengths of the visible region (400 to 700 nm), but also those that are shorter (e.g., the ultraviolet and X-ray regions) and longer (e.g., the infrared and radio wave regions). See Figure D3.1. From Equation D3.1, shorter wavelength EM radiation has higher energy.

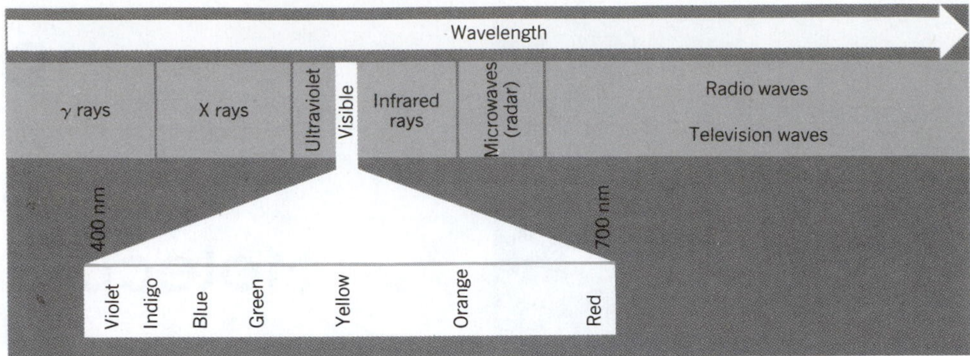

Figure D3.1 Visible light (400–700 nm) is a small part of the electromagnetic spectrum.

Spectrophotometer: an instrument used to detect and monitor the interaction of electromagnetic radiation with matter. The instrument has an EM radiation source, a grating to sort wavelengths, a sample cell, and an EM radiation detector.

White light: EM radiation containing all wavelengths of visible light

The unique sets of energy states for atoms of elements and molecules of compounds will absorb correspondingly "unique" wavelengths from the incident radiation; as a result a unique transmission of wavelengths occurs. This results in a transmitted or emission **spectrum,** the range of wavelengths *not* absorbed by the sample, that is sensed by the EM detector of a **spectrophotometer.**

When an atom or molecule absorbs EM radiation from the visible region of the spectrum, it is usually an electron that is excited from a lower to a higher energy state. When **white light** passes through a sample, our eyes (and the EM detector of a spectrophotometer) detect the wavelengths of visible light *not* absorbed, i.e., the transmitted light. Therefore, the color we see is complementary to the one absorbed. If, for example, the atom or molecule absorbs energy from the violet region of the visible spectrum, the transmitted light (and the substance) appears yellow (violet's complementary color)—the higher the concentration of violet absorbing atoms or molecules, the more intense is the yellow (Figure D3.2).

Table D3.1 lists the colors corresponding to wavelength regions of light (and their complements) in the visible region of the EM spectrum.

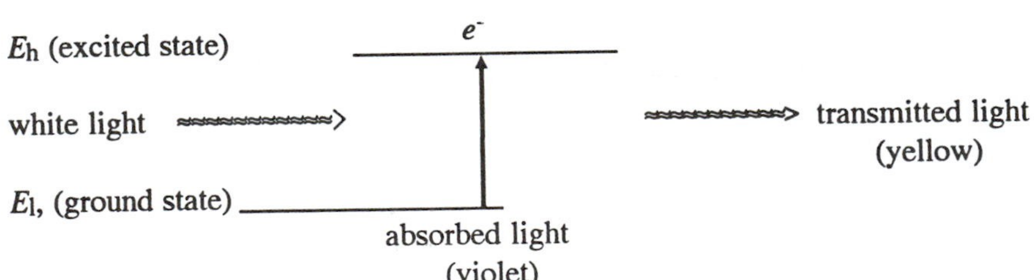

Figure D3.2 White light that is *not* absorbed is the transmitted light that we detect with our eyes.

Table D3.1 Color and Wavelengths in the Visible Region of the Electromagnetic Spectrum

Color Absorbed	Wavelength (nm)	Color Transmitted
red	750–610	green-blue
orange	610–595	blue-green
yellow	595–580	violet
green	580–500	red-violet
blue	500–435	orange-yellow
violet	435–380	yellow

When an atom of an element absorbs EM radiation, it is the electrons that absorb quanta of energy to reach excited states. When the electrons return to the lowest energy states **(the ground state of the atom),** the same amount of energy absorbed is now emitted as **photons.** The photons have unique energies and wavelengths that represent the difference in the energy states of the atom.

Photon: a particlelike quantity of electromagnetic radiation, often associated with electron transitions

E_h (excited state)

E_l (ground state)

e^-

absorption of energy

e^-

$E_{photon\ 1}$

$E_{photon\ 2}$

emission of energy

Since electrons can have a large number of excited states, a large collection of excited state electrons returning to the ground state produces an array of photons. When these emitted photons pass through a prism, an emission **line spectrum** is produced; each line in the spectrum corresponds to photons of fixed energy and wavelength. The line spectrum for hydrogen is shown in Figure D3.3.

Each element exhibits its own characteristic line spectrum because of the unique electronic energy states in its atoms. For example, the 11 electrons in sodium have a different set of electron energy states than do the 80 electrons in mercury. Therefore when an electron in an excited state of a sodium atom moves to a lower energy state, the emitted photon has a different energy and wavelength from one that is emitted when an electron de-excites (i.e., moves to a lower energy state) in a mercury atom.

The different wavelengths of the emitted photons produce different, yet characteristic, colors of light. Light emitted from an excited sodium atom is characteristically yellow-orange but mercury emits a blue light. Flame tests (Experiments 35 and 37) and exploding, aerial fireworks attest to the uniqueness of the electronic energy states of the atoms for different elements.

Much of the modern theory of atomic structure, which we call quantum theory or quantum mechanics, is based on the emission spectra of the elements.

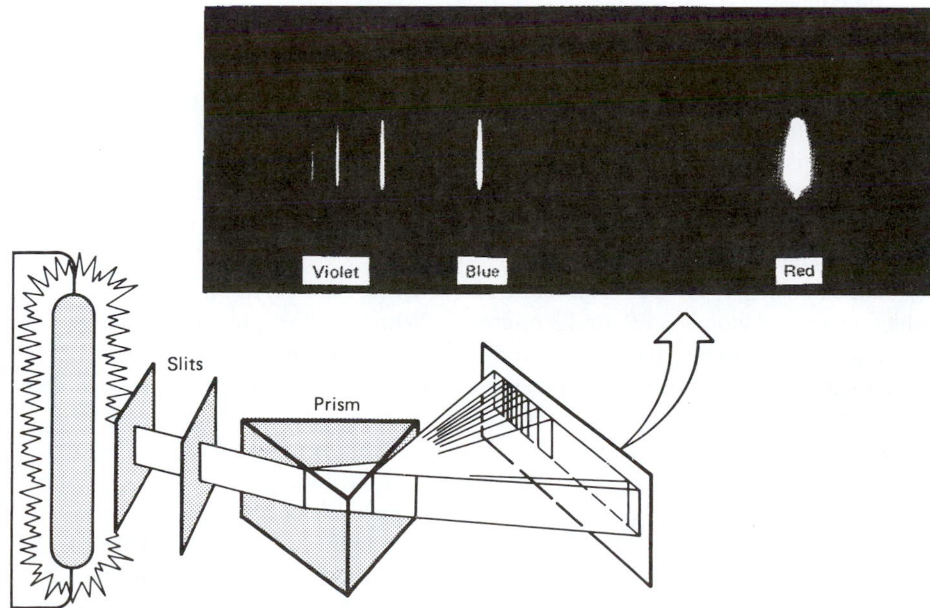

Figure D3.3 Formation of the line spectrum for hydrogen.

Table D3.2 Infrared Absorption Bands for Specific Atom Arrangements in Molecules

Atom Arrangement	Wavenumbers	Wavelengths
O—H	3700 to 3500 cm^{-1}	2.7 to 2.9 μm
C—H	3000 to 2800 cm^{-1}	3.3 to 3.6 μm
C═O	1800 to 1600 cm^{-1}	5.6 to 6.2 μm
C—O	1200 to 1050 cm^{-1}	8.3 to 9.5 μm
C—C	1670 to 1640 cm^{-1}	6.0 to 6.1 μm

Molecular Structure

The structures of atoms are much simpler than the structures of molecules. While the structures of atoms are one-dimensional, those of molecules are three-dimensional with atoms bonded in an almost endless number of configurations. Additionally, there are only a relatively few elements, but over 22 million compounds (see Dry Lab 2A). Using EM radiation to elucidate the three-dimensional structure of a molecule can be painstakingly tedious, especially the structures of the "living" (biochemical) molecules, such as DNA and hemoglobin. Oftentimes, EM radiation is only a partial tool for determining the molecular structure of a compound.

For simpler molecules, infrared EM radiation is used as a "probe" for the identification of specific atom arrangements in molecules. For example, the O—H bond, as in water and alcohols, absorbs infrared radiation for a principal vibrational energy transition between 3700 and 3500 **cm^{-1}** or wavelengths of 2.7 to 2.9 μm. Other characteristic infrared absorption bands for specific atom arrangements in molecules are listed in Table D3.2.

cm^{-1}: Infrared spectroscopists often indicate absorption bands in units of reciprocal centimeters, called ***wavenumbers,*** *rather than as wavelengths.*

For other molecules and molecular ions, other regions of EM radiation are most effective. For example, the $FeNCS^{2+}$ ion has a major absorption of radiation at 447 nm, and this property is used to measure its concentration in an aqueous solution in Experiment 25—the higher the concentration, the more EM radiation that is absorbed!

Lewis Theory

Valence electrons: electrons in the highest energy state (outermost shell) of an atom in the ground state

In 1916, G. N. Lewis developed a theory that focused on the significance of **valence electrons** in chemical reactions and in bonding. He proposed the "octet rule" in which atoms form bonds by losing, gaining, or sharing valence electrons until each atom of the molecule has the same number of valence electrons (eight) as the nearest noble gas in the periodic table. The resulting arrangement of atoms formed the Lewis structure of the compound.

The Lewis structure for water shows that by sharing the one valence electron on each of the hydrogen atoms with the six valence electrons on the oxygen atom, all three atoms obtain the same number of valence electrons as the nearest noble gas. Thus, in water the hydrogen atoms are **isoelectronic** with helium atoms and the oxygen atom is isoelectronic with the neon atom.

Isoelectronic: two atoms are isoelectronic if they have the same number of electrons.

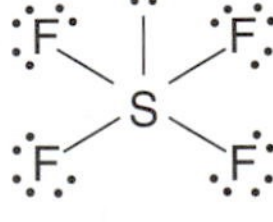

The Lewis structures for molecular ions are written similarly, except that electrons are removed from (for cations) or added to (for anions) the structure to account for the molecular charge.

An extension of the Lewis structure also exists for molecules or molecular ions in which its central atom is of Period 3 or greater. While the "peripheral" atoms retain the noble gas configuration, the central atom often "extends" its valence electrons to accommodate additional bonds. As a consequence the central atom accommodates more than eight valence electrons, an extension of the octet rule. For example, the six valence electrons of sulfur bond to four fluorine atoms in forming SF_4. To do so, four of the six valence electrons on sulfur share with the four fluorine atoms and two remain "nonbonding"—now, ten valence electrons exist in the bonded sulfur atom.

Although a Lewis structure accounts for the bonding based on the valence electrons on each atom, it does not predict the three-dimensional structure for a molecule. The development of the *v*alence *s*hell *e*lectron *p*air *r*epulsion (VSEPR) theory provides some insight into the nature of the three-dimensional structure of the molecule.

Gilbert Newton Lewis (1875–1946).

Valence Shell Electron Pair Repulsion (VSEPR) Theory of the Structures of Molecules and Molecular Ions

VSEPR theory proposes that the structure of a molecule is determined by the repulsive interaction of electron pairs in the valence shell of its central atom. The orientation is such that the distance between the electron pairs is maximized, or such that the electron pair–electron pair interactions are minimized. A construction of the Lewis structure of a molecule provides the first link in predicting the molecular structure.

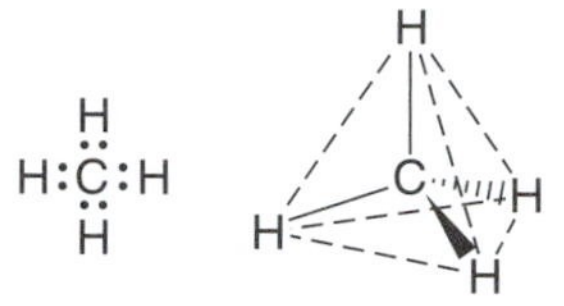

Methane, CH_4, has four bonding electron pairs in the valence shell of its carbon atom (the central atom in the molecule). Repulsive interactions between these four electron pairs are minimized when the electron pairs are positioned at the vertices of a tetrahedron with H—C—H bond angles of 109.5°. On the basis of the VSEPR theory, one can generalize that *all* molecules (or molecular ions) having four electron pairs in the valence shell of its central atom have a tetrahedral arrangement of these electron pairs with *approximate* bond angles of 109.5°. The nitrogen atom in ammonia, NH_3, and the oxygen atom in water, H_2O, also have four electron pairs in their valence shell!

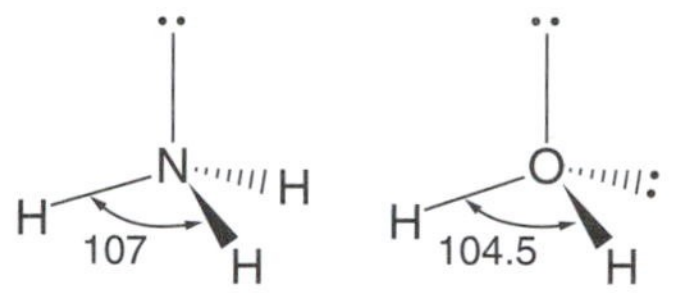

The preferred arrangement of the bonding and nonbonding electron pairs around the central atom gives rise to the corresponding structure of a molecule. The three-dimensional structures for numerous molecules and molecular ions can be grouped into a few basic structures. Based upon a correct Lewis structure, a VSEPR formula summarizes the number and type (bonding and nonbonding) of electron pairs in the compound or ion. The VSEPR formula uses the following notations:

A refers to the central atom

X_m refers to m number of *bonding* pairs of electrons on A

E_n refers to n number of *nonbonding* pairs of electrons on A

If a molecule has the formula AX_mE_n, it means there are $m + n$ electron pairs in the valence shell of A, the central atom of the molecule; m are bonding and n are nonbonding electron pairs. For example, CH_4, SiF_4, $GeCl_4$, PH_4^+, and PO_4^{3-} all have a VSEPR formula of AX_4. Thus, they all have the same three-dimensional structure, that of a tetrahedral structure.

It should be noted here that valence electrons on the central atom contributing to a multiple bond do *not* affect the geometry of a molecule. For example in SO_2, the VSEPR formula is AX_2E, and the geometric shape of the molecule is "V-shaped." Further applications are presented in more advanced chemistry courses.

Table D3.3 VSEPR and Geometric Shapes of Molecules and Molecular Ions

Valence Shell Electron Pairs	Bonding Electron Pairs	Nonbonding Electron Pairs	VSEPR Formula	Three-Dimensional Structure	Bond Angle	Geometric Shape	Examples
2	2	0	AX_2	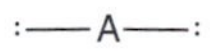 :—A—:	180°	Linear	$HgCl_2$, $BeCl_2$
3	3	0	AX_3	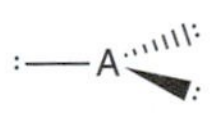	120°	Planar triangular	BF_3, $In(CH_3)_3$
	2	1	AX_2E		<120°	V-shaped	$SnCl_2$, $PbBr_2$
4	4	0	AX_4	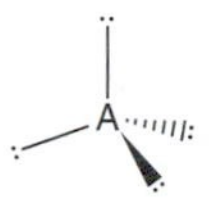	109.5°	Tetrahedral	CH_4, $SnCl_4$
	3	1	AX_3E		<109.5°	Trigonal pyramidal	NH_3, PCl_3, H_3O^+
	2	2	AX_2E_2		<109.5°	Bent	H_2O, OF_2, SCl_2
5	5	0	AX_5	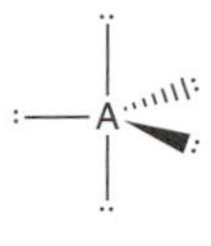	90°/120°	Trigonal bipyramidal	PCl_5, $NbCl_5$
	4	1	AX_4E		>90°	Irregular tetrahedral	SF_4, $TeCl_4$
	3	2	AX_3E_2		<90°	T-shaped	ICl_3
	2	3	AX_2E_3		180°	Linear	ICl_2^-, XeF_2
6	6	0	AX_6		90°	Octahedral	SF_6
	5	1	AX_5E		>90°	Square pyramidal	BrF_5
	4	2	AX_4E_2		90°	Square planar	ICl_4^-, XeF_4

Table D3.3 presents a summary of the VSEPR theory for predicting the geometric shape and approximate bond angles of a molecule or molecular ion based on the five VSEPR three-dimensional structures of molecules and molecular ions.

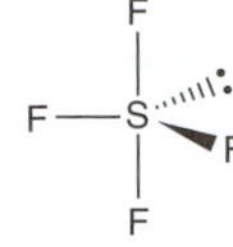

Let us refer back to SF_4, the molecule with the "extended" valence shell on the central atom. The sulfur atom has four bonding electron pairs ($m = 4$) and one nonbonding pair ($n = 1$). This gives a VSEPR formula of AX_4E_1, predicting a geometric shape of "irregular tetrahedral" (sometimes also called "seesaw") with bond angles greater than 90°. Molecular models will enable you to envision these properties of SF_4.

In Part A of the Dry Lab Procedure, the visible spectra of a number of elements are studied (see color plate on back cover of this manual). The wavelengths of the spectra are to be calibrated relative to the mercury spectrum at the bottom of color plate. The most intense lines of the mercury spectrum are listed in Table D3.4.

In Part B, a spectrum from the color plate will be assigned, and, with reference to Table D3.5, the element producing the line spectrum will be identified.

In Part C, several infrared spectra will be analyzed using the data from Table D3.2. After analysis, the spectra will be matched with the molecular structures of a list of compounds.

In Part D, a number of simple molecules and molecular ions will be assigned, and their three-dimensional structure and approximate bond angles will be determined. The Lewis structure and the VSEPR adaptation of the Lewis structure are used for analysis.

Dry Lab Procedure

Procedure Overview: The experimental atomic spectra for a number of elements appearing on the color plate on the back cover of this manual are calibrated using the mercury spectrum as the reference spectrum. An element is identified from an experimental emission spectrum and the data in Table D3.5. The molecular structures of a number of compounds will be paired with their infrared spectra followed by a prediction of molecular structures based on the VSEPR theory.

The Mercury Spectrum

1. **The Color Plate.** Notice the various experimental emission line spectra on the color plate. A continuous spectrum appears at the top, the line spectra for various elements appear in the middle, and the Hg spectrum appears at the bottom.
2. **Calibrate the Spectra of the Color Plate.** Use a ruler to mark off a linear wavelength scale across the bottom of the color plate such that the experimental wavelengths of the mercury spectrum correlate with those in Table D3.4. Extend the linear wavelength scale perpendicularly and upward across spectra on the color plate, thus creating the same wavelength scale for all of the emission spectra. A wax marker or "permanent" felt tip pen may be required for marking the wavelength scale along the axis.

   Have your instructor approve your calibration of the spectra on the color plate. See the Report Sheet.

Table D3.4 Wavelengths of the Visible Lines in the Mercury Spectrum

Violet	404.7 nm
Violet	407.8 nm
Blue	435.8 nm
Yellow	546.1 nm
Orange	577.0 nm
Orange	579.1 nm

B. The Spectra of Elements

Use Figure D3.3 to identify which of the emission spectra on the color plate on the back cover is that of hydrogen. Justify your selection.

Your instructor will assign to you one or two emission spectra from the color plate. Analyze each spectrum by locating the most intense wavelengths in the assigned

Table D3.5 Wavelengths and Relative Intensities of the Emission Spectra of Several Elements

Element	Wavelength (nm)	Relative Intensity
Argon	451.1	100
	560.7	35
	591.2	50
	603.2	70
	604.3	35
	641.6	70
	667.8	100
	675.2	150
	696.5	10000
	703.0	150
	706.7	10000
	706.9	100
Barium	435.0	80
	553.5	1000
	580.0	100
	582.6	150
	601.9	100
	606.3	200
	611.1	300
	648.3	150
	649.9	300
	652.7	150
	659.5	3000
	665.4	150
Cadmium	467.8	200
	479.9	300
	508.6	1000
	610.0	300
	643.8	2000
Cesium	455.5	1000
	459.3	460
	546.6	60
	566.4	210
	584.5	300
	601.0	640
	621.3	1000
	635.5	320
	658.7	490
	672.3	3300

Element	Wavelength (nm)	Relative Intensity
Helium	388.9	500
	396.5	20
	402.6	50
	412.1	12
	438.8	10
	447.1	200
	468.6	30
	471.3	30
	492.2	20
	501.5	100
	587.5	500
	587.6	100
	667.8	100
Neon	585.2	500
	587.2	100
	588.2	100
	594.5	100
	596.5	100
	597.4	100
	597.6	120
	603.0	100
	607.4	100
	614.3	100
	616.4	120
	618.2	250
	621.7	150
	626.6	150
	633.4	100
	638.3	120
	640.2	200
	650.7	150
	660.0	150
Potassium	404.4	18
	404.7	17
	536.0	14
	578.2	16
	580.1	17
	580.2	15
	583.2	17
	691.1	19

Element	Wavelength (nm)	Relative Intensity
Rubidium	420.2	1000
	421.6	500
	536.3	40
	543.2	75
	572.4	60
	607.1	75
	620.6	75
	630.0	120
Sodium	466.5	120
	466.9	200
	497.9	200
	498.3	400
	568.2	280
	568.8	560
	589.0	80000
	589.6	40000
	616.1	240
Thallium	377.6	12000
	436.0	2
	535.0	18000
	655.0	16
	671.4	6
Zinc	468.0	300
	472.2	400
	481.1	400
	507.0	15
	518.2	200
	577.7	10
	623.8	8
	636.2	1000
	647.9	10
	692.8	15

emission spectrum. Compare the wavelengths of the most intense lines with the data in Table D3.5. Identify the element having the shown spectrum.

C. Infrared Spectra of Compounds

Identify the atom arrangement from Table D3.2 that is responsible for the major absorption bands appearing in the six infrared spectra (Figures D3.5A-F). Thereafter, match the six infrared spectra with the six compounds appearing at the bottom of the page.

D. Structure of Molecules and Molecular Ions

Using an appropriate set of molecular models, construct the five basic three-dimensional structures shown in Table D3.3. Because of the possible limited availability of molecular models, some sharing with other chemists may be necessary. Consult with your laboratory instructor.

On the basis of the Lewis structure of the molecule or molecular ion and the selection of the corresponding basic three-dimensional structure, determine the geometric shape and approximate bond angles of the molecules/ions shown on the Report Sheet.

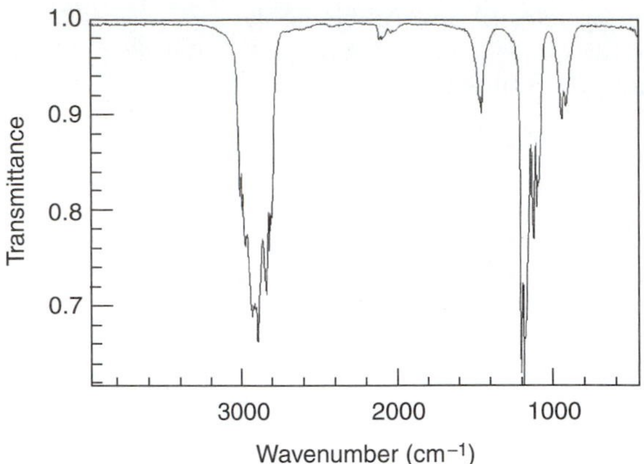

Figure D3.5A

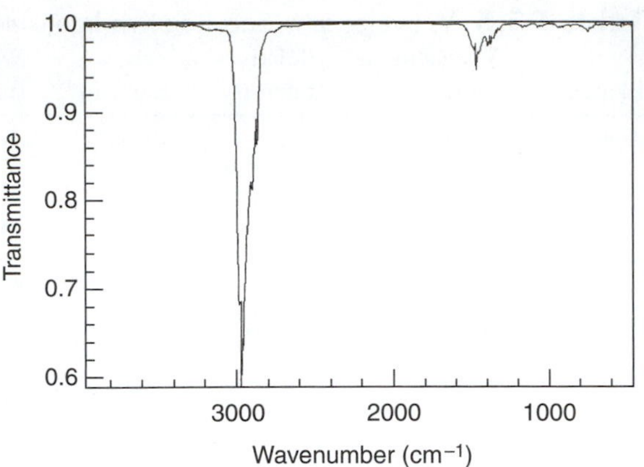

Figure D3.5B

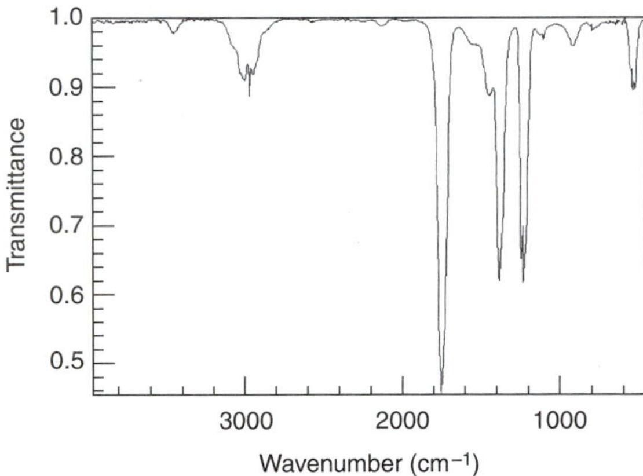

Figure D3.5C

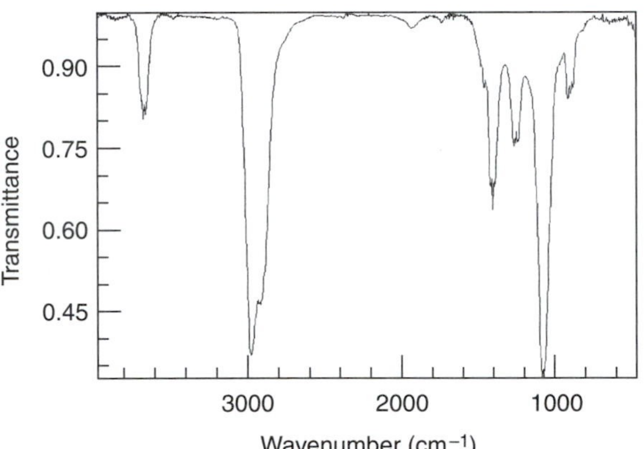

Figure D3.5D

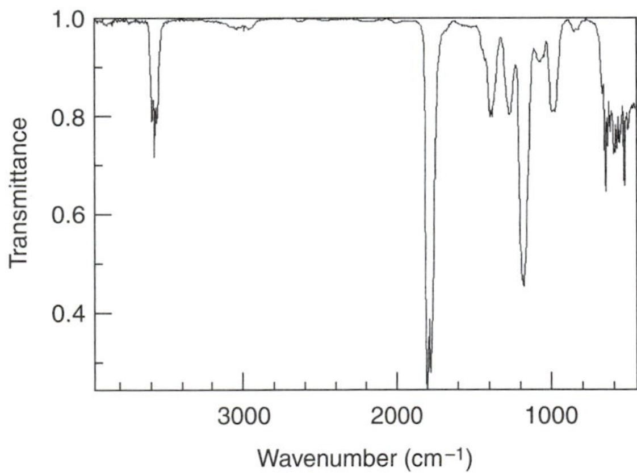

Figure D3.5E

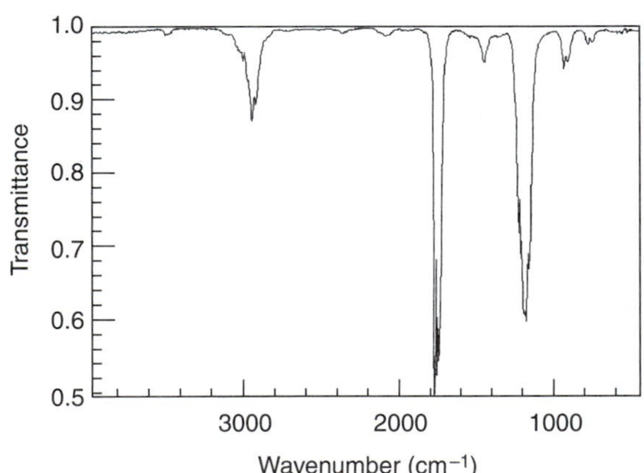

Figure D3.5F

$H_3C{-}CH_2{-}CH_3$
propane

$H_3C{-}CH_2{-}O{-}H$
ethanol

$H_3C{-}O{-}CH_3$
dimethyl ether

$H{-}C({=}O){-}O{-}CH_3$
methyl formate

$H_3C{-}C({=}O){-}O{-}H$
acetic acid

$H_3C{-}C({=}O){-}CH_3$
acetone

Dry Lab 3 *Report Sheet*

Atomic and Molecular Structure

Date __________ Lab Sec. ______ Name ______________________________ Desk No. __________

A. The Mercury Spectrum

Instructor's approval of the calibration of the color plate (back cover) _______________

B. The Spectra of Elements

1. Spectrum number _____ is the emission line spectrum for hydrogen on the color plate.

   What are the wavelengths and colors of the emission lines of its visible spectrum?

2. Identification of Spectra
   a. Unknown spectrum number _____

      Intense lines in the spectrum: _____ nm; _____ nm; _____ nm; _____ nm; _____ nm; _____ nm; _____ nm

      Element producing the spectrum _____

   b. Unknown spectrum number _____

      Intense lines in the spectrum: _____ nm; _____ nm; _____ nm; _____ nm; _____ nm; _____ nm; _____ nm

      Element producing the spectrum _____

C. Infrared Spectra of Compounds

On a spearate sheet of paper, set up the following table for each of the six infrared spectra. For each spectrum, list the wavenumbers (cm^{-1}) for the major absorption bands and identify the corresponding atom arrangement (Table D3.2). From the data analysis of the spectra, match the six compounds with the six spectra. Follow each assignment with a brief explanation.

Spectrum	Absorption Band (cm^{-1})/ Atom Arrangment	Compound/Explanation
D3.5A		______________________

D. Structure of Molecules and Molecular Ions

On a separate sheet of paper, set up the following table (with eight columns) for each of the molecules/molecular ions that are assigned to you/your group. The central atom of the molecule/molecular ion is italicized.

Molecule or Molecular Ion	Lewis Structure	Valence Shell Electron Pairs	Bonding Electron Pairs	Nonbonding Electron Pairs	VSEPR Formula	Approx. Bond Angle	Geometric Shape
1. $\mathit{C}H_4$	H H:C:H H	4	4	0	AX_4	109.5°	tetrahedral
2. $\mathit{S}F_4$							
3. $H_2\mathit{O}$							

1. Complete the table (as outlined above) for the following molecules/molecular ions, all of which obey the Lewis octet rule. Complete those that are assigned by your laboratory instructor.

   a. $H_3\mathit{O}^+$ **d.** $\mathit{C}H_3^-$ **g.** $\mathit{P}O_4^{3-}$ **j.** $\mathit{Si}F_4$

   b. $\mathit{N}H_3$ **e.** $\mathit{Sn}H_4$ **h.** $\mathit{P}F_3$ **k.** $H_2\mathit{S}$

   c. $\mathit{N}H_4^+$ **f.** $\mathit{B}F_4^-$ **i.** $\mathit{As}H_3$ **l.** $\mathit{N}H_2^-$

2. Complete the table (as outlined above) for the following molecules/molecular ions, *none* of which obey the Lewis octet rule. Complete those that are assigned by your laboratory instructor.

   a. $\mathit{Ga}I_3$ **d.** $\mathit{Xe}F_2$ **g.** $\mathit{Xe}OF_4$ **j.** $\mathit{Sn}F_6^{2-}$

   b. $\mathit{P}Cl_2F_3$ **e.** $\mathit{Xe}F_4$ **h.** $\mathit{Sb}F_6^-$ **k.** $\mathit{I}F_4^-$

   c. $\mathit{Br}F_3$ **f.** $\mathit{Xe}OF_2$ **i.** $\mathit{S}F_6$ **l.** $\mathit{I}F_4^+$

3. Complete the table (as outlined above) for the following molecules/molecular ions. No adherence to the Lewis octet rule is indicated. Complete those that are assigned by your laboratory instructor.

   a. $\mathit{As}F_3$ **d.** $\mathit{Sn}F_2$ **g.** $\mathit{P}F_5$ **j.** $\mathit{Kr}F_2$

   b. $\mathit{Cl}O_2^-$ **e.** $\mathit{Sn}F_4$ **h.** $\mathit{S}O_4^{2-}$ **k.** $\mathit{Te}F_6$

   c. $\mathit{C}F_3Cl$ **f.** $\mathit{P}F_4^+$ **i.** $\mathit{C}N_2^{2-}$ **l.** $\mathit{As}F_5$

4. Complete the table (as outlined above) for the following molecules/molecular ions. For molecules or molecular ions with two or more atoms considered as central atoms, consider each atom separately in the analysis according to Table D3.3. Complete those that are assigned by your laboratory instructor.

   a. $O\mathit{P}Cl_3$ **d.** $Cl_3\mathit{CC}F_3$ **g.** $\mathit{C}OCl_2$ **j.** O_3

   b. H_2CCH_2 **e.** $Cl_2\mathit{O}$ **h.** OCCCO **k.** $\mathit{N}O_3^-$

   c. $\mathit{C}H_3\mathit{N}H_3^+$ **f.** ClCN **i.** $\mathit{C}O_2$ **l.** $\mathit{Te}F_2(CH_3)_4$

Dry Lab Questions

Circle the questions that have been assigned.

1. What experimental evidence leads scientists to believe that only "quantized" electronic energy states exist in atoms?

2. a. What is the wavelength range of the visible spectrum for electromagnetic radiation?

   b. What is the color of the "short" wavelength end of the visible spectrum?

   c. If a substance absorbed the wavelengths from the "short" wavelength end of the visible spectrum, what would be its color?

3. Explain why an aqueous solution containing copper(II) ion appears blue.

4. Why does the mercury light from public lighting appear blue, even though yellow and orange lines appear in the spectrum?

5. a. Is the energy absorption associated with "bands" in an infrared spectrum of higher or lower energy than the "lines" appearing in a visible line spectrum? Explain.

   b. Identify the type of energy transition causing a band to appear in an infrared spectrum?

   c. Identify the type of energy transition causing a line to appear in a visible line spectrum?

6. Since $FeNCS^{2+}$ has an absorption maximum at 447 nm, what is the color of the $FeNCS^{2+}$ ion in solution?

7. a. Write the Lewis structure for XeF_6.

   b. Write the VSEPR formula for XeF_6.

   c. Sketch (or describe) the three dimensional structure (or geometric shape) of XeF_6.

   d. What are the approximate F–Xe–F bond angles in XeF_6?

8. a. Write the Lewis structure for $TeF_2(CH_3)_4$, where tellurium is the central atom.

   b. Predict the geometric shape of $TeF_2(CH_3)_4$.

9. a. Write the Lewis structure for XeO_3.

   b. Predict the geometric shape of XeO_3.

10. a. Why do different elements have different line spectra? Explain.

    b. What effect do you suppose the nucleus might have on the atomic spectra of the different elements? Explain.

Experiment 12

Inorganic Compounds and Metathesis Reactions

Drops of a potassium hydroxide solution added to an iron(III) chloride solution produces insoluble iron(III) hydroxide.

Objectives

- To observe the physical appearance of common laboratory chemicals
- To systematically observe and express the stages of a **metathesis reaction**
- To determine the solubilities of some salts by studying metathesis reactions

Techniques

The following techniques are used in the Experimental Procedure

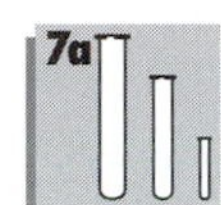

Introduction

By now you have learned to name and write the formulas for a large number of inorganic compounds. In addition you should also be familiar with balanced equations and **stoichiometry.** But names, formulas, and equations have little meaning unless there is some tangible relationship to chemicals and chemical reactions. To a chemist, sulfur is not just an element with the symbol S that reacts with oxygen to form sulfur dioxide, but rather a yellow solid that can be held in the hand and burns in air with a blue flame, producing a choking irritant called sulfur dioxide.

Stoichiometry: a study of a chemical reaction using a balanced equation

In this experiment you will "look" at some laboratory chemicals with the intent that a mental association develops between a compound's formula and its physical appearance. You will also conduct a number of chemical reactions, observe some of the chemical properties of chemicals, and make conclusions on the basis of your observations.

In Part A, a number of compounds are identified by their formulas, names, physical states, colors, crystal characteristics (if a solid), and solubilities in water.

In Part B, the chemical reactions of some ionic compounds are studied in an aqueous solution. In these chemical reactions two ionic compounds, as reactants, are dissolved in solution whereby the ions are free to exchange partners to form products. A chemical reaction (an exchange of ionic partners) is observed if a chemical change is evident, such as the formation of a precipitate or the evolution of a gas. These chemical reactions are called **metathesis** (or double displacement) **reactions.**

To systematically interpret the progress of a metathesis reaction, a sequence of observations and equation writing steps are followed. Using the **salts** silver nitrate and sodium chloride as reactants, these steps are as follows:

Salt: an ionic compound

1. **Define the Reactants.** Write the formulas of the reactants.

$$AgNO_3 + NaCl \quad (12.1)$$

Figure 12.1 The progression of a reaction between solutions of silver nitrate and sodium chloride.

2. **Determine the Products and Write the Balanced Molecular Equation.** The formulas of the products are written by exchanging the cations with the anions of the reactants. A *balanced* **molecular equation** for the proposed metathesis reaction is

$$AgNO_3 + NaCl \rightarrow AgCl + NaNO_3 \qquad (12.2)$$

It is this molecular equation that is studied in detail in the laboratory—what is the nature of each of these substances in aqueous solution?

Species: any ion or molecule, generally in an aqueous solution

3. **Formulas of Reactant *Species* in the Solution.** The physical states of the reactants in solution *before mixing* are observed; for example, are the salts soluble or insoluble? What is their *actual* existence in solution? The formulas of the salts are written accordingly, either as separated hydrated ions or as a solid.

Both silver nitrate and sodium chloride are soluble salts (Figure 12.1a). The cations and anions of the two salts move about the aqueous solution as separate hydrated species; each is written as (*aq*).

$$Ag^+(aq) + NO_3^-(aq) + Na^+(aq) + Cl^+(aq) \rightarrow \qquad (12.3)$$

4. **Evidence of Reaction.** On combination of the ions from the two aqueous solutions into a single system, an observation determines whether or not a reaction has occurred between the ions. A chemical reaction is observed when

- a precipitate forms
- a gas is evolved
- heat is evolved or absorbed
- a color change occurs
- a change in acidity (or basicity) occurs
- light is emitted

In this system, the formation of the white precipitate of silver chloride is observed (Figure 12.1b).

5. **Formulas of Product *Species* in Solution.** The formulas of the products are then written as they appear in the aqueous solution after the mixing (Figure 12.1c). A precipitate is written as (*s*), and an ion remaining in solution is written as (*aq*).

$$\rightarrow AgCl(s) + Na^+(aq) + NO_3^-(aq) \qquad (12.4)$$

6. **Ionic Equation.** The reactants (from Step 3) and the products (from Step 5) are combined to form a balanced **ionic equation** that represents each species as it actually exists in the aqueous solution.

$$Ag^+(aq) + NO_3^-(aq) + Na^+(aq) + Cl^-(aq) \rightarrow AgCl(s) + Na^+(aq) + NO_3^-(aq) \qquad (12.5)$$

7. **Net Ionic Equation.** A final equation, called the **net ionic equation,** includes only those ions responsible for the observed reaction. The ions present, but not involved in the observed reaction, are considered **spectator ions** and do not appear in a net ionic equation. Only the silver and chloride ions are involved in any observable reaction, the sodium and nitrate ions are not.

Spectator ions: ions that do not participate in an observable/detectable reaction

$$Ag^+(aq) + Cl^-(aq) \rightarrow AgCl(s) \qquad (12.6)$$

Similar molecular equations, ionic equations, and net ionic equations are written for a number of chemical systems in Part B of this experiment.

EXPERIMENTAL PROCEDURE

Procedure Overview: An array of laboratory chemicals are observed and described. Eleven metathesis reactions are observed, and appropriate formulas and equations are *systematically* (7 step format) written to express the nature of each species in solution.

A. Identification

Crystal characteristics: shape (powder, granular, etc.), brightness, and wet or dry appearance

A broad selection of chemical compounds in test tubes or laboratory dishes (Figure 12.2) are located on the display table. Describe the color and **crystal characteristics** (if a solid) of each chemical; also predict the water solubility[1] of each chemical. Use the Report Sheet as a format for reporting your observations and predictions.

Attempt to group some of the compounds on the basis of color, crystal characteristics, and/or solubility.

B. Systematic Study of Metathesis Reactions

This experiment requires either a set of 12 clean, small test tubes or a 24-well plate. Consult with your instructor. If you use test tubes, label them in accordance with the letter/number designations in Table 12.1.

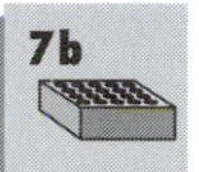

1. **Test Solutions.** Set up the test tubes (Figure 12.3) or the 24-well plate (Figure 12.4) with the reactant solutions or preparations listed in Table 12.1. Volumes of solutions only need to be approximate.[2] Your instructor may substitute, add, or delete chemicals from the table. **Caution:** *Avoid skin contact with any of the solutions used in this experiment. Wash the skin immediately with copious amounts of water.*

Figure 12.2 A view of the nature of substances.

Table 12.1 An Organization of the Reactants for a Series of Metathesis Reactions

Test Tube No. or Well No.	Reactant Solution or Preparation
A1	Several crystals of $CaCO_3$
A2	2 mL of 3.0 *M* HCl and several drops of universal indicator[3]
A3	2 mL of 3.0 *M* NaOH
A4	Several crystals of $FeCl_3•6H_2O$ in 2 mL of water *or* 2 mL of 0.1 *M* $FeCl_3$
A5	Several crystals of $CoCl_2•6H_2O$ in 2 mL of water *or* 2 mL of 0.1 *M* $CoCl_2$
A6	Several crystals of $AgNO_3$ in 2 mL of water *or* 2 mL of 0.1 *M* $AgNO_3$
B3	Several crystals of NH_4Cl
B6	Several crystals of Na_2CO_3 in 2 mL of water *or* 2 mL of 0.1 *M* Na_2CO_3
C4	Several crystals of $NiCl_2•6H_2O$ in 2 mL of water *or* 2 mL of 0.1 *M* $NiCl_2$
C5	Several crystals of $Na_3PO_4•12H_2O$ in 2 mL of water *or* 2 mL of 0.1 *M* Na_3PO_4
C6	Several crystals of $CuSO_4•5H_2O$ in 2 mL of water *or* 2 mL of 0.1 *M* $CuSO_4$
D6	Several crystals of $BaCl_2•2H_2O$ in 2 mL of water *or* 2 mL of 0.1 *M* $BaCl_2$

[1]The predicted solubility is based on the rules in Appendix G.
[2]The volume of a small (75-mm) test tube is about 3 mL; the volume of each well is about 3.4 mL.
[3]Universal indicator is a mixture of acid–base indicators that gradually changes color, depending on the acidity of the solution.

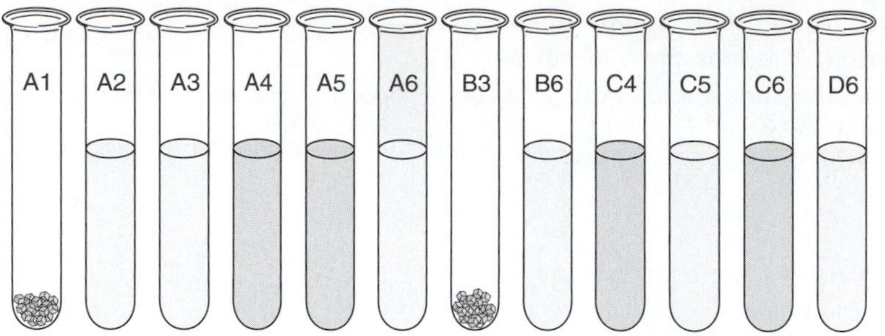

Figure 12.3 The arrangement of 12 small, labeled test tubes for test solutions.

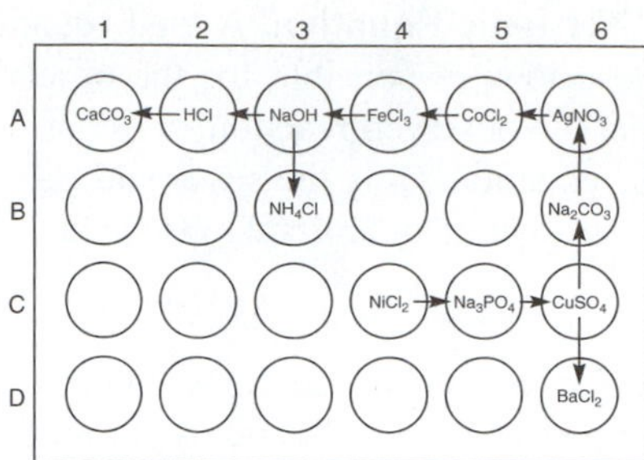

Figure 12.4 The arrangement of the test solutions in a 24-well plate.

2. **Systematic Procedure for Studying Reactions.** Pairs of reactant solutions are combined according to Figure 12.4. Record your stepwise analysis of each reaction on the Report Sheet using the seven steps outlined in the Introduction. Be sure to look for any evidence of a chemical change.

   Complete the seven (7) step format of writing equations as suggested in the Introduction and on the Report Sheet for each reaction mixture before, during, and after *each* solution transfer. Use clean dropping pipets or Beral pipets to slowly make the following solution transfers:

   a. Transfer approximately one-half of solution A2 to A1 in (test tube/well) A1.

   b. Transfer one-third of solution A3 to the remainder of solution A2 in (test tube/well) A2.

Appendix G

   c. Transfer one-third of solution A3 to B3 in (test tube/well) B3. (Warm the test tube in your hand or the bottom of the well with your finger and smell cautiously. Test the fumes with wet, red litmus.)

   d. Transfer one-half of solution A4 to the remainder of solution A3 in (test tube/well) A3.

   e. Transfer one-half of solution A5 to the remainder of solution A4 in (test tube/well) A4.

   f. Transfer one-half of solution A6 to the remainder of solution A5 in (test tube/well) A5.

   g. Transfer one-half of solution B6 to the remainder of solution A6 in (test tube/well) A6.

   h. Transfer one-third of solution C6 to the remainder of solution B6 in (test tube/well) B6.

   i. Transfer one-third of solution C6 to solution D6 in (test tube/well) D6.

   j. Transfer one-half of solution C5 to the remainder of solution C6 in (test tube/well) C6.

   k. Transfer one-half of solution C4 to the remainder of solution C5 in (test tube/well) C5.

Disposal: Dispose of the waste solutions, as identified by your instructor, in the "Waste Salts" container.

CLEANUP: Rinse the test tubes or well plate with tap water twice and with deionized water twice. Discard each rinse in the "Waste Salts" container.

Experiment 12 *Prelaboratory Assignment*

Inorganic Compounds and Metathesis Reactions

Date __________ Lab Sec. ______ Name ____ Desk No. __________

1. a. Nickel(II) chloride dissolves in water to form a light green solution. What species are present in solution, i.e., what is "swimming around" in the aqueous solution? Write the formulas.

   b. Sodium phosphate dissolves in water to form a colorless solution. What species are in solution? Write the formulas.

   c. When solutions of nickel(II) chloride and sodium phosphate are combined, a green nickel(II) phosphate precipitate forms. What species remain in solution (spectator ions)? What is the color of the solution (assume stoichiometric amounts of nickel(II) chloride and sodium phosphate). Explain.

2. Complete the seven (7) step format for writing equations for the reaction mixture of $AgNO_3$ and ZnI_2.
   1) Define the reactants (formulas in *molecular* form)

   2) Determine the products and write the balanced *molecular* equation

   3) Formulas of reactant *species* in the two solutions (in the test tubes prior to mixing)

   4) Evidence of reaction
      A yellow precipitate forms (refer to Appendix G)

   5) Formulas of product *species* in solution (in the test tube after mixing)

   6) Ionic equation (balanced)

   7) Net ionic equation (balanced) . . . species responsible for observed reaction

3. Refer to Table 12.1. What species are present in the solution for test tube #C6 (i.e., several crystals of $CuSO_4 \cdot 5H_2O$ in 2 mL of water)?

4. For the metathesis reaction, $K_2CO_3 + FeCl_3$,
   a. Write the balanced molecular equation.

   b. What evidence of reaction occurs in the mixture? (Appendix G)

   c. Write the balanced ionic equation.

   d. Write the balanced net ionic equation for the reaction that occurs.

5. List five observations, appealing to your senses, that indicate a chemical reaction has occurred. In each case, identify an example where you might have experienced that observation.

	Observation	Example
a.		
b.		
c.		
d.		
e.		

Inorganic Compounds and Metathesis Reactions

Date ________ Lab Sec. ______ Name ______________________________ Desk No. ________

A. Identification

On a separate sheet of paper, construct a table with the headings shown below. Fill in the table as described in the Experimental Procedure. Submit this data sheet along with your Report Sheet.

	Formula	Name	Physical State (g, l, s)	Color	Characteristics	Predicted Solubility in Water (Appendix G)
Ex.	NaCl	Sodium chloride	Solid	White	Small, shiny, dry	Soluble
1	____	____	____	____	____	____
2	____	____	____	____	____	____
3	____	____	____	____	____	____

B. Metathesis Reactions (Double-Displacement Reactions)

Use the seven steps presented in the Introduction and the Prelaboratory Assignment (Q#2) as a guide to writing formulas and equations as you observe the stages of a chemical reaction.

Step	Formulas, Observations, Equations	Step	Formulas, Observations, Equations
1	Formulas of reactants (below)	5	Formulas of products in solution
2	Molecular equation	6	Ionic equation
3	Formulas of reactants in solution	7	Net ionic equation
4	Evidence of reaction		

On a separate sheet of paper, use this 7-step format to write the formulas, observations, and equations for the systems described in Part B.2. Submit this data sheet along with the Report Sheet.

a. $CaCO_3 + HCl \rightarrow$

b. $HCl + NaOH \rightarrow$

c. $NH_4Cl + NaOH \rightarrow$

d. $NaOH + FeCl_3 \rightarrow$

e. $FeCl_3 + CoCl_2 \rightarrow$

f. $CoCl_2 + AgNO_3 \rightarrow$

g. $AgNO_3 + Na_2CO_3 \rightarrow$

h. $Na_2CO_3 + CuSO_4 \rightarrow$

i. $BaCl_2 + CuSO_4 \rightarrow$

j. $CuSO_4 + Na_3PO_4 \rightarrow$

k. $Na_3PO_4 + NiCl_2 \rightarrow$

On the basis of your observations in Part B.2, summarize your data in the following table:

anions / cations	SO_4^{2-}	CO_3^{2-}	OH^-	PO_4^{3-}	NO_3^-	Cl^-
Cations forming insoluble salts						
Cations forming soluble salts						

Laboratory Questions

Circle the questions that have been assigned.

1. Part A and Part B. A wide range of chemicals was observed and studied in this experiment. General conclusions can be made regarding the following:
   a. Identify the color of salts or salt solutions containing the copper(II) ion.
   b. Identify the color of salts or salt solutions containing the nickel(II) ion.
   c. Identify the color of salts or salt solutions containing the cobalt(II) ion.
   d. What can be concluded about the color of most anions? Explain and/or cite some examples.
2. Part A. Distinguish between the general appearance of those salts that are hydrated and those that are not. In which section of the periodic table are cations found that are commonly hydrated?
3. Part B.2. Review the observed reactions for the mixtures created in this experiment. Identify the *general* solubility (soluble or insoluble) of salts containing the following ions:
   a. Cl^-
   b. CO_3^{2-}
   c. PO_4^{3-}
   d. Na^+
   e. NO_3^-
4. Part B.2. a. Identify the chemicals necessary to produce $CO_2(g)$.
   b. Write a net ionic equation for the generation of $CO_2(g)$.
5. Part B.2. a. Identify the chemicals necessary to produce $NH_3(g)$.
   b. Write a net ionic equation for the generation of $NH_3(g)$.
6. Part B.2. Write the net ionic equation for an acid–base neutralization reaction.

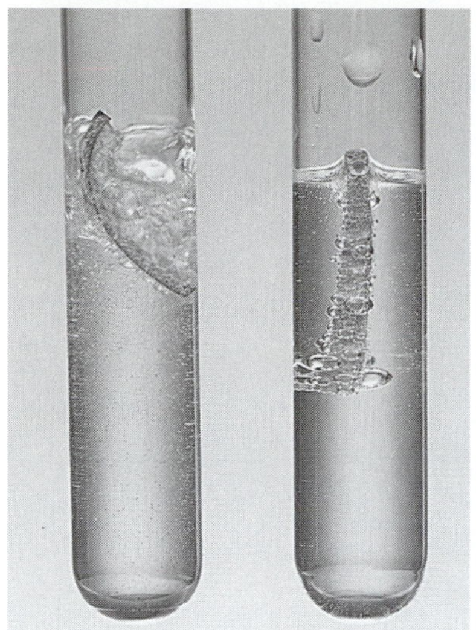

Hydrogen gas is produced at different rates from the reaction of zinc metal with hydrochloric acid (left) and phosphoric acid (right).

Experiment 13

Acids, Bases, and Salts; pH

Objectives

- To become familiar with the chemical and physical properties of acids and bases
- To measure the pH of laboratory and household acid, base, and salt solutions

Techniques

The following techniques are used in the Experimental Procedure

   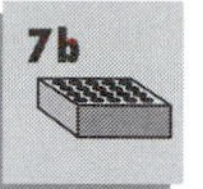 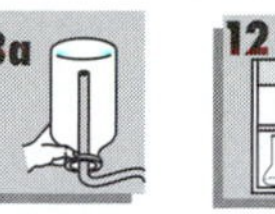 

Introduction

Acids, both inorganic and organic, are important laboratory and industrial chemicals primarily because, in aqueous solutions, they produce hydronium ions or because they are oxidizing agents. Sulfuric acid, also called **oil of vitriol,** is perhaps the most versatile of all inorganic industrial chemicals. Presently, so many industries around the world use sulfuric acid that it is often called the "old horse of chemistry." In 2002, when it ranked number one in chemical usage, over 36 million metric tons were produced in the United States alone (Table 13.1 and Figure 13.1). Chemically, a dilute solution of sulfuric acid acts as a strong acid, providing hydronium ion to an aqueous system, whereas the concentrated acid acts as a vigorous **oxidizing agent** and strong **dehydrating agent.**

Oxidizing agent: a substance that removes electrons from another substance

Dehydrating agent: a substance that absorbs water

Table 13.1 Top Ten Inorganic Chemicals Produced in the United States in 2002

1998 Rank	Chemical	Thousands of Metric Tons
1	Sulfuric acid	36,567
2	Ammonia	13,171
3	Chlorine	11,362
4	Phosphoric acid	10,789
5	Sodium hydroxide	8,984
6	Nitric acid	6,752
7	Ammonium nitrate	6,655
8	Hydrochloric acid	4,033
9	Ammonium sulfate	2,567
10	Titanium dioxide	1,409

Courtesy of *Chemical and Engineering News*, July 7, 2003

Top 10 Inorganic Chemicals

Thousands of Metric Tons

Chemical

Year

Figure 13.2 Phosphoric acid is the additive that delivers the tart taste in many soft drinks.

Year	H_2SO_4	NH_3	Cl_2	H_3PO_4	NaOH	HNO_3	NH_4SO_4	HCl	$(NH_4)_2SO_4$	TiO_2
1992	40383	16257	10664	11633	11105	7379	7239	3274	2169	1136
	36134	15596	10956	10444	11307	7486	7510	3167	2206	1160
1994	40645	16207	11054	11602	11373	7904	7771	3405	2344	1252
	43100	15785	11242	11913	10347	8018	7700	3541	2401	1253
1996	43327	16256	11301	11981	10488	8349	7708	3733	2414	1226
	43472	16227	11720	11935	9953	8556	7804	4145	2451	1330
1998	44000	16757	11647	12599	11893	8421	8235	4226	2528	1323
	40594	15725	12111	12433	11971	8113	6920	4081	2357	1354
2000	39584	14339	12698	11320	10451	7898	7237	4278	2547	1403
	36999	11372	11513	10490	9660	6416	6276	3972	2318	1327
2002	36567	13171	11362	10789	8984	6752	6665	4033	2567	1409

Figure 13.1 Top Ten Inorganic Chemicals Produced in the United States, 1992–2002, *Chemical and Engineering News,* July 7, 2003

Figure 13.3 Common household acids.

Figure 13.4 Common household bases.

Other prominent inorganic acids are hydrochloric acid (also called **muriatic acid**), nitric acid, and phosphoric acid (Figure 13.2). This experiment shows that hydrochloric acid is a nonoxidizing acid, whereas nitric acid, like sulfuric acid, has excellent oxidizing properties.

Some common "organic" acids are acetic acid found in vinegar, citric acid found in citrus fruits, and ascorbic acid, the vitamin C acid (Figure 13.3). Other organic acids, such as formic acid, adipic acid, terephthalic acid, oxalic acid, and oleic acid, are commercially useful and important.

The more familiar bases are aqueous ammonia (found in household "sudsy" ammonia) and the soluble metal hydroxides: sodium hydroxide (found in oven cleaner and drain cleaner), called **caustic soda** or **lye;** calcium hydroxide, called **slaked lime;** magnesium hydroxide, called **milk of magnesia** (Figure 13.4); and potassium hydroxide, called **caustic potash.** Other bases used commercially are hydrated sodium carbonate, called soda ash and **washing soda,** used in detergent formulations, and calcium oxide, or **quicklime,** used in cement.

Appendix D lists the formulas and the common and chemical names of some familiar acids and bases.

Acidic Solutions

Acidic solutions have a sour or tart taste, cause a pricking sensation on the skin, and turn blue litmus (an acid-base indicator) red. Nearly all of the foods we eat and the drinks we drink are acidic, some more acidic than others. Aqueous solutions become acidic because of the presence of hydronium ions, H_3O^+.

Acidic aqueous solutions exist from the reactions of

- a nonmetallic hydride with water

$$HCl(g) + H_2O(l) \rightarrow \boldsymbol{H_3O^+}(aq) + Cl^-(aq) \qquad (13.1)$$

- a nonmetallic oxide with water

$$SO_3(g) + 2\,H_2O(l) \rightarrow \boldsymbol{H_3O^+}(aq) + HSO_4^-(aq) \qquad (13.2)$$
$$CO_2(g) + 2\,H_2O(l) \rightarrow \boldsymbol{H_3O^+}(aq) + HCO_3^-(aq) \qquad (13.3)$$

- a molecular species with water

$$CH_3COOH(aq) + H_2O(l) \rightleftharpoons \boldsymbol{H_3O^+}(aq) + CH_3CO_2^-(aq) \qquad (13.4)$$

- some cations of salts with water (discussed later)

In each case, hydronium ion is produced in the aqueous solution.

Basic Solutions

Basic solutions (such as the commercial antacids) have a bitter taste, are slippery to the touch, and turn red litmus blue. Aqueous solutions become basic because of the presence of hydroxide ions, OH^-. Most household cleaning agents, such as soaps, detergents, and oven cleaners, are (or form) basic solutions. The most common foodstuff that is basic is the wide collection of commercial products called *antacids* (see Experiment 26 for further information on antacids).

Basic aqueous solutions exist from the reactions of

- a metallic hydride with water

$$2\,LiH(s) + H_2O(l) \rightarrow 2\,Li^+(aq) + 2\,\boldsymbol{OH^-}(aq) \qquad (13.5)$$

- a soluble metallic oxide or hydroxide with water

$$NaOH(s) + H_2O(l) \rightarrow Na^+(aq) + \boldsymbol{OH^-}(aq) + H_2O(l) \qquad (13.6)$$
$$BaO(s) + H_2O(l) \rightarrow Ba^{2+}(aq) + 2\,\boldsymbol{OH^-}(aq) \qquad (13.7)$$

- a molecular species with water

$$NH_3(aq) + H_2O(l) \rightleftharpoons NH_4^+(aq) + \boldsymbol{OH^-}(aq) \qquad (13.8)$$

- some anions of salts with water (discussed later)

In each case hydroxide ion is a product of the reaction.

Other soluble carbonates, metallic oxides, and substituted ammonia molecules behave similarly.

Salts

When aqueous solutions of an acid and a base are mixed, a reaction occurs producing water and a salt as products. This is a **neutralization reaction.** For example a reaction mixture of stoichiometric amounts of sulfuric acid, H_2SO_4, and sodium hydroxide, NaOH, produce the salt sodium sulfate, Na_2SO_4, and water:

$$H_2SO_4(aq) + 2\,NaOH(aq) \rightarrow Na_2SO_4(aq) + H_2O(l) \qquad (13.9)$$

A **salt** therefore can be defined as *any* ionic compound that is a neutralization product of an acid–base reaction.

pH

The acidity or basicity of most aqueous solutions is frequently the result of *low* concentrations of hydronium ion or hydroxide ion. To express these low concentrations conveniently, Søren P. L. Sørensen, a Danish biochemist, introduced **pH,** the negative logarithm of the molar concentration of hydronium ion.

[ ] Brackets placed around an ion indicate the molar concentration of that ion

$$pH = -\log[H_3O^+] \qquad (13.10)$$

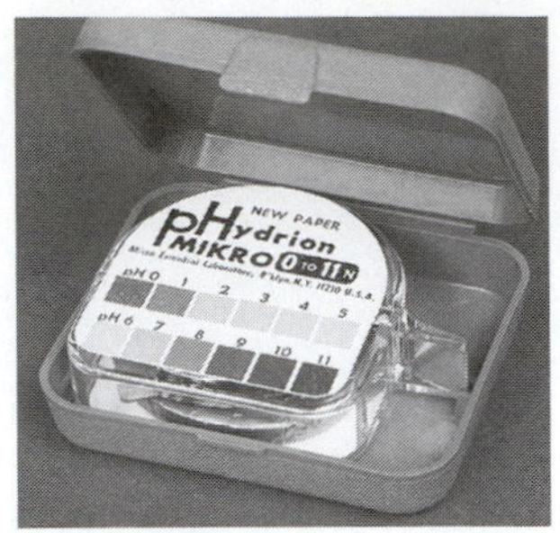

Figure 13.6 The pH of a solution can be estimated with pH paper, a strip of paper that is impregnated with a mixture of acid–base indicators. A gradation of colors on the pH paper approximates the pH of the solution.

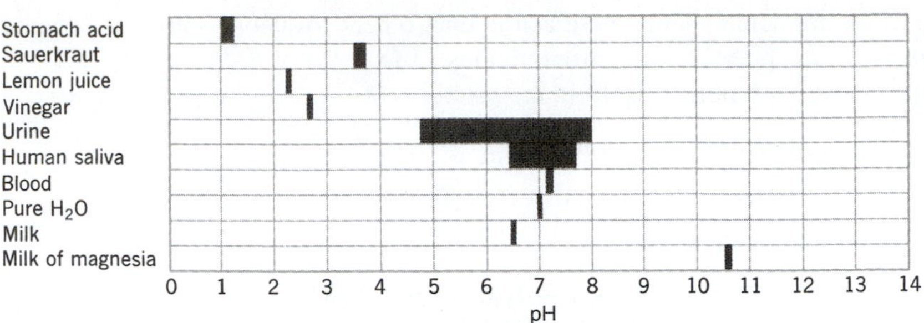

Figure 13.5 pH ranges of various common solutions.

Water dissociates very slightly, producing equal concentrations of hydronium ion and hydroxide ion.

$$2\,H_2O(l) \rightleftharpoons H_3O(aq) + OH^-(aq) \qquad (13.11)$$

$pH = -\log[1.0 \times 10^{-7}] = 7$

At 25°C, the molar concentration of hydronium ion, $[H_3O^+]$, equals the molar concentration of hydroxide ion, $[OH^-]$; both equal 1.0×10^{-7} mol/L. This results in a pH of 7. Addition of an acid increases the $[H_3O^+]$, causing a pH less than 7. A base decreases the $[H_3O^+]$ in solution, producing a pH greater than 7. The pH scale at 25°C can be summarized as follows:

pH | 1 | 2 | 3 | 4 | 5 | 6 **7** 8 | 9 | 10 | 11 | 12 | 13 | 14 |

acidic neutral basic

In this experiment, the pH of some acids and bases are measured. The pH ranges of some familiar solutions are shown in Figure 13.5.

Acid–base indicators are generally organic compounds whose color depends on the pH of the solution. Litmus is red in an acidic solution, but blue in a basic solution. Mixtures of acid–base indicators can be used to approximate the pH of a solution. Various pH test papers (Figure 13.6) and universal indicator are examples of mixed indicators.

Ions of Salts as Acids and Bases

$$\oplus \cdots {}^{\delta-}O(H^{\delta+})_2$$

a hydrated cation

$$\ominus \cdots (H^{\delta+})_2 O^{\delta-}$$

a hydrated anion

Salts dissociate in aqueous solution to form cations and anions. Because of ion–dipole interactions these ions are surrounded to varying degrees by polar water molecules (the ions are said to be **hydrated**). The $\delta-$ end of the water molecule is attracted to the cation of the salt in solution, and the $\delta+$ end of the water molecule is attracted to the anion of the salt.

Generally the larger the charge and the smaller the size of the ion (charge and size are collectively referred to as the **charge density** of the ion), the more strongly hydrated is the ion. For example, the Fe^{3+} and Al^{3+} cations are strongly hydrated because of their large charge and small ionic size; the larger cations Ba^{2+} and Cs^+ are very weakly hydrated.

A strongly hydrated cation, such as Fe^{3+}, weakens the O—H bond of a hydrated water molecule. This liberates a H^+ ion to bulk water molecules, producing the H_3O^+ ion and an acidic solution. The pH of a solution containing a cation with a large charge density is less than 7.

$$Fe^{3+}\cdots{}^{\delta-}O(H^{\delta+})_2 + H^{\delta+}\text{—}O^{\delta-}\text{—}H^{\delta+} \longrightarrow Fe^{3+}\text{—}{}^{\delta-}O(H^{\delta+})_2 \;\; H^{\delta+}\text{—}O^{\delta-}\text{—}H^{\delta+} \longrightarrow FeOH^{2+} + H_3O^+ \qquad (13.12)$$

A strongly hydrated anion, such as PO_4^{3-}, also weakens the O—H bond in a hydrated water molecule. But in this case H^+ ion attaches to the PO_4^{3-} ion and OH^- ion is liberated, producing a basic solution. The pH of a solution containing an anion with a large charge density is greater than 7.

$$\left[O_3P{-}O\right]^{3-}\cdots\overset{\delta+}{H}{-}\overset{\delta-}{O}{-}\overset{\delta+}{H} \longrightarrow \left[O_3P{-}O{-}H\cdots\overset{\delta-}{O}{-}\overset{\delta+}{H}\right]^{3-} \longrightarrow \left[O_3P{-}OH\right]^{2-} + OH^- \qquad (13.13)$$

Cations and anions having a very weak attraction for water molecules do not affect the pH of a solution. These ions typically have a low charge density and are called **spectator ions** with respect to pH effects. The common spectator ions are

Spectator ions (acid/base): cations or anions that have no affect on the pH of an aqueous solution

- cations: Group 1A (Na^+, K^+, Rb^+, Cs^+), Group 2A (Mg^{2+}, Ca^{2+}, Sr^{2+}, Ba^{2+}), and all other metal cations having an oxidation number of +1
- anions: the halides (Cl^-, Br^-, I^-), NO_3^-, ClO_4^-, and ClO_3^-.

Conversely this implies that *all* other metal cations and anions, exclusive of these spectator ions, affect the pH of a solution to some degree.

Experimental Procedure

Procedure Overview: The chemical properties of a range of acids and bases are observed. The pH values of selected acids and bases are estimated with pH test paper or universal indicator. Included is the practice of following laboratory safety rules in handling acids and bases.

Perform the experiment with a partner. At each circled superscript (1–19) in the procedure, *stop,* and record your observation on the Report Sheet. Discuss your observations with your lab partner and your instructor. **Caution:** *Dilute and concentrated (conc) acids and bases cause severe skin burns and irritation to mucous membranes. Be very careful in handling these chemicals. Clean up all spills immediately with excess water, followed by a covering of baking soda,* $NaHCO_3$. *Refer to the Laboratory Safety section at the beginning of this manual.*

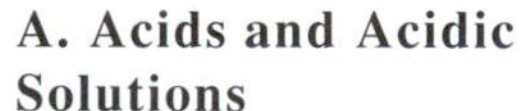

A. Acids and Acidic Solutions

1. **Action of Acids on Metals.** a. Place a small (~1 cm) polished (with steel wool) strip of Mg, Zn, and Cu into separate small test tubes. To each test tube add just enough 6 *M* HCl to submerge the metal and observe for several minutes. Record your observations on the Report Sheet.①

   b. Repeat the test of the three metals with 6 *M* $HNO_3$② and then again with 6 *M* CH_3COOH.③

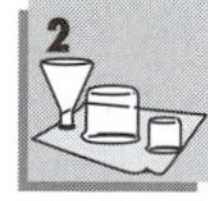

2. **Effect of Acid Concentration on Reaction Rate.** Set up six small clean test tubes containing about 1½ mL of the acid solutions shown in Figure 13.7. Add a small

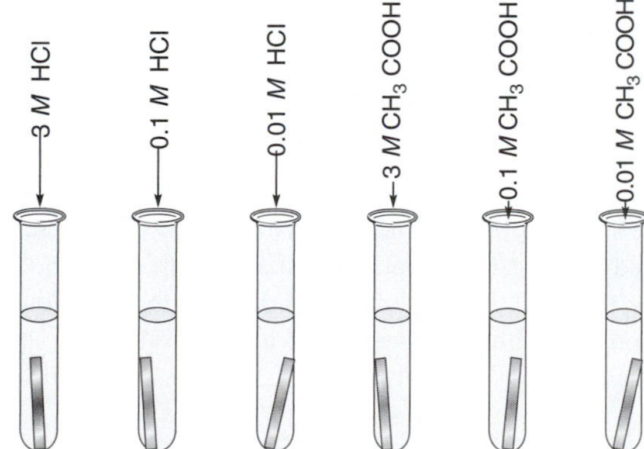

Figure 13.7 A setup for testing the effect of different acids and acid strengths on their reactivity with a metal.

Figure 13.8 Sugar only (left). Sugar after the addition of concentrated sulfuric acid (right).

(~1 cm) polished strip of magnesium to each solution and observe. Explain your observations. Account for any similarities and differences between HCl and CH_3COOH. What effect do differences in the acids *and* the concentrations of the acids have on the reaction rate?④⑤

3. **Oxidizing Strength of Acids.** The iodide ion is readily oxidized to iodine with a mild oxidizing agent. Place a **pinch** of NaI into two small test tubes. Add several drops of conc H_2SO_4 ***(Careful!)*** to one test tube; add several drops of conc H_3PO_4 to the other. What color changes, if any, occur?⑥ Hold moistened blue litmus paper over each test tube to test for any escaping gases. What are the products of these reactions?⑦

> *Disposal:* Discard the acid solutions in the "Waste Acids" container.

CLEANUP: Rinse all of the used test tubes twice with tap water and twice with deionized water.

4. **Dehydrating Effect of "Concentrated" Acids.** A demonstration is suggested; consult with your laboratory instructor. Do this test in the fume hood because of the choking gas that is evolved. (**Caution:** *Do not allow these concentrated acids to touch the skin.*) Test the dehydrating effects of conc H_2SO_4 by placing a few drops on a wood splint and on a small amount (≈1g) of sugar (Figure 13.8). Repeat the test with conc HCl. Record and explain your observations.⑧

B. Bases and Basic Solutions

1. **Reaction of Aqueous Sodium Hydroxide with Acid.** Place about 1 mL of 1 *M* NaOH in a small test tube. Test the solution with litmus. Add and count drops of 6 *M* HCl—after each drop, agitate the mixture and test the solution with litmus—until the litmus changes color. Record your data.⑨

Appendix D

Anhydrous salt: a salt having no hydrated water molecules in its solid structure

2. **Dissolution of Solid Bases.** a. *Slaking of Quicklime.* Place a small "BB-sized" sample of "fresh" CaO, NaOH, and **anhydrous** Na_2CO_3 in separate, clean small test tubes. Hold the test tube containing the CaO in your hand and add drops of water to cover the CaO sample (called **slaking**).[1] Record your observations and test the solution with litmus paper. Write a balanced equation for the slaking of quicklime.⑩

[1]A saturated solution of calcium hydroxide is called **limewater.**

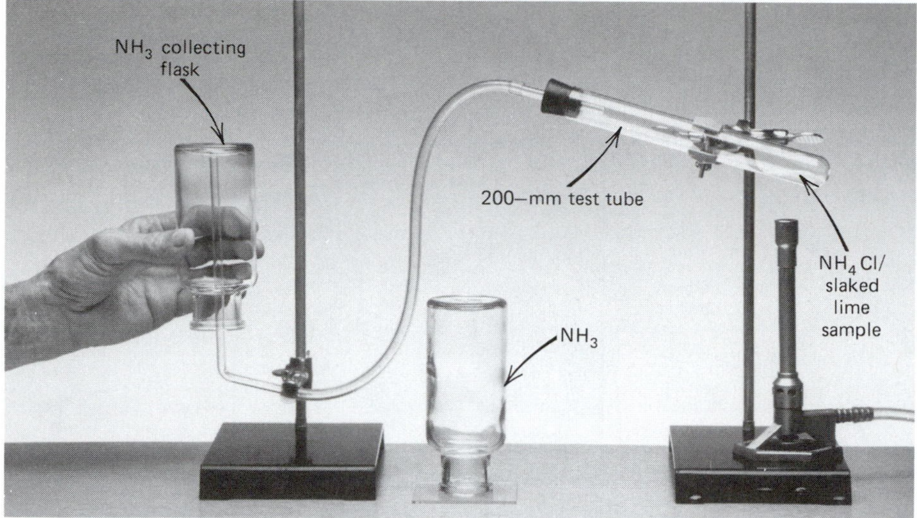

Figure 13.9 Apparatus for the generation and collection of ammonia gas by air displacement. A 200–m test tube can be substituted for the collecting flask.

b. *Other Bases with Water.* Repeat the addition of water to the test tubes containing the NaOH and the Na_2CO_3. Record your observations and test the solutions with litmus paper.⑪

Disposal: Discard the solutions from Parts B.1 and B.2 in the "Waste Bases" container.

3. **Production of Ammonia.**

a. *Set Up the Apparatus for the Collection of Ammonia.* Set up the apparatus shown in Figure 13.9. Place a 2 : 1 mixture (by volume of solid) of 4 g (±0.01 g) of ammonium chloride, NH_4Cl, and fresh slaked lime, $Ca(OH)_2$, into the 200-mm test tube. Stir the solid mixture.

b. *Generation and Collection of Ammonia.* Heat the mixture *gently;* begin near the mouth of the test tube and gradually extend the heating over the entire test tube. As soon as gas evolves freely and the apparatus is free of air, collect two to four bottles of ammonia gas by air displacement.[2] Cover each bottle with a glass plate or stopper each bottle with a rubber stopper. Record your data.⑫

c. *Test for the Flammability of Ammonia.* Note *carefully* the odor and color of the NH_3 gas. Test for its flammability by inserting a glowing wood splint into one of the gas bottles. Consult your instructor on this technique.⑬

d. *Test for the Solubility of Ammonia.* Place about 250 mL of water in a 600-mL beaker. Add several drops of phenolphthalein to the water and stir. Submerge the mouth of a bottle of ammonia beneath the water surface (Figure 13.11). Allow it to remain there for 5 minutes; be sure the mouth of the bottle remains submerged. Note the water level in the bottle and any color change of the water. Account for your observations.⑭

[2]One of two tests can be used to determine if the bottle is filled with ammonia: at the mouth of the ammonia-filled bottle either (1) place a strip of wet red litmus paper or (2) place a drop of conc HCl suspended on a glass rod (fumes of NH_4Cl will appear). See Figure 13.10. Consult with your laboratory instructor.

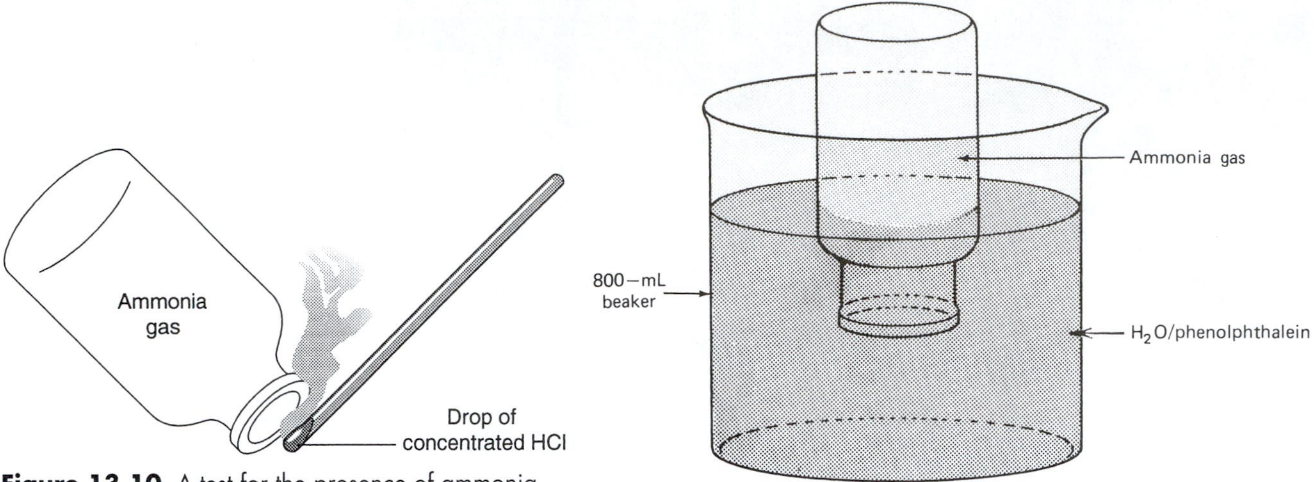

Figure 13.10 A test for the presence of ammonia gas.

Figure 13.11 Testing the solubility of ammonia gas in water.

C. pH Measurements

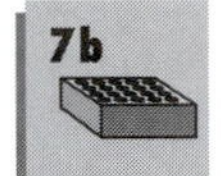

Boil 10 mL of deionized water for about 10 minutes to expel any dissolved gases, specifically carbon dioxide. Set the water aside to cool.

1. **pH of Water.** Clean a 24-well plate (small test tubes may also be used) with soap and water and rinse with deionized water. Half-fill wells A1–A3 with boiled (but cooled) deionized water, unboiled deionized water, and tap water, respectively. Add 1–2 drops of universal indicator to each. Account for any differences in the pH.[15]
2. **Estimate the pH of Acids and Bases.** Place 1 mL of each solution listed on the Report Sheet into wells B1–B6 and add 1–2 drops of universal indicator to each. Estimate the pH of each solution. Write an equation showing the origin of the free H_3O^+ or OH^- in each solution.[16]
3. **Estimate the pH of "Common" Solutions.** By the same method as in Part C.2, use wells C1–C6 to determine the pH of vinegar, lemon juice, household ammonia, detergent solution, or substitutes designated by your instructor.[17]

D. Ions of Salts as Acids and Bases

1. **Estimate the pH of Each Solution.** Transfer 1 mL of each solution listed on the Report Sheet into wells D1–D6 and add 1–2 drops of universal indicator to each. Estimate the pH of each solution as in Part C. Identify the ion that causes the acidity/basicity of the solution.[18]

Disposal and Cleanup: Discard the test solutions into the "Waste Salts" container. Rinse the 24-well plate twice with tap water and twice with deionized water.

E. Acids and Bases

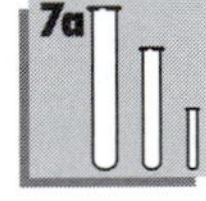

1. **Neutralization of Acids.** Into three test tubes, successively pipet 2 mL of 0.10 *M* HCl, 0.10 *M* H_2SO_4, and 0.10 *M* CH_3COOH. Add 1–2 drops of phenolphthalein indicator to each solution. Add (and count) drops of 0.5 *M* NaOH to each acid until a color change occurs (agitate the solution after each drop). Record the drops added. Compare the available hydronium ion of the three acids.[19]

Disposal and Cleanup: Discard the test solutions into the "Waste Salts" container. Rinse the test tubes twice with tap water and twice with deionized water.

Experiment 13 *Prelaboratory Assignment*

Acids, Bases, and Salts; pH

Date __________ Lab Sec. ______ Name ______________________________ Desk No. __________

1. In an aqueous solution,
   a. identify the "species" that makes a solution acidic.

   b. identify the "species" that makes a solution basic.

2. Aqueous salt solutions often are *not* neutral with respect to pH. Explain.

3. a. Milk of magnesia is used as a laxative and to treat upset stomachs. What is the formula of milk of magnesia?

   b. Washing soda is often added to detergent formulations to make the wash water more basic. What is the formula of the anhydrous form of washing soda? Does it increase or decrease the pH of the wash water? Explain.

   c. The composition of "ordinary" Portland cement is approximately 63.82% CaO, 20.56% SiO_2, 5.50% Al_2O_3, 2.99% Fe_2O_3, 0.76% K_2O, 0.32% Na_2O, 0.95% other, and 0.52% insolubles. When mixed with water the resulting cement slurry is basic. Identify three bases of Portland cement and the compounds they form when mixed with water.

4. Three solutions have the following pH:
   - Solution 1: pH 12.1
   - Solution 2: pH 6.2
   - Solution 3: pH 10.2

   a. Which solution contains the highest H_3O^+ ion concentration? ______________

   b. Which solution is the most acidic? ______________

   c. Which solution is the most basic? ______________

5. Experimental Procedure, Part B.3. Describe the technique for testing for the presence of $NH_3(g)$.

6. a. Which ion is more strongly hydrated in an aqueous solution, Co^{2+} or Co^{3+}? Explain.

   b. Which salt solution has a lower pH, 0.1 *M* $CoCl_2$ or 0.1 *M* $CoCl_3$?

7. When a solution resulting from a stoichiometric reaction mixture of an unknown acid, HA, and sodium hydroxide, NaOH, is tested with litmus paper, the litmus is blue.
   a. Why is the litmus blue?

   b. What *salt* (write its formula) is present from the reaction between the unknown acid and the sodium hydroxide? Write a balanced equation.

   c. Which ion of the salt is responsible for the observation of the litmus test?

Experiment 13 *Report Sheet*

Acids, Bases, and Salts; pH

Date __________ Lab Sec. ______ Name ______________________________ Desk No. __________

A. Acids and Acidic Solutions

1. **Action of Acids on Metals.** In the following table, state whether or not a reaction is observed. Indicate the relative reaction rate (i.e., fast, moderate, slow) of the metals with the corresponding acid.

	Mg	Zn	Cu
① 6 *M* HCl	______	______	______
② 6 *M* HNO_3	______	______	______
③ 6 *M* CH_3COOH	______	______	______

2. **Effect of Acid Concentration on Reaction Rate.** ④ Explain how changes in HCl concentration affect the reaction rate with Mg.

⑤ Explain how changes in CH_3COOH concentration affect the reaction rate with Mg.

Discuss the similarities and differences of HCl and CH_3COOH in the observed reactions with magnesium.

What effect do concentration changes have on the reaction rate?

3. **Oxidizing Strength of Acids.** ⑥ Describe and differentiate between your observations for the reaction of NaI with H_2SO_4 and with H_3PO_4.

⑦What are the products of the reactions?

Compare the relative oxidizing strengths of H_2SO_4 and H_3PO_4.

4. **Dehydrating Effect of "Concentrated" Acids.**⑧a. Why is the product of the reaction black?

   b. Compare the relative dehydrating effects of H_2SO_4 and HCl.

B. Bases and Basic Solutions

1. **Reaction of Aqueous Sodium Hydroxide with Acid.**⑨On the basis of your observations of the litmus test, describe the progress of the reaction of NaOH with HCl.

2. **Dissolution of Solid Bases.**⑩ a. *Slaking of Quicklime.* Describe the observed effect of H_2O on quicklime, CaO, and write a balanced equation for the reaction.

   Is limewater acidic or basic? Explain.

   b. ⑪*Other Bases with Water* Describe and distinguish your observations for the dissolution of NaOH and Na_2CO_3.

3. **Production of Ammonia.**⑫Write a balanced equation for the production of NH_3 from NH_4Cl and $Ca(OH)_2$.

   ⑬On the basis of your observed data, characterize NH_3 gas with respect to odor, color, and flammability.

   ⑭On the basis of your observed data, characterize NH_3 gas with respect to solubility and basicity.

C. pH Measurements

15 Water	Approximate pH	Briefly account for the pH if *not* equal to 7
Boiled, deionized		
Deionized		
Tap		

Solution	Approximate pH	Balanced Equation
16 0.10 *M* HCl		
0.00010 *M* HCl		
0.10 *M* CH_3COOH		
0.10 *M* NaOH		
0.00010 *M* NaOH		
0.10 *M* NH_3		
17 Vinegar		
Lemon juice		
Household ammonia		
Detergent solution		

D. Ions of Salts as Acids and Bases

Salt Solution	Approximate pH	Ion Affecting pH	Spectator Ion (Ions)
18 0.10 *M* NaCl			
0.10 *M* Na_2CO_3			
0.10 *M* Na_3PO_4			
0.10 *M* $NaHSO_4$			
0.10 *M* NH_4Cl			
0.10 *M* $KAl(SO_4)_2$			
0.10 *M* $AlCl_3$			

E. Acids and Bases

1. ⑲**Neutralization of Acids**

	0.10 *M* HCl	0.10 *M* H_2SO_4	0.10 *M* CH_3COOH
Drops of NaOH for color change	________	________	________

Discuss the relative acidity (availability of hydronium ion) of the three acids.

Laboratory Questions

Circle the questions that have been assigned

1. Part A.1. The observation for the reaction of 6 *M* HCl with copper metal was obviously different from that of 6 *M* HNO_3. What were the contrasting observations? How do the two acids differ? Explain.

2. Part A.2. As the molar concentration of an acid decreases, the reaction rate with an active metal, such as magnesium, is expected to __________. Explain.

3. Part A.3. Based upon your observations in Part A.1, would you expect HNO_3 to have properties more like those of H_2SO_4 or H_3PO_4? Explain.

*4. Part A.4. Concentrated sulfuric acid blackens wood and sugar because of its ability to extract hydrogen and oxygen atoms in a 2:1 mole ratio from compounds. The formula of table sugar is $C_{12}H_{22}O_{11}$. Write a balanced equation for the dehydration of table sugar, using concentrated sulfuric acid. (*Hint:* The concentrated sulfuric acid merely absorbs the water and becomes more dilute.)

5. Part B.1. 6 *M* H_2SO_4 is substituted for 6 *M* HCl. Will more or less drops of 6 *M* H_2SO_4 be required for the litmus to change color? Explain.

*6. Part B.2. When quicklime slakes in the atmosphere, a mixture of calcium carbonate, $CaCO_3$, and calcium hydroxide, $Ca(OH)_2$, forms because of the reaction of quicklime with carbon dioxide and water in the atmosphere. This is the time-consuming reaction that occurs when mortar sets. Write balanced equations that represent the setting of mortar.

7. Part B.3. Gaseous HCl is evolved from a conc HCl solution (an acid) and ammonia is a gas (a base). What is the chemical formula of the "fumes" that form in the ammonia test? Write a balanced equation for the reaction.

8. Part C.1. The unboiled deionized water has a measured pH less than seven. Explain.

9. Part E.1. In Part A.2 there were obvious differences in the chemical reactivity of 0.1 *M* HCl and 0.1 *M* CH_3COOH. However in Part E.1, there seemed to be no difference. Distinguish between the acidic properties of HCl and CH_3COOH.

Experiment 14

Oxidation–Reduction Reactions

Zinc metal displaces the blue copper(II) ion from solution (left). Copper metal collects on the zinc strip and the colorless zinc(II) ion goes into solution (right).

Objectives

- To observe and predict products of oxidation–reduction reactions
- To determine the relative reactivity of a series of metallic elements

Techniques

The following techniques are used in the Experimental Procedure

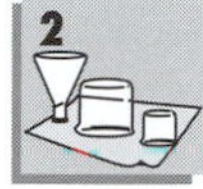

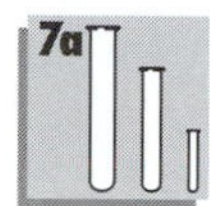

Introduction

Most chemical reactions are classified as being either acid–base (Experiment 13) or oxidation–reduction (redox). Reactions may also be classified as single displacement, double displacement (metathesis, Experiment 12), or decomposition.

Redox reactions are often accompanied by spectacular color changes, generally more so than what is observed in acid–base reactions. The changing of the color of the leaves each fall season, the rusting of our automobiles, the detonation of explosives, the formation of "brown" smog, and the commercial production of copper, aluminum, and iron are all examples of redox reactions.

In an acid–base reaction, protons (H^+) are transferred; in a redox reaction, electrons (e^-) are transferred from one substance to another, resulting in changes in oxidation numbers (see Dry Lab 2A) of two or more elements in the chemical reaction.

In a redox reaction the substance that experiences an increase in oxidation number is said to be **oxidized**—to do so it must have donated or lost electrons, a process of **oxidation.** Conversely, the substance that experiences a decrease in oxidation number is said to be **reduced**—as a result it must have accepted or gained electrons, a process of **reduction.**

Oxidation: a process whereby a substance loses electrons, but increases in oxidation number

Reduction: a process whereby a substance gains electrons, but decreases in oxidation number

In all redox reactions the substance that is oxidized must lose its electrons to the substance that is reduced (or gains the electrons). Therefore the substance being oxidized is causing a reduction and therefore called a **reducing agent.** Conversely, the substance that is reduced in the reaction must gain electrons from the substance that is oxidized, thereby causing oxidation and called an **oxidizing agent** (Figure 14.1).

Reducing agent: a substance that donates electrons causing reduction to occur (and is therefore oxidized)

Oxidizing agent: a substance that accepts electrons causing oxidation to occur (and is therefore reduced)

Electrons are never considered a reactant or product in a chemical reaction—the total number of electrons donated must equal the total number of electrons gained in a redox reaction.

For example, zinc metal reacts with copper(II) ion (see opening photo):

$$Zn(s) + Cu^{2+}(aq) \rightarrow Zn^{2+}(aq) + Cu(s) \quad (14.1)$$

In the reaction, zinc has increased its oxidation number from 0 to 2^+, by releasing 2 moles of electrons per mole of zinc; zinc has therefore been oxidized:

$$Zn(s) \rightarrow Zn^{2+}(aq) + 2\,e^- \quad \text{an } \textit{oxidation} \text{ half-reaction} \quad (14.2)$$

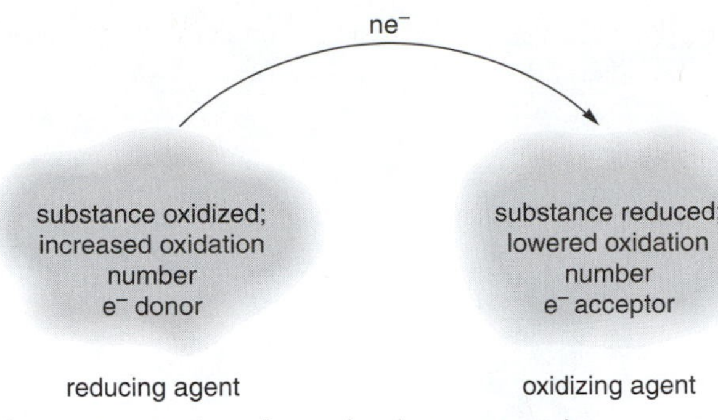

Figure 14.1 The relationship between oxidation, reduction, oxidizing agent, and reducing agent.

In addition, copper(II) ion has decreased in oxidation number from 2^+ to 0 by accepting 2 moles of electrons per mole of copper(II) ion; copper(II) ion has therefore been reduced:

$$Cu^{2+}(aq) + 2\,e^- \rightarrow Cu(s) \quad \text{a } \textit{reduction} \text{ half-reaction} \qquad (14.3)$$

An overview of the reaction suggests that the presence of zinc caused the reduction of copper(II) ion—*zinc is the reducing agent;* the presence of copper(II) ion caused the oxidation of zinc—*copper(II) ion is the oxidizing agent.* The "total" balanced redox reaction (Equation 14.1) is a sum of the two half-reactions, the oxidation half-reaction and the reduction half-reaction.

Reducing Agents and Oxidizing Agents

A common laboratory reducing agent (in addition to zinc metal) is the thiosulfate ion, $S_2O_3^{2-}$. Its half-reaction for oxidation in an aqueous solution is

$$2\,S_2O_3^{2-}(aq) \rightarrow S_4O_6^{2-}(aq) + 2\,e^- \qquad (14.4)$$

Reducing agents with their corresponding half-reactions used in this experiment are:

$Cu(s$, copper-colored$) \rightarrow Cu^{2+}(aq$, blue$) + 2\,e^-$

$3\,I^-(aq$, colorless$) \rightarrow I_3^-(aq$, purple$) + 2\,e^-$

$Fe^{2+}(aq$, light green to colorless$) \rightarrow Fe^{3+}(aq$, red-brown$) + e^-$

$C_2O_4^{2-}(aq$, colorless$) \rightarrow 2\,CO_2(g$, colorless$) + 2\,e^-$

A common laboratory oxidizing agent is the permanganate ion, MnO_4^-. Its half-reaction for reduction in an acidic solution is

$$MnO_4^-(aq) + 8\,H^+(aq) + 5\,e^- \rightarrow Mn^{2+}(aq) + 4\,H_2O(l) \qquad (14.5)$$

Oxidizing agents with their corresponding half-reactions used in this experiment are:

$NO_3^-(aq$, colorless$) + 2\,H^+(aq) + e^- \rightarrow NO_2(g$, red-brown$) + H_2O(l)$
$NO_3^-(aq$, colorless$) + 4\,H^+(aq) + 3\,e^- \rightarrow NO(g$, colorless$) + 2\,H_2O(l)$
$MnO_4^-(aq$, purple$) + 8\,H^+(aq) + 5\,e^- \rightarrow Mn^{2+}(aq$, light pink to colorless$) + 4\,H_2O(l)$
$O_2(g$, *colorless*$) + 4\,H^+(aq) + 4\,e^- \rightarrow 2\,H_2O(l)$
$H_2O_2(aq$, *colorless*$) + 2\,H^+(aq) + 2\,e^- \rightarrow 2\,H_2O(l)$
$ClO^-(aq$, colorless$) + 2H^+(aq) + 2\,e^- \rightarrow Cl^-(aq$, colorless$) + H_2O(l)$

In Part A, several redox reactions involving these reducing and oxidizing agents are observed and studied, reactants and products are recorded. You may have to consult with your instructor when writing the balanced redox equation for the reactions that are studied.

Displacement Reactions (Activity Series)

Activity series: a listing of (generally) metals in order of decreasing chemical reactivity

Parts B and C of the experiment seek to establish the relative chemical reactivity of several metals and hydrogen. The metals and hydrogen, listed in order of decreasing activity (decreasing tendency to react), constitute an abbreviated **activity series.**

The result of one metal being placed into a solution containing the cation of another metal establishes the relative reactivity of the metals. A displacement reaction occurs if there is evidence of a chemical change. For example, if metal A is more reactive (has a greater tendency to lose electrons to form its cation) than metal B, then metal A displaces B^{n+} from an aqueous solution. Metal A is oxidized to A^{m+} in solution and B^{n+} is reduced to metal B.

$$A(s) + B^{n+}(aq) \rightarrow A^{m+}(aq) + B(s) \quad (14.6)$$

A specific example is the relative reactivity of iron versus lead: experimentally, when iron metal is placed in a solution containing lead(II) ion, iron is oxidized (loses two electrons) to form the iron(II) ion and the lead(II) ion is reduced (gains two electrons) to form lead metal (Figure 14.2). The equation for the reaction is

$$Fe(s) + Pb^{2+}(aq) \rightarrow Fe^{2+}(aq) + Pb(s) \quad (14.7)$$

On the other hand, when lead metal is placed into a solution containing iron(II) ion, *no reaction* occurs:

$$Pb(s) + Fe^{2+}(aq) \rightarrow \text{no reaction} \quad (14.8)$$

Consequently, iron prefers to be ionic, whereas lead prefers the metallic state—iron loses electrons more easily than lead. Iron, therefore, is more easily oxidized than lead and is said to have a greater **activity** (chemical reactivity) than lead.

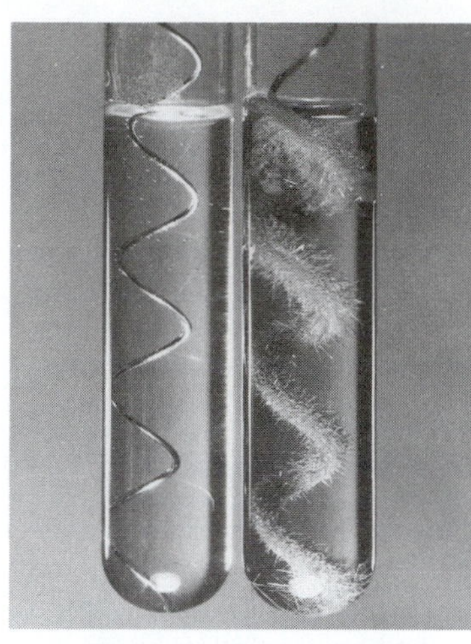

Figure 14.2 Lead metal does *not* react in a solution containing iron(II) ion (left). Iron metal does react in a solution containing lead(II) ion (right).

Experimental Procedure

Procedure Overview: Observations of a number of redox reactions are analyzed and equations are written. The relative reactivity of several metals is determined from a series of oxidation–reduction/displacement reactions.

Perform the experiment with a partner. At each circled superscript (1–7) in the procedure, *stop*, and record your observation on the Report Sheet. Discuss your observations with your lab partner and your instructor.

A. Oxidation–Reduction Reactions

1. **Oxidation of Magnesium.** Half-fill a 50-mL beaker with deionized water and heat to near boiling—test the water with litmus. Grip a 2-cm piece of magnesium ribbon with crucible tongs. Heat it directly in a Bunsen burner flame until it ignites. (**Caution:** *Do not look directly at the flame.*) The white-ash product is magnesium oxide. Allow the ash to drop into the 50-mL beaker. Swirl and again test the solution with litmus at the end of the laboratory period. Compare your observations for the two litmus tests. Record your observations on the Report Sheet. (1)

2. **Oxidation of Copper.** In three separate small test tubes add approximately 1 mL of 6 *M* HCl, 1 mL of 6 *M* HNO_3, and 1 mL of conc HNO_3. (**Caution:** *Avoid skin contact. Flush affected areas with large amounts of water.*) Place the three test tubes in a fume hood and add a 1-cm wire strip of copper metal to each. Record your observations on the Report Sheet. (2)

3. **A Series of Redox Reactions.** Refer to Table 14.1. Clean and label eight small test tubes. Place about 1 mL of each solution listed as Solution 1 in Table 14.1 into the test tubes. (**Caution:** *Avoid skin contact with all solutions!*) Slowly add up to 1 mL of Solution 2 until a permanent change is observed. For test solutions 5 and 6, after the addition of solution 2, stopper the test tube and shake vigorously. Some reactions may be slow to develop. Record your observations. (3)

Disposal: Dispose of the test solutions in the "Waste Salts" container.

CLEANUP: Rinse the test tubes plate twice with tap water and twice with deionized water. Discard the rinses in the "Waste Salts" container.

Figure 14.3 The deep purple permanganate ion, added to an acidic iron(II) solution, is reduced to the nearly colorless manganese(II) ion.

Table 14.1. Preparation of Several Oxidation–Reduction Reactions

Test Tube No.	Solution 1	Addition of Solution 2
1	Chlorine bleach and 2 drops of 6 *M* HCl (the active ingredient in chlorine bleach is ClO^-)	Drops of 0.1 *M* $Fe(NH_4)_2SO_4$
2	Chlorine bleach and 2 drops of 6 *M* HCl	*Add* 1 drop of starch solution followed by drops of 0.1 *M* KI
3	0.01 *M* $KMnO_4$ and 2 drops of 6 *M* H_2SO_4	Drops of 0.1 *M* $Fe(NH_4)_2SO_4$ (Figure 14.3)
4	0.01 *M* $KMnO_4$ and 2 drops of 6 *M* H_2SO_4	Drops of 1 *M* $K_2C_2O_4$
5	Deionized water and 2 drops of 6 *M* H_2SO_4	Drops of 0.1 *M* $Fe(NH_4)_2SO_4$
6	Deionized water and 2 drops of 6 *M* H_2SO_4	Add 1 drop of starch solution followed by drops of 0.1 *M* KI
7	0.1 *M* H_2O_2 and 2 drops of 6 *M* H_2SO_4	Drops of 0.1 *M* $Fe(NH_4)_2SO_4$
8	0.1 *M* H_2O_2 and 2 drops of 6 *M* H_2SO_4	Add 1 drop of starch solution followed by drops of 0.1 *M* KI

B. Reactions with Hydronium Ion

1. **Reactivity of Ni, Cu, Zn, Fe, Al, and Mg with H_3O^+.** Obtain 1-cm strips of Ni, Cu, Zn, Fe, Al, and Mg. Place about 1.5 mL of 6 *M* HCl into each of six small test tubes (Figure 14.4). Polish each metal (with steel wool) to remove any oxide coating and immediately place it in a test tube of hydrochloric acid.[1] Allow 10–15 minutes for any reactions to occur. Record what you see—*look closely!*④

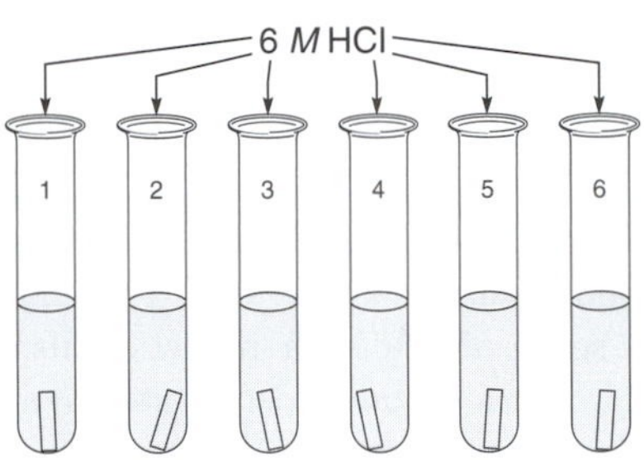

Figure 14.4 Setup for observing the reactivity of six metals in hydrochloric acid.

C. Displacement Reactions between Metals and Metal Cations

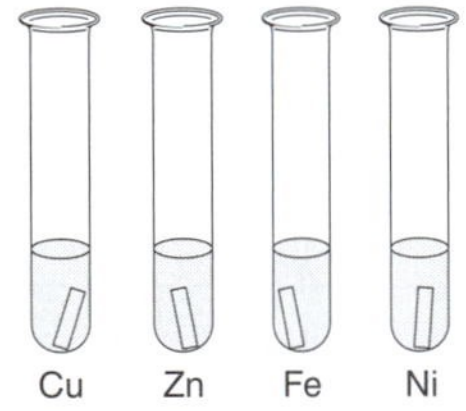

Figure 14.5 Setup for the observing the relative reactivity of metal and metal ions.

1. **Reactivity of Cu, Zn, and Fe with Ni^{2+}.** Obtain 1-cm strips of Cu, Zn, Fe, and Ni. Place about 1 mL of 0.1 *M* $NiSO_4$ in each of three test tubes. Place a short strip of freshly polished Cu, Zn, and Fe in successive test tubes (Figure 14.5). Observation of a tarnishing or dulling of the metal or color change of the solution indicates that a reaction has occurred. Allow 5–10 minutes to observe the reaction. Record.⑤
2. **Reactivity of Cu, Zn, Fe, and Ni with Other Cations.** Test the reactivity of the metals in the following 0.1 *M* test solutions: $Cu(NO_3)_2$, $Zn(NO_3)_2$, and $Fe(NH_4)_2(SO_4)_2$. This series of tests will fill the table on the Report Sheet.⑥ The metal strip may be reused if it is unreacted after the previous test and is rinsed with deionized water or if it is again freshly polished. Record your observations.
3. **Relative Activity of the Metals.** List the four metals in order of decreasing activity.⑦

Disposal: Dispose of the waste test solutions and one rinse of the test tubes in the "Waste Salts" container. Place the unreacted metals in the "Waste Solids" container.

CLEANUP: Rinse the test tubes twice with tap water and twice with deionized water. Discard the rinses in the sink.

[1]The polishing of the metal and quick placement in HCl are especially critical for aluminum and magnesium as they form quick, tough protective oxide coatings.

Experiment 14 *Prelaboratory Assignment*

Oxidation–Reduction Reactions

Date __________ Lab Sec. ______ Name ______________________________ Desk No. __________

1. a. Oxygen is a common oxidizing agent in nature. What changes in the oxidation number of oxygen must occur if it is to be an oxidizing agent? Explain.

   b. If oxygen gas were to oxidize copper metal, what change in oxidation number must occur of the copper metal?

2. Zinc metal is a common reducing agent in analytical chemistry. What does it mean for a substance to be a reducing agent?

3. Each of the following processes is a likely change in a redox reaction. Label the chemical change as an oxidation process, a reduction process, or neither:

   a. $ClO_2^- \rightarrow ClO_3^-$ ______________________

   b. $Co^{3+} \rightarrow Co^{2+}$ ______________________

   c. $Cr_2O_7^{2-} \rightarrow Cr^{3+}$ ______________________

   d. $SO_3 \rightarrow SO_4^{2-}$ ______________________

   e. $H^+ \rightarrow H_2$ ______________________

4. Based upon your experiences, would you predict a reaction to occur when
   a. zinc metal is placed into a solution containing silver ion, Ag^+? Explain.

   b. gold metal is placed into a solution containing zinc ion, Zn^{2+}? Explain.

5. List the generic chemicals A, E, F, and G in order of decreasing activity on the basis of the following reactions:

   $A + E^+ \rightarrow A + E$

   $G + A^+ \rightarrow$ no reaction

   $F + A^+ \rightarrow F^+ + A$

   $E + G^+ \rightarrow$ no reaction

   most active _____, _____, _____, _____, least active

*6. Zinc metal is used as a protective covering on *galvanized* iron. Explain its function in terms of chemical activity.

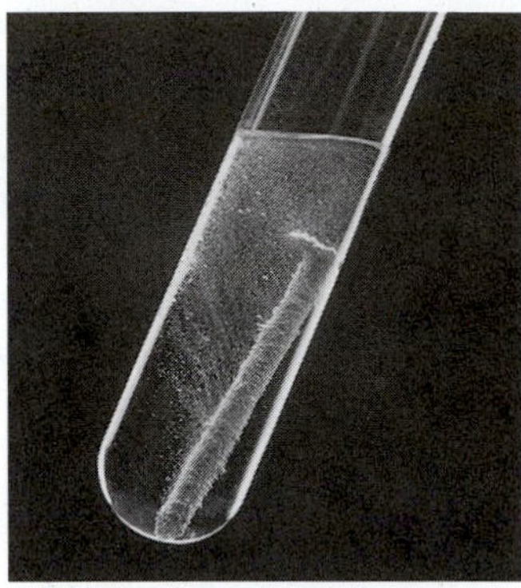

The zinc layer of the galvanized iron nail reacts with sulfuric acid.

Experiment 14 *Report Sheet*

Oxidation–Reduction Reactions

Date __________ Lab Sec. ______ Name ____ Desk No. __________

A. Oxidation–Reduction Reactions

1. ①**Oxidation of Magnesium.** Write a description of the reaction. What did the litmus tests reveal?

   Write a balanced equation for the oxidation of magnesium in air. [Box] the oxidizing agent in the equation.

2. ②**Oxidation of Copper.** Describe your observations of each test tube.

   Comment on the relative oxidizing strengths of the three acids in the test tubes.

3. ③**A Series of Redox Reactions.** On a separate sheet of paper, organize your data to record the test tube number and your observations for each mixture that shows a reaction. Write a balanced redox equation for each observed reaction: [Box] the oxidizing agent in each written equation.

B. Reactions with Hydronium Ion

1. ④**Reactivity of Ni, Cu, Zn, Fe, Al, and Mg with H_3O^+**

   Which metals show a definite reaction with HCl?

   Record this information on the table in Part C of the Report Sheet.

   Arrange the metals that *do react* in order of decreasing activity.

   Write a balanced equation for the reaction that occurs between Mg and H_3O^+.

C. Displacement Reactions between Metals and Metal Cations

Complete this table with NR (no reaction) or R (reaction) where appropriate. For all observed reactions, write a balanced *net ionic* equation. Use additional paper if necessary. Box the oxidizing agent in each written equation.

	Ni	Cu	Zn	Fe	Al	Mg
HCl	______	______	______	______	______	______
⑤$NiSO_4$	NR	______	______	______		
⑥$Cu(NO_3)_2$	______	NR	______	______		
$Zn(NO_3)_2$	______	______	NR	______		
$Fe(NH_4)_2(SO_4)_2$	______	______	______	NR		

⑦The four metals along with Al, Mg, and hydrogen in order of decreasing activity are

______, ______, ______, ______, ______, ______, ______.

Laboratory Questions

Circle the questions that have been assigned.

1. Part A.1. Sodium metal is readily oxidized by oxygen gas. If the product of the reaction were dissolved in water, what would be the color of the litmus for a litmus test? Explain. What is the product?
2. Part A.2. Oxygen gas has an oxidizing strength comparable to that of nitric acid. Patina is a green or greenish-blue coating that forms on copper metal in the environment. Account for its formation.
3. Part A.3. The presence of starch enhances the appearance of the formation of the I_3^- ion in an aqueous solution. What color would appear in the solution if the starch were omitted?
4. Part A.3. Does the iron in the $Fe(NH_4)_2SO_4$ solutions function as an oxidizing agent or a reducing agent? Explain.
5. Part B.1. 6 *M* HCl was not to be found for the experiment, the available 6 *M* HNO_3 was used instead. Explain how the results of the experiment might have been different.
6. Part C. Single displacement, double displacement, and decomposition reactions may all be redox reactions. Identify the type of redox reactions in Part C. Explain.
7. Part C. a. On the basis of your intuitive understanding of the chemical properties of sodium and gold, *where* in the activity series would you place sodium and gold?

   b. Will hydrochloric acid react with gold metal to produce gold(III) ions and hydrogen gas? Explain.

*8. Part C. Magnesium metal is used as a sacrificial anode to reduce the corrosion of underground storage tanks (made of iron, or more accurately, steel which is an iron/carbon alloy). As a result the sacrificial metal corrodes instead of the iron. Explain the "chemistry" of the process.

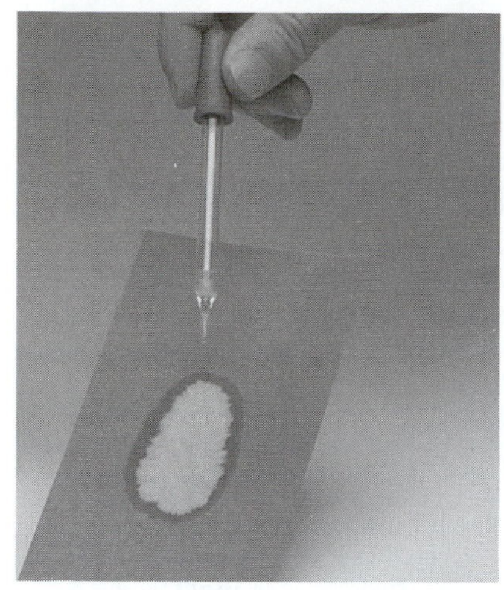

Sodium hypochlorite, the bleaching agent of bleach, oxidizes the dye and "whitens" the fabric.

Experiment 15

Bleach Analysis

Objectives

- To determine the "available chlorine" in commercial bleaching agents
- To prepare a standardized sodium thiosulfate solution

Techniques

The following techniques are used in the Experimental Procedure

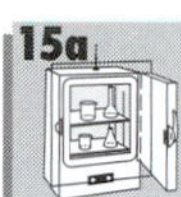

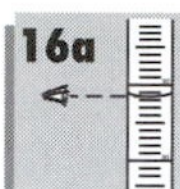

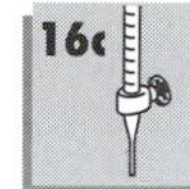

Introduction

Natural fibers, such as cotton and wool, and paper products tend to retain some of their original color during the manufacturing process. For example, brown paper is more "natural" than white paper, and original cotton fibers tend *not* to be "white."

Natural dyes and, in general, all of nature's colors exist because of the presence of molecules with multiple bonds. Similarly, stains and synthetic colors are due to molecules with one or more double and/or triple bonds. The electrons of the multiple bond readily absorb visible radiation, reflecting (or transmitting) the colors of the visible radiation that are *not* absorbed.[1] These electrons of the double/triple bonds are readily removed by oxidizing agents. Because the oxidized form[2] of the molecules causing the colors and stains no longer absorb visible radiation, the fabric or paper appears whiter.

All of the halogens and solutions of the oxyhalides are excellent oxidizing agents. Chlorine, as an **oxidant,** is an excellent bactericide for use in swimming pools, drinking water, and sewage treatment plants.

Oxidant: oxidizing agent, a substance that causes the oxidation of another substance

Most commercial **bleaching agents** contain the hypochlorite ion, ClO^-. The ClO^- ion removes light-absorbing electrons from the multiple bonds of molecules. Therefore, the absorption of visible radiation does not occur; all visible light is reflected (or transmitted), giving the object a "white" appearance.

Bleaching agent: oxidizing agent

$$ClO^-(aq) + H_2O(l) + 2\,e^- \rightarrow Cl^-(aq) + 2\,OH^-(aq) \qquad (15.1)$$

The textile and paper industries, as well as home and commercial laundries, use large quantities of hypochlorite solutions to oxidize the color- and stain-causing molecules. The hypochlorite ion is generally in the form of a sodium or a calcium salt in the bleaching agent.

[1] In Dry Lab 3, the explanation of light absorption in the visible region is discussed in terms of an absorption of energy (from the visible region) by electrons.

[2] The stains or impurities may be oxidized to form several products.

Strengths of Bleaching Agents

Mass ratio: mass of Cl_2/mass of sample (solution or solid)

Industry expresses the strength of the bleaching agents as the mass of molecular *chlorine,* Cl_2, per unit mass of solution (or mass of powder) regardless of the actual chemical form of the oxidizing agent. This **mass ratio,** expressed as a percent, is called the percent "available chlorine" in the bleach solution.

As an example, "ultra" liquid laundry bleach is generally a 6.00% NaClO aqueous solution by mass. Liquid laundry bleach is manufactured by the electrolysis of a cold, stirred, concentrated sodium chloride solution. The chlorine gas and the hydroxide ion that are generated at the electrodes in the electrolysis cell combine to form the hypochlorite ion, ClO^-, and in the presence of Na^+, a sodium hypochlorite solution.

$$Cl_2(aq) + 2\,OH^-(aq) \rightarrow ClO^-(aq) + Cl^-(aq) + H_2O(l) \qquad (15.2)$$

The percent "available chlorine" in a 6.00% NaClO solution is calculated as

$$\frac{\text{grams } Cl_2}{\text{grams soln}} \times 100 = \frac{6.00 \text{ g NaClO}}{100 \text{ g soln}} \times \frac{\text{mol NaClO}}{74.44 \text{ g NaClO}} \times \frac{1 \text{ mol } Cl_2}{1 \text{ mol NaClO}} \cdot$$
$$\times \frac{70.90 \text{ g } Cl_2}{\text{mol } Cl_2} \times 100 = 5.71\% \; Cl_2$$

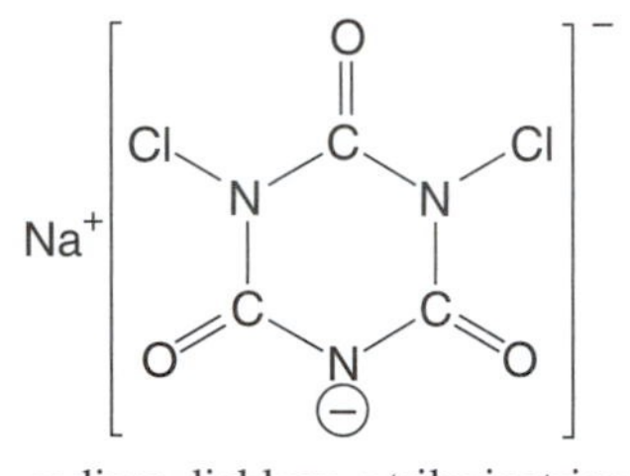

sodium dichloro-*s*-trilazinetrione

Solid bleaching powders and water disinfectants (e.g., for home swimming pools or hot tubs) contain a variety of compounds that serve as oxidizing agents. Solids such as $Ca(ClO)Cl/Ca(ClO)_2$ mixtures and sodium dichloro-*s*-triazinetrione, commonly found in household cleansers and water disinfectants, are substances that release ClO^- in solution. Solid household bleaches often contain sodium perborate, $NaBO_2{\bullet}H_2O_2{\bullet}H_2O$, in which the bleaching agent is hydrogen peroxide, a compound that is also found in hair bleaches.

Analysis of Bleach with Potassium Iodide and Sodium Thiosulfate Solutions

The oxidizing property of the hypochlorite ion is used to perform an analysis for the amount of "available chlorine" in a bleach sample for this experiment. In the analysis, the hypochlorite ion oxidizes iodide ion to molecular iodine—the hypochlorite ion is the limiting reactant in the analysis. In the presence of the excess amount of added iodide ion, the molecular iodine actually exists as a water-soluble complex, I_3^- (or $I_2 \bullet I^-$).

$$ClO^-(aq) + 2\,I^-(aq) + H_2O(l) \rightarrow I_2(aq) \text{ [or } I_3^-\text{] } Cl^-(aq) + 2\,OH^-(aq) \quad (15.3)$$

The amount of molecular iodine produced in the reaction is determined by its reaction with a standard sodium thiosulfate, $Na_2S_2O_3$, solution.

$$I_2(aq) + 2\,S_2O_3^{2-}(aq) \rightarrow 2\,I^-(aq) + S_4O_6^{2-}(aq) \qquad (15.4)$$

Stoichiometric point: point at which the mole ratio of reactants in a titration is the same as the mole ratio in the balanced equation

The **stoichiometric point** for this redox titration is detected by adding a small amount of starch. Starch forms a rather strong bond to molecular iodine; whereas molecular iodine has a yellow-brown color, the I_2•starch complex is a deep, and very intense, blue. Just prior to the disappearance of the yellow-brown color of I_2 in the titration, the starch is added to intensify the presence of the molecular iodine still to be titrated.

$$I_2(aq) + \text{starch}(aq) \rightarrow I_2{\bullet}\text{starch}(aq, \text{deep blue}) \qquad (15.5)$$

Titrant: the solution being added from the buret

The addition of the sodium thiosulfate **titrant** is then resumed until the blue color of the $I_2 \bullet$ starch complex disappears.

$$I_2 \bullet \text{starch}(aq) + 2\,S_2O_3^{2-}(aq) \rightarrow 2\,I^-(aq) + S_4O_6^{2-}(aq) + \text{starch}(aq, \text{colorless}) \quad (15.6)$$

Figure 15.1 I_3^- without starch added (left); I_3^- with starch added (right).

From the stoichiometries of Equations 15.4 (2 mol $S_2O_3^{2-}$ to 1 mol I_2), 15.3 (1 mol I_2 from 1 mol ClO^-), and 15.2 (1 mol ClO^- from 1 mol Cl_2) in that sequence, the moles of "available chlorine" in the sample can be calculated. From the molar mass of chlorine and the mass of the sample used for the analysis, the percent "available chlorine" is calculated.

A Standard Solution of Sodium Thiosulfate, $Na_2S_2O_3$

The concentration of a standard $Na_2S_2O_3$ solution (used in Equation 15.4)[3] is determined by its reaction with solid potassium iodate, KIO_3, a **primary standard** (Figure 15.2). A measured mass of potassium iodate is dissolved in an acidic solution containing an excess of potassium iodide, KI. Potassium iodate, a strong oxidizing agent, oxidizes I^- to a water-soluble, yellow-brown $I_2(aq)$ in solution:

$$IO_3^-(aq) + 5\,I^-(aq) + 6\,H^+(aq) \rightarrow 3\,I_2(aq) + 3\,H_2O(l) \qquad (15.7)$$

The molecular iodine is then titrated with a prepared sodium thiosulfate solution, where the I_2 is reduced to I^- and the thiosulfate ion, $S_2O_3^{2-}$, is oxidized to the tetrathionate ion, $S_4O_6^{2-}$ (Equation 15.4).

The stoichiometric point is detected using a starch solution. Just prior to the disappearance of the yellow-brown iodine in solution, starch is added to form a deep blue complex with iodine (Equation 15.5).

The addition of the thiosulfate solution is then continued until all of the I_2 is converted to I^- and the blue color disappears (Equation 15.6).

Note that the reaction of IO_3^- with I^- (Equation 15.7) for the standardization of the sodium thiosulfate solution substitutes for the reaction of ClO^- with I^- (Equation 15.3) in the analysis of the bleaching agent.

Primary standard: a substance that has a known high degree of purity, a relatively large molar mass, is nonhygroscopic, and reacts in a predictable way

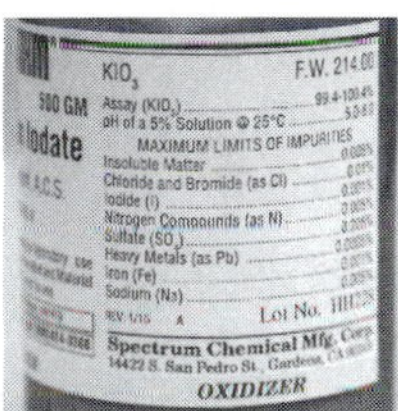

Figure 15.2 Potassium iodate has the necessary properties of a primary standard.

Experimental Procedure

Procedure Overview: A solution of $Na_2S_2O_3$ is prepared and standardized using solid KIO_3 as a primary standard. A bleach sample is mixed with an excess of iodide ion, generating iodine. The amount of iodine generated is analyzed by titration with the standard $Na_2S_2O_3$ solution.

Quantitative, reproducible data are the objective of this experiment. Practice good laboratory techniques. Three trials are necessary. Ask your instructor about the completion of both Parts C and D in this experiment. Nearly 800 mL of boiled, deionized water is required for this experiment. Begin preparing this at the beginning of the laboratory period.

[3]The $Na_2S_2O_3$ solution is a secondary standard solution in that its concentration can only be determined from an analysis; it cannot be prepared directly from solid $Na_2S_2O_3$.

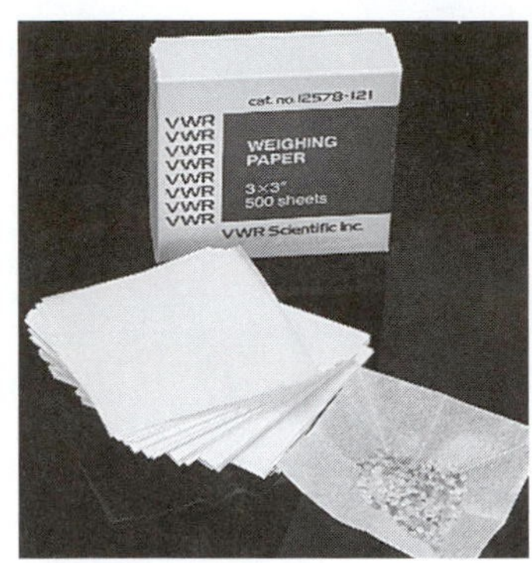

Figure 15.3 Weighing paper for measuring the KIO_3 primary standard.

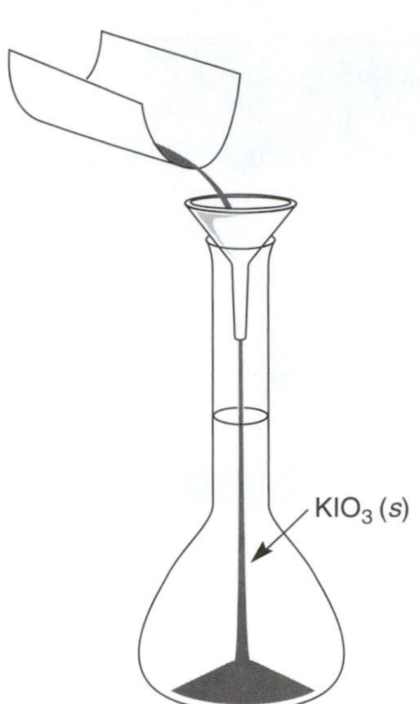

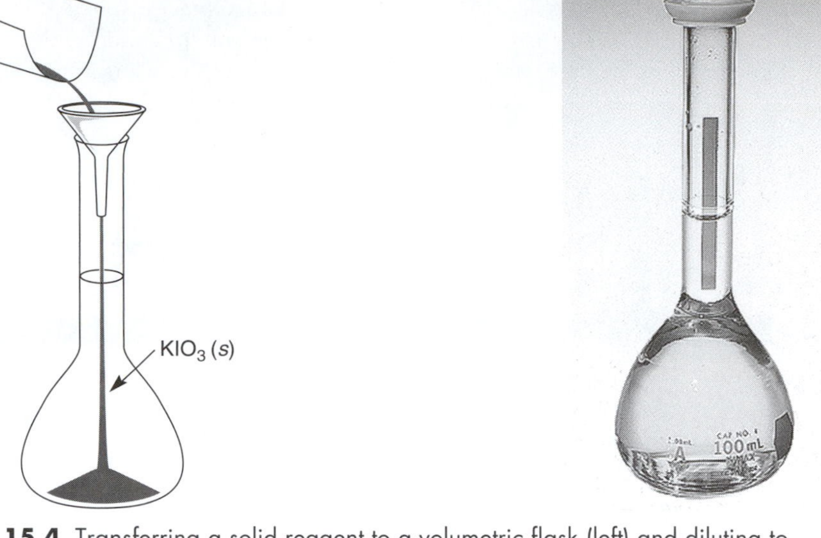

Figure 15.4 Transferring a solid reagent to a volumetric flask (left) and diluting to volume (right).

A. Preparation of the KIO_3 Primary Standard

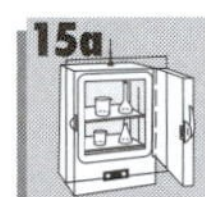

1. **Prepare a Primary Standard Solution.** Use weighing paper (Figure 15.3) to measure the tared mass (±0.001 g) of KIO_3 (previously dried[4] at 110°C) that is required to prepare 100 mL of 0.01 *M* KIO_3. *Show this calculation on the Report Sheet.* Dissolve the solid KIO_3 and dilute to volume with freshly boiled, deionized water in a 100-mL volumetric flask (Figure 15.4). Calculate and record its exact molar concentration.

B. A Standard 0.1 *M* $Na_2S_2O_3$ Solution

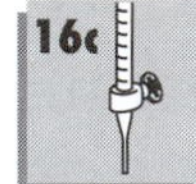

1. **Prepare a $Na_2S_2O_3$ Solution.**[5] This solution should be prepared no more than 1 week in advance because unavoidable decomposition occurs. Measure the mass (±0.01 g) of $Na_2S_2O_3{\bullet}5H_2O$ needed to prepare 250 mL of a 0.1 *M* $Na_2S_2O_3$ solution. *Show this calculation on the Report Sheet.* Dissolve the solid $Na_2S_2O_3{\bullet}5H_2O$ with freshly boiled, deionized water and dilute to 250 mL; an Erlenmeyer flask can be used to prepare and store the solution. Agitate until the salt dissolves.
2. **Prepare a Buret.** Prepare a *clean,* 50-mL buret for titration. Fill the buret with the $Na_2S_2O_3$ solution, drain the tip of air bubbles, and after 10–15 seconds, record the volume "using all certain digits (from the labeled calibration marks on the glassware) *plus* one uncertain digit (the last digit which is the best estimate between the calibration marks)."
3. **Titrate the KIO_3 Solution.** Pipet 25 mL of the standard KIO_3 solution into a 125-mL Erlenmeyer flask, add about 2 g (±0.01 g) of solid KI, and swirl to dissolve. Add about 10 mL of 0.5 *M* H_2SO_4. Begin titrating the KIO_3 solution immediately. When the yellow iodine nearly disappears (pale yellow color),[6] add 2 mL of starch solution.[7] Stirring constantly, continue titrating *slowly*[8] until the blue color completely (and suddenly) disappears.

---

[4]The KIO_3 should have been dried prior to your coming to the laboratory.
[5]Ask your instructor about this solution; it may have already been prepared.
[6]Consult your instructor at this point.
[7]If the deep blue color of $I_2{\bullet}$starch does *not* appear, you have passed the stoichiometric point in the titration and you will need to discard the sample and prepare another.
[8]The blue $I_2{\bullet}$starch complex is slow to dissociate.

4. **Repeat the Titration with Another KIO_3 Sample.** Repeat twice the procedure in Part B.3 by rapidly adding the $Na_2S_2O_3$ titrant until about 1 mL before the yellow-brown color of iodine disappears. Add the starch solution and continue titrating until the solution is colorless. Three trials for the standardization of $Na_2S_2O_3$ should yield molar concentrations that agree to within $\pm 1\%$.

Disposal: Dispose of the test solutions as directed by your instructor.

5. **Prepare the Buret for Bleach Analysis.** Prepare the 50-mL buret with the $Na_2S_2O_3$ solution for titration as described in Part B.2. This is to be ready for immediate use in Part E. Ten to fifteen seconds after filling the buret, read and record (Report Sheet, E.2) the volume of $Na_2S_2O_3$ solution.

C. Preparation of Liquid Bleach for Analysis

1. **Obtain Bleach Samples.** Two trials for each bleach are suggested. Obtain about 12 mL of each liquid bleach from your instructor.
2. **Prepare the Bleach Solution for Analysis.** Pipet 10 mL of bleach into a 100-mL volumetric flask and dilute to the 100-mL mark with boiled, deionized water. Mix the solution thoroughly. Pipet 25 mL of this diluted bleach solution into a 250-mL Erlenmeyer flask; add 20 mL of deionized water, 2 g of KI, and 10 mL of 0.5 *M* H_2SO_4. A yellow color indicates the presence of free I_2 that has been generated by the ClO^- ion. Proceed *immediately* to Part E.

D. Preparation of Solid Bleach for Analysis

1. **Obtain Bleach Samples.** Two trials for each bleach are suggested. Obtain about 5 g of each powdered bleach from your instructor.
2. **Prepare a Solution from the Powdered Bleach.** Place about 5 g of a powdered bleach sample into a mortar and grind (Figure 15.5). Measure the tared mass (± 0.001 g) of the finely pulverized sample on weighing paper and transfer it to a 100-mL volumetric flask fitted with a funnel. Dilute the sample to 100 mL with deionized water. Mix thoroughly. If an abrasive is present in the powder, the sample may *not* dissolve completely. Allow the insolubles to settle.
3. **Prepare the Bleach Solution for Analysis.** Pipet 25 mL of this bleach solution (be careful to *not* draw any solids into the pipet!) into a 250-mL Erlenmeyer flask and add 20 mL of deionized water, 2 g of KI, and 10 mL of 0.5 *M* H_2SO_4. A yellow color indicates the presence of free I_2 that has been generated by the ClO^- ion. Proceed *immediately* to Part E.

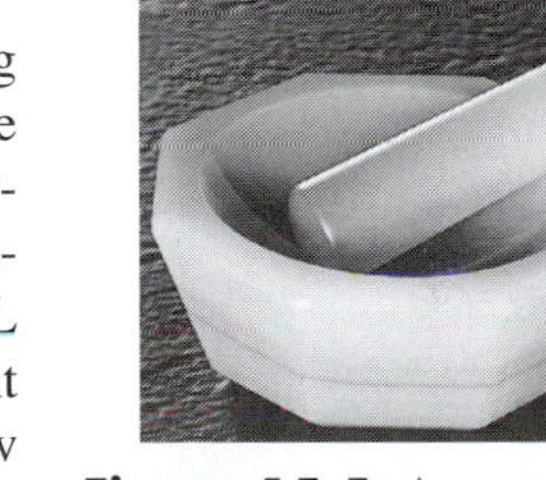

Figure 15.5 A mortar and pestle is used to grind the powder.

E. Analysis of Bleach Samples

1. **Titrate the Bleach Sample.** *Immediately* titrate the liberated I_2 with the standardized $Na_2S_2O_3$, prepared in Part B.5, until the yellow-brown I_2 color is almost discharged.
2. **Perform the Final Steps in the Analysis.** Add 2 mL of starch solution to form the blue I_2•starch complex.[9] While swirling the flask, continue titrating *slowly* until

[9] If you are analyzing a powdered bleach, the 2 mL of starch may be added immediately after the 0.5 *M* H_2SO_4 in Part D.2.

the deep blue color disappears. Refer to the titration procedure in Part B.3. Record the final volume of the titrant in the buret.

3. **Repeat with Another Sample.** Repeat the experiment with a second sample of the same bleach.
4. **Repeat the Analysis for Another Bleach.** Analyze another bleaching agent (liquid, Part C, or solid, Part D) to determine the amount of "available chlorine." Compare the available chlorine content of the two bleaches.

Disposal: Dispose of the KIO_3 solution and the remaining bleach solutions in the "Waste Oxidants" container. Dispose of the $Na_2S_2O_3$ solution in the buret and the Erlenmeyer flask in the "Waste Thiosulfate" container. Dispose of the sulfuric acid in the "Waste Acids" container. Dispose of all test solutions as directed by your instructor.

CLEANUP: Flush the buret twice with tap water and twice with deionized water. Repeat the cleaning operation with the pipets. Discard all washings as directed by your instructor. Clean up the balance area, discarding any solids in the "Waste Solids" container.

Experiment 15 *Prelaboratory Assignment*

Bleach Analysis

Date __________ Lab Sec. ______ Name ____ Desk No. __________

1. A standardized solution of sodium thiosulfate, prepared in this experiment, is a common reducing agent. Write a balanced half-reaction showing its reducing properties.

2. What is the color change at the stoichiometric point in the titration of the bleach solution with the standardized sodium thiosulfate solution? Explain.

3. Write the net-ionic equation for the standardization of the sodium thiosulfate solution using Equations 15.7 and 15.4.

4. A 25.0 mL volume of 0.010 *M* KIO_3 is pipetted into a 250-mL Erlenmeyer flask. The solution is titrated to the stoichiometric point with 14.7 mL of a sodium thiosulfate solution. What is the molar concentration of the sodium thiosulfate solution?

5. a. Experimental Procedure, Part B.3. Calculate the mass of KI that is necessary to react with 25.0 mL of a 0.010 *M* KIO_3 solution.

   b. Comment on the mass of KI that is used in the standardization of the $Na_2S_2O_3$ solution.

6. Write the net-ionic equation for the analysis of the "available chlorine" in a sample using sodium thiosulfate as the reducing agent using Equations 15.2, 15.3, and 15.4.

7. A 10.0-mL volume of "Bleach White" is diluted to 100 mL in a volumetric flask. A 25.0-mL sample of this solution is analyzed according to the procedure in this experiment. Given that 30.9 mL of 0.138 *M* $Na_2S_2O_3$ is needed to reach the stoichiometric point, answer the following questions.
   a. How many grams of "available" Cl_2 are in the titrated sample?

   b. How many grams of "Bleach White" bleach solution are analyzed? Assume that the density of bleach is 1.084 g/mL.

   c. Calculate the percent "available chlorine" in the "Bleach White."

*8. A shower stall is laden with mildew and has accumulated lime deposits from the use of hard water. If a one-step cleaning operation is used, that is, the use of a mixture of liquid bleach and tub and tile cleaner (which is usually acidic in order to react with the lime deposits), a potentially lethal gas is generated. Explain how this occurs (refer to Equation 15.2).

Experiment 15 *Report Sheet*

Bleach Analysis

Date __________ Lab Sec. ______ Name ______________________________ Desk No. __________

A. Preparation of the KIO_3 Primary Standard

Calculation for mass of KIO_3, Part A.1.

1. Tared mass of KIO_3 (g) ____________
2. Moles of KIO_3 (mol) ____________
3. Molar concentration of KIO_3 solution (mol/L) ____________

B. A Standard 0.1 *M* $Na_2S_2O_3$ Solution

Calculation for mass of $Na_2S_2O_3{\cdot}5H_2O$, Part B.1.

	Trial 1	*Trial 2*	*Trial 3*
1. Volume of KIO_3 solution (mL)	25.0	25.0	25.0
2. Moles of KIO_3 titrated (mol)			
3. Buret reading, *initial* (mL)			
4. Buret reading, *final* (mL)			
5. Volume of $Na_2S_2O_3$ added (mL)			
6. Moles of $Na_2S_2O_3$ added (mol)			
7. Molar concentration of $Na_2S_2O_3$ (mol/L)			
8. Average molar concentration of $Na_2S_2O_3$ (mol/L)			

	Trial 1	*Trial 2*	*Trial 1*	*Trial 2*
C. Preparation of Liquid Bleach for Analysis				
Sample Name or Description				
1. Volume of original bleach titrated (mL)	2.5	2.5	2.5	2.5
2. Mass of bleach titrated (g) (assume the density of liquid bleach is 1.084 g/mL)				

	Trial 1	Trial 2	Trial 1	Trial 2
D. Preparation of Solid Bleach for Analysis				
Sample Name or Description	________		________	
1. Tared mass of bleach (g)	____	____	____	____
E. Analysis of Bleach Samples				
1. Buret reading, *initial* (mL)	____	____	____	____
2. Buret reading, *final* (mL)	____	____	____	____
3. Volume of $Na_2S_2O_3$ added (mL)	____	____	____	____
F. Data Analysis				
1. $Na_2S_2O_3$ added (mol)	____	____	____	____
2. ClO^- reacted (mol)	____	____	____	____
3. "Available chlorine" (mol)	____	____	____	____
4. Mass of "available chlorine" (g)	____	____	____	____
5. Percent "available chlorine" (liquid) (assume the density of liquid bleach is 1.084 g/mL) (%)	____	____	____	____
6. Average percent "available chlorine" (liquid) (%)	________		________	
7. Percent "available chlorine" (solid) (%)	____	____	____	____
8. Average percent "available chlorine" (solid) (%)	________		________	

Laboratory Questions

Circle the questions that have been assigned.

1. Part A.1. The solid KIO_3 is diluted shy of the 100-mL mark in the volumetric flask.
   a. How does this technique error affect the molar concentration of the KIO_3 solution?
   b. Will the reported molar concentration of the standardized $Na_2S_3O_3$ solution be too high, too low, or unaffected? Explain.
2. Part B.1. The solid $Na_2S_3O_3$ is diluted shy of the 250-mL mark in the volumetric flask. Will the reported molar concentration of the standardized $Na_2S_3O_3$ solution be too high, too low, or unaffected by this technique error? Explain.
3. Part B.3. The buret is dirty and some of the $Na_2S_3O_3$ titrant adhered to the buret wall between the initial and final buret readings. Will the reported molar concentration of the standardized $Na_2S_3O_3$ solution be too high, too low, or unaffected by this technique error? Explain.
4. Part B.3. Deionized water from a wash bottle was added to wash down the wall of the receiving flask during the titration. How does this affect the reported molar concentration of the sodium thiosulfate solution? Explain.
5. Part C.1. The bleach is diluted but beyond the 100-mL mark in the volumetric flask. Will the percent "available chlorine" calculated after the analysis in Part E be reported as being too high or too low? Explain.
6. Part C.2 (or D.3). The starch solution is added immediately after the sulfuric acid. What is observed? Because the I_2•starch complex is slow to dissociate, what difficulties might be encountered in Part E.2 if the starch were added too early in the analysis? Explain.
7. Part E.2. The endpoint of the titration is surpassed. How will this technique error affect the reported percent "available chlorine" in the bleach sample?
8. Part E.2. An air bubble initially trapped in the tip of the buret is released during the titration. Will the reported percent "available chlorine" in the bleach sample be reported too high or too low? Explain.

Experiment 16

Vitamin C Analysis

Objective

- To determine the amount of vitamin C in a vitamin tablet, a fresh fruit, or a fresh vegetable sample

Techniques

The following techniques are used in the Experimental Procedure

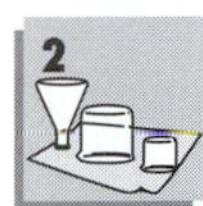

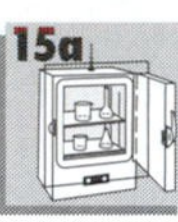

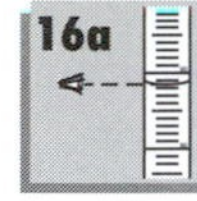

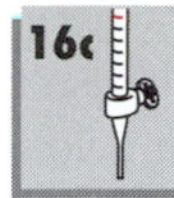

Principles

The human body does not synthesize vitamins; therefore the vitamins that we need for catalyzing specific biochemical reactions are gained only from the food that we eat. Vitamin C can be obtained from a variety of fresh fruits (most notably, citrus fruits) and vegetables. The storage and processing of fruits and vegetables generally decrease their vitamin C content. Cooking (boiling or steaming) leaches the water-soluble vitamin C from the fruits and vegetables.

Vitamin C plays a vital role for the proper growth and development of teeth, bones, gums, cartilage, skin, and blood vessels. It is an important anti-oxidant (reducing agent) that helps protect against cancers, heart disease, and stress. Vitamin C is also critical in enabling the body to absorb iron and folic acid.

The recommended daily allowance (RDA) of vitamin C for an adult is 60 mg, however levels as high as 200 mg are proven to be beneficial. Table 16.1 lists the concentration ranges of vitamin C for various fruits and vegetables (Figure 16.1).

Vitamin C, also called **ascorbic acid**, is one of the more abundant and easily obtained vitamins in nature. It is a colorless, water-soluble acid that, in addition to its acidic properties, is a powerful biochemical reducing agent, meaning it readily undergoes oxidation, even from the oxygen of the air.

Even though ascorbic acid is an acid, its *reducing* properties are used in this experiment to analyze its concentration in various samples. There are many other acids present in foods (e.g., citric acid) that could interfere with an acid analysis and not permit

Table 16.1 Vitamin C in Foods

<10 mg/100 g	beets, carrots, eggs, milk
10–25 mg/100 g	asparagus, cranberries, cucumbers, green peas, lettuce, pineapple
25–100 mg/100 g	Brussels sprouts, citrus fruits, tomatoes, spinach
100–350 mg/100 g	chili peppers, sweet peppers, turnip, greens, kiwi

Figure 16.1 Vegetables are a good supply of vitamin C

a selective determination of the ascorbic acid content. The equation for the oxidation of ascorbic acid is

$$C_6H_8O_6 \text{ (structure)} + H_2O \leftrightharpoons C_6H_8O_7 \text{ (structure)} + 2\,H^+ + 2\,e^- \quad (16.1)$$

or $C_6H_8O_6(aq) + H_2O(l) \rightarrow C_6H_8O_7(aq) + 2H^+(aq) + 2\,e^-$

Analysis of Vitamin C

A sample containing ascorbic acid is prepared and dissolved in an acidic solution. The sample is treated with a measured amount of a *standard* potassium iodate, KIO_3, solution to which is added an excess of I^-. From the IO_3^-/I^- reaction a stoichiometric amount of yellow-brown molecular iodine is generated in solution (in the presence of the excess I^- the molecular iodine actually exists as a water-soluble complex, I_3^-, in solution).

$$IO_3^-(aq) + 5\,I^-(aq) + 6\,H^+(aq) \rightarrow 3\,I_2(aq)\ [\text{or } I_3^-] + 3\,H_2O(l) \qquad (16.2)$$

For the analysis, ascorbic acid from the sample reduces only a portion of the quantitative amount of I_2 present in solution.

$$C_6H_8O_6(aq) + I_2(aq) + H_2O(l) \rightarrow C_6H_8O_7(aq) + 2\,I^-(aq) + 2\,H^+(aq) \qquad (16.3)$$

The remainder of the I_2 (or the excess, "xs") is titrated with a standard thiosulfate, $S_2O_3^{2-}$, solution, producing the colorless I^- and $S_4O_6^{2-}$ ions:

$$(\text{xs})\ I_2(aq) + 2\,S_2O_3^{2-}(aq) \rightarrow 2\,I^-(aq) + S_4O_6^{2-}(aq) \qquad (16.4)$$

Therefore the difference between the I_2 generated (from the IO_3^- in Equation 16.2) and that which is titrated as an excess (in Equation 16.4) is a measure of the ascorbic acid content of the sample.

A small amount of starch is used as an indicator to detect the stoichiometric point. Just prior to the disappearance of the yellow-brown I_2 in the titration (Equation 16.4), starch is added; starch forms a deep-blue complex with molecular iodine, I_2•starch, that is more easily to view than the yellow-brown color of molecular iodine.

$$I_2(aq,\ \textit{yellow-brown}) + \text{starch}(aq) \rightarrow I_2\bullet\text{starch}(aq,\ \textit{deep blue}) \qquad (16.5)$$

The addition of the $S_2O_3^{2-}$ titrant is continued until the I_2•starch is reduced to colorless I^- ion; the solution appears colorless at the endpoint.

$$I_2\bullet\text{starch}(aq,\ \textit{deep blue}) + 2\,S_2O_3^{2-}(aq) \rightarrow 2\,I^-(aq,\ \textit{colorless}) + S_4O_6^{2-}(aq) + \text{starch}\ (aq,\ \textit{colorless}) \qquad (16.6)$$

A Standard Solution of Sodium Thiosulfate, $Na_2S_2O_3$

Primary standard: a substance that has a known high degree of purity, a relatively large molar mass, is nonhygroscopic, and reacts in a predictable way

The preparation of a standardized solution of sodium thiosulfate follows the same procedure as for the analysis in vitamin C. For the standardization procedure, a measured mass of solid potassium iodate, KIO_3, used as a **primary standard** (Figure 16.2), is diluted to prepare a standard KIO_3 solution. An aliquot of this solution is mixed with an acidified solution containing an excess of potassium iodide (Equation 16.2).

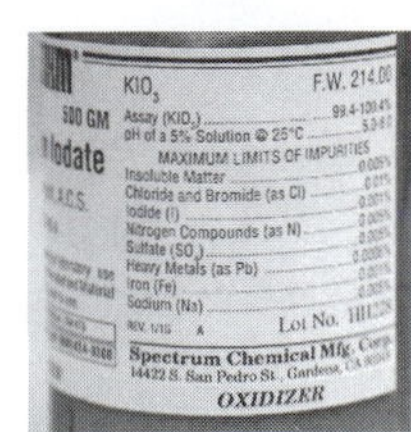

Figure 16.2 Potassium iodate has the necessary properties of a primary standard

The molecular iodine generated from the reaction is titrated with a prepared (unknown concentration) solution of sodium thiosulfate (Equation 16.4) using starch as the indicator (Equations 16.5 and 16.6). The standardization procedure of $Na_2S_2O_3$ differs from the analysis procedure of vitamin C by the omission of the vitamin C (Equation 16.3).

Experimental Procedure

A standard solution of potassium iodate is prepared from solid KIO_3, a primary standard. The solution is first used to standardize a sodium thiosulfate solution. Secondly, the standard KIO_3 solution in conjunction with KI is used to analyze a sample to determine its ascorbic acid content. The excess molecular iodine (not having reacted with ascorbic acid) is back-titrated with the standardized $Na_2S_2O_3$ solution, using starch as an indicator.

Three trials are necessary to standardize the $Na_2S_2O_3$ solution and three trials are necessary to analyze the ascorbic sample assigned by your instructor. Quantitative, reproducible data are objectives of this experiment; practice good laboratory techniques. Nearly 600 mL of boiled, deionized water is required for this experiment. Begin preparing this at the beginning of the laboratory period.

A. Preparation of the KIO_3 Primary Standard

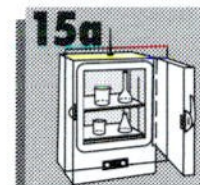

1. **Prepare a Primary Standard Solution.** Use weighing paper to measure the tared mass ($\pm$0.001 g) of KIO_3 (previously dried at 110°C) that is required to prepare 250 mL of 0.01 *M* KIO_3. *Show this calculation on the Report Sheet.* Dissolve the solid KIO_3 and dilute to volume with freshly boiled, deionized water in a 250-mL volumetric flask. Calculate and record its exact molar concentration.

B. A Standard 0.1 *M* $Na_2S_2O_3$ Solution

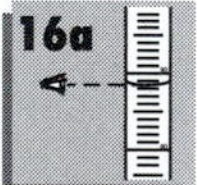

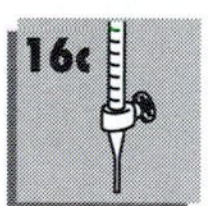

1. **Prepare a $Na_2S_2O_3$ Solution.** This solution should be prepared no more than 1 week in advance because unavoidable decomposition occurs. Measure the mass ($\pm$0.01 g) of $Na_2S_2O_3{\cdot}5H_2O$ needed to prepare 250 mL of a 0.1 *M* $Na_2S_2O_3$ solution. *Show this calculation on the Report Sheet.* Dissolve the solid $Na_2S_2O_3{\cdot}5H_2O$ with freshly boiled, deionized water and dilute to 250 mL; an Erlenmeyer flask can be used to prepare and store the solution. Agitate until the salt dissolves.
2. **Prepare a Buret.** Prepare a *clean,* 50-mL buret for titration. Fill the buret with the $Na_2S_2O_3$ solution, drain the tip of air bubbles, and after 10–15 seconds, record the volume.
3. **Titrate the KIO_3 Solution.** Pipet 25 mL of the standard KIO_3 solution into a 125-mL Erlenmeyer flask, add about 2 g ($\pm$0.01 g) of solid KI, and swirl to dissolve. Add about 10 mL of 0.5 *M* H_2SO_4. Begin titrating the KIO_3 solution immediately. When the yellow iodine nearly disappears (pale yellow color), add 2 mL of starch solution. Stirring constantly, continue titrating *slowly* until the blue color completely (and suddenly) disappears.
4. **Repeat the Titration with Another KIO_3 Sample.** Repeat twice the procedure in Part B.3 by rapidly adding the $Na_2S_2O_3$ titrant until about 1 mL before the yellow-brown color of iodine disappears. Add the starch solution and continue titrating until the solution is colorless.
5. **Prepare the Buret for the Ascorbic Analysis.** Prepare the 50-mL buret with the $Na_2S_3O_3$ solution for titration as described in Part B.2. This is to be ready for immediate use in Part D. Ten to fifteen seconds after filling the buret, read and record the volume of the $Na_2S_3O_3$ solution.

C. Sample Preparation

1. **Vitamin C tablet.** Read the label on the bottle to determine the approximate mass of vitamin C in each tablet. Measure (±0.001 g) the fraction of the total mass of a tablet that corresponds to 100 mg of ascorbic acid. Dissolve the sample in a 250 mL Erlenmeyer flask with 40 mL of 0.5 *M* H_2SO_4[1] and then about 0.5 g $NaHCO_3$.[2] Kool-Aid™, Tang™, or Gatorade™ may be substituted as dry samples, even though their ascorbic acid concentrations are much lower. Fresh Fruit™ has a very high concentration of ascorbic acid. Proceed immediately to Part D.

2. **Fresh fruit sample.** Filter 125–130 mL of freshly squeezed juice through several layers of cheesecloth (or vacuum-filter). Measure the mass (±0.01 g) of a clean, dry 250 mL Erlenmeyer flask. Add about 100 mL of filtered juice and again determine the mass. Add 40 mL of 0.5 *M* H_2SO_4 and 0.5 g $NaHCO_3$. Concentrated fruit juices may also be used as samples. Proceed immediately to Part D.

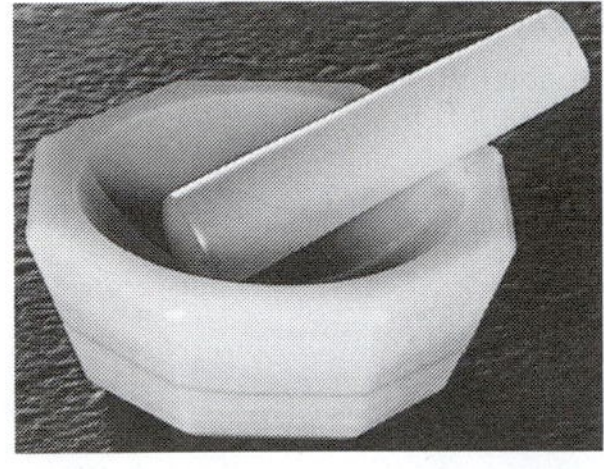

Figure 16.3 A mortar and pestle is used to grind the sample.

3. **Fresh vegetable sample.** Measure about 100 g (±0.01 g) of a fresh vegetable. Transfer the sample to a mortar[3] (Figure 16.3) and grind. Add 5 mL of 0.5 *M* H_2SO_4 and continue to pulverize the sample. Add another 15 mL of 0.5 *M* H_2SO_4, stir, and filter through several layers of cheesecloth (or vacuum-filter). Add 20 mL of 0.5 *M* H_2SO_4 to the mortar, stir, and pour through the same filter. Repeat the washing of the mortar with 20 mL of freshly boiled, deionized water. Combine all of the washings in a 250 mL Erlenmeyer flask and add 0.5 g $NaHCO_3$. Proceed immediately to Part D.

D. Vitamin C Analysis

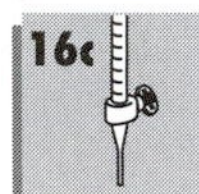

1. **Titrate the Vitamin C Sample.** Pipet 25.0 mL of the standard KIO_3 solution into the sample from Part C and add 1 g of KI. Add about 5 mL of 0.5 *M* H_2SO_4 and 0.1 g $NaHCO_3$. Immediately titrate the excess I_2 in the sample with the standard $Na_2S_2O_3$ solution until the yellow-brown color is almost discharged.
2. **Perform the Final Steps in the Analysis.** Add 2 mL of starch solution to form the deep-blue I_2•starch complex.[4] While swirling the flask, continue titrating *slowly* until the deep-blue color disappears. Refer to the titration procedure in Part B.3. Record the final volume of titrant in the buret.
3. **Repeat the Analysis.** Repeat the analysis twice in order to complete the three trials. Repeated analyses of the ascorbic acid content of the sample should be ±1%.

Disposal: Dispose of the remaining KIO_3 solution in the "Waste Oxidants" container. Dispose of the excess $Na_2S_2O_3$ solution in the "Waste Reducing Agent" container.

CLEANUP: Flush the buret twice with tap water and twice with deionized water. Repeat the cleaning operation with the pipets. Discard all washings as advised by your instructor. Clean up the balance area, discarding any solids in the "Waste Solids" container.

[1]Remember that vitamin tablets contain binders and other material that may be insoluble in water—*do not heat* in an attempt to dissolve the tablet!

[2]The $NaHCO_3$ reacts in the acidic solution to produce $CO_2(g)$, providing an inert atmosphere above the solution, minimizing the possibility of the air oxidation of the ascorbic acid.

[3]A blender may be substituted for the mortar and pestle.

[4]If you are analyzing a vegetable, the 2 mL of starch may be added immediately after the 0.5 *M* H_2SO_4 in Part C.3.

Experiment 16 *Prelaboratory Assignment*

Vitamin C Analysis

Date __________ Lab Sec. ______ Name ____ Desk No. __________

1. Explain why cooked fruits and vegetables have a lower vitamin C content than fresh fruits and vegetables.

2. Vitamin C is an acid (ascorbic acid) and a reducing agent. Which property is utilized for its analysis in the experiment?

3. a. What is the oxidizing agent in Experimental Procedure, Part D?

   b. What is the color change of the starch indicator that signals the stoichiometric point in Part D?

4. Six ounces (1 fl. oz = 29.57 mL) of a well-known vegetable juice contains 35% of the recommended daily allowance of vitamin C (RDA = 60 mg). How many milliliters of the vegetable juice will provide 100% of the recommended daily allowance?

5. A 25.0-mL volume of 0.010 *M* KIO_3, containing an excess of KI, is added to a 0.246 g sample of a Real Lemon™ solution containing vitamin C. The yellow-brown solution, caused by the presence of excess I_2, is titrated to a colorless starch endpoint with 10.7 mL of 0.100 *M* $Na_2S_2O_3$.

   a. How many moles of KIO_3 were added to the Real Lemon™ solution?

   b. Calculate the moles of I_2 that are generated from the KIO_3?

   c. How many moles of I_2 reacted with the 0.100 *M* $Na_2S_2O_3$ in the titration?

   d. How many moles of I_2 reacted with the vitamin C in the Real Lemon™ sample?

   e. Calculate the moles and grams of vitamin C, $C_6H_8O_6$, in the sample.

   f. Calculate the percent (by mass) of vitamin C in the Real Lemon™ sample.

Experiment 16 *Report Sheet*

Vitamin C Analysis

Date __________ Lab Sec. ______ Name ____ Desk No. __________

A. Preparation of the KIO_3 Primary Standard

Calculation for mass of KIO_3, Part A.1.

1. Tared mass of KIO_3 (g) ____________________
2. Moles of KIO_3 (mol) ____________________
3. Molar concentration of KIO_3 solution (mol/L) ____________________

B. A Standard 0.1 *M* $Na_2S_2O_3$ Solution

Calculation for mass of $Na_2S_2O_3 \cdot 5H_2O$, Part B.1.

C. Sample Preparation

Sample Name: ____________________

1. Mass of sample (*g*) ____________________

D. Vitamin C Analysis

	Trial 1	Trial 2	Trial 3
1. Volume of KIO_3 added (*mL*)	25.0	25.0	25.0
2. Moles of IO_3^- added (*mol*)			
3. Moles of I_2 generated, total (*mol*)			
4. Buret reading, *initial* (*mL*)			
5. Buret reading, *final* (*mL*)			
6. Volume of $Na_2S_2O_3$ added (*mL*)			
7. Moles of $S_2O_3^{2-}$ added (*mol*)			
8. Moles of I_2 reduced by $S_2O_3^{2-}$ (*mol*)			
9. Moles of I_2 reduced by $C_6H_8O_6$ (*mol*)			
10. Moles of $C_6H_8O_6$ in sample (*mol*)			
11. Mass of $C_6H_8O_6$ in sample (*g*)			
12. Mass ratio (*mg $C_6H_8O_6$/100 g sample*)			
13. Average mass ratio of $C_6H_8O_6$ in sample (mg/100 g)			

Laboratory Questions

Circle the questions that have been assigned.

1. Part A.1. The KIO_3 sample was not sufficiently dried. Will the reported molar concentration of the $Na_2S_2O_3$ solution (Part B.3) be too high, too low, or unaffected? Explain.

2. Part B.1. The mass used to prepare the $Na_2S_2O_3$ solution was erroneously calculated too low. Will the reported molar concentration of the $Na_2S_2O_3$ (determined in Part B.3) be too high, too low, or unaffected? Explain.

3. Part B.1. The mass used to prepare the $Na_2S_2O_3$ solution was calculated correctly. The calculation was based on the availability $Na_2S_2O_3$•$5H_2O$. Only anhydrous $Na_2S_2O_3$ is available. If the calculated mass is used to measure the mass of $Na_2S_2O_3$ instead of $Na_2S_2O_3$•$5H_2O$ for the preparation of the solution, will the molar concentration of the $Na_2S_2O_3$ be higher or lower *than expected?* Explain.

4. Part C.1. a. What is the purpose of adding sodium bicarbonate to the ascorbic acid sample? Write a balanced equation for its reaction.
   b. The $NaHCO_3$ solution is omitted in an analysis of a sample. Will the reported amount of ascorbic acid in the sample be too high, too low, or unaffected? Explain.

5. Part D.1. After adding the standard solution of KIO_3, the KI, H_2SO_4, and $NaHCO_3$ solutions, the sample solution remains colorless! What modification of the Experimental Procedure can be made to correct for this unexpected observation in order to complete the analysis?

6. Part D.2. The deep-blue color of the I_2•starch complex does not appear! What next? Should you continue titrating with the standard $Na_2S_2O_3$ solution or discard the sample? Explain.

7. Part D.2. The final buret reading is read and recorded as 27.43 mL instead of the correct 28.43 mL. Will the reported amount of ascorbic acid in the sample be too high or too low? Explain.

Experiment 17

Synthesis of an Alum

A crystal of potassium alum, $KAl(SO_4)_2 \cdot 12H_2O$.

OBJECTIVES

- To prepare alums from an aluminum can and a salt
- To test the purity of an alum using a melting point measurement

TECHNIQUES

The following techniques are used in the Experimental Procedure

INTRODUCTION

An alum is a hydrated double sulfate salt with the general formula

$$M^{+}M'^{3+}(SO_4)_2 \cdot 12H_2O.$$

M^+ is a **univalent** cation, commonly Na^+, K^+, Tl^+, NH_4^+, or Ag^+; M'^{3+} is a trivalent cation, commonly Al^{3+}, Fe^{3+}, Cr^{3+}, Ti^{3+}, or Co^{3+}. A common household alum is ammonium aluminum sulfate dodecahydrate (Figure 17.1).

Univalent: an ion that has a charge of one

Dodeca: Greek prefix meaning "12"

Some common alums and their uses are listed in Table 17.1.

This experiment describes the preparation of two alums—potassium alum, chrome alum, and ferric alum.

Preparation of Potassium Alum

Sizing: to effect the porosity of paper or fabrics

Potassium alum, commonly just called **alum,** is widely used in the chemical industry for home and commercial uses. It is extensively used in the pulp and paper industry for **sizing** paper, also for sizing fabrics in the textile industry. Alum is also used in municipal water treatment plants for purifying drinking water. Potassium aluminum sulfate dodecahydrate (potassium alum), $KAl(SO_4)_2 \cdot 12H_2O$, is prepared from an aluminum can and potassium hydroxide. Aluminum metal rapidly reacts with hot aqueous KOH producing a soluble potassium aluminate salt solution and hydrogen gas.

$$2\,Al(s) + 2\,K^+(aq) + 2\,OH^-(aq) + 6\,H_2O(l) \rightarrow 2\,K(aq) + 2\,Al(OH)_4^-(aq) + 3\,H_2(g) \quad (17.1)$$

When treated with sulfuric acid, the aluminate ion, $Al(OH)_4^-$, precipitates as aluminum hydroxide, but redissolves with the application of heat.

$$2\,K(aq) + 2\,Al(OH)_4^-(aq) + 2\,H^+(aq) + SO_4^{2-}(aq) \rightarrow 2\,Al(OH)_3(s) + 2\,K^+(aq) + SO_4^{2-}(aq) + 2\,H_2O(l) \quad (17.2)$$

$$2\,Al(OH)_3(s) + 6\,H^+(aq) + 3\,SO_4^{2-}(aq) \xrightarrow{\Delta} 2\,Al^{3+}(aq) + 3\,SO_4^{2-}(aq) + 6\,H_2O(l) \quad (17.3)$$

Figure 17.1 Ammonium aluminum sulfate dodecahydrate.

Table 17.1 Common Alums

Alum	Formula	Uses
Sodium aluminum sulfate dodecahydrate (sodium alum)	$NaAl(SO_4)_2 \cdot 12H_2O$	Baking powders: hydrolysis of Al^{3+} releases H^+ in water to react with the HCO_3^- in baking soda—this produces CO_2, causing the dough to rise
Potassium aluminum sulfate dodecahydrate (alum or potassium alum)	$KAl(SO_4)_2 \cdot 12H_2O$	Water purification, sewage treatment, fire extinguishers, and "sizing" paper
Ammonium aluminum sulfate dodecahydrate (ammonium alum)	$NH_4Al(SO_4)_2 \cdot 12H_2O$	Pickling cucumbers
Potassium chromium(III) sulfate dodecahydrate (chrome alum)	$KCr(SO_4)_2 \cdot 12H_2O$	Tanning leather and waterproofing fabrics
Ammonium ferric sulfate dodecahydrate (ferric alum)	$NH_4Fe(SO_4)_2 \cdot 12H_2O$	Mordant in dying and printing textiles

The potassium aluminum sulfate dodecahydrate alum forms octahedral-shaped crystals when the nearly saturated solution cools.

$$K^+(aq) + Al^{3+}(aq) + 2\ SO_4^{2-}(aq) + 12\ H_2O(l) \rightarrow KAl(SO_4)_2 \cdot 12H_2O(s) \quad (17.4)$$

Preparation of Ferric Alum

Ammonium iron(III) sulfate dodecahydrate (ferric alum), $NH_4Fe(SO_4)_2 \cdot 12H_2O$, is prepared by oxidizing iron(II) ion to iron(III) ion with nitric acid in the presence of ammonium sulfate.

$$2\ H^+(aq) + NO_3^-(aq) + Fe^{2+}(aq) \rightarrow Fe^{3+}(aq) + NO_2(g) + H_2O(l) \quad (17.5)$$

Ammonium and sulfate ions from ammonium sulfate, $(NH_4)_2SO_4$, crystallize the iron(III) ion as ferric alum.

$$NH_4^+(aq) + Fe^{3+}(aq) + 2\ SO_4^{2-}(aq) + 12\ H_2O(l) \rightarrow NH_4Fe(SO_4)_2 \cdot 12H_2O(s) \quad (17.6)$$

For each preparation, some solvent is removed by evaporation to produce a supersaturated solution from which the alum ultimately crystallizes.

EXPERIMENTAL PROCEDURE

Procedure Overview: A known mass of starting material is used to synthesize one or more alums. The synthesis requires the careful transfer of solutions and some evaporation and cooling techniques.

Ask your instructor for instructions as to which alums you are to prepare. Prepare an ice bath by half-filling a 600-mL beaker with ice.

A. Potassium Alum

1. **Prepare the Aluminum Sample.** Cut an approximate 2-inch square of scrap aluminum from a beverage can and clean both sides (to remove the plastic coating on the inside, a paint covering on the outside) with steel wool. Rinse the aluminum with deionized water. Cut the "clean" aluminum into small pieces[1]. Tare a 100-mL beaker and measure about 0.5 g (±0.01 g) of aluminum pieces.

   Add 10–12 mL of 4 *M* KOH to the aluminum pieces (**Caution:** *Wear safety glasses; do not splatter the solution, KOH is caustic.*) and swirl the reaction mixture.

Cool flame: a nonluminous Bunsen flame with a reduced flow of natural gas

2. **Dissolve the Aluminum Pieces.** Move the beaker to a well-ventilated area, such as a fume hood. Warm the beaker *gently* with a **cool flame** or hot plate to initiate the reaction. As the reaction proceeds hydrogen gas is being evolved as is evidenced by the "fizzing" at the edges of the aluminum pieces.

[1]The smaller the aluminum pieces, the more rapid is the reaction. Aluminum foil may be used in place of the scrap aluminum pieces.

The dissolution of the aluminum pieces may take up to 20 minutes—it is important to maintain the solution at a level that is one-half to three-fourths of its original volume by adding small portions of deionized water during the dissolution process.[2]

3. **Gravity Filter the Reaction Mixture.** When no further reaction is evident, return the reaction mixture to the laboratory desk. Gravity filter the warm reaction mixture through a cotton plug (although filter paper will suffice) into a 100-mL beaker to remove the insoluble impurities (see Figures T.11d and T.11e). If solid particles appear in the filtrate, repeat the filtration. Rinse the filter with 2–3 mL of deionized water.

4. **Allow the Formation of Aluminum Hydroxide.** Allow the clear solution (the filtrate) to cool in the 100-mL beaker. While stirring, *slowly* add, in 5-mL increments (because the reaction is exothermic!), approximately 15 mL of 6 M H_2SO_4 (**Caution:** *Avoid skin contact!*).

5. **Dissolve the Aluminum Hydroxide.** If the solution shows evidence of the white, gelatinous $Al(OH)_3$ precipitate in the acidified filtrate, gently heat the mixture until the $Al(OH)_3$ dissolves.
6. **Crystallize the Alum.** Remove the solution from the heat. Cool the solution in an ice bath. Alum crystals should form within 20 minutes. If crystals do *not* form, use a hot plate (Figure 17.3) to gently reduce the volume by one-half (*do not boil!*) and return to the ice bath. For larger crystals and a higher yield, allow the crystallization process to continue until the next laboratory period.

7. **Test the Purity of the Alum.** Proceed to Part C.

B. Ferric Alum

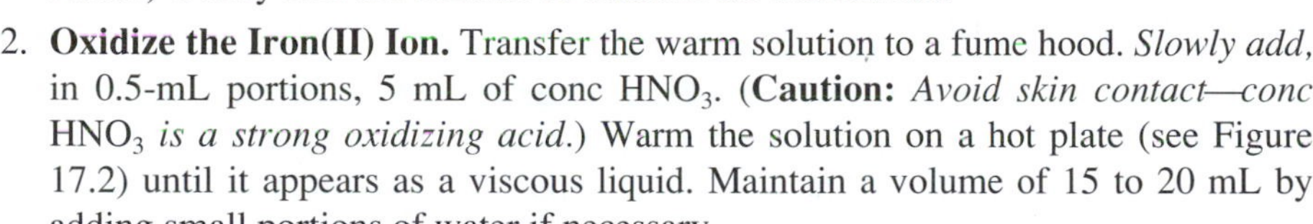

1. **Dissolve the Iron(II) Salt.** Dissolve 10 g (±0.01 g) of $FeSO_4 \cdot 7H_2O$ in 15 mL of 2.0 M H_2SO_4 in a 150-mL beaker. (**Caution:** *Avoid skin contact—it causes severe burns.*) Gently heat the mixture to dissolve the ferrous salt.

2. **Oxidize the Iron(II) Ion.** Transfer the warm solution to a fume hood. *Slowly add,* in 0.5-mL portions, 5 mL of conc HNO_3. (**Caution:** *Avoid skin contact—conc* HNO_3 *is a strong oxidizing acid.*) Warm the solution on a hot plate (see Figure 17.2) until it appears as a viscous liquid. Maintain a volume of 15 to 20 mL by adding small portions of water if necessary.

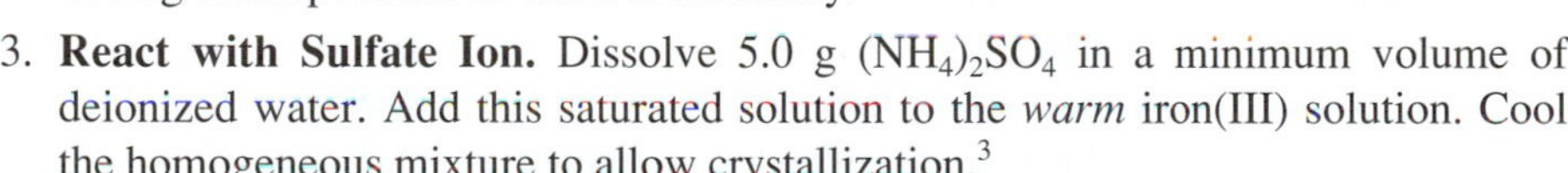

3. **React with Sulfate Ion.** Dissolve 5.0 g $(NH_4)_2SO_4$ in a minimum volume of deionized water. Add this saturated solution to the *warm* iron(III) solution. Cool the homogeneous mixture to allow crystallization.[3]

4. **Nurture the Crystallization.** If crystallization does not occur after 24 hours, again warm the solution on a hot plate (Figure 17.3) to evaporate one-third of the original volume (*do not boil!*); cool for the alum to crystallize. An ice bath can hasten the crystallization process, although the crystals may be small.
5. **Test the Purity of the Alum.** Proceed to Part C.

C. Melting Point of the Synthesized Alum

The melting point of the alum sample can be determined with either a commercial melting point apparatus or with the apparatus shown in Figure 17.6 and described in Parts C.3 and C.4. Consult with your instructor.

1. **Isolate and Wash the Alum Crystals.** Vacuum filter the alum crystals from the solution. Wash the crystals on the filter paper with two (cooled-to-ice temperature) 5-mL portions of a 50% (by volume) ethanol–water solution.[4] Maintain the suc-

---

[2]Some impurities, such as the label or the plastic lining of the can, may remain undissolved.

[3]An ice bath may encourage the crystallization process.

[4]The alum crystals are marginally soluble in a 50% (by volume) ethanol–water solution.

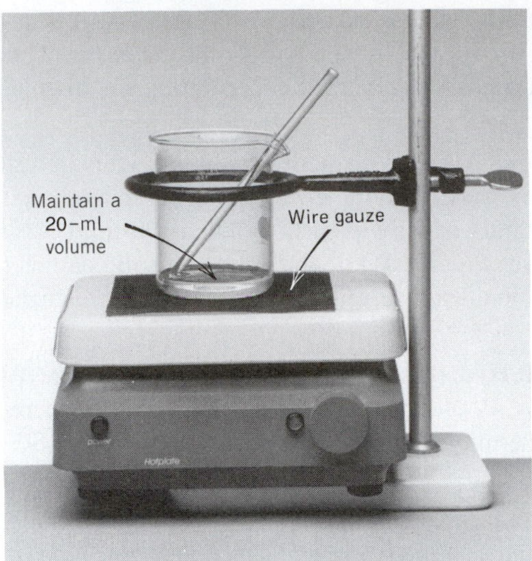

Figure 17.2 Flameless heating using a hot plate.

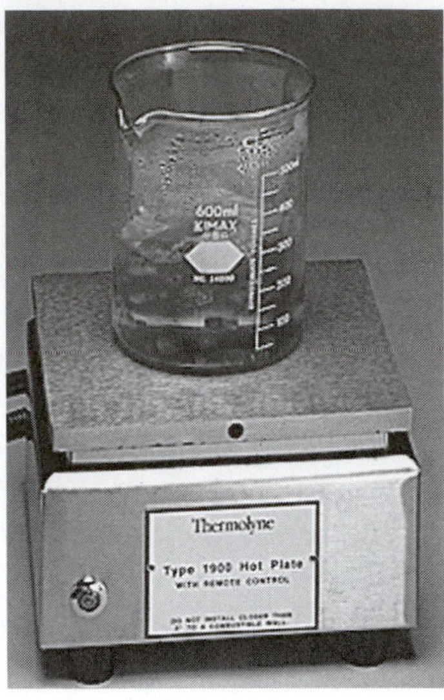

Figure 17.3 Reduce the volume of the solution on a hot plate.

tion until the crystals appear dry. Determine the mass (±0.01 g) of the crystals. Have your laboratory instructor approve the synthesis of your alum.

2. **Percent Yield.** Calculate the percent yield of your alum crystals.

Disposal: Discard the filtrate in the "Waste Salts" container.

CLEANUP: Rinse all glassware twice with tap water and twice with deionized water. All rinses can be discarded as advised by your instructor, followed by a generous amount of tap water.

3. **Prepare the Alum in the Melting Point Tube.** Place finely ground, dry alum to a depth of about 0.5 cm in the bottom of a melting point capillary tube. To do this, place some alum on a piece of dry filter paper and "tap–tap" the open end of the capillary tube into the alum until the alum is at a depth of about 0.5 cm (Figure 17.4). Invert the capillary tube and compact the alum at the bottom of the tube—either drop the tube onto the lab bench through a 25-cm piece of glass tubing (Figure 17.5a) or vibrate the capillary tube with a triangular file (Figure 17.5b).

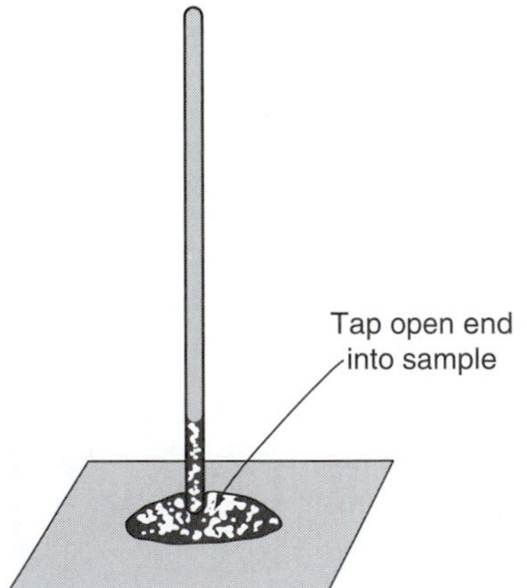

Figure 17.4 Invert the capillary melting point tube into the sample and "tap."

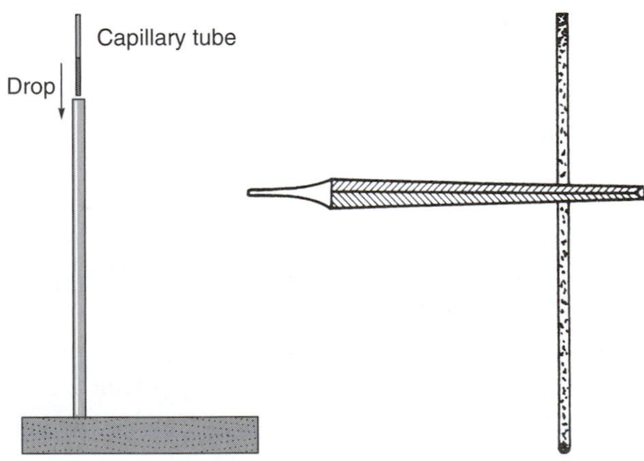

Figure 17.5 Compact the sample to the bottom of the capillary melting point tube by (a) dropping the capillary tube into a long piece of glass tubing or (b) vibrating the sample with a triangular file.

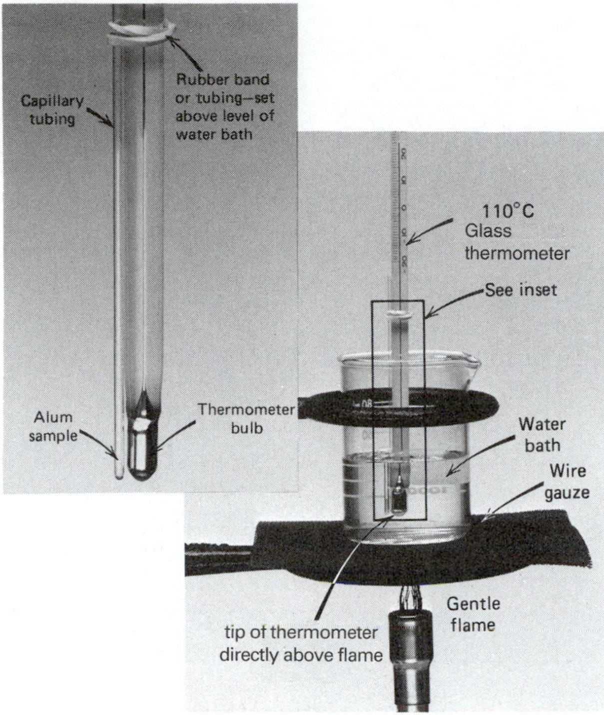

Figure 17.6 Melting point apparatus for an alum.

4. **Determine the Melting Point of the Alum.** Mount the capillary tube *beside* the thermometer bulb with a rubber band or tubing (Figure 17.6, insert). Slowly heat the water in the beaker at about 3°C per minute; carefully watch the solid. When the solid melts, note the temperature. Allow the bath to cool to just below this *approximate* melting point; at a 1°C per minute heating rate, heat again until it melts. Repeat the cooling/heating cycle until reproducibility is obtained—this is the melting point of your alum. Record this on the Report Sheet.
5. **Compare Your Melting Point with Known Values.** Use the values in Table 17.2 to check the melting point of your alum.

Table 17.2 Melting Points of Various Alums

Alum	Melting Point
Potassium alum	92.5°C
Ferric alum	40°C

Disposal: Dispose of the melting point tube in the "Glass Only" container.

Experiment 17 *Prelaboratory Assignment*

Synthesis of an Alum

Date __________ Lab Sec. ______ Name ____ Desk No. __________

1. An alum is a "double salt" consisting of a monovalent cation, a trivalent cation, and two sulfate ions with twelve waters of hydration (waters of crystallization) as part of the crystalline structure.
   a. Are the twelve waters of hydration used to calculate the theoretical yield of the alum? Explain.

   b. The 12 waters of hydration are "hydrated (strongly attracted)" to the metal ions in the crystalline alum structure. Are the water molecules more strongly hydrated to the monovalent cation of the trivalent cation? Explain.

   c. What might you expect to happen to the alum if it were heated to a high temperature? Explain.

2. Potassium alum, synthesized in this experiment, has the formula $KAl(SO_4)_2•12\ H_2O$, but written as a "double salt" its formula is $K_2SO_4•Al_2(SO_4)_3•24\ H_2O$. Refer to Table 17.1 and write the formula of
   a. chrome alum as a double salt

   b. ferric alum as a double salt

3. Potassium dichromate, $K_2Cr_2O_7$, is the starting material in the synthesis of chrome alum (Table 17.1). With a starting mass of 15 g of $K_2C_2O_7$ (molar mass = 294.19 g/mol), how many grams of chrome alum could be synthesized?

4. For the synthesis of potassium alum, advantage is taken by the fact that aluminum hydroxide is amphoteric.
   a. What is the property of amphoterism?

   b. Write equations showing the amphoteric properties of $Al(OH)_3$.

5. An aluminum can is cut into small pieces. A 0.62-g sample of the aluminum "chips" is used to prepare potassium alum according to the procedure described in this experiment. Calculate the theoretical yield (in grams) of potassium alum that could be obtained in the reaction.

6. A mass of 12.62 g of $(NH_4)_2SO_4$ is dissolved in water. After the solution is heated, 26.42 g of $Al_2(SO_4)_3 \bullet 18H_2O$ is added. Calculate the theoretical yield of the resulting alum (refer to Table 17.1 for the formula of the alum). *Hint:* This is a limiting reactant problem.

7. Equation 17.5 and Experimental Procedure, Part B.2. How many milliliters of conc (16 *M*) HNO_3 is required to reaction with 9.81 g of $FeSO_4 \bullet 7H_2O$ (molar mass = 278.02 g/mol)?

8. Experimental Procedure, Part C.4. To measure the melting point of the alum, the water bath temperature is *slowly* increased. Why does this procedure ensure a more accurate melting point measurement?

Experiment 17 *Report Sheet*

Synthesis of an Alum

Date __________ Lab Sec. ______ Name ______________________________ Desk No. __________

Alum prepared ______________________________; Limiting Reactant ______________________________

1. Mass of limiting reactant (g) ______________________________
2. Mass of alum synthesized (g) ______________________________
3. Instructor's approval of alum ______________________________
4. Theoretical yield (g) ______________________________
5. Percent yield (%) ______________________________
6. Melting point (°C) ______________________________

Calculations:

Laboratory Questions

Circle the questions that have been assigned

1. Part A.1. The aluminum sample is not cut into small pieces, but rather left as one large piece.
   a. How will this oversight affect the progress of completing the experimental procedure? Explain.
   b. Will the percent yield of the alum be too high, too low, or unaffected by the oversight? Explain.
2. Part A.3. Aluminum pieces inadvertently collect on the filter.
   a. If left on the filter, will the percent yield of the alum be reported too high or too low? Explain.
   b. If the aluminum pieces are detected on the filter, what steps would be used to remedy the observation? Explain.
3. Part A.3. Identify the ions present in the filtrate.
4. Part A.4. Too much sulfuric acid is added. What observation would be expected in Part A.6? Explain.
5. Part A.5. Identify the ions that are present in the solution.
6. Part B.2. A volume of 5 mL of *conc* HNO_3 has the same number of moles of HNO_3 as 90 mL of 1 *M* HNO_3.
   a. Explain why the *conc* HNO_3 is preferred in the synthesis.
   b. If the 1 *M* HNO_3 is used will the percent yield of the ferric alum be reported too high, too low, or unaffected? Explain.
7. Part B.3. Ammonium sulfate is not present on the reagent shelf, but ammonium nitrate is—both provide the ammonium ion for the ferric alum. What observation would be expected in Part B.4? Explain.
8. Part C.2. Explain why a percent yield of greater than 100% might be reported for the synthesis of the alum.
9. Part C.5. Explain why the melting point of your prepared alum must either equal to or be less than the actual melting point of the alum. Consult with your laboratory instructor.

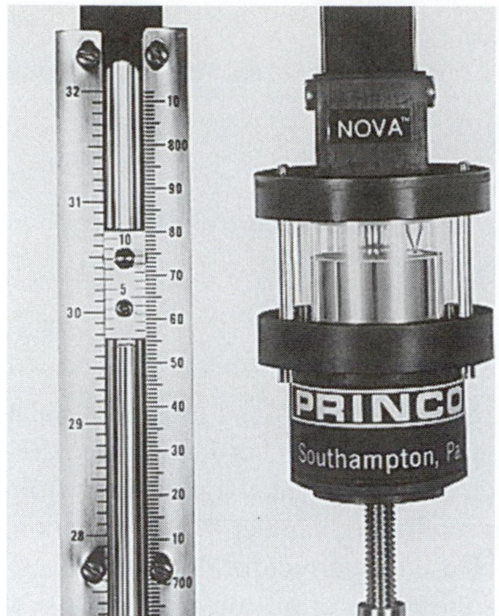

The mercury barometer accurately measures atmospheric pressure in mmHg (or torr).

Experiment 18

Molar Mass of a Volatile Liquid

Objectives

- To measure the physical properties of pressure, volume, and temperature for a gaseous substance
- To determine the molar mass (molecular weight) of a volatile liquid

Techniques

The following techniques are used in the Experimental Procedure

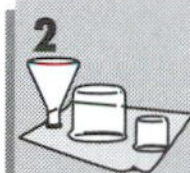

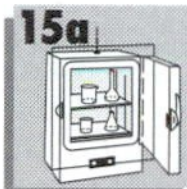

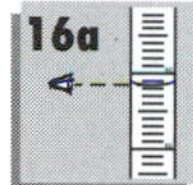

Introduction

Chemists in academia, research, and industry synthesize new compounds daily. To identify a new compound a chemist must determine its properties; physical properties such as melting point, color, density, and elemental composition are all routinely measured. The molar mass of the compound, also one of the most fundamental properties, is often an initial determination.

A number of analytical methods can be used to measure the molar mass of a compound; the choice of the analysis depends on the properties of the compound. For example, the molar masses of "large" molecules, such as proteins, natural drugs, and enzymes found in biochemical systems, are often determined with an osmometer, an instrument that measures changes in osmotic pressure of the solvent in which the molecule is soluble. For smaller molecules a measurement of the melting point change of a solvent (Experiment 20) in which the molecule is soluble can be used. The molar mass of a molecule stable in the gas phase can be measured by **mass spectrometry.**

Mass spectrometry: an instrumental method for identifying a gaseous ion according to its mass and charge (an ionized gas phase molecule travels a specific arc of a curved path in a magnetic or electric field depending on its mass)

For **volatile** liquids, molecular substances with low boiling points and relatively low molar masses, the Dumas method (John Dumas, 1800–1884) of analysis can provide a fairly accurate determination of molar mass. In this analytical procedure the liquid is vaporized into a fixed-volume vessel at a measured temperature and barometric pressure. From the data and the use of the ideal gas law equation (assuming ideal gas behavior), the number of moles of vaporized liquid, n_{vapor}, is calculated:

Volatile: readily vaporizable

$$n_{\text{vapor}} = \frac{PV}{RT} = \frac{P(\text{atm}) \times V(\text{L})}{(0.08206\ \text{L•atm/mol•K}) \times T(\text{K})} \qquad (18.1)$$

In this equation, R is the universal gas constant, P is the barometric pressure in atmospheres, V is the volume in liters of the vessel into which the liquid is vaporized, and T is the temperature in kelvins of the vapor.

R = 0.08206 L•atm/mol • K

The mass of the vapor, m_{vapor}, is determined from the mass difference between the "empty" vessel and the vapor-filled vessel.

$$m_{vapor} = m_{flask + vapor} - m_{flask} \tag{18.2}$$

The molar mass of the compound, $M_{compound}$, is then calculated from the acquired data.

$$M_{compound} = \frac{m_{vapor}}{n_{vapor}} \tag{18.3}$$

Gases and liquids with relatively large intermolecular forces and large molecular volumes do *not* behave according to the ideal gas law equation; in fact, some compounds that we normally consider as liquids, such as H_2O, deviate significantly from ideal gas behavior in the vapor state. Under these conditions, van der Waals' equation, a modification of the ideal gas law equation, can be used to correct for the intermolecular forces and molecular volumes in determining the moles of gas present in the system:

$$\left(P + \frac{n^2a}{V^2}\right)(V - nb) = nRT \tag{18.4}$$

In this equation, $P, V, T, R,$ and n have the same meanings as in Equation 18.1. a is an experimental value that is representative of the intermolecular forces of the vapor, and b is an experimental value that is representative of the volume (or size) of the molecules.

If a more accurate determination of the moles of vapor, n_{vapor}, in the flask is required, van der Waals' equation can be used instead of the ideal gas law equation. Some values of a and b for a number of low-boiling-point liquids are listed in Table 18.1. Others may be found in your textbook.

Table 18.1 Van der Waals' Constants for Some Low-Boiling-Point Compounds

Name	a $\left(\frac{L^2 \cdot atm}{mol^2}\right)$	b (L/mol)	Boiling Point (°C)
methanol	9.523	0.06702	65.0
ethanol	12.02	0.08407	78.5
acetone	13.91	0.0994	56.5
propanol	14.92	0.1019	82.4
hexane	24.39	0.1735	69.0
cyclohexane	22.81	0.1424	80.7
pentane	19.01	0.1460	36.0
water	5.46	0.0305	100.0

Experimental Procedure

Procedure Overview: A boiling water bath of measured temperature is used to vaporize an unknown liquid into a flask. The volume of the flask is measured by filling the flask with water. As the flask is open to the atmosphere, you will record a barometric pressure.

You are to complete three trials in determining the molar mass of your low-boiling-point liquid. Initially obtain 15 to 20 mL of liquid from your instructor. The same apparatus is used for each trial.

A. Preparing the Sample

1. **Prepare the Flask for the Sample.** Clean a 125-mL Erlenmeyer flask and dry it either in a drying oven or by allowing it to air-dry. Do *not* wipe it dry or heat it over a direct flame. Cover the dry flask with a small piece of aluminum foil (Figure 18.1) and secure it with a rubber band. Determine the mass ($\pm$0.001 g) of the *dry* flask, aluminum foil, and rubber band.

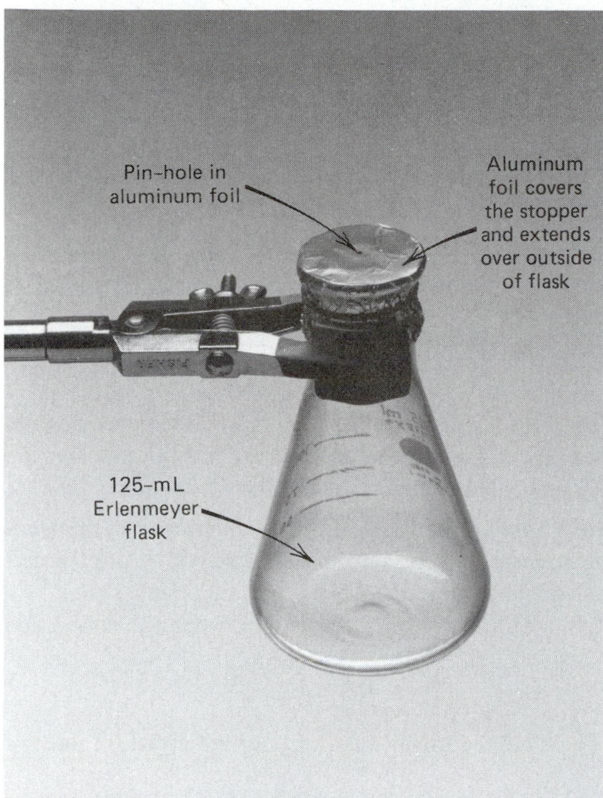

Figure 18.1 Preparation of a flask for the placement of the volatile liquid.

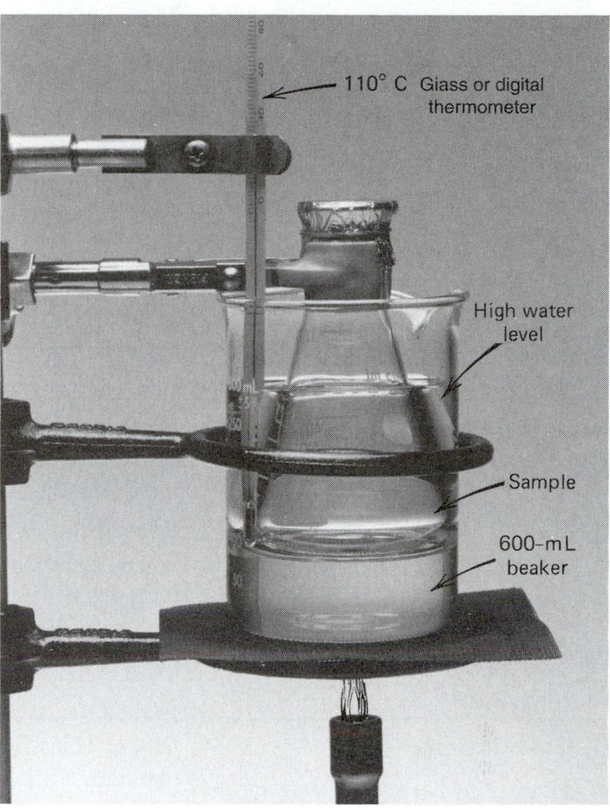

Figure 18.2 Apparatus for determining the molar mass of a volatile liquid.

2. **Place the Sample in the Flask.** Transfer *about* 5 mL of the unknown liquid into the flask; again cover the flask with the aluminum foil and secure the foil with a rubber band. You do *not* need to conduct a mass measurement. With a pin, pierce the aluminum foil several times.

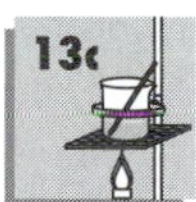

3. **Prepare a Boiling Water Bath.** Half-fill a 600-mL beaker with water; support it on a wire gauze and, near the top, with a support ring. Add one or two **boiling chips** to the water. The heat source may be a hot plate or a Bunsen flame—consult with your instructor.

Boiling chip: a piece of porous ceramic that releases air when heated (the bubbles formed prevent water from becoming superheated)

B. Collecting the Data

1. **Place the Flask/Sample in the Bath.** Lower the flask/sample into the bath and secure it with a utility clamp. Be certain that neither the flask nor the clamp touches the beaker wall. Adjust the water level *high* on the neck of the flask (Figure 18.2).[1]

2. **Heat the Sample to the Temperature of Boiling Water.** Gently heat the water until it reaches a *gentle* boil. (**Caution:** *Most unknowns are flammable; use a moderate flame for heating.*) When the liquid in the flask or the vapors escaping from the holes in the aluminum foil are no longer visible, continue heating for another 5 minutes. Read and record the temperature of the boiling water, "using all certain digits (from the labeled calibration marks on the thermometer) *plus* one uncertain digit (the last digit which is the best estimate between the calibration marks)."

[1]You may choose to wrap the upper portion of the flask and beaker with aluminum foil; this will maintain the upper portion of the flask *not* in the boiling water bath at the same temperature as the boiling water.

3. **Measure the Mass of the Flask/Sample.** Remove the flask and allow it to cool to room temperature. Sometimes the remaining vapor in the flask condenses; that's O.K. *Dry the outside of the flask* and determine the mass (±0.001 g, use the same balance!) of the flask, aluminum foil, rubber band, and the vapor.
4. **Do It Again and Again.** Repeat the experiment for Trials 2 and 3. You only need to transfer another 5 mL of liquid to the flask (i.e., begin with Part A.2) and repeat Parts B.1–B.3.

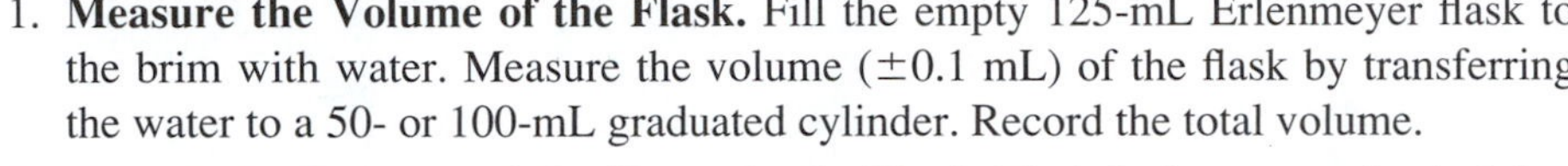

Disposal: Dispose of the leftover unknown liquid in the "Waste Organics" container.

C. Determine the Volume and Pressure of the Vapor

1. **Measure the Volume of the Flask.** Fill the empty 125-mL Erlenmeyer flask to the brim with water. Measure the volume (±0.1 mL) of the flask by transferring the water to a 50- or 100-mL graduated cylinder. Record the total volume.

2. **Record the Pressure of the Vapor in the Flask.** Find the barometer in the laboratory. Read and record the atmospheric pressure in atmospheres, "using all certain digits (from the labeled calibration marks on the barometer) *plus* one uncertain digit (the last digit which is the best estimate between the calibration marks)."

D. Calculations

1. **Molar Mass from Data.** Calculate the molar mass of your unknown for each of the three trials.

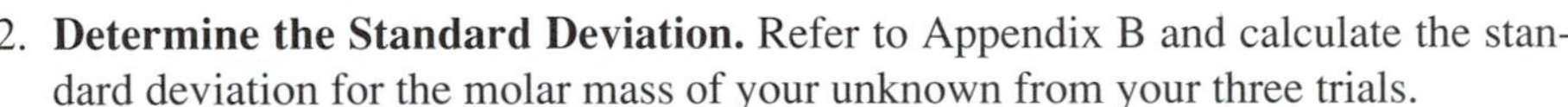

Appendix B

2. **Determine the Standard Deviation.** Refer to Appendix B and calculate the standard deviation for the molar mass of your unknown from your three trials.

3. **Obtain Group Data.** Obtain the values of molar mass for the same unknown from other chemists. Calculate the standard deviation for the molar mass of the unknown.

Experiment 18 *Prelaboratory Assignment*

Molar Mass of a Volatile Liquid

Date __________ Lab Sec. ______ Name ____ Desk No. __________

1. A 125-mL Erlenmeyer flask has a measured volume of 152 mL. A 0.177-g sample of an unknown vapor occupies the flask at 97.8°C a pressure of 748 torr. Assume ideal gas behavior.
   a. How many moles of vapor are present?

   b. What is the molar mass of the vapor?

2. a. If the volume of the flask is mistakenly recorded as 125 mL in Question 1, what is the reported molar mass of the vapor?

   b. What is the percent error caused by the error in the recording of the volume of the flask?

   $$\%\ \text{error} = \frac{M_{\text{difference}}}{M_{\text{actual}}} \times 100$$

3. The ideal gas law equation (Equation 18.1) is an equation used for analyzing "ideal gases." According to the kinetic molecular theory that defines an ideal gas, no ideal gases exist in nature, only "real" gases. Van der Waals' equation is an attempt to make corrections to real gases that do not exhibit ideal behavior. Describe the type of gaseous molecules that are most susceptible to nonideal behavior.

4. a. How is the pressure of the vaporized liquid determined in the experiment?

   b. How is the volume of the vaporized liquid determined in the experiment?

   c. How is the temperature of the vaporized liquid determined in the experiment?

   d. How is the mass of the vaporized liquid determined in the experiment?

5. The molar mass of a compound is measured to be 33.4, 35.2, 34.1, and 33.9 g/mol in four trials.
   a. What is the average molar mass of the compound?

   b. Calculate the standard deviation (see Appendix B) for the determination of the molar mass.

Experiment 18 *Report Sheet*

Molar Mass of a Volatile Liquid

Date _________ Lab Sec. ______ Name ______________________________ Desk No. _________

Unknown Number ____________________	*Trial 1**	*Trial 2*	*Trial 3*
1. Mass of dry flask, foil, and rubber band (g)	________	________	________
2. Temperature of boiling water (°C, K)	________	________	________
3. Mass of dry flask, foil, rubber band, and vapor (g)	________	________	________
4. Volume of 125-mL flask (L)			
_____ + _____ + _____ = total volume	________________		
5. Atmospheric pressure (torr, atm)	________________		

Calculations

1. Amount of vapor, n_{vapor} (mol)	________	________	________
2. Mass of vapor, m_{vapor} (g)	________	________	________
3. Molar mass of compound (g/mol)	________	________	________
4. Average molar mass (g/mol)	________________		
5. Standard deviation of molar mass	________________		

*Calculation of Trial 1. Show work here.

Class Data/Group	1	2	3	4	5	6
Molar mass						
Sample Unknown No. _________						

What is the standard deviation of the molar mass of the unknown for the class?

(Optional) Ask your instructor for the name of your unknown liquid. Using van der Waals' equation and the values of a and b for your compound, repeat the calculation for the moles of vapor, n_{vapor} (show for Trial 1 below), to determine a more accurate molar mass of the compound.

Calculations (van der Waals' equation)

Unknown Number ______________	*Trial 1**	*Trial 2*	*Trial 3*
1. Amount of vapor, n_{vapor} (mol)	______	______	______
2. Mass of vapor, m_{vapor} (g)	______	______	______
3. Molar mass of compound (g/mol)	______	______	______

4. Average molar mass (g/mol) ______________

*Calculation of n_{vapor} from van der Waals' equation for Trial 1. Show work here.

Laboratory Questions

Circle the questions that have been assigned.

1. Part A.1. The mass of the flask is measured when the outside of the flask is wet. However in Part B.3 the outside of the flask is dried before its mass is measured.
   a. Will the mass of vapor in the flask be reported too high, too low, or unaffected? Explain.
   b. Will the molar mass of vapor in the flask be reported too high, too low, or unaffected? Explain.
2. Part A.1. From the time the mass of the flask is first measured in Part A.1 until the time it is finally measured in Part B.3, it is handled a number of times with oily fingers. How does this lack of proper technique affect the reported mass of the vapor in the flask? Explain.
3. Part A.2. The aluminum foil is pierced several times with pencil-size holes rather than pin-size holes.
   a. How will this oversight in the procedure affect the mass of vapor measured in Part B.3? Explain.
   b. Will the reported molar mass of the liquid be reported too high, too low, or unaffected? Explain.
4. Part B.2. The flask is not only completely filled with vapor, but some liquid also remains in the bottom of the flask when it is removed from the hot water bath and cooled in Part B.3. How will this oversight affect the reported molar mass of the liquid . . . too high, too low, or unaffected? Explain.
5. Part B.2. The flask is *only* completely filled with vapor when it is removed from the hot water bath and cooled in Part B.3. However when the flask cools some of the vapor condenses in the flask. As a result of this observation, will the reported molar mass of the liquid be too high, too low, or unaffected? Explain.
6. Part C.2. The pressure reading from the barometer is recorded higher than it actually is. How does this affect the reported molar mass of the liquid . . . too high, too low, or unaffected? Explain.

Experiment 19

Calcium Carbonate Analysis; Molar Volume of Carbon Dioxide

The reaction of hydrochloric acid on calcium carbonate produces carbon dioxide gas.

OBJECTIVES

- To determine the percent calcium carbonate in a **heterogeneous mixture**
- To determine the molar volume of carbon dioxide gas at 273 K and 760 torr

Heterogeneous mixture: a nonuniform mix of two or more substances, oftentimes in different phases

TECHNIQUES

The following techniques are used in the Experimental Procedure

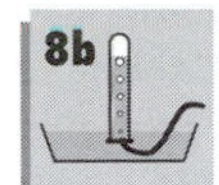

INTRODUCTION

Calcium carbonate is perhaps the most prevalent simple inorganic compound in the Earth's crust. More commonly known as limestone, calcium carbonate is found in many forms and formulations. Chalk, marble (a dense form of calcium carbonate), shells of shellfish, stalactites, stalagmites (Figure 19.1), caliche, and the minerals responsible for hard water have their origin from fossilized remains of marine life.

Figure 19.1 Stalagmites and stalactites are primarily calcium carbonate.

Calcium carbonate readily reacts in an acidic medium to produce carbon dioxide gas:

$$CaCO_3(s) + 2\,H_3O^+(aq) \rightarrow Ca^{2+}(aq) + 3\,H_2O(l) + CO_2(g) \qquad (19.1)$$

In this experiment, the percent by mass of calcium carbonate in a heterogeneous mixture is determined. The calcium carbonate of the sample is treated with an excess of hydrochloric acid, and the carbon dioxide gas is collected over water. The moles of carbon dioxide generated in the reaction are measured and, from the stoichiometry of Equation 19.1, the moles and mass of $CaCO_3$ in the sample are calculated. Because carbon dioxide is relatively soluble in water, the water over which the CO_2 is collected is pretreated to saturate it with carbon dioxide.

In addition to analyzing the unknown mixture for percent calcium carbonate and impurities, the **molar volume** of carbon dioxide is also determined. At **standard temperature and pressure** (STP), one mole of an *ideal* gas occupies 22.4 L; that is, its molar volume is 22.4 L at STP. Because carbon dioxide is not an ideal gas, we may expect its molar volume to vary slightly from this number.

Molar volume: volume occupied by one mole (6.023×10^{23} atoms or molecules) of gas at defined temperature and pressure conditions

Standard temperature and pressure: 273 K (0°C) and 1 atmosphere (760 torr)

To make these two determinations in the experiment, two important measurements are made: (1) the CO_2 gas evolved from the reaction is collected and its volume is measured and (2) the mass difference of the $CaCO_3$ mixture, before and after reaction, is measured.

Molar Volume of Carbon Dioxide

The mass loss of the $CaCO_3$ mixture is due to the mass of CO_2 gas evolved in the reaction; the mass of CO_2 is converted to moles of CO_2 evolved. The volume of CO_2 gas, collected over water under the temperature and pressure conditions of the experiment, is adjusted to STP conditions. Knowing the number of moles and the volume at STP, the molar volume of CO_2 gas is calculated.

$$\frac{V_{CO_2}(\text{STP})}{n_{CO_2}} = \text{molar volume of } CO_2 \qquad (19.2)$$

Volume of Collected Carbon Dioxide at STP

Dalton's law of partial pressures: the total pressure, P_T, exerted by a mixture of gases is the sum of the individual pressures (called ***partial pressures****) exerted by each of the constituent gases*

Leveling tank: a large container of water in which the level of water in a flask can be adjusted to equal the water level in the container

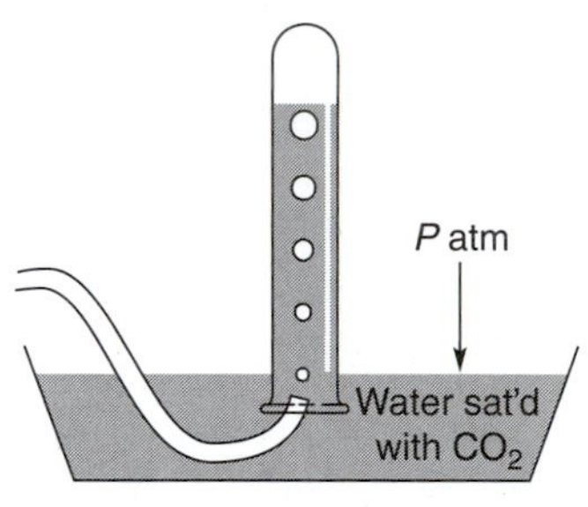

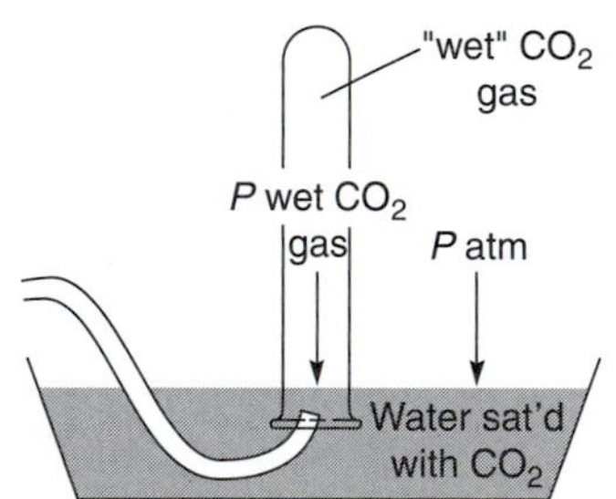

Figure 19.2 The collection of CO_2 gas over water (top) and the pressure measurement of "wet" CO_2 gas (bottom).

The CO_2 gas evolved in the reaction is collected by displacing an equal volume of water (Figure 19.2 (top)). Because the CO_2 gas is bubbled through the water, it is considered "wet," meaning that the volume occupied by the CO_2 is also saturated with water vapor at the temperature of the water over which it is collected. Therefore, the total pressure, P_T, in this volume is due the combined pressures of the CO_2 gas, p_{CO_2}, and the water vapor, p_{H_2O}. The pressure of the "dry" CO_2 is calculated using **Dalton's law of partial pressures.**

For carbon dioxide gas,

$$p_{CO_2} = P_T - p_{H_2O} \qquad (19.3)$$

The pressure of the water vapor, p_{H_2O}, at the gas-collecting temperature (obtained from Appendix E) is subtracted from the total pressure of the gases, P_T, in the gas-collecting vessel. Experimentally, the total pressure of the gaseous mixture ($CO_2 + H_2O$) is adjusted to atmospheric pressure ($P_T = P_{atm}$) by using a **leveling tank** (Figure 19.2 (bottom)). Atmospheric pressure is read from the laboratory barometer.

Once the experimental values for the volume, pressure, and temperature of the CO_2 gas are determined, the volume of CO_2 gas at STP conditions is calculated using a combination of Boyle's law ($P \propto 1/V$) and Charles' law ($V \propto T$).

$$V_{CO_2}(\text{at STP}) = V_{CO_2\,\text{expt}} \times \underbrace{\left(\frac{p_{CO_2\,\text{expt}}(\text{torr})}{760 \text{ torr}}\right)}_{\text{Boyle's law correction}} \times \underbrace{\left(\frac{273 \text{ K}}{T_{CO_2\,\text{expt}}(\text{K})}\right)}_{\text{Charles' law correction}} \qquad (19.4)$$

This value is used in Equation 19.2.

In addition to the fact that the collected CO_2 gas is "wet," CO_2 also has an appreciable solubility in water. When the CO_2 gas is bubbled through pure water, some of the CO_2 dissolves in the water and is not measured as gas evolved in the reaction. To minimize any loss of evolved CO_2 as a result of its water solubility, the water is saturated with CO_2 *before* the experiment is conducted. This saturation is completed with the addition of either sodium bicarbonate to slightly acidified water or an antacid tablet that evolves CO_2, such as Alka-Seltzer®.

Percent Calcium Carbonate in Mixture

By use of Equation 19.1, once the moles of CO_2 gas evolved in the reaction is known, the moles and mass of $CaCO_3$ in the mixture can be calculated. The percent $CaCO_3$ in the mixture is calculated by dividing this mass of $CaCO_3$ by that of the original mixture and multiplying by 100.

$$\frac{\text{mass of } CaCO_3}{\text{mass of mixture}} \times 100 = \% \; CaCO_3 \qquad (19.5)$$

EXPERIMENTAL PROCEDURE

Procedure Overview: A gas generator apparatus is constructed to collect the CO_2 gas evolved from a reaction. The masses of the sample in the gas generator before and after the reaction are measured; the volume of gas evolved in the reaction is measured.

Two trials are required in this experiment. To hasten the analyses, prepare two samples for Part A and perform the experiment with a partner—a partner you can trust!

A. Sample Preparation and Setup of Apparatus

1. **Water Saturated with CO_2.** Fill a 1-L beaker with tap water and saturate the water with CO_2, using one Alka-Seltzer® tablet.[1] Save for Part A.3.
2. **Sample Preparation.** a. *Mass of sample.* Calculate the mass of $CaCO_3$ that would produce 50 mL of CO_2. *Show the calculation on the Report Sheet.* On weighing paper, measure this calculated mass (±0.001 g) of the sample mixture, one that contains $CaCO_3$ and a noncarbonate impurity. Record the number of the unknown sample. Carefully transfer the sample to a 75-mm test tube.

   b. *Set up the CO_2 generator.* Place 10 mL of 3 *M* HCl in a 200-mm test tube; carefully slide the 75-mm test tube into the 200-mm test tube without splashing any of the acid into the sample. **Important:** The HCl(*aq*) is in the 200-mm test tube and separately, but inserted, is the $CaCO_3(s)$ sample in the 75-mm test tube . . . do not mix the two substances until Part B.1!

   c. *Mass of CO_2 generator.* Measure the combined mass (±0.001 g) of this CO_2 generator (Figure 19.3).

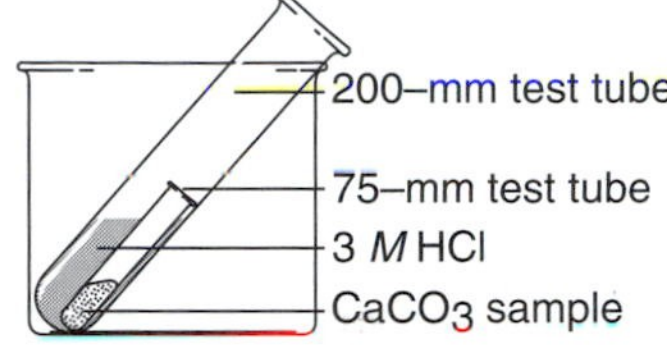

Figure 19.3 Apparatus for measuring the combined mass of the CO_2 generator.

3. **Setup of CO_2 Collection Apparatus.** a. *Saturated Water with CO_2.* Fill the pan (Figure 19.4) about two-thirds full with the water that is saturated with CO_2 (Part A.1). Do not proceed in the experiment until the CO_2 from the Alka-Seltzer® is no longer evolved.

   b. *Fill CO_2 collection tubes.* Use the CO_2-saturated water in the pan to fill the 200-mm test tube that will collect the CO_2 from the sample. Fill the 200-mm test tube by laying it horizontal in the water, and then, without removing the mouth of the test tube from the water, set it upright.[2]

   c. *Connect the gas inlet tube.* Place the gas inlet tube that connects to the CO_2 generator into the mouth of the 200-mm CO_2 collection test tube. Support the test tube with a ring stand and clamp.
4. **Setup of CO_2 Generator.** Prepare a one-hole rubber stopper (check to be certain there are no "cracks" in the rubber stopper) fitted with a short piece of glass tubing and insert it into the 200-mm test tube. Clamp the CO_2 generator (200-mm test

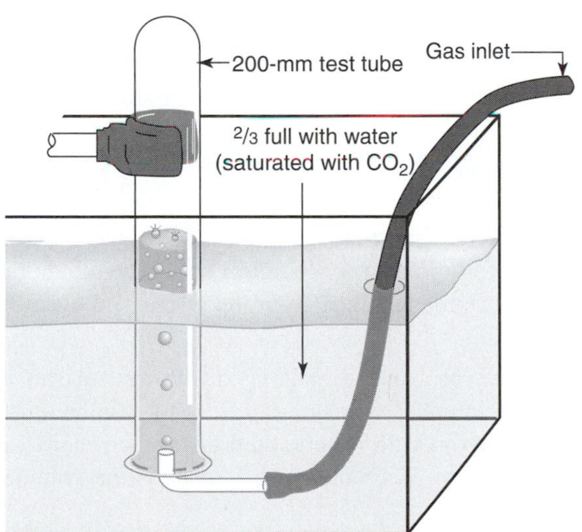

Figure 19.4 A CO_2 gas-collection apparatus.

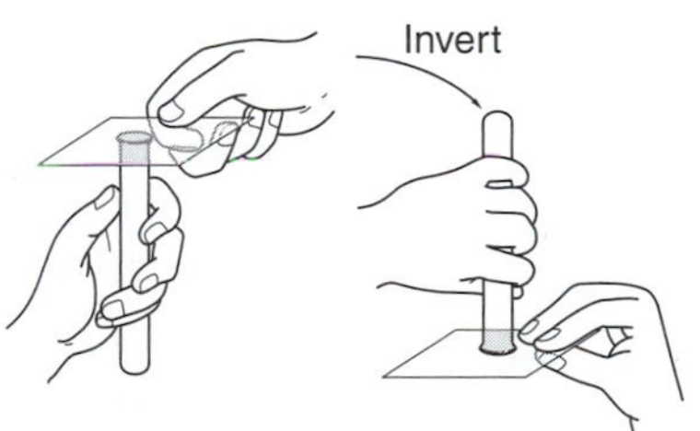

Slide glass plate over top of test tube. Do not allow an air bubble to enter the test tube.

Figure 19.5 Inverting a test tube filled with CO_2-saturated water.

[1] An acidic solution of sodium bicarbonate can also be used to saturate the water with CO_2.
[2] If this procedure is not possible, fill the test tube with CO_2-saturated water and invert the test tube into the pan as shown in Figure 19.5.

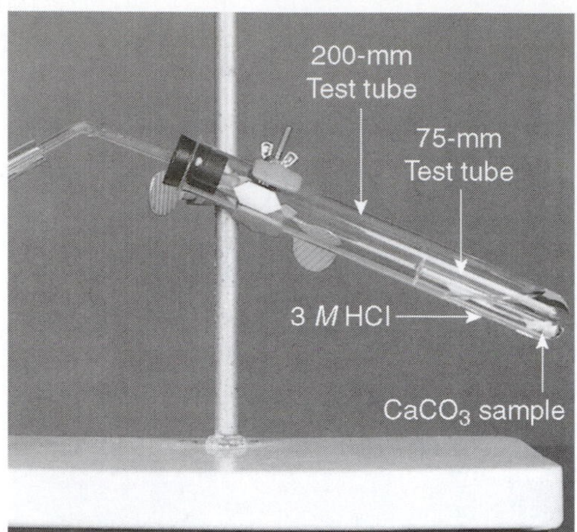

Figure 19.6 A CO_2 gas generator.

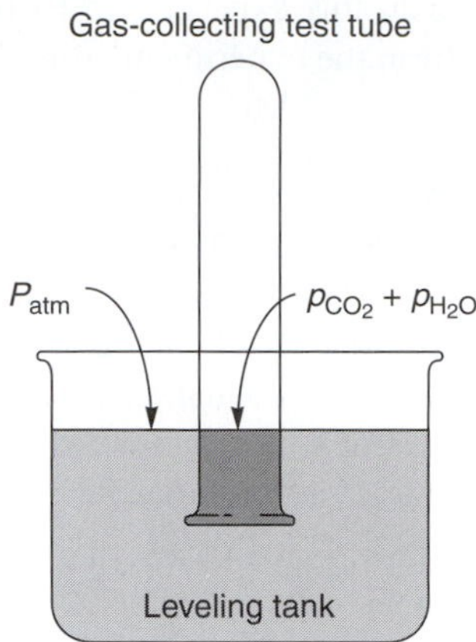

Figure 19.7 Equalizing the pressure in the test tube to atmospheric pressure.

tube) to the ring stand at a 45° angle from the horizontal (Figure 19.6). Connect the gas delivery tube (using rubber or Tygon tubing) from the CO_2 collection apparatus (Figure 19.4) to the CO_2 generator (Figure 19.6).

5. **Obtain Instructor's Approval.** Obtain your instructor's approval before continuing.

B. Collection of the Carbon Dioxide Gas

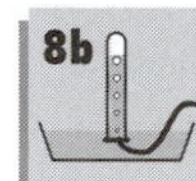

1. **Generate and Collect the CO_2 Gas.** a. *Initiate the reaction. Gently* agitate the generator (Figure 19.6) to allow some of the HCl solution to contact the sample mixture. As the evolution rate of CO_2 gas decreases, agitate again and again until CO_2 gas is no longer evolved.

   b. *Complete the reaction.* When no further generation of CO_2 gas is evident (Figure 19.4), remove the CO_2-collecting test tube from the gas inlet tube. Same for Part C.2.

C. Determination of the Temperature, Volume, and Pressure of the Carbon Dioxide Gas

Appendix E

1. **Determine the Temperature of the CO_2 Gas.** Record the temperature of the water in the pan.
2. **Determine the Pressure of the CO_2 Gas.** Adjust the pressure inside the CO_2-collecting 200-mm test tube to atmospheric pressure by raising or lowering it until the water levels inside and out are the same (Figure 19.7). You may need to transfer the test tube to the leveling tank in the laboratory.

   Read and record the barometric pressure in the laboratory. Obtain the vapor pressure of water at the gas-collecting temperature in Appendix E.
3. **Determine the Volume of CO_2 Gas.** a. *Invert the 200-mm CO_2 collection test tube.* Place a glass plate, a piece of filter paper, or Parafilm under the mouth of the test tube (from Part C.2), remove the test tube from the water, and place it right side up in your test tube rack.

   b. *Volume of CO_2 collected.* Transfer the water remaining in the test tube to a graduated cylinder and record its volume. Now, measure the total volume of the test tube—fill the test tube to the rim with water, and transfer the water to a graduated cylinder. Record this volume. The volume difference is the volume of CO_2 collected from the reaction.

D. Mass of Carbon Dioxide Evolved

1. **Determine a Mass Difference.** Determine the mass (±0.001 g) of the 200-mm CO_2 generator and its remaining contents. Compare this mass with that in Part A.2c. After subtracting for the mass of the generator, calculate the mass loss of the sample.

Experiment 19 *Prelaboratory Assignment*

Calcium Carbonate Analysis; Molar Volume of Carbon Dioxide

Date __________ Lab Sec. ______ Name ____ Desk No. __________

1. In some solid calcium carbonate samples, calcium bicarbonate, $Ca(HCO_3)_2$, is also present. Write a balanced equation for its reaction with hydrochloric acid.

2. There is a movement afoot in the scientific community to change current STP conditions (1 atmosphere pressure and 273 K) to STP conditions using SI base units. This would set STP conditions to be 1 bar and 273 K. What would be the value of the "new" molar volume of an ideal gas under the "new" STP conditions? 1 atm = 101.325 kPa = 1.10325 bars.

3. A mixture of gases collected over water at 14°C has a total pressure of 0.981 atm and occupies 55 mL.
   a. How many grams of water escaped into the vapor state?

   b. What volume does the mixture of gases, exclusive of the water vapor, occupy?

   c. What volume does the mixture of gases, exclusive of the water vapor, occupy at STP conditions?

4. A 37.7-mL volume of an unknown gas is collected *over water* at 19°C and a total pressure of 770 torr. The mass of the gas is 68.6 mg.
   a. What is the pressure of the gas?

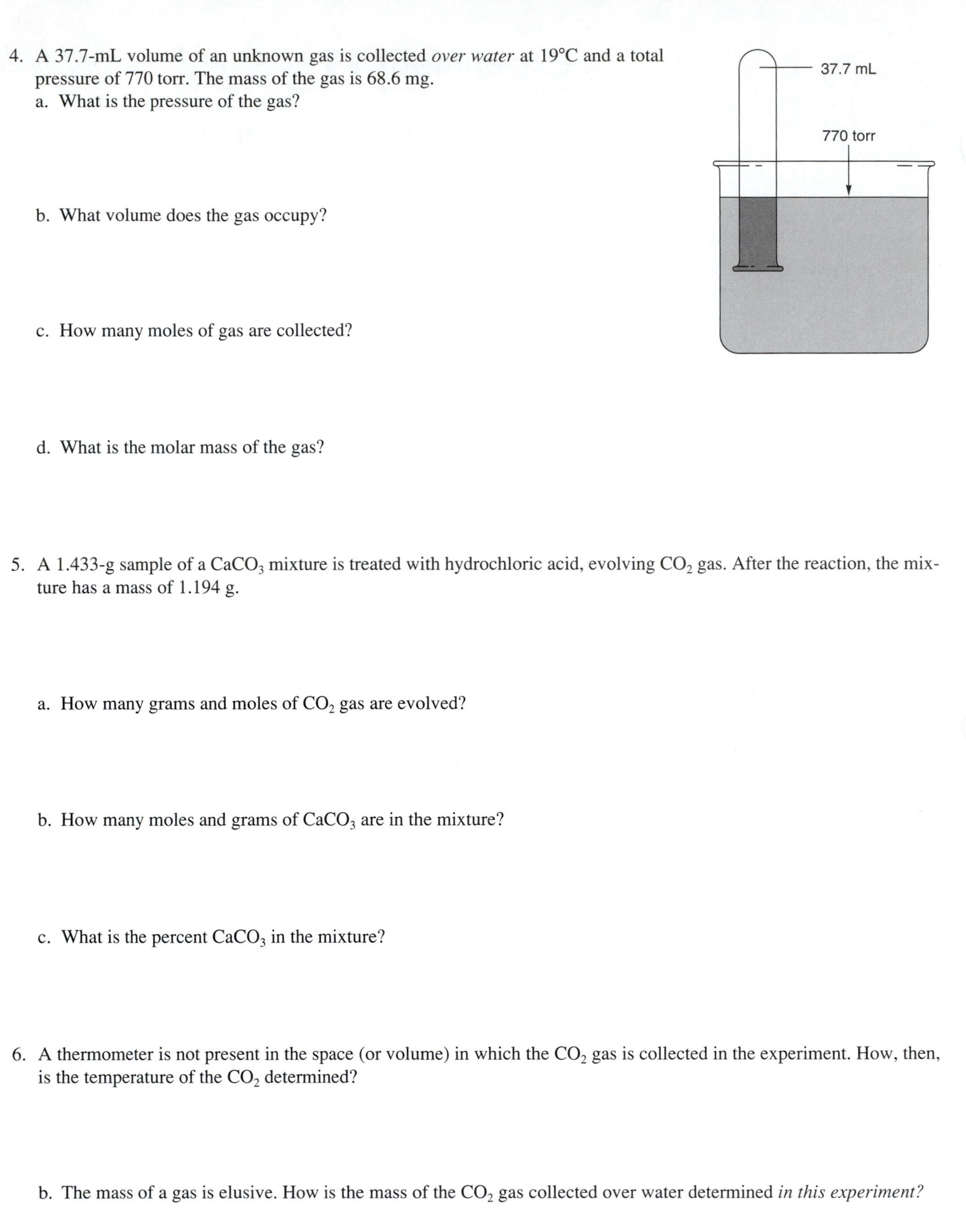

   b. What volume does the gas occupy?

   c. How many moles of gas are collected?

   d. What is the molar mass of the gas?

5. A 1.433-g sample of a $CaCO_3$ mixture is treated with hydrochloric acid, evolving CO_2 gas. After the reaction, the mixture has a mass of 1.194 g.

   a. How many grams and moles of CO_2 gas are evolved?

   b. How many moles and grams of $CaCO_3$ are in the mixture?

   c. What is the percent $CaCO_3$ in the mixture?

6. A thermometer is not present in the space (or volume) in which the CO_2 gas is collected in the experiment. How, then, is the temperature of the CO_2 determined?

   b. The mass of a gas is elusive. How is the mass of the CO_2 gas collected over water determined *in this experiment?*

Calcium Carbonate Analysis; Molar Volume of Carbon Dioxide

Date ________ Lab Sec. ______ Name ______________________________ Desk No. ________

Calculation of Mass of $CaCO_3$ Sample for Analysis

Unknown Sample No. __________

Data	*Trial 1*	*Trial 2*
1. Mass of sample (g)		
2. Mass of generator + sample before reaction (g)		
3. Instructor's approval of apparatus		
4. Temperature of water (°C)		
5. Vapor pressure of H_2O at ____°C (Appendix E, torr)		
6. Barometric pressure (torr)		
7. Volume of water remaining in "CO_2-collecting" test tube (mL)		
8. Volume of "CO_2-collecting" test tube (mL)		
9. Volume of CO_2 gas collected (L)		
10. Mass of generator and sample after reaction (g)		
11. Mass of CO_2 generated (g)		

Molar Volume of CO_2 Gas

1. Pressure of "dry" CO_2 gas (torr) ______________ ______________
2. Volume of CO_2 gas at STP (L) ______________ ______________
3. Amount of CO_2 gas generated (mol) ______________ ______________
4. Molar volume of CO_2 gas at STP (L/mol) ______________ ______________
5. Average molar volume of CO_2 gas at STP (L/mol) ______________

Percent $CaCO_3$ in Mixture

1. Calculated amount of $CaCO_3$ in sample (mol) ______________ ______________
2. Mass of $CaCO_3$ in sample (g) ______________ ______________
3. Mass of original sample (g) ______________ ______________
4. Percent of $CaCO_3$ in sample (%) ______________ ______________
5. Average percent of $CaCO_3$ in sample (%) ______________

Laboratory Questions

Circle the questions that have been assigned.

1. Part A.1. The water for the pan (Part A.3) is not saturated with CO_2. Will the reported percent $CaCO_3$ in the original sample be too high, too low, or unaffected? Explain.
2. Part A.3. A few drops of HCl(*aq*) spilled over into the $CaCO_3$ sample. What error in the reported percent $CaCO_3$ will result from this poor laboratory technique? Explain.
3. Part A.4. A small crack is present in the rubber stopper. How does this affect the calculated molar volume of the CO_2 . . . too high, too low, or unaffected? Explain.
4. Part C.2. The water level in the CO_2-gas collecting tube is *higher* than the water level in the leveling tank.
   a. Is the "wet" CO_2 gas pressure greater or less than atmospheric pressure? Explain.
   b. An adjustment is made to equilibrate the water levels. Will the volume of the "wet" CO_2 gas increase or decrease? Explain.
   c. The student chemist chooses *not* to equilibrate the inside and outside water levels. Will the reported number of moles of CO_2 generated in the reaction be too high, too low, or unaffected by this carelessness? Explain.

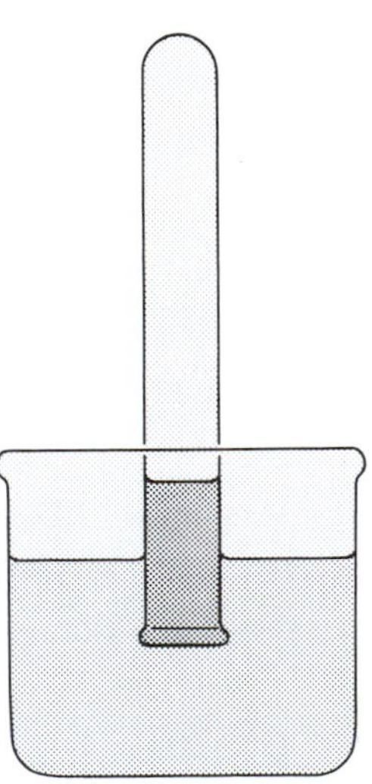

5. Part C.3a. During the inversion of the test tube, water from the pan entered the test tube. How does this poor technique affect the reported moles of CO_2 collected? Explain.
6. Part D.1. The handling of the CO_2 gas-generator with oily fingers affects the final mass determination. How does this (nearly unavoidable) error affect the reported percent $CaCO_3$ in the original sample? Explain.

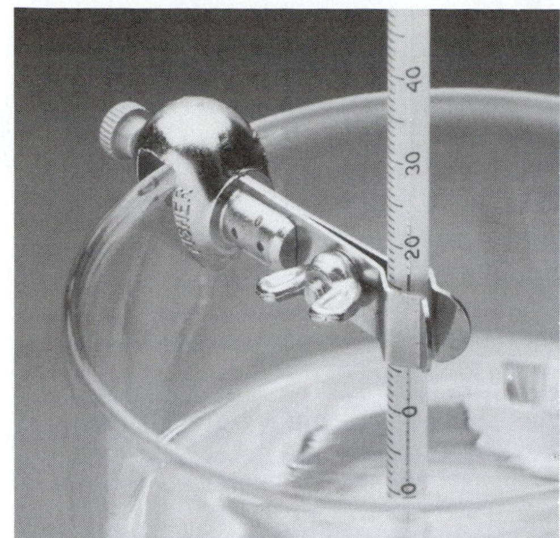

A thermometer is secured with a thermometer clamp to guard against breakage.

Experiment 20

Molar Mass of a Solid

Objectives

- To observe and measure the effect of a solute on the freezing point of a solvent
- To determine the molar mass (molecular weight) of a nonvolatile, nonelectrolyte solute

Techniques

The following techniques are used in the Experimental Procedure

Introduction

A pure liquid, such as water or ethanol, has characteristic physical properties: the melting point, boiling point, density, vapor pressure, viscosity, surface tension, and additional data are listed in handbooks of chemistry. The addition of a soluble solute to the liquid forms a homogeneous mixture, called a **solution.** The solvent of the solution assumes physical properties that are no longer definite, but dependent on the amount of solute added. The vapor pressure of the solvent decreases, the freezing point of the solvent decreases, the boiling point of the solvent increases, and the osmotic pressure of the solvent increases. The degree of the change depends on the *number* of solute particles that have dissolved, *not* on the chemical identity of the solute. These four physical properties that depend on the number of solute particles dissolved in a solvent are called **colligative properties.**

For example, one mole of glucose or urea (neither of which dissociates in water) lowers the freezing point of one kilogram of water by 1.86°C; whereas one mole of sodium chloride lowers the freezing point of one kilogram of water by nearly twice that amount (~3.72°C) because, when dissolved in water, it dissociates into Na^+ and Cl^- providing *twice* as many moles of solute particles per mole of solute as do glucose or urea.

When freezing ice cream at home, a salt/ice/water mixture provides a lower temperature bath than an ice/water mixture alone. Antifreeze (ethylene glycol, Figure 20.1) added to the cooling system of an automobile reduces the probability of freeze-up in the winter and boiling over in the summer because the antifreeze/water solution has a lower freezing point and a higher boiling point than pure water.

These changes in the properties of pure water that result from the presence of a **nonvolatile solute** are portrayed by the phase diagram in Figure 20.2, a plot of vapor pressure versus temperature. The solid lines refer to the equilibrium conditions between the respective phases for pure water; the dashed lines represent the same conditions for an aqueous solution.

ZEREX PROTECTION CHART

Freeze-up/Boil-over Protection	Protects against freeze-ups down to	Protects against boil-overs up to
For Extreme Conditions Install a 70/30 Mix of ZEREX and Water	−84°F	276°F*
For Year Round Operation Install a 50/50 Mix of ZEREX and Water	−34°F	265°F*

*Using a 15 PSI radiator cap

WARNING:

HARMFUL OR FATAL IF SWALLOWED. Do not drink antifreeze coolant or solution. If swallowed, induce vomiting immediately. Call a physician. Contains Ethylene Glycol which caused birth defects in animal studies. Do not store in open or unlabeled containers.

KEEP OUT OF REACH OF CHILDREN AND ANIMALS.

Ingredients: Ethylene glycol (107-21-1), Diethylene glycol (111-46-6), Dipotassium phosphate (7758-11-4), Water (7732-18-5), Corrosion inhibitors, Silicone silicate, Defoamer, Dyes.

Figure 20.1 Ethylene glycol is a major component of most antifreeze solutions.

Colligative properties: properties of a solvent that result from the presence of the number of solute particles in the solution and not their chemical composition

Nonvolatile solute: a solute that does not have a measurable vapor pressure

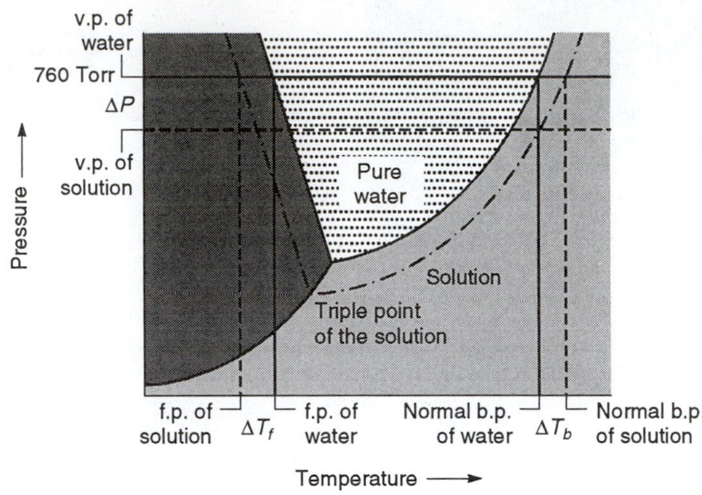

Figure 20.2 Phase diagram (not shown to scale) for water (—) and for an aqueous solution (- - -).

Vapor pressure: pressure exerted by a vapor when the vapor is in a state of dynamic equilibrium with its liquid

Boiling point: the temperature at which the vapor pressure of a liquid equals atmospheric pressure

Freezing point: the temperature at which the liquid and solid phases of a substance co-exist

The **vapor pressure** of water is 760 torr at its boiling point of 100°C. When a nonvolatile solute dissolves in water to form a solution, solute molecules occupy a part of the surface area. This inhibits movement of some water molecules into the vapor state, causing a **vapor pressure lowering** of the water (ΔP in Figure 20.2), lower than 760 torr. With the vapor pressure less than 760 torr, the solution (more specifically, the water in the solution) no longer boils at 100°C. For the solution to boil, the vapor pressure must be increased to 760 torr; boiling can resume only if the temperature is increased above 100°C. This **boiling point elevation** (ΔT_b in Figure 20.2) of the water is due to the presence of the solute.

A solute added to water also affects its freezing point. The normal freezing point of water is 0°C, but in the presence of a solute, the temperature must be lowered below 0°C before freezing occurs (the energy of the water molecules must be lowered to increase the magnitude of the intermolecular forces, so that the water molecules "stick" together to form a solid); this is called a **freezing point depression** of the water (ΔT_f in Figure 20.2).

The changes in the freezing point, ΔT_f, and the boiling point, ΔT_b, are directly proportional to the molality, m, of the solute in solution. The proportionality is a constant, characteristic of the actual solvent. For water the freezing point constant, k_f, is 1.86°C•kg/mol and the boiling point constant, k_b, is 0.512°C•kg/mol.

$$\Delta T_f = |T_{\text{f, solvent}} - T_{\text{f, solution}}| = k_f m \quad (20.1)$$

$$\Delta T_b = |T_{\text{b, solvent}} - T_{\text{b, solution}}| = k_b m \quad (20.2)$$

Table 20.1 Molal Freezing Point and Boiling Point Constants for Several Solvents

Substance	Freezing Point (°C)	$k_f\left(\frac{°C \cdot kg}{mol}\right)$	Boiling Point (°C)	$k_b\left(\frac{°C \cdot kg}{mol}\right)$
H_2O	0.0	1.86	100.0	0.512
cyclohexane	—	20.0	80.7	2.69
naphthalene	80.2	6.9	—	—
camphor	179	39.7	—	—
acetic acid	17	3.90	118.2	2.93
t-butanol	25.5	9.1	—	—

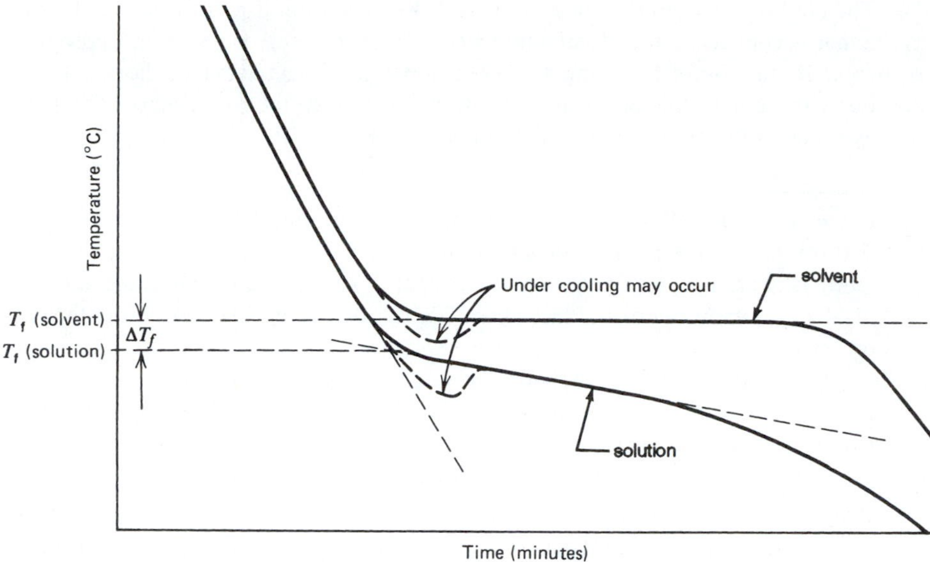

Figure 20.3 Cooling curves for a solvent and solution.

In Equations 20.1 and 20.2, T_f represents the freezing point and T_b represents the boiling point of the respective system. $|T_{f,\text{ solvent}} - T_{f,\text{ solution}}|$ represents the absolute temperature difference in the freezing point change. **Molality** is defined as

$$\text{molality, } m = \frac{\text{mol solute}}{\text{kg solvent}} = \frac{(\text{mass/molar mass})}{\text{kg solvent}} \qquad (20.3)$$

k_f and k_b values for various solvents are listed in Table 20.1.

In this experiment the freezing points of a selected pure solvent and of a solute/solvent mixture are measured. The freezing point lowering (difference), the k_f data from Table 20.1 for the solvent, and Equations 20.1 and 20.3 are used to calculate the moles of solute dissolved in solution and, from its measured mass, the molar mass of the solute.

The freezing points of the solvent and the solution are obtained from a **cooling curve**—a plot of temperature versus time. An ideal plot of the data appears in Figure 20.3. The cooling curve for a pure solvent reaches a plateau at its freezing point; extrapolation of the plateau to the temperature axis determines its freezing point. The cooling curve for the solution does *not* reach a plateau, but continues to decrease slowly as the solvent freezes out of solution. Its freezing point is determined at the intersection of two straight lines drawn through the data points above and below the freezing point (Figure 20.3).

Cooling curve: a data plot of temperature vs. time showing the rate of cooling of a substance before, during, and after its phase changes

Experimental Procedure

Procedure Overview: Cyclohexane is the solvent selected for this experiment although other solvents are just as effective for the determination of the molar mass of a solute. Another solvent[1] may be used at the discretion of the laboratory instructor. Consult with your instructor. The freezing points of cyclohexane and a cyclohexane *solution* are determined from plots of temperature versus time. The mass of the solute is measured before it is dissolved in a known mass of cyclohexane.

Obtain about 15 mL of cyclohexane. You'll use the cyclohexane throughout the experiment. Your laboratory instructor will issue you about 1 g of unknown solute. Record the unknown number of the solute on the Report Sheet.

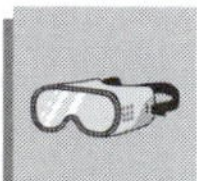

[1] *t*-Butanol is a suitable substitute for cyclohexane in this experiment.

The cooling curve to be plotted in Part A.4 can be established by using a thermal probe that is connected directly to either a calculator or computer with the appropriate software. If this thermal sensing/recording apparatus is available in the laboratory, consult with your instructor for its use and adaptation to the experiment. The probe merely replaces the glass or digital thermometer in Figure 20.4.

A. Freezing Point of Cyclohexane (Solvent)

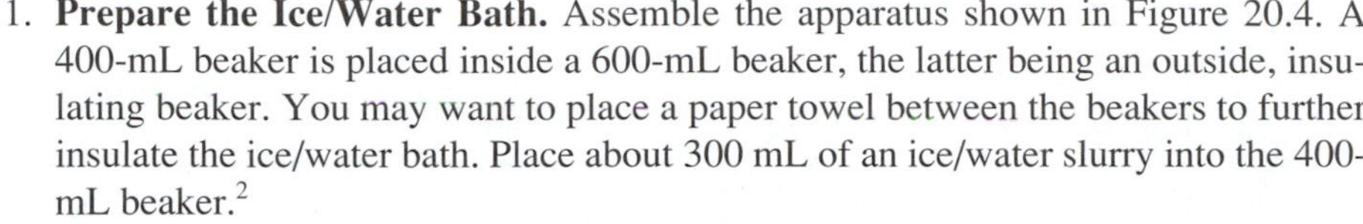

1. **Prepare the Ice/Water Bath.** Assemble the apparatus shown in Figure 20.4. A 400-mL beaker is placed inside a 600-mL beaker, the latter being an outside, insulating beaker. You may want to place a paper towel between the beakers to further insulate the ice/water bath. Place about 300 mL of an ice/water slurry into the 400-mL beaker.[2]

   Obtain a glass or digital thermometer and thermometer clamp, mount it to the ring stand, and position the thermometer in the test tube. (**Caution:** *If the thermometer is a glass thermometer, handle the thermometer carefully. If the thermometer is accidentally broken, notify your instructor immediately.*)

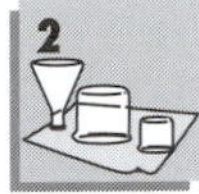

2. **Prepare the Cyclohexane.** Determine the mass (±0.01 g)[3] of a *clean, dry* 200-mm test tube in a 250-mL beaker (Figure 20.5). Add approximately 12 mL of cyclohexane (**Caution:** *Cyclohexane is flammable—keep away from flames; cyclohexane is a mucous irritant—do not inhale*) to the test tube and again measure the mass. Record the mass of the cyclohexane. Place the test tube containing the cyclohexane into the ice/water bath (Figure 20.5). Secure the test tube with a utility clamp. Insert the thermometer and a wire stirrer into the test tube. *Secure the thermometer* so that the thermal sensor extends into the cyclohexane.

3. **Record Data for the Freezing Point of Cyclohexane.** While stirring with the wire stirrer, record the temperature at timed intervals (15 or 30 seconds) on the Report Sheet. The temperature remains virtually constant at the freezing point until the solidification is complete. Continue collecting data until the temperature begins to drop again.

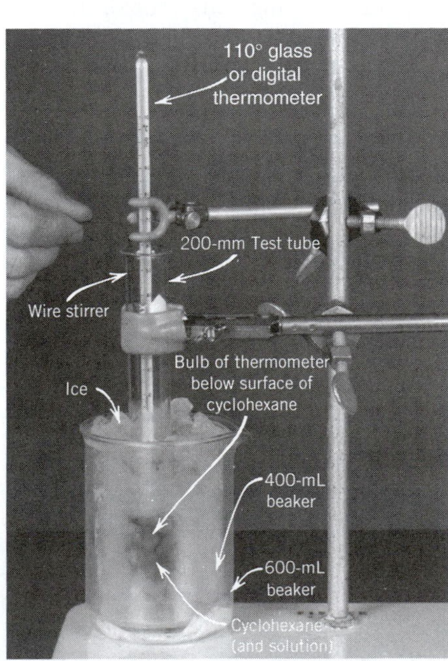

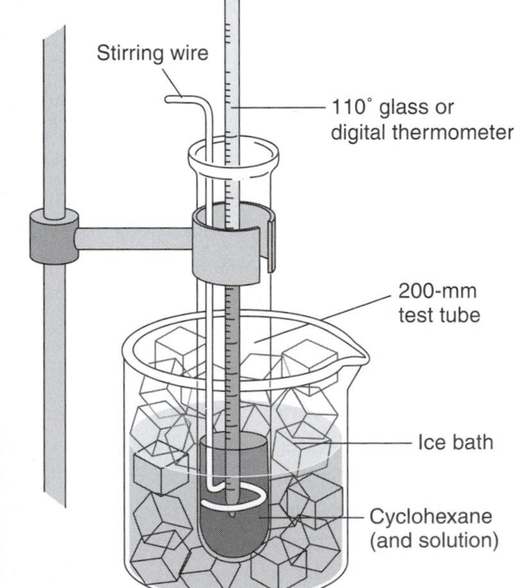

Figure 20.4 Freezing point apparatus.

---

[2]Rock salt may be added to further lower the temperature of the ice/water bath.
[3]Use a balance with ±0.001 g sensitivity, if available.

Figure 20.5 Determining the mass of cyclohexane.

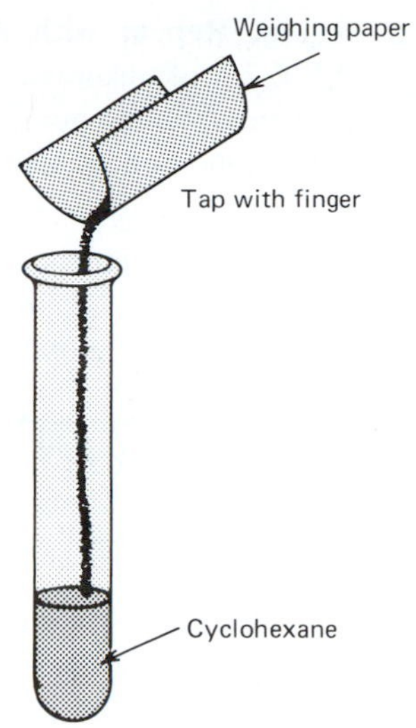

Figure 20.6 Transfer of the unknown solid solute to the test tube containing cyclohexane.

4. **Plot the Data.** On linear graph paper, plot the temperature (°C, vertical axis) versus time (sec, horizontal axis) to obtain the cooling curve for cyclohexane. Have your instructor approve your graph.

Appendix C

B. Freezing Point of Cyclohexane plus Unknown Solute

Three freezing point trials for the cyclohexane *solution* are to be completed. Successive amounts of unknown sample are added to the cyclohexane in Parts B.4 and B.5.

1. **Measure the Mass of Solvent and Solid Solute.** Dry the outside of the test tube containing the cyclohexane and again measure its mass in the same 250-mL beaker with the same balance as in Part A.2. On weighing paper tare the mass of 0.1–0.3 g of unknown solid solute (ask your instructor for the approximate mass to use) and record. *Quantitatively* transfer the solute to the cyclohexane in the 200-mm test tube (Figure 20.6).[4]

2. **Record Data for the Freezing Point of Solution.** Determine the freezing point of this solution in the same way as that of the solvent (Part A.3). Record the time/temperature data on the Report Sheet. When the solution nears the freezing point of the pure cyclohexane, record the temperature at more frequent time intervals (~15 seconds). A "break" in the curve occurs as the freezing begins, although it may not be as sharp as that for the pure cyclohexane.
3. **Plot the Data on the Same Graph.** Plot the temperature versus time data on the *same* graph (and same coordinates) as those for the pure cyclohexane (Part A.4). Draw straight lines through the data points above and below the freezing point (see Figure 20.3); the intersection of the two straight lines is the freezing point of the solution.

Appendix C

4. **Repeat with Additional Solute.** Remove the test tube/solution from the ice/water bath. Add an additional 0.1–0.3 g of unknown solid solute, using the *same procedure* as in Part B.1. Repeat the freezing point determination and, again, plot the temperature versus time data on the same graph (Parts B.2 and B.3). The total mass of solute in solution is the sum from the first and second trials.

[4]In the transfer be certain that *none* of the solid solute adheres to the test tube wall. If some does, roll the test tube until the solute dissolves.

5. **Again, Repeat with Additional Solute.** Repeat Part B.4 with an additional 0.1–0.2 g of unknown solid solute, using the *same procedure* as in Part B.1. Repeat the freezing point determination and again plot the temperature versus time data on the same graph (Parts B.2–4). You now should have four plots on the same graph.
6. **Obtain Instructor's Approval.** Have your instructor approve the three temperature versus time graphs (Parts B.3–5) that have been added to your first temperature versus time graph (Part A.4) for the pure cyclohexane.

Disposal: Dispose of the waste cyclohexane and cyclohexane solution in the "Waste Organic Liquids" container.

CLEANUP: Safely store/return the thermometer. Rinse the test tube once with acetone; discard the rinse in the "Waste Organic Liquids" container.

C. Data Analysis

1. From the plotted data, determine ΔT_f for Trial 1, Trial 2, and Trial 3.
2. From k_f, the mass (in kg) of the cyclohexane, and the measured ΔT_f, calculate the moles of solute for each trial.
3. Determine the molar mass of the solute for each trial (remember the mass of the solute for each trial is different).
4. What is the average molar mass of your unknown solute?

Experiment 20 *Prelaboratory Assignment*

Molar Mass of a Solid

Date __________ Lab Sec. ______ Name ____ Desk No. __________

1. This experiment is more about understanding the colligative properties of a solution rather than the determination of the molar mass of a solid.
   a. Define colligative properties.

   b. Vapor pressure lowering is a colligative property. Explain "how" the vapor pressure of a solvent decreases when a nonvolatile solute is added to the solvent.

   c. Which of the following solutes has the greatest affect on the colligative properties of a given mass of pure water? Explain.
      i) 0.01 mol of $CaCl_2$ (an electrolyte)
      ii) 0.01 mol of KNO_3 (an electrolyte)
      iii) 0.01 mol of $CO(NH_2)_2$ (a nonelectrolyte)

2. A 0.442 g sample of a nonvolatile solid solute dissolves in 15.0 g of *t*-butanol. The freezing point of the solution is 23.9°C.
   a. What is the molality of the solute in the solution.

   b. Calculate the molar mass of the solute.

   c. The same mass of solute is dissolved in 15.0 g of cyclohexane instead of *t*-butanol. What is the expected freezing point *change* of this solution?

3. Explain why ice cubes formed from water of a glacier freeze at a higher temperature than ice cubes formed from water of an underground aquifer.

4. Two students prepare two cyclohexane solutions having the same mass of cyclohexane. Student 1 uses 0.22 g of a solute and Student 2 uses 0.17 g of the same solute. Which student will observe the larger freezing point change? Explain.

5. Two solutions are prepared using the same solute:
   Solution A: 0.14 g of the solute dissolves in 15.4 g of *t*-butanol
   Solution B: 0.17 g of the solute dissolves in 12.7 g of cyclohexane
   Which solution has the greatest freezing point change? Explain.

6. See Figure 20.3. The temperature of a solvent remains fixed over time during the phase change from liquid to solid. However, the temperature of a solution continually decreases over time during the phase change from liquid to solid. Explain.

Experiment 20 *Report Sheet*

Molar Mass of a Solid

Date __________ Lab Sec. ______ Name ______________________________ Desk No. __________

A. Freezing Point of Cyclohexane (Solvent)

1. Mass of beaker, test tube, cyclohexane (g) ______________
2. Mass of beaker, test tube (g) ______________
3. Mass of cyclohexane (g) ______________
4. Freezing point, from cooling curve (°C) ______________
5. Instructor's approval of graph ______________

B. Freezing Point of Cyclohexane plus Unknown Solute

Unknown Solute No. ______________	*Trial 1* (Parts B.1, B.3)	*Trial 2** (Part B.4)	*Trial 3* (Part B.5)
1. Mass of beaker, test tube, cyclohexane (g)			
2. Mass of cyclohexane (g)			
3. Mass of solute, *total* (g)			
4. Freezing point, from cooling curve (°C)			
5. Instructor's approval of graph			

Calculations

1. k_f for cyclohexane (pure solvent)		20.0 °C • kg/mol	
2. Freezing point *change*, ΔT_f (°C)			
3. Mass of cyclohexane in solution (kg)			
4. Moles of solute, *total* (mol)			
5. Mass of solute in solution, *total* (g)			
6. Molar mass of solute (g/mol)			
7. Average molar mass of solute			

*Show calculations for Trial 2 on the next page.

Calculations for Trial 2.

A. Cyclohexane		B. Cyclohexane + Unknown Solute					
Time	Temp	*Trial 1*		*Trial 2*		*Trial 3*	
		Time	Temp	Time	Temp	Time	Temp

Continue recording data on your own paper and submit it with the Report Sheet.

Laboratory Questions

Circle the questions that have been assigned.

1. Part A.2. Some of the cyclohexane solvent vaporized during the temperature vs. time measurements in Part A.3. How will this loss of solvent affect its freezing point determination? Explain.
2. Part A.3. The digital thermometer is miscalibrated by +0.15°C over its entire range. If the same thermometer is used in Part B.2, will the reported moles of solute in the solution be too high, too low, or unaffected? Explain.
3. Part B.1. Some of the solid solute adheres to the side of the test tube during the freezing point determination of the solution in Part B.2. As a result of the oversight, will the reported molar mass of the solute be too high, too low, or unaffected? Explain.
4. Part B.2. Some of the cyclohexane solvent vaporized during the temperature vs. time measurement. How will loss of solvent affect the freezing point determination of the solution? Explain.
5. Part B.2. The solute dissociates slightly in the solvent. How will the slight dissociation affect the reported molar mass of the solute . . . too high, too low, or unaffected? Explain.
6. Part C.1. Interpretation of the data plots consistently reports that the freezing points of three solutions are too low. As a result of this "misreading of the data," will the reported molar mass of the solute be too high, too low, or unaffected? Explain.

Experiment 21

Calorimetry

A set of nested coffee cups is a good constant pressure calorimeter.

Objectives

- To determine the specific heat of a metal and its approximate molar mass
- To determine the enthalpy of neutralization for a strong acid-strong base reaction
- To determine the quantity and direction of heat flow for the dissolution of a salt

Techniques

The following techniques are used in the Experimental Procedure

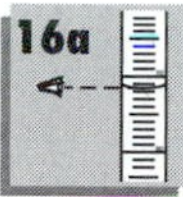

Introduction

Accompanying all chemical and physical changes is a transfer of heat (energy); heat may be either evolved (exothermic) or absorbed (endothermic). A **calorimeter** is the laboratory apparatus that is used to measure the quantity and direction of heat flow accompanying a chemical or physical change. The heat change in chemical reactions is quantitatively expressed as the **enthalpy (or heat) of reaction, ΔH,** at constant pressure. ΔH values are negative for exothermic reactions and positive for endothermic reactions.

ΔH values are often expressed as J/mol or kJ/mol

Three quantitative measurements of heat are detailed in this experiment: measurements of the specific heat of a metal, the heat accompanying an acid–base reaction, and the heat associated with the dissolution of a salt in water.

Specific Heat and the Molar Mass of a Metal

The energy (heat, expressed in joules, J) required to change the temperature of one gram of a substance by 1°C is the **specific heat** of that substance:

$$\text{specific heat}\left(\frac{\text{J}}{\text{g}\bullet{}^\circ\text{C}}\right) = \frac{\text{energy (J)}}{\text{mass (g)} \times \Delta T\ (^\circ\text{C})} \tag{21.1}$$

or, rearranging for energy,

$$\text{energy (J)} = \text{specific heat}\left(\frac{\text{J}}{\text{g}\bullet{}^\circ\text{C}}\right) \times \text{mass (g)} \times \Delta T\ (^\circ\text{C}) \tag{21.2}$$

ΔT is the temperature change of the substance. Although the specific heat of a substance changes slightly with temperature, for our purposes, we assume it is constant over the temperature changes of this experiment.

The specific heat of a metal that does not react with water is determined by (1) heating a measured mass of the metal, M, to a known (higher) temperature, (2) placing it into a measured amount of water at a known (lower) temperature, and (3) measuring the final equilibrium temperature of the system after the two are combined.

The following equations, based on the law of conservation of energy, show the calculations for determining the specific heat of a metal. Considering the direction of energy flow by the conventional sign notation of energy loss being "negative" and energy gain being "positive," then

$$-\text{energy (J) lost by metal}_{M} = \text{energy (J) gained by water}_{H_2O} \tag{21.3}$$

Substituting from Equation 21.2,

Equation 21.4 is often written as $-c_{p,M} \times m_M \times \Delta T_M = c_{p,H_2O} \times m_{H_2O} \times \Delta T_{H_2O}$

$$-\text{specific heat}_{M} \times \text{mass}_{M} \times \Delta T_{M} = \text{specific heat}_{H_2O} \times \text{mass}_{H_2O} \times \Delta t_{H_2O} \tag{21.4}$$

Rearranging Equation 21.4 to solve for the specific heat of the metal$_M$ gives

$$\text{specific heat}_{M} = \frac{\text{specific heat}_{H_2O} \times \text{mass}_{H_2O} \times \Delta T_{H_2O}}{\text{mass}_{M} \times \Delta T_{M}} \tag{21.5}$$

In the equation, the temperature change for either substance is defined as the difference between the final temperature, T_f, and the initial temperature, T_i, of the substance:

$$\Delta T = T_f - T_i \tag{21.6}$$

These equations assume no heat loss to the calorimeter when the metal and the water are combined. The specific heat of water is 4.18 J/g • °C.

Pierre Dulong and Alexis Petit in 1819 proposed that one mole of all pure metals has the same capacity of absorbing heat, approximately 25 J/(mol • °C).

$$\text{specific heat}_{M}\left(\frac{\text{J}}{\text{g} \bullet {}^\circ\text{C}}\right) \times \text{molar mass}_{M}\left(\frac{\text{g}}{\text{mol}}\right) \approx 25\left(\frac{\text{J}}{\text{mol} \bullet {}^\circ\text{C}}\right) \tag{21.7}$$

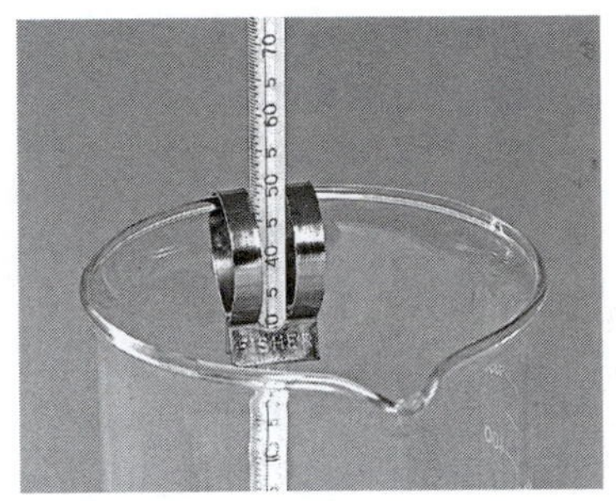

Figure 21.1 Acid–base reactions are exothermic.

Although their "constant" varies by 10% or more, a molar mass can and will be estimated from the specific heat of a metal.

Enthalpy (Heat) of Neutralization of an Acid–Base Reaction

Enthalpy of neutralization: energy released per mole of water formed in an acid–base reaction—an exothermic quantity

The reaction of a strong acid with a strong base is an exothermic reaction (Figure 21.1) that produces water and heat as products.

$$H_3O^+(aq) + OH^-(aq) \rightarrow 2\,H_2O(l) + \text{heat} \tag{21.8}$$

The **enthalpy (heat) of neutralization,** ΔH_n, is determined by (1) assuming the density and the specific heats of the acid and base solutions are equal to that of water and (2) measuring the temperature change, ΔT (Equation 21.6), when the two are mixed.

$$\Delta H_n = -\text{specific heat}_{H_2O} \times \textit{combined}\ \text{masses}_{\text{acid + base}} \times \Delta T \tag{21.9}$$

ΔH_n is generally expressed in units of kJ/mol of acid (or base) reacted. The mass (grams) of the solution equals the *combined* masses of the acid and base solutions.

Enthalpy (Heat) of Solution for the Dissolution of a Salt

Lattice energy: energy required to vaporize one mole of salt into its gaseous ions—an endothermic quantity

Hydration energy: energy released when one mole of a gaseous ion is attracted to and surrounded by water molecules forming one mole of hydrated ion in aqueous solution—an exothermic quantity

When a salt dissolves in water, energy is either absorbed or evolved depending upon the magnitude of the salt's lattice energy and the hydration energy of its ions. For the dissolution of KI:

$$KI(s) \xrightarrow{H_2O} K^+(aq) + I^-(aq) \qquad \Delta H_s = +\,13\ \text{kJ/mol} \tag{21.10}$$

The **lattice energy** (an endothermic quantity) of a salt, ΔH_{LE}, and the **hydration energy** (an exothermic quantity), ΔH_{hyd}, of its composite ions account for the amount of heat evolved or absorbed when one mole of the salt dissolves in water. The **enthalpy (heat) of solution,** ΔH_s, is the sum of these two terms (for KI, see Figure 21.2).

$$\Delta H_s = \Delta H_{LE} + \Delta H_{hyd} \tag{21.11}$$

Whereas ΔH_{LE} and ΔH_{hyd} are difficult to measure in the laboratory, ΔH_s is easily measured.

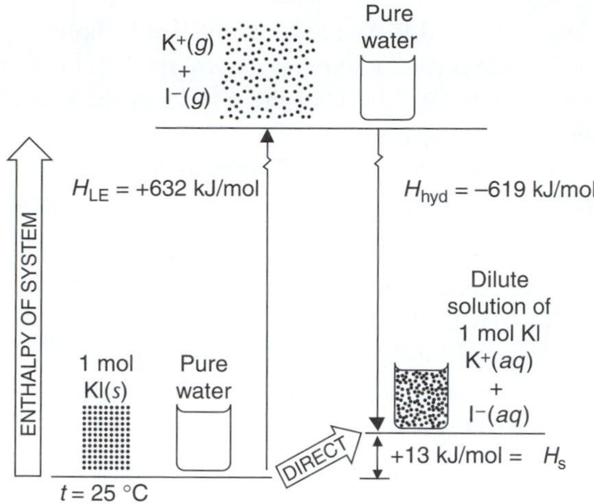

Figure 21.2 Energy changes in the dissolving of solid KI in water.

The enthalpy of solution for the dissolution of a salt, ΔH_s, is determined experimentally by adding the heat changes of the salt and the water when the two are mixed. ΔH_s is expressed in units of kilojoules per mole of salt.

$$\Delta H_s = (-\text{heat change}_{H_2O}) + (\text{heat change}_{salt}) \tag{21.12}$$

$$\Delta H_s = \left(-\frac{\text{specific heat}_{H_2O} \times \text{mass}_{H_2O} \times \Delta T_{H_2O}}{\text{mole}_{salt}}\right) + \left(-\frac{\text{specific heat}_{salt} \times \text{mass}_{salt} \times \Delta T_{salt}}{\text{mole}_{salt}}\right) \tag{21.13}$$

Refer to Equation 21.6 for an interpretation of ΔT. A temperature rise for the dissolution of a salt, indicating an exothermic process, means that the ΔH_{hyd} is greater than the ΔH_{LE} for the salt; conversely, a temperature decrease in the dissolution of the salt indicates that ΔH_{LE} is greater than ΔH_{hyd} and ΔH_s is positive.

The specific heats of some salts are listed in Table 21.1.

Table 21.1 Specific Heat of Some Salts

Salt	Formula	Specific Heat (J/g•°C)
ammonium chloride	NH_4Cl	1.57
ammonium nitrate	NH_4NO_3	1.74
ammonium sulfate	$(NH_4)_2SO_4$	1.409
calcium chloride	$CaCl_2$	0.657
sodium carbonate	Na_2CO_3	1.06
sodium hydroxide	NaOH	1.49
sodium sulfate	Na_2SO_4	0.903
sodium thiosulfate pentahydrate	$Na_2S_2O_3{\bullet}5H_2O$	1.45
potassium bromide	KBr	0.439
potassium nitrate	KNO_3	0.95

Experimental Procedure

Procedure Overview: Three different experiments are completed in a "double" coffee cup calorimeter. Each experiment requires careful mass/volume measurements and temperature measurements before and after the mixing of the respective components. Calculations are based on an interpretation of plotted data.

Ask your instructor which parts of this experiment you are to complete.

You and a partner are to complete at least two trials for each part assigned.

- Part A: The same metal sample is used. A boiling water bath is required.
- Part B: New solutions are required for each trial; therefore initially obtain about 110 mL of 1.1 *M* HCl, 110 mL of 1.1 *M* HNO_3, and 210 mL of 1.0 *M* NaOH from the stock reagents and plan your work schedule carefully.
- Part C: Two trials for a single salt are required; therefore determine the mass of a sample for each trial while using the balance.

The temperature vs. time curves to be plotted in Parts A.5, B.4, and C.4 can be established by using a thermal probe that is connected directly to either a calculator or computer with the appropriate software. If this thermal sensing/recording apparatus is available in the laboratory, consult with your instructor for its use and adaptation to the experiment. The probe merely replaces the glass or digital thermometer in Figure 21.5.

A. Specific Heat and the Molar Mass of a Metal

1. **Prepare the Metal.** Obtain 10–30 g of an unknown metal[1] from your instructor. Record the number of the unknown metal on the Report Sheet. Use weighing paper to measure its mass on your assigned balance. Transfer the metal to a dry, 200-mm test tube. Place the 200-mm test tube in a 400-mL beaker filled with water, well above the level of the metal sample in the test tube (Figure 21.3). Heat the water to boiling and maintain this temperature for at least 5 minutes so that the metal reaches thermal equilibrium with the boiling water. Proceed to Part A.2 while the water is heating.
2. **Prepare the Water in the Calorimeter.** Obtain two 6- or 8-oz styrofoam coffee cups, a plastic lid, and a 110° glass or digital thermometer. Thoroughly clean the styrofoam cups with several rinses of deionized water. Assemble the apparatus shown in Figure 21.4.[2] Determine the mass of the calorimeter. Using a graduated cylinder, add about 20.0 mL of water. Measure the combined mass of the calorimeter and water. Secure the thermometer with a clamp and position the bulb or thermal sensor below the water surface. (**Caution:** *Carefully handle a glass thermometer. If the thermometer is accidentally broken, notify your instructor immediately.*)
3. **Measure and Record the Temperatures of the Metal and Water.** Once thermal equilibrium has been reached in Parts A.1 and A.2, measure the temperatures of the *boiling* water from Part A.1 and of the water in the calorimeter from Part A.2. Record the temperature "using all certain digits (from the labeled calibration marks on the thermometer) *plus* one uncertain digit (the last digit which is the best estimate between the calibration marks)."

The temperature is to be recorded with the correct number of significant figures

4. **Transfer the Hot Metal to the Cool Water and Record the Data.** Remove the test tube from the boiling water and *quickly* transfer *only* the metal to the water in the calorimeter.[3] Replace the lid and swirl the contents gently. Record the water temperature as a function of time (about 5-second intervals for 1 minute and then 15-second intervals) on the table at the end of the Report Sheet.

Use the stirrer to assist in the gentle transfer of the metal into the water of the calorimeter

Appendix C

5. **Plot the Data.** Plot the temperature (y axis) versus time (x axis) on the top half of a sheet of linear graph paper. The maximum temperature is the intersection point of

[1]Ask your instructor to determine the approximate mass of metal to use for the experiment.
[2]This setup is called your **calorimeter** for the remainder of the experiment. Use a clamp to mount the glass or digital thermometer. Be sure that the thermometer is secure!
[3]Be careful *not* to splash out any of the water in the calorimeter. If you do, you will need to repeat the entire procedure.

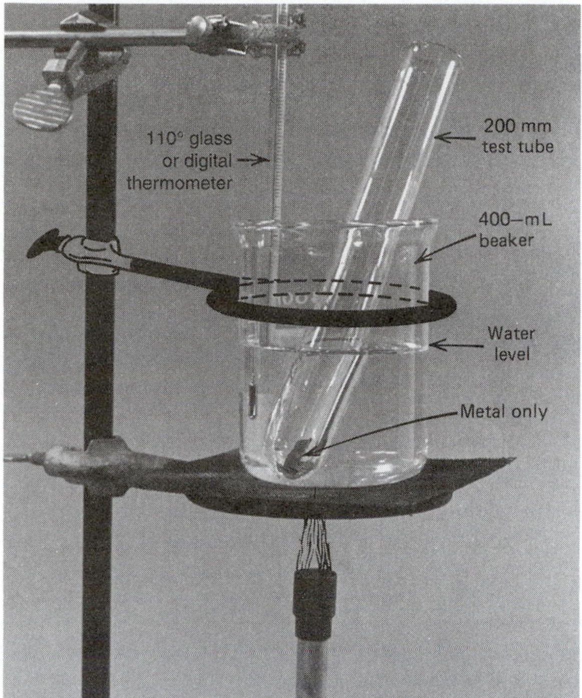

Figure 21.3 Placement of the metal in the dry test tube below the water surface in the beaker.

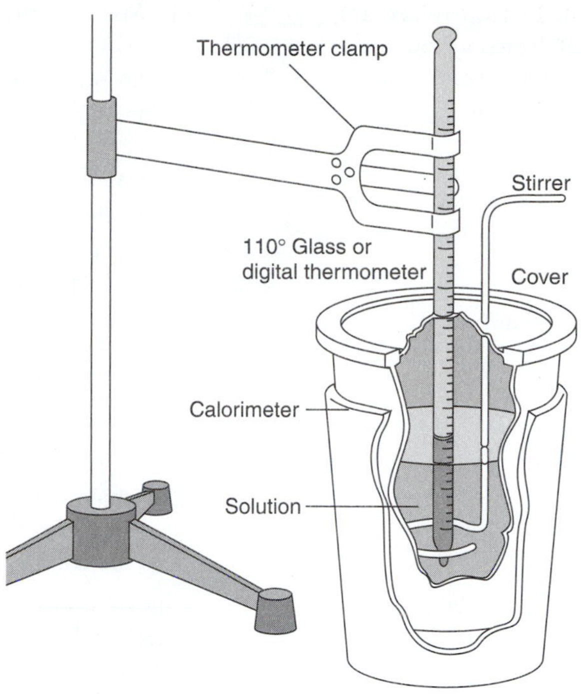

Figure 21.4 Schematic of a "coffee cup" calorimeter (see opening photo).

two lines: (1) the best line drawn through the data points on the cooling portion of the curve and (2) a line drawn perpendicular to the time axis at the mixing time [when the metal is added to the water (Figure 21.5)].[4] Have your instructor approve your graph.

6. **Do It Again.** Repeat Parts A.1 through A.5 for the same metal sample. Plot the data on the bottom half of the same sheet of linear graph paper.

Disposal: Return the metal to the appropriately labeled container, as advised by your instructor.

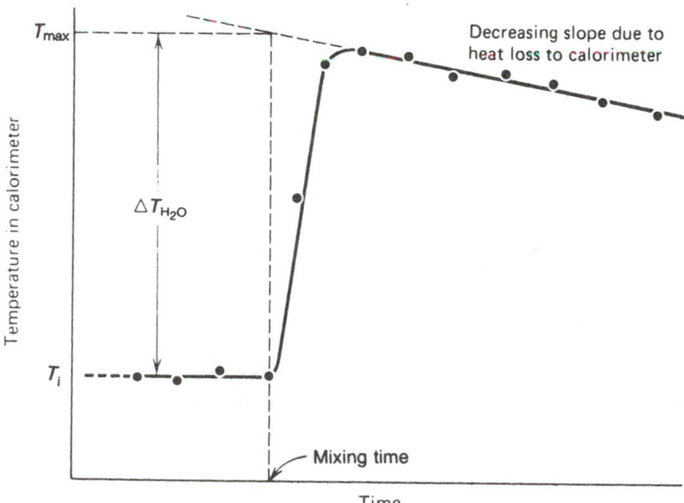

Figure 21.5 Extrapolation of temperature vs. time data (not to scale) for an exothermic process.

[4]The maximum temperature is never recorded because of some, albeit very small, heat loss to the calorimeter wall.

B. Enthalpy (Heat) of Neutralization for an Acid–Base Reaction

Standard solution: a solution with a very accurately measured concentration of a solute

Appendix C

1. **Measure the Volume and Temperature of the HCl.** Measure 50.0 mL of 1.1 *M* HCl in a *clean,* graduated cylinder. Measure and record its temperature.
2. **Measure the Volume and Temperature of the NaOH.** Using another *clean,* graduated cylinder transfer 50.0 mL of a **standard** 1.0 *M* NaOH **solution** to the *dry* calorimeter (see Figure 21.4). Record the temperature and exact molar concentration of the NaOH solution.
3. **Collect the Data.** Carefully, but quickly, add the acid to the base, replace the calorimeter lid, and swirl gently. Read and record the temperature and time every 5 seconds for 1 minute and thereafter every 15 seconds. Record the data.
4. **Plot the Data.** Plot the temperature (*y* axis) versus time (*x* axis) on the top half of a sheet of linear graph paper. Determine the maximum temperature as was done in Part A.5. Have your instructor approve your graph.
5. **Do It Again.** Repeat the acid-base experiment, Parts B.1 through B.4. Plot the data on the bottom half of the same sheet of graph paper.
6. **Change the Acid; Repeat the Neutralization Reaction.** Repeat Parts B.1 through B.5, substituting 1.1 *M* HNO_3 for 1.1 *M* HCl. On the Report Sheet, compare the ΔH_n values for the two strong acid-strong base reactions.

> *Disposal:* Discard the neutralized solutions contained in the calorimeter into the "Waste Acids" container. Rinse the calorimeter twice with deionized water.

C. Enthalpy (Heat) of Solution for the Dissolution of a Salt

Appendix C

1. **Prepare the Salt.** On weighing paper, measure about 5.0 g (±0.001 g) of the assigned salt. Record the name of the salt on the Report Sheet.
2. **Prepare the Calorimeter.** Measure the mass of the *dry* calorimeter. Using your clean, graduated cylinder, add about 20.0 mL of deionized water to the calorimeter (see Figure 21.4) and record its temperature. Measure the combined mass of the calorimeter and water. Secure the thermometer with a clamp and position the bulb or thermal sensor below the water surface.
3. **Collect the Temperature Data.** Carefully add (do not spill) the salt to the calorimeter, replace the lid, and swirl gently. Read and record the temperature and time at 5-second intervals for 1 minute and thereafter every 15-seconds. Record the data.
4. **Plot the Data.** Plot the temperature (*y* axis) versus time (*x* axis) on the top half of a sheet of linear graph paper. Determine the maximum (for an exothermic process) or minimum (for an endothermic process) temperature as was done in Part A.5. Have your instructor approve your graph.
5. **Do It Again.** With a fresh sample, repeat the dissolution of your assigned salt, Parts C.1 through C.4. Plot the data on the bottom half of the same sheet of linear graph paper.

> *Disposal:* Discard the salt solution into the "Waste Salts" container, followed by additional tap water. Consult with your instructor.

CLEANUP: Rinse the coffee cups twice with tap water and twice with deionized water, insert the thermometer into its carrying case, and return them.

Experiment 21 *Prelaboratory Assignment*

Calorimetry

Date __________ Lab Sec. ______ Name ____ Desk No. __________

1. An 18.50-g sample of aluminum is heated to 99.9°C in a hot water bath until thermal equilibrium is reached. The aluminum sample is quickly transferred to 50.0 mL of water at 23.8°C contained in a calorimeter. The thermal equilibrium temperature of the aluminum sample plus water mixture is 29.4°C. What is the specific heat of aluminum?

2. Will the recorded temperature change for an exothermic reaction performed in a glass calorimeter be greater or less than that in a styrofoam "coffee cup" calorimeter? Explain. Assume glass to be a better conductor of heat than styrofoam.

3. An acid–base neutralization reaction is exothermic. For the measurement of the enthalpy of neutralization, ΔH_n, for the reaction, heat is inevitably lost to the calorimeter (beaker or styrofoam cup). How will this unaccounted for heat loss affect the reported value for the enthalpy of neutralization for the reaction? Explain.

4. Three student chemists measured 50.0 mL of 1.00 *M* NaOH in separate styrofoam "coffee cup" calorimeters (Part B). Andrew added 50.0 mL of 1.10 *M* HCl to his solution of NaOH; Lyndsay added 45.5 mL of 1.10 *M* HCl (equal moles) to her NaOH solution. Dale added 50.0 mL of 1.00 *M* HCl to his NaOH solution. Each student recorded the temperature change and calculated the enthalpy of neutralization.
   Two of the chemists will report, within experimental error, the same temperature change for the HCl/NaOH reaction. Identify the two students and explain.

5. A student chemist observes a temperature decrease when her salt dissolves in water (Part C).
   a. Is the lattice energy for the salt greater or less than the hydration energy for the salt? Explain.

   b. Will the solubility of the salt increase or decrease with temperature increases? Explain.

6. A 4.50-g sample of NaOH at 25.0°C dissolves in 75.0 mL of water also at 25.0°C. The final equilibrium temperature of the resulting solution is 40.5°C. What is the enthalpy of solution, ΔH_s, of NaOH expressed in kilojoules per mole.

Experiment 22

Factors Affecting Reaction Rates

Iron reacts slowly in air to form iron(III) oxide, commonly called rust. When pure iron is heated and thrust into pure oxygen, the reaction is rapid.

OBJECTIVE

- To study the various factors that affect the rates of chemical reactions

TECHNIQUES

The following techniques are used in the Experimental Procedure

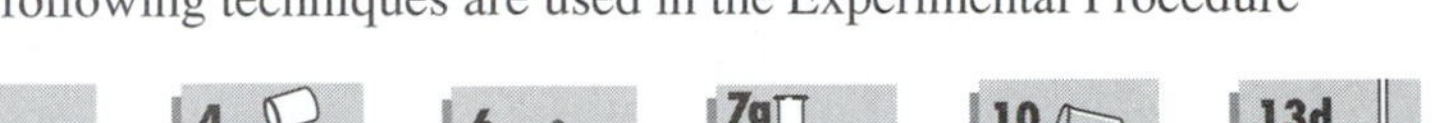

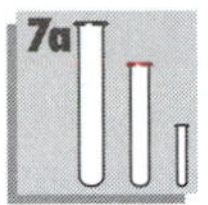

INTRODUCTION

Chemical kinetics is the study of chemical reaction rates, how reaction rates are controlled, and the pathway or mechanism by which a reaction proceeds from its reactants to its products.

Reaction rates vary from the very fast, in which the reaction, such as the explosion of a hydrogen/oxygen mixture, is essentially complete in microseconds or even **nanoseconds,** to the very slow, in which the reaction, such as the setting of concrete, requires years to complete.

The rate of a chemical reaction may be expressed as a *change* in the concentration of a reactant (or product) as a function of time (e.g., per second)—the greater the change in the concentration per unit of time, the faster the rate of the reaction. Other parameters that can follow the change in concentration of a **species** as a function of time in a chemical reaction are color (expressed as absorbance, Figure 22.1), temperature, pH, odor, and conductivity. The parameter chosen for following the rate of a particular reaction depends on the nature of the reaction and the species of the reaction.

We will investigate four of five factors that can be controlled to affect the rate of a chemical reaction. The first four factors listed below are systematically studied in this experiment:

- Nature of the reactants
- Temperature of the chemical system
- Presence of a catalyst
- Concentration of the reactants
- Surface area of the reactants

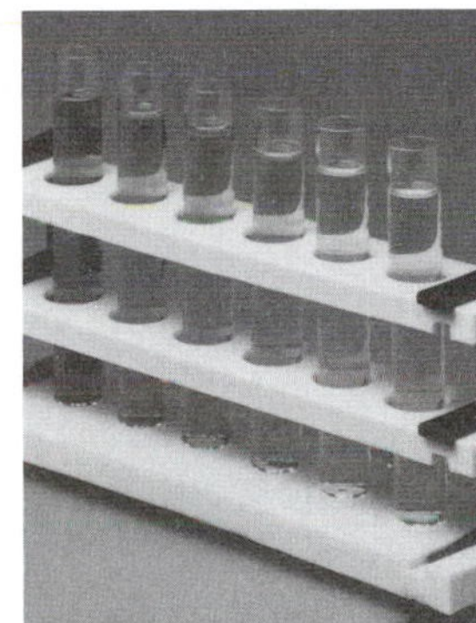

Figure 22.1 The higher concentration of light-absorbing species, the more intense is the color of the solution.

Nanosecond: 1×10^{-9} second

Species: any atom, molecule, or ion that may be a reactant or product of a chemical reaction

Nature of the Reactants

Some substances are naturally more reactive than others and, therefore, undergo rapid chemical changes. For example, the reaction of **sodium metal and water** is a very rapid, exothermic reaction (see Experiment 11, Part E), whereas the corrosion of iron is much slower. Plastics, reinforced with fibers such as carbon or glass, are now being substituted for iron and steel in specialized applications where corrosion has historically been a problem.

Sodium metal and water: the reaction releases $H_2(g)$ which ignites with the oxygen in the air to produce a yellow/blue flame, the yellow resulting from the presence of Na^+ in the flame

Temperature of the Chemical System

Internal energy: the energy contained within the molecules/ions when they collide

As a rule of thumb, a 10°C rise in temperature doubles (increases by a factor of 2) the rate of a chemical reaction. The added heat not only increases the number of collisions[1] between reactant molecules, but also, and more importantly, increases their kinetic energy. On collision of the reactant molecules, this kinetic energy is converted into an **internal energy** that is distributed throughout the collision system. This increased internal energy increases the probability for the weaker bonds to be broken and the new bonds to be formed.

Presence of a Catalyst

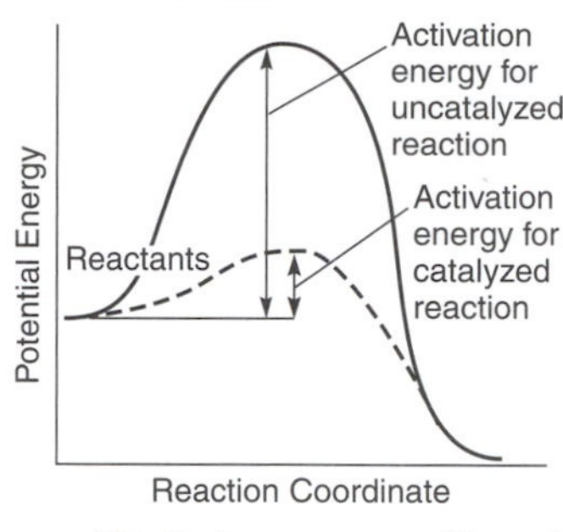

Figure 22.2 Reaction profiles of an uncatalyzed and a catalyzed reaction.

A **catalyst** increases the rate of a chemical reaction without undergoing any *net* chemical change. Some catalysts increase the rate of only one specific chemical reaction without affecting similar reactions. Other catalysts are more general and affect an entire set of similar reactions. Catalysts generally reroute the pathway of a chemical reaction so that this "alternate" path, although perhaps more circuitous, has a lower activation energy for reaction than the uncatalyzed reaction (Figure 22.2).

ncentration of the ıctants

An increase in the concentration of a reactant generally increases the reaction rate. See the opening photo. The larger concentration of reactant molecules increases the probability of an "effective" collision between reacting molecules for the formation of product. On occasion, such an increase may have no effect or may even decrease the reaction rate. A quantitative investigation on the affect of concentration changes on reaction rate is undertaken in Experiment 23.

ıce Area of the :ants

Generally speaking, the greater the exposed surface area of the reactant, the greater the reaction rate. For example, a large piece of coal burns very slowly, but coal *dust* burns rapidly, a consequence of which can be a disastrous coal mine explosion; solid potassium iodide reacts very slowly with solid lead nitrate, but when both are dissolved in solution, the formation of lead iodide is instantaneous.

E :RIMENTAL F :EDURE

Procedure Overview: A series of qualitative experiments are conducted to determine how various factors affect the rate of a chemical reaction.

Caution: *A number of strong acids are used in the experiment. Handle with care; do not allow them to touch the skin or clothing.*

Perform the experiment with a partner. At each circled superscript (1–19) in the procedure, *stop,* and record your observation on the Report Sheet. Discuss your observations with your lab partner and your instructor.

Ask your instructor which parts of the Experimental Procedure you are to complete. Prepare the hot water bath needed for Parts B and C. Use a 250-mL beaker.

A. N of the Reactants

1. **Different Acids Affect Reaction Rates.** Half-fill a set of four, labeled small test tubes (Figure 22.3) with 3 *M* H_2SO_4, 6 *M* HCl, 6 *M* CH_3COOH, and 6 *M* H_3PO_4, respectively. (**Caution:** *Avoid skin contact with the acids.*) Submerge a 1-cm strip of magnesium ribbon into each test tube. Compare the reaction rates and record your observations.(1)

[1]A 10°C temperature rise only increases the collision frequency between reactant molecules by a factor of 1.02—nowhere near the factor of 2 that is normally experienced in a reaction rate.

2. **Different Metals Affect Reaction Rates.** Half-fill a set of three, labeled small test tubes (Figure 22.4) with 6 *M* HCl. Submerge 1-cm strips of zinc, magnesium, and copper separately into the test tubes. Compare the reaction rates of each metal in HCl and record your observations.② Match the relative reactivity of the metals with the photos in Figure 22.5.③

Disposal: Dispose of the reaction solutions in the "Waste Inorganic Test Solutions" container.

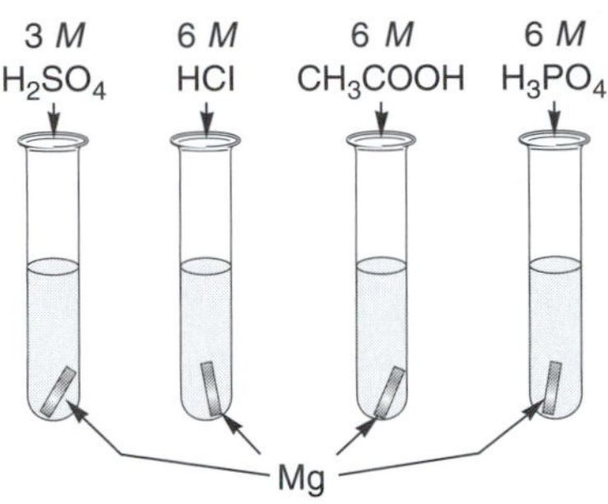

Figure 22.3 Setup for the effect of acid type on reaction rate.

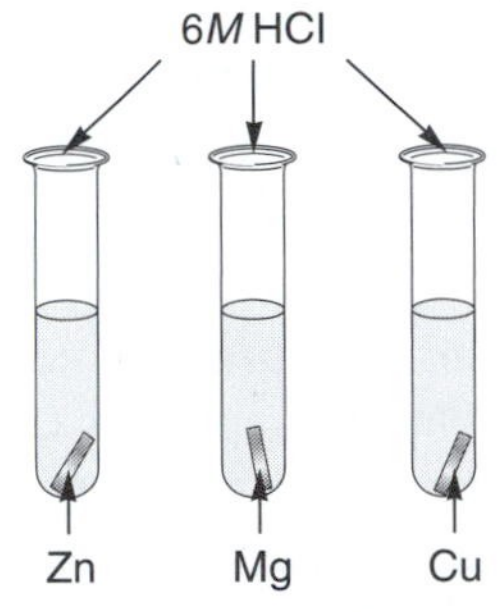

Figure 22.4 Setup for the effect of metal type on reaction rate.

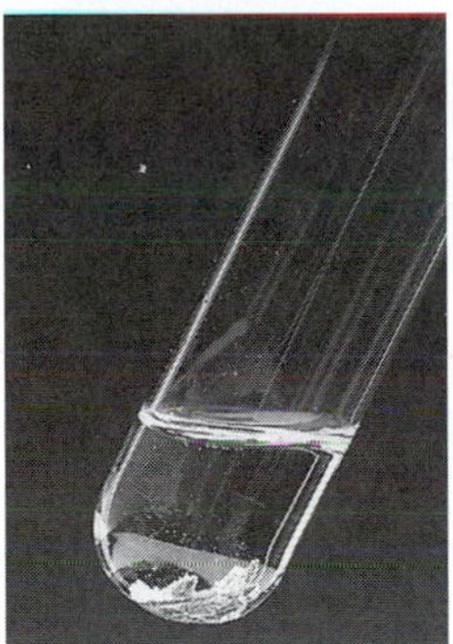

Figure 22.5 Zinc, copper, and magnesium react at different rates with 6 *M* HCl. Identify the metals in the photo according to their reactivity.③

B. Temperature of the Reaction: Hydrochloric Acid–Sodium Thiosulfate Reaction System

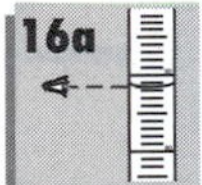

Ask your instructor to determine if *both* Parts B and C are to be completed. You should perform the experiment with a partner; as one student combines the test solutions, the other notes the time.

The oxidation–reduction reaction that occurs between hydrochloric acid and sodium thiosulfate, $Na_2S_2O_3$, produces insoluble sulfur as a product.

$$2\ HCl(aq) + Na_2S_2O_3(aq) \rightarrow S(s) + SO_2(g) + 2\ NaCl(aq) + H_2O(l) \quad (22.1)$$

The time required for the cloudiness of sulfur to appear is a measure of the reaction rate. Measure each volume of reactant with separate graduated pipets.

1. **Prepare the Solutions.** Pipet 2 mL of 0.1 *M* $Na_2S_2O_3$ into each of a set of three 150-mm, *clean* test tubes. Into a second set of three 150-mm test tubes pipet 2 mL of 0.1 *M* HCl. Label the test tubes.
2. **Record the Time for Reaction at the "Lower" Temperature.** Place a $Na_2S_2O_3$–HCl pair of test tubes in an ice water bath until thermal equilibrium is established (approximately 5 minutes). Pour the HCl solution into the $Na_2S_2O_3$ solution, START TIME, agitate the mixture for several seconds, and return the reaction mixture to the ice bath. STOP TIME when the cloudiness of the sulfur

appears. Record the time lapse for the reaction and the temperature of the bath, "using all certain digits (from the labeled calibration marks on the thermometer) *plus* one uncertain digit (the last digit which is the best estimate between the calibration marks)."④

3. **Record the Time for Reaction at the "Higher" Temperature.** Place a second $Na_2S_2O_3$–HCl pair of test tubes in a warm water (<60°C) bath until thermal equilibrium is established (approximately 5 minutes). Pour the HCl solution into the $Na_2S_2O_3$ solution, START TIME, agitate the mixture for several seconds, and return the reaction mixture to the warm water bath. STOP TIME when the cloudiness of the sulfur appears. Record the temperature of the bath.⑤

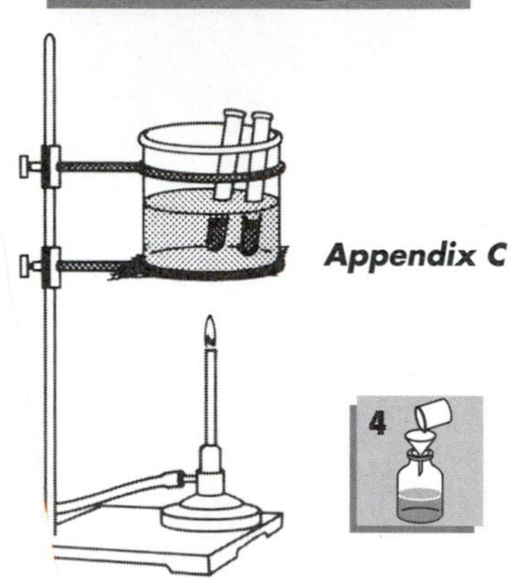

Appendix C

4. **Record the Time for Reaction at "Room" Temperature.** Combine the remaining set of $Na_2S_2O_3$–HCl test solutions at room temperature and proceed as in Parts B.2 and B.3. Record the appropriate data.⑥ Repeat any of the above reactions as necessary.
5. **Plot the Data.** Plot temperature (*y* axis) versus time (*x* axis) on one-half of a sheet of linear graph paper for the three data points. Have the instructor approve your graph.⑦ Further interpret your data as suggested on the Report Sheet.

Disposal: Dispose of the reaction solutions in the "Waste Inorganic Test Solutions" container.

Temperature of the ...ction: Oxalic ...-Potassium ...nanganate Reaction ...m

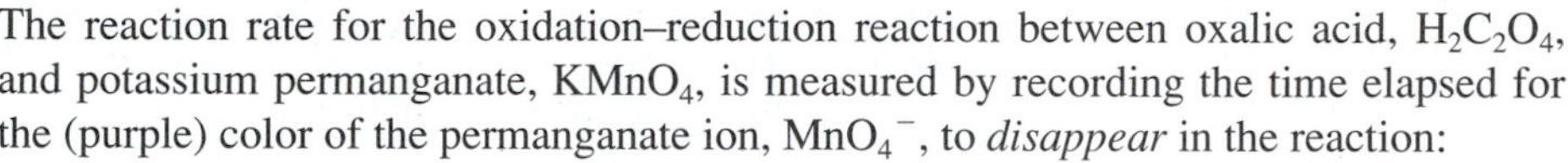

The reaction rate for the oxidation–reduction reaction between oxalic acid, $H_2C_2O_4$, and potassium permanganate, $KMnO_4$, is measured by recording the time elapsed for the (purple) color of the permanganate ion, MnO_4^-, to *disappear* in the reaction:

$$5\,H_2C_2O_4(aq) + 2\,KMnO_4(aq) + 3\,H_2SO_4(aq) \rightarrow 10\,CO_2(g) + 2\,MnSO_4(aq) + K_2SO_4(aq) + 8\,H_2O(l) \quad (22.2)$$

Measure the volume of each solution with separate, *clean* graduated pipets. As one student pours the test solutions together, the other notes the time.

1. **Prepare the Solutions.** Into a set of three, clean 150-mm test tubes, pipet 1 mL of 0.01 *M* $KMnO_4$ (in 3 *M* H_2SO_4) and 4 mL of 3 *M* H_2SO_4. (**Caution:** *$KMnO_4$ is a strong oxidant and causes brown skin stains; H_2SO_4 is a severe skin irritant and is corrosive. Do not allow either chemical to make skin contact.*) Into a second set of three, clean 150-mm test tubes pipet 5 mL of 0.33 *M* $H_2C_2O_4$.

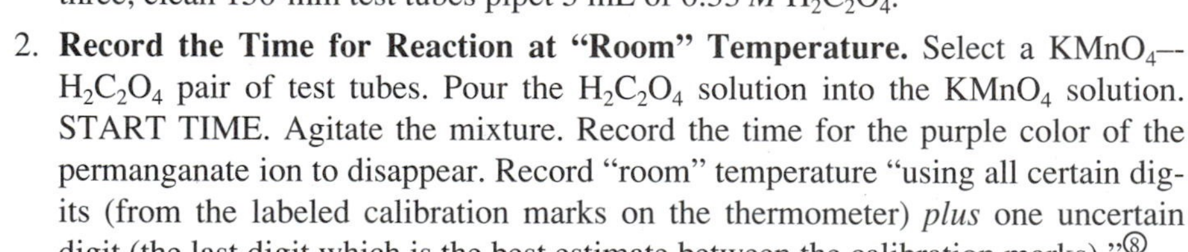

2. **Record the Time for Reaction at "Room" Temperature.** Select a $KMnO_4$–$H_2C_2O_4$ pair of test tubes. Pour the $H_2C_2O_4$ solution into the $KMnO_4$ solution. START TIME. Agitate the mixture. Record the time for the purple color of the permanganate ion to disappear. Record "room" temperature "using all certain digits (from the labeled calibration marks on the thermometer) *plus* one uncertain digit (the last digit which is the best estimate between the calibration marks)."⑧

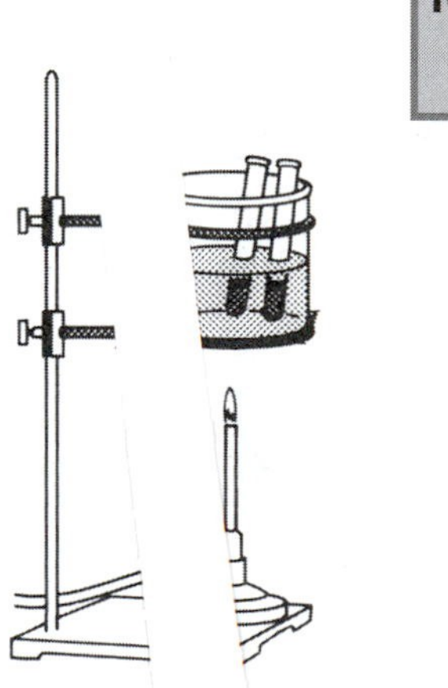

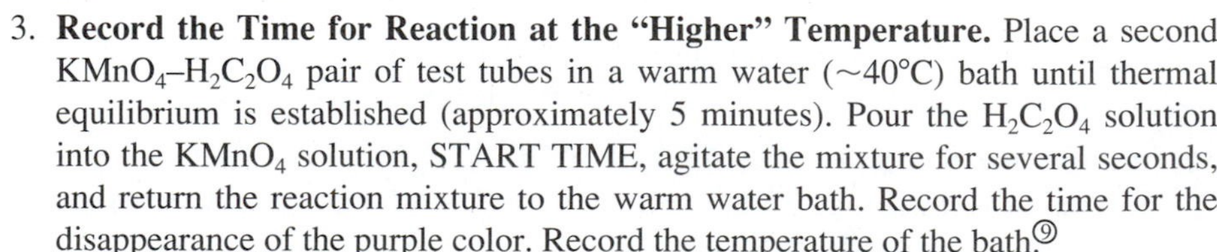

3. **Record the Time for Reaction at the "Higher" Temperature.** Place a second $KMnO_4$–$H_2C_2O_4$ pair of test tubes in a warm water (~40°C) bath until thermal equilibrium is established (approximately 5 minutes). Pour the $H_2C_2O_4$ solution into the $KMnO_4$ solution, START TIME, agitate the mixture for several seconds, and return the reaction mixture to the warm water bath. Record the time for the disappearance of the purple color. Record the temperature of the bath.⑨
4. **Record the Time for Reaction at the "Highest" Temperature.** Repeat Part C.3, but increase the temperature of the bath to about 60°C. Record the appropriate data.⑩ Repeat any of the preceding reactions as necessary.

Appendi...

5. **Plot the Data.** Plot temperature (*y* axis) versus time (*x* axis) on one-half of a sheet of linear graph paper for the three data points. Have the instructor approve your graph.⑪

Disposal: Dispose of the reaction solutions in the "Waste Inorganic Test Solutions" container.

D. Presence of a Catalyst

Hydrogen peroxide is relatively stable, but it readily decomposes in the presence of a catalyst.

1. **Add a Catalyst.** Place approximately 2 mL of a 3% H_2O_2 solution in a small test tube. Add 1 or 2 crystals of MnO_2 to the solution and observe. Note its instability.(12)

E. Concentration of Reactants: Magnesium–Hydrochloric Acid System

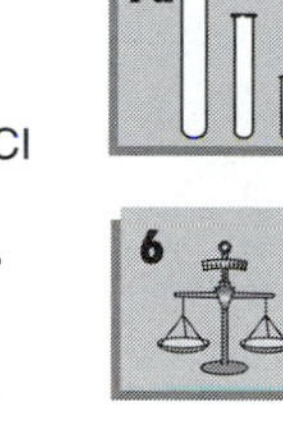

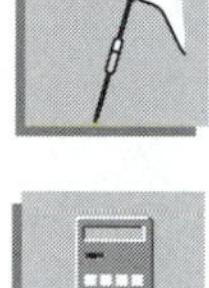

Ask your instructor for advice in completing *both* Parts E and F.

1. **Prepare the Reactants.** Into a set of four, clean, labeled test tubes, pipet 5 mL of 6 *M* HCl, 4 *M* HCl, 3 *M* HCl, and 1 *M* HCl, respectively (Figure 22.6).[2] Determine the mass (±0.001 g)—separately (for identification)—of four 1-cm strips of *polished* (with steel wool) magnesium. Calculate the number of moles of magnesium in each strip.(13)

6 *M* HCl 4 *M* HCl 3 *M* HCl 1 *M* HCl

Figure 22.6 Setup for the effect of acid concentration on reaction rate.

2. **Record the Time for Completion of the Reaction.** Add the first magnesium strip to the 6 *M* HCl solution, START TIME, and record the time for all traces of the magnesium strip to disappear. Repeat the experiment with the remaining three magnesium strips and the 4 *M* HCl, 3 *M* HCl, and 1 *M* HCl, solutions.(14)

3. **Plot the Data.** Plot $\frac{\text{mol HCl}}{\text{mol Mg}}$ (*y* axis) versus time in seconds (*x* axis) for the four tests on one-half of a sheet of linear graph paper. Have the instructor approve your graph.(15)

Appendix C

Disposal: Dispose of the reaction solutions in the test tubes in the "Waste Inorganic Test Solutions" container.

CLEANUP: Rinse the test tubes twice with tap water and twice with deionized water. Discard each rinse in the sink; flush the sink with water.

F. Concentration of Reactants: Iodic Acid–Sulfurous Acid System

A series of interrelated oxidation–reduction reactions occur between iodic acid, HIO_3, and sulfurous acid, H_2SO_3, that ultimately lead to the formation of iodine, I_2, and sulfuric acid, H_2SO_4, as the final products. The net equation for the reaction is

$$2\,HIO_3(aq) + 5\,H_2SO_3(aq) \rightarrow I_2(aq) + 5\,H_2SO_4(aq) + H_2O(l) \qquad (22.3)$$

The free iodine appears *only* after all of the sulfurous acid is consumed in the reaction. Once the iodine forms, its presence is detected by its reaction with starch, forming a deep blue complex.

$$I_2(aq) + \text{starch}(aq) \rightarrow I_2{\bullet}\text{starch}(aq) \qquad (22.4)$$

[2]Remember to rinse the pipet properly with the appropriate solution before dispensing it into the test tube.

Table 22.1 Reactant Concentration and Reaction Rate

	Solution in Test Tube			Add to Test Tube
Test Tube	0.01 *M* HIO_3	Starch	H_2O	0.01 *M* H_2SO_3
1	3 drops	1 drop	17 drops	1.0 mL
2	6 drops	1 drop	14 drops	1.0 mL
3	12 drops	1 drop	8 drops	1.0 mL
4	15 drops	1 drop	5 drops	1.0 mL
5	20 drops	1 drop	0 drops	1.0 mL

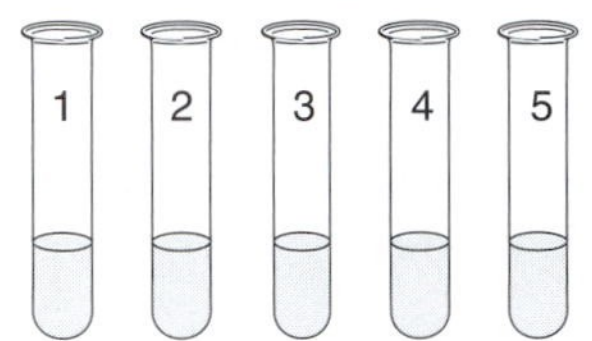

Figure 22.7 Setup for changes in HIO_3 concentration on reaction rate.

1. **Prepare the Test Solutions.** Review the preparation of the test solutions in Table 22.1. Set up five, clean and labeled test tubes (Figure 22.7). Measure the volumes of the 0.01 *M* HIO_3, starch, and water with dropping (or Beral) pipets.[3] Calibrate the HIO_3 dropping pipet to determine the volume (mL) per drop.(16) Calibrate a second dropping (or Beral) pipet with water to determine the number of milliliters per drop.(17)

   Calibrate a third dropping (or Beral) pipet for the 0.01 *M* H_2SO_3 solution that delivers 1 mL; mark the level on the pipet so that quick delivery of 1 mL of the H_2SO_3 solution to each test tube can be made. Alternatively, use a calibrated 1-mL Beral pipet.

2. **Record the Time for the Reaction.** Place a sheet of white paper beside the test tube (Figure 22.8). As one student quickly transfers 1.0 mL of the 0.01 *M* H_2SO_3 to the respective test tube, the other notes the time. *Immediately* agitate the test tube; record the time lapse (seconds) for the deep blue I_2•starch complex to appear.[4] Repeat any of the trials as necessary.(18)

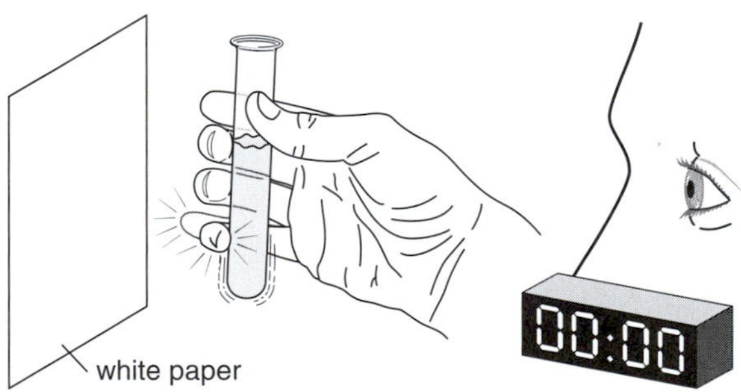

Figure 22.8 Viewing the reaction rate in a test tube.

Appendix C

3. **Plot the Data.** On one-half of a sheet of linear graph paper, plot for each solution the initial concentration of iodic acid,[5] $[HIO_3]_0$ (*y* axis), versus the time in seconds (*x* axis) for the reaction.(19)

Disposal: Dispose of all test solutions in the "Waste Inorganic Test Solutions" container.

CLEANUP: Rinse the test tubes twice with tap water and discard each into the "Waste Inorganic Test Solutions" container. Two final rinses with deionized water can be discarded in the sink.

[3] Be careful! Do not intermix the dropping pipets between solutions. This error in technique causes a significant error in the data.
[4] Be ready! The appearance of the deep blue solution is sudden.
[5] Remember that in calculating $[HIO_3]_0$, its total volume is the *sum* of the volumes of the two solutions.

Experiment 22 *Prelaboratory Assignment*

Factors Affecting Reaction Rates

Date __________ Lab Sec. ______ Name ____ Desk No. __________

1. Identify the major factor affecting reaction rate that accounts for the following observations:
   a. Tadpoles grow more rapidly near the cooling water discharge from a power plant.

   b. Enzymes accelerate certain biochemical reactions, but are not consumed.

   c. Gold and silver jewelry are common whereas iron jewelry is not.

   d. Iron and steel corrode more rapidly near the coast of an ocean than in the desert.

   e. Hydrogen peroxide antiseptic rapidly decomposes in an open wound.

2. Explain "why or how" the rate of a chemical reaction is increased by an increase in the concentration of a reactant.

3. A 20-mg strip of magnesium metal reacts in 5.0 mL of 4.0 *M* HCl over a given time period. Evaluate the $\frac{\text{mol Mg}}{\text{mol HCl}}$ ratio for the reaction.

4. A 1.0-mL volume of 0.010 *M* H_2SO_3 is added to a mixture of 12 drops of 0.01 *M* HIO_3, 8 drops of deionized water, and 1 drop of starch solution. A color change in the reaction mixture occurred in 40 seconds.
   a. Assuming 25 drops per milliliter, determine the initial molar concentration of IO_3^- after the mixing but before any reaction occurs (at time = 0).

   b. At the same point in time (see Part **a**, time = 0), what is the molar concentration of H_2SO_3?

   c. For the reaction (Experimental Procedure, Part F), the rate of the reaction is measured by the disappearance of H_2SO_3. For the reaction mixture in this question, what is the reaction rate? Express the reaction rate with appropriate units.

5. The reactions in Parts B, C, E, and F are timed. Indicate the "signal" to stop timing in each reaction.
   a. Part B.

   b. Part C.

   c. Part E.

   d. Part F.

Experiment 22 *Report Sheet*

Factors Affecting Reaction Rates

Date __________ Lab Sec. ______ Name ______________________________ Desk No. __________

A. Nature of the Reactants

1. ①List the acids in order of decreasing reaction rate with magnesium: _______, _______, _______, _______
2. ②List the metals in order of decreasing reaction rate with 6 *M* HCl: _______, _______, _______
3. ③Identify the metals reacting in Figure 22.5 (from left to right). _____, _____, _____

B. Temperature of the Reaction: Hydrochloric Acid–Sodium Thiosulfate Reaction System

1. Time for Sulfur to Appear — Temperature of the Reaction

Time for Sulfur to Appear	Temperature of the Reaction
④_____ seconds	_____ °C
⑤_____ seconds	_____ °C
⑥_____ seconds	_____ °C

2. ⑦Plot temperature (*y* axis) versus time (*x* axis) for the three trials. Instructor's approval of graph: _______________
3. From the plotted data, interpret the effect of temperature on reaction rate.

4. From your graph, estimate the temperature at which the appearance of sulfur should occur in 20 seconds. Assume no changes in concentration.

C. Temperature of the Reaction: Oxalic Acid–Potassium Permanganate Reaction System

1. Time for Permanganate Ion to Disappear — Temperature of the Reaction

Time for Permanganate Ion to Disappear	Temperature of the Reaction
⑧_____ seconds	_____ °C
⑨_____ seconds	_____ °C
⑩_____ seconds	_____ °C

2. ⑪Plot temperature (*y* axis) versus time (*x* axis) for the three trials. Instructor's approval of graph: _______________
3. From your plotted data, interpret the effect of temperature on reaction rate.

4. From your graph, estimate the time for the disappearance of the purple permanganate ion at 15°C. Assume no changes in concentration.

D. Presence of a Catalyst

1. ⑫What effect does the MnO_2 catalyst have on the rate of evolution of O_2 gas?

2. Write a balanced equation for the decomposition of H_2O_2.

E. Concentration of Reactants: Magnesium–Hydrochloric Acid System

Concentration of HCl	mol HCl	mass of Mg	⑬mol Mg	$\frac{\text{mol HCl}}{\text{mol Mg}}$	⑭Time (sec)
6 *M*					
4 *M*					
3 *M*					
1 *M*					

1. ⑮Plot $\frac{\text{mol HCl}}{\text{mol Mg}}$ (*y* axis) versus time (*x* axis). Instructor's approval of graph: ______________

2. From your graph, predict the time, in seconds, for 5 mg of Mg to react in 5 mL of 2.0 *M* HCl.

F. Concentration of Reactants: Iodic Acid–Sulfurous Acid System

Molar concentration of HIO_3 _____ Molar concentration of H_2SO_3 _____

⑯ What is the volume (mL) per drop of the HIO_3 solution? _____ ⑰What is the volume (mL) per drop of the water? _____

Test Tube	mL HIO_3	mL H_2O	mL H_2SO_3	$[HIO_3]_0$[6] (diluted)	Time (sec)
⑱1			1.0		
2			1.0		
3			1.0		
4			1.0		
5			1.0		

1. ⑲Plot $[HIO_3]_0$ (*y* axis) versus time (*x* axis). Instructor's approval of graph: ______________

2. How does a change in the molar concentration of HIO_3 affect the time required for the appearance of the deep blue I_2•starch complex?

3. Estimate the time, in seconds, for the deep blue I_2•starch complex to form when 10 drops of 0.01 *M* HIO_3 are used for the reaction. Assume all other conditions remain constant.

[6]See footnote 5.

Experiment 23

Determination of a Rate Law

Drops of blood catalyze the decomposition of hydrogen peroxide to water and oxygen gas.

OBJECTIVES

- To determine the rate law for a chemical reaction
- To utilize a graphical analysis of experimental data

TECHNIQUES

The following techniques are used in the Experimental Procedure

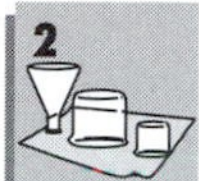

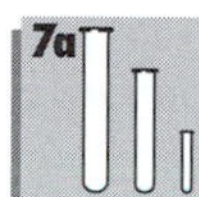

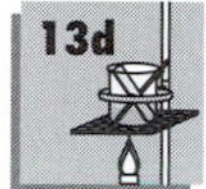

INTRODUCTION

The rate of a chemical reaction is affected by a number of factors, most of which were observed in Experiment 22. The rate of a reaction can be expressed in a number of ways, depending on the nature of reactants being consumed or the products being formed. The rate may be followed as a change in concentration (mol/L) of one of the reactants or products per unit of time, the volume of gas produced per unit of time (Figure 23.1), or the change in color (measured as absorbance) per unit of time, just to cite a few examples.

In this experiment, a quantitative statement as to *how* changes in reactant concentrations affect reaction rate is expressed in an experimentally derived rate law.

To assist in understanding the relationship between reactant concentration and reaction rate, consider the general reaction, $A_2 + 2\ B_2 \rightarrow 2\ AB_2$. The rate of this reaction

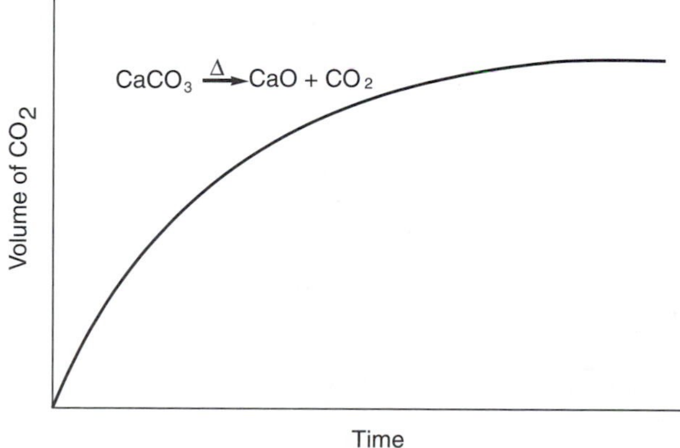

Figure 23.1 The rate of thermal decomposition of calcium carbonate is determined by measuring the volume of evolved carbon dioxide gas versus time.

is related, by some exponential power, to the initial concentration of each reactant. For this reaction, we can write the relationship as

$$\text{rate} = k\,[A_2]^p[B_2]^q \qquad (23.1)$$

Rate constant: a proportionality constant relating the rate of a reaction to the initial concentrations of the reactants

Order: the exponential factor by which the concentration of a substance affects reaction rate

This expression is called the **rate law** for the reaction. The value of k, the reaction **rate constant,** varies with temperature but is independent of reactant concentrations.

The superscripts p and q designate the **order** with respect to each reactant and are *always* determined experimentally. For example, if tripling the molar concentration of A_2 while holding the B_2 concentration constant increases the reaction rate by a factor of 9, then $p = 2$. In practice, a large excess of B_2 makes an insignificant change in its concentration during the course of the reaction; therefore, the change in reaction rate results from the more significant changes in the smaller amounts of A_2 in the reaction. An experimental study of the kinetics of any reaction involves determining the values of k, p, and q.

In this experiment the rate law for the reaction of hydrogen peroxide, H_2O_2, with potassium iodide, KI, is determined.[1] When these reactants are mixed, hydrogen peroxide slowly oxidizes iodide ion to elemental iodine, I_2.

$$2\,I^-(aq) + H_2O_2(aq) + 2\,H_3O^+\,(aq) \rightarrow I_2(aq) + 4\,H_2O(l) \qquad (23.2)$$

The rate of the reaction, governed by the molar concentrations of I^-, H_2O_2, and H_3O^+, is expressed by the rate law,

$$\text{rate} = k\,[I^-]^p[H_2O_2]^q[H_3O^+]^r \qquad (23.3)$$

Buffer: a solution that resists changes in acidity or basicity in the presence of added H^+ or OH^+ (Buffer solutions are studied in Experiment 24.)

When the $[H_3O^+]$ is greater than 1×10^{-3} mol/L, the reaction rate is too rapid to measure in the general chemistry laboratory; however, if the $[H_3O^+]$ is *less than* 1×10^{-3} mol/L, the reaction proceeds at a measurable rate. An acetic acid–sodium acetate **buffer** maintains a nearly constant $[H_3O^+]$ at about 1×10^{-5} mol/L during the experiment.[2] As H_3O^+ ion does not affect the reaction rate at this lower concentration, the rate law for the reaction becomes more simply

$$\text{rate} = k'\,[I^-]^p[H_2O_2]^q \qquad (23.4)$$

where $k' = k\,[H_3O^+]^r$.

In this experiment we will determine p, q, and k' for the hydrogen peroxide–iodide ion system. Two sets of experiments are required: One set of experiments is designed to determine the value of p and the other to determine the value of q.

Determination of *p*, the Order of the Reaction with Respect to Iodide Ion

In the first set of experiments, the effect that iodide ion has on the reaction rate is observed in several kinetic trials. A "large" excess of hydrogen peroxide in a buffered system maintains the H_2O_2 and H_3O^+ concentrations essentially constant during each trial. Therefore, for this set of experiments, the rate law, Equation 23.4, reduces to the form

$$\text{rate} = k'\,[I^-]^p\bullet c \qquad (23.5)$$

c, a constant, equals $[H_2O_2]^q$.

In logarithmic form, Equation 23.5 becomes

$$\log(\text{rate}) = \log k' + p\log[I^-] + \log c \qquad (23.6)$$

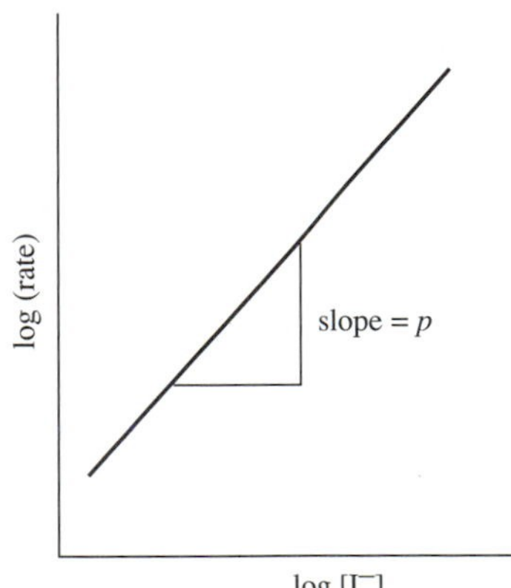

[1]Your laboratory instructor may substitute $K_2S_2O_8$ for H_2O_2 for this experiment. The balanced equation for the reaction is
$S_2O_8^{2-}(aq) + 2\,I^-(aq) \rightarrow 2\,SO_4^{2-}(aq) + I_2(aq)$

[2]In general, a combined solution of H_2O_2 and I^- is only very slightly acidic and the acidity changes little during the reaction. Therefore, the buffer solution may not be absolutely necessary for the reaction. However, to ensure that change in H_3O^+ concentrations is *not* a factor in the reaction rate, the buffer is included as a part of the experiment.

Combining constants, we have the equation for a straight line:

$$\log (\text{rate}) = p \log [\text{I}^-] + C$$
$$y = mx + b \qquad (23.7)$$

C equals $\log k' + \log c$ or $\log k' + \log [H_2O_2]^q$.

Therefore, a plot of log (rate) versus log $[\text{I}^-]$ produces a straight line with a slope equal to p, the order of the reaction with respect to the molar concentration of iodide ion. See margin figure on page 268.

Determination of q, the Order of the Reaction with Respect to Hydrogen Peroxide

In the second set of experiments, the effect that hydrogen peroxide has on the reaction rate is observed in several kinetic trials. A "large" excess of iodide ion in a buffered system maintains the I^- and H_3O^+ concentrations essentially constant during each trial. Under these conditions the logarithmic form of the rate law (Equation 23.4) becomes

$$\log (\text{rate}) = q \log [H_2O_2] + C'$$
$$y = mx + b \qquad (23.8)$$

C' equals $\log k' + \log [\text{I}^-]^p$.

A second plot, log (rate) versus log $[H_2O_2]$, produces a straight line with a slope equal to q, the order of the reaction with respect to the molar concentration of hydrogen peroxide.

Determination of Activation Energy

Reaction rates are temperature dependent. Higher temperatures increase the kinetic energy of the (reactant) molecules, such that when two reacting molecules collide, they do so with a much greater force (more energy is dispersed within the collision system) causing bonds to rupture, atoms to rearrange, and new bonds (products) to form more rapidly. The energy required for a reaction to occur is called the **activation energy** for the reaction.

The relationship between the reaction rate constant, $\boldsymbol{k'}$, at a measured temperature, **T(K)**, and the activation energy, $\mathbf{E_a}$, is expressed in the Arrhenius equation.

$$k' = Ae^{-E_a/RT} \qquad (23.9)$$

A is a collision parameter for the reaction and **R** is the gas constant (=8.314 J/mol K). The logarithmic form of Equation 23.9 is

$$\ln k' = \ln A - \frac{E_a}{RT} \quad \text{or} \quad \ln k' = \ln A - \frac{E_a}{R}\left[\frac{1}{T}\right] \qquad (23.10)$$

The latter equation of 23.10 conforms to the equation for a straight line, $y = b + mx$, where a plot of $\ln k'$ vs. $1/T$ yields a straight line with a slope of $-E_a/R$ and a y-intercept of $\ln A$.

Therefore, the activation energy for a reaction can be calculated from the slope of the plotted data.

The orders of the reactants of the reaction are determined in Part C and the room temperature rate constant is calculated in Part D. In Part E, the temperature of the solutions for kinetic trial 4 (Table 23.1) will be increased/decreased to determine rate constants at these new temperatures.

Observing the Rate of the Reaction

To follow the progress of the rate of the reaction, two solutions are prepared:

- Solution A: a diluted solution of iodide ion, starch, thiosulfate ion ($S_2O_3^{2-}$), and the acetic acid–sodium acetate buffer
- Solution B: the hydrogen peroxide solution

When Solutions A and B are mixed, the H_2O_2 reacts with the I^- (Equation 23.2).

$$2\,\text{I}^-(aq) + H_2O_2(aq) + 2\,H_3O^+(aq) \rightarrow I_2(aq) + 4\,H_2O(l) \qquad \text{(repeat of Equation 23.2)}$$

To prevent an equilibrium (a back reaction) from occurring in Equation 23.2, the presence of thiosulfate ion removes I_2 as it is formed:

$$2\ S_2O_3^{2-}(aq) + I_2(aq) \rightarrow 2\ I^-(aq) + S_4O_6^{2-}(aq) \qquad (23.11)$$

As a result, iodide ion is regenerated in the reaction system; this maintains a constant iodide ion concentration during the course of the reaction until the thiosulfate ion is consumed. When the thiosulfate ion has completely reacted in solution, the generated I_2 combines with starch, forming a deep blue I_2•starch complex. Its appearance signals a length of time for the reaction (Equation 23.2) to occur, and the length of time for the disappearance of the thiosulfate ion.

$$I_2(aq) + \text{starch}\ (aq) \rightarrow I_2 \bullet \text{starch}\ (aq, \text{deep blue}) \qquad (23.12)$$

The time required for a quantitative amount of thiosulfate ion to react is the time lapse for the appearance of the deep blue solution. During that period a quantitative amount of I_2 is generated; therefore, the rate of I_2 production (mol I_2/time), and thus the rate of the reaction, is affected *only* by the initial concentrations of H_2O_2 and I^-.

Therefore, the rate of the reaction is followed by measuring the time required to produce a preset number of moles of I_2, *not* the time required to deplete the moles of reactants.

Experimental Procedure

Procedure Overview: Measured volumes of several solutions having known concentrations of reactants are mixed in a series of trials. The time required for a visible color change to appear in the solution is recorded for the series of trials. The data are collected and plotted (two plots); from the plotted data, the order of the reaction with respect to each reactant is calculated and the rate law for the reaction is derived. After the rate law for the reaction is established, the reaction rate is observed at non-ambient temperatures. The plotted data produces a value for the activation energy of the reaction.

Read the entire procedure before beginning the experiment. Student pairs should gather the kinetic data.

A. Solutions of Reactants

1. **Prepare Solutions A for the Kinetic Trials.** Table 23.1 summarizes the preparation of the solutions for the kinetic trials. Use previously-boiled deionized water. Measure the volumes of KI and $Na_2S_2O_3$ solutions with *clean*[3] pipets.[4] Burets or pipets can be used for the remaining solutions. Prepare, at the same time, all of the Solutions A for Kinetic Trials 1–8 in either clean and labeled 20-mL beakers or 150-mm test tubes. Trial 8 is to be of your design.
2. **Prepare Solutions for Kinetic Trial 1.**
   Solution A. Stir the solution in a small 20-mL beaker or 150-mm test tube.
   Solution B. Pipet 3.0 mL of 0.1 *M* H_2O_2 into a clean 10-mL beaker or 150-mm test tube.
3. **Prepare for the Reaction.** The reaction begins when the H_2O_2 (Solution B) is added to Solution A; be prepared to start timing the reaction in seconds. Place the beaker on a white sheet of paper so the color change is more easily detected (Figure 23.2). As one student mixes the solutions, the other notes the time. All of the solutions should be at ambient temperature before mixing.
4. **Time the Reaction.** Rapidly add Solution B to Solution A—START TIME and swirl (once) the contents of the mixture. Continued swirling is unnecessary. The appearance of the deep blue color is sudden. Be ready to STOP TIME. Record the time lapse to the nearest second on the Report Sheet.

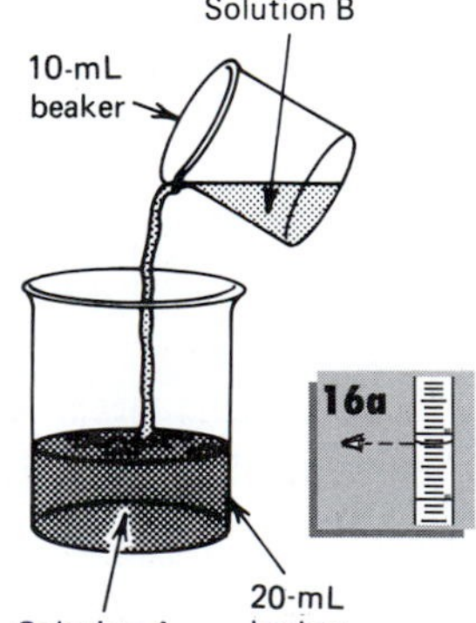

[3]Cleanliness is important in preparing these solutions because H_2O_2 readily decomposes in the presence of foreign particles. Do *not* dry glassware with paper towels.

$$2\ H_2O_2 \xrightarrow{\text{catalyst}} 2\ H_2O + O_2$$

[4]5-mL graduated (±0.1 mL) pipets are suggested for measuring these volumes.

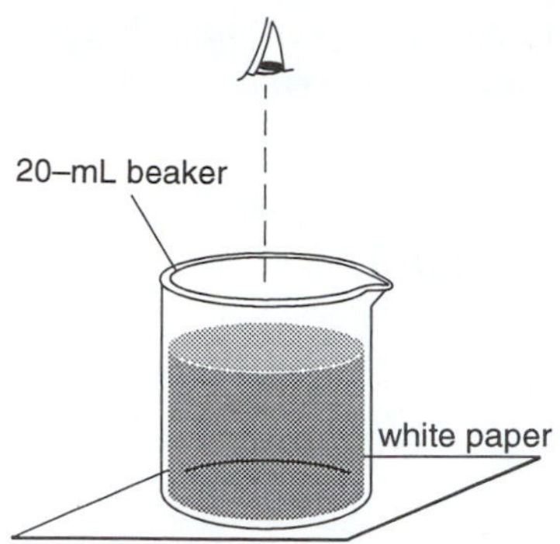

Figure 23.2 Viewing the appearance of the I_2•starch complex.

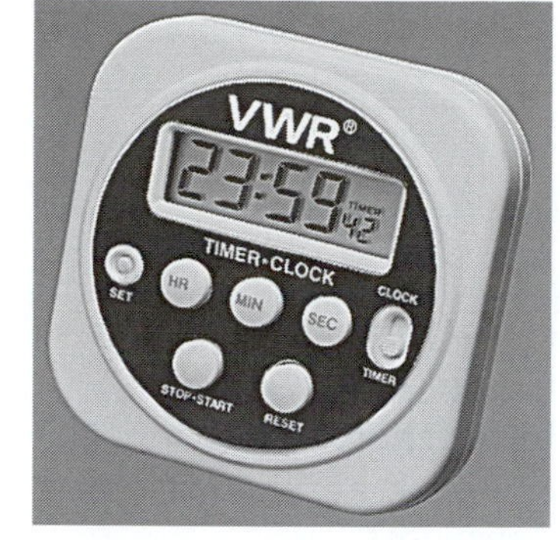

5. **Repeat for the Remaining Kinetic Trials.** Mix and time the test solutions for the remaining seven kinetic trials. If the instructor approves, conduct additional kinetic trials, either by repeating those in Table 23.1 or by preparing other combinations of KI and H_2O_2. Make sure that the total diluted volume remains constant at 10 mL.

Disposal: Dispose of the solutions from the kinetic trials in the "Waste Iodide Salts" container.

CLEANUP: Rinse the beakers/test tubes twice with tap water and discard in the "Waste Iodide Salts" container. Dispose of two final rinses with deionized water in the sink.

Table 23.1 Composition of Test Solutions

	Solution A					Solution B*
Kinetic Trial	Boiled, Deionized Water	Buffer**	0.3 *M* KI	0.02 *M* $Na_2S_2O_3$	Starch	0.1 *M* H_2O_2
1	4.0 mL	1.0 mL	1.0 mL	1.0 mL	5 drops	3.0 mL
2	3.0 mL	1.0 mL	2.0 mL	1.0 mL	5 drops	3.0 mL
3	2.0 mL	1.0 mL	3.0 mL	1.0 mL	5 drops	3.0 mL
4	1.0 mL	1.0 mL	4.0 mL	1.0 mL	5 drops	3.0 mL
5	2.0 mL	1.0 mL	1.0 mL	1.0 mL	5 drops	5.0 mL
6	0.0 mL	1.0 mL	1.0 mL	1.0 mL	5 drops	7.0 mL
7	5.0 mL	1.0 mL	1.0 mL	1.0 mL	5 drops	2.0 mL
8†	—	1.0 mL	—	1.0 mL	5 drops	—

*0.1 *M* $K_2S_2O_8$ may be substituted.
**0.5 *M* CH_3COOH and 0.5 *M* $NaCH_3CO_2$.
†You are to select the volumes of solutions for the trial.

B. Calculations for Determining the Rate Law

Perform the calculations, one step at a time. As you read through this section, complete the appropriate calculation and record it for each test solution on the Report Sheet.

1. **Moles of I_2 Produced.** Calculate the moles of $S_2O_3^{2-}$ consumed in each reaction. From Equation 23.9 the moles of I_2 that form in the reaction equals one-half the moles of $S_2O_3^{2-}$ that react. This also equals the change in the moles of I_2, starting with none at time zero up until a final amount that was produced at the time of the color change. This is designated as "$\Delta(\text{mol } I_2)$" produced.
2. **Reaction Rate.** The reaction rate for this reaction is calculated as the ratio of the moles of I_2 produced, $\Delta(\text{mol } I_2)$, to the time lapse, Δt, for the appearance of the deep blue color.[5] Compute these reaction rates, $\frac{\Delta(\text{mol } I_2)}{\Delta t}$, and the logarithms of

[5]The moles of I_2 present initially, at time zero, is zero.

the reaction rates for each reaction and enter them on the Report Sheet. Because the total volume is a constant for all kinetic trials, we do *not* need to calculate the molar concentrations of the I_2 produced.

3. **Initial Iodide Concentrations.** Calculate the initial molar concentration, $[I^-]_0$, and the logarithm of the initial molar concentration, log $[I^-]_0$, of iodide ion for each solution.[6]
4. **Initial Hydrogen Peroxide Concentrations.** Calculate the initial molar concentration, $[H_2O_2]_0$, and the logarithm of the initial molar concentration, log $[H_2O_2]_0$, of hydrogen peroxide for each solution.[7]

C. Determination of the Reaction Order, p and q, for Each Reactant

Appendix C

1. **Determination of p from Plot of Data.** Plot on the top half of a sheet of linear graph paper log (Δmol $I_2/\Delta t$), which is log (rate) (y axis), versus log $[I^-]_0$ (x axis) at constant hydrogen peroxide concentration. Kinetic Trials 1, 2, 3, and 4 have the same H_2O_2 concentration. Draw the best straight line through the four points. Calculate the slope of the straight line. The slope is the order of the reaction, p, with respect to the iodide ion.

Appendix C

2. **Determination of q from Plot of Data.** Plot on the bottom half of the same sheet of linear graph paper log (Δmol $I_2/\Delta t$) (y axis) versus log $[H_2O_2]_0$ (x axis) at constant iodide ion concentration using Kinetic Trials 1, 5, 6, and 7. Draw the best straight line through the four points and calculate its slope. The slope of the plot is the order of the reaction, q, with respect to the hydrogen peroxide.
3. **Approval of Graphs.** Have your instructor approve both graphs.

D. Determination of k', the Specific Rate Constant for the Reaction

Appendix B

1. **Substitution of p and q into Rate Law.** Use the values of p and q (from Part C) and the rate law, rate $= \dfrac{\Delta(\text{mol } I_2}{\Delta t} = k' \, [I^-]^p \, [H_2O_2]^q$, to determine k' for the seven solutions. Calculate the average value of k' with proper units. Also determine the standard deviation of k' from your data.
2. **Class Data.** Obtain average k' values from other groups in the class. Calculate a standard deviation of k' for the class.

E. Determination of Activation Energy

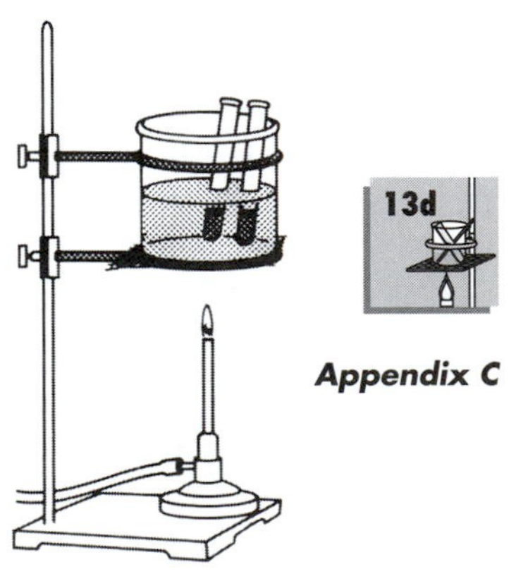

Appendix C

1. **Prepare Test Solutions.** Refer to Table 23.1, kinetic trial 4. In separate, clean 150-mm test tubes prepare *two* sets of Solution A and Solution B. Place one (Solution A/Solution B) set in an ice bath. Place the other set in a warm water (~35°C) bath. Allow thermal equilibrium to be established for each set, about 5 minutes.

   Test solutions prepared at other temperatures are encouraged for additional data points.
2. **Mix Solutions A and B.** When thermal equilibrium has been established, quickly pour Solution B into Solution A, START TIME, and agitate the mixture. When the deep blue color appears STOP TIME. Record the time lapse as before. Record the temperature of the water bath and use this time lapse for your calculations. Repeat for the other set(s) of solutions.
3. **The Reaction Rates and "New" Rate Constants.** The procedure for determining the reaction rates is described in Part B.2. Calculate and record the reaction rates for the (at least) two trials from Part E.2 and re-record the reaction rate for the (room temperature) Kinetic Trial 4 in Part A.5.

[6]Remember this is *not* 0.3 M I^- because the total volume of the solution is 10 mL after mixing.
[7]Remember, too, this is *not* 0.1 M H_2O_2 because the total volume of the solution is 10 mL after mixing.

Use the reaction rates at the three temperatures (ice, room, and ~35°C temperatures) and the established rate law from Part C to calculate the rate constants, k' at these temperatures. Calculate the natural logarithm of these rate constants.

4. **Plot the Data.** Plot ln k' vs. $1/T(K)$ for the (at least) three trials at which the experiment was performed. Remember to express temperature in kelvins and R = 8.314 J/mol•K.

Appendix C

5. **Activation Energy.** From the data plot, determine the slope of the linear plot ($= -E_a/R$) and calculate the activation energy for the reaction.

Experiment 23 *Prelaboratory Assignment*

Determination of a Rate Law

Date __________ Lab Sec. ______ Name ______________________________ Desk No. __________

1. The following data were collected in a kinetics experiment for the disproportionation of A_2 in a buffered solution:

   $3\ A_2 \rightarrow$ products

   The "generic" rate law for the reaction is rate $= k[A_2]^p$.

$[A_2]_0$	Time for Color Change	Rate (min^{-1})
0.10	7.50 minutes	______________
0.20	2.00 minutes	______________
0.30	1.30 minutes	______________
0.40	0.90 minutes	______________
0.50	0.50 minutes	______________

   Plot the data according to the method described in this experiment to determine p, the order of the reaction with respect to A_2. (*Hint:* The reaction order is *not* a whole number. Round the order to 2 significant figures.)

2. a. In the collection of the rate data for the experiment, when does START TIME and when does STOP TIME occur for each kinetic trial in Table 23.1?

   b. What is the color of the solution at STOP TIME?

   c. Account for the color of the solution at STOP TIME.

3. In the kinetic analysis of this experiment for the reaction of iodide ion with hydrogen peroxide, state the purpose for each of the following solutions (see Table 23.1):
   a. deionized water

   b. buffer solution (acetic acid, sodium acetate mixture)

4. a. If 4.0 mL of 0.30 *M* KI is mixed with 1.0 mL of deionized water, 1.0 mL of buffer solution, 1.0 mL of 0.02 *M* $Na_2S_2O_3$, and 3.0 mL of H_2O_2, what is the *initial* molar concentration of I^- (at time = 0, before reaction)?

   b. How many moles of I_2 are generated in the reaction during the time lapse of the rate measurement? (See Equations 23.2 and 23.11.)

5. How is the order of the reaction with respect to H_2O_2 determined in the experiment
   a. chemically?

   b. graphically?

6. Explain how the rate constant, k', is determined for the rate law in the experiment.

7. Explain how the activation energy, E_a, is determined for the reaction in the experiment.

Experiment 23 *Report Sheet*

Determination of a Rate Law

Date ________ Lab Sec. ______ Name ______________________________ Desk No. ________

Kinetic Trial	1*	2	3	4	5	6	7	8
Ambient temperature ______ °C								
1. Time for color change, Δt (sec)								
2. Amount of $S_2O_3^{2-}$ reacted (mol)								
3. Δ(mol I_2) produced								
4. $\frac{\Delta(\text{mol } I_2)}{\Delta t}$								
5. $\log \frac{\Delta(\text{mol } I_2)}{\Delta t}$								
6. $[I^-]_0$ (mol/L)								
7. $\log [I^-]_0$								
8. $[H_2O_2]_0$ (mol/L)								
9. $\log [H_2O_2]_0$								

*Calculations for Kinetic Trial 1.

C. Determination of the Reaction Order, *p* and *q*, for Each Reactant

Instructor's approval of graphs

log (Δmol $I_2/\Delta t$) versus log $[I^-]_0$ ________________________

log (Δmol $I_2/\Delta t$) versus log $[H_2O_2]_0$ ________________________

value of p from graph __________; value of q from graph __________

Write the rate law for the reaction.

D. Determination of k', the Specific Rate Constant for the Reaction

Kinetic Trial	1	2	3	4	5	6	7	8
Value of k'	____	____	____	____	____	____	____	____

Average value of k' __________________

Standard deviation of k' __________________

Class Data/Group	1	2	3	4	5	6
Average value of k'						

From Appendix B, determine the standard deviation of the reaction rate constant for the class.

E. Determination of Activation Energy

Time for color change	Rate	Calc. k'	ln k'	Temperature	$1/T(K)$

Instructor's Approval of Data Plot ________________________

Activation Energy, E_a, from data plot. Show calculation.

Laboratory Questions

Circle the questions that have been assigned.

1. Part A.4. Describe the chemistry that was occurring in the experiment between the time when Solution A and Solution B were mixed and STOP TIME?

2. Part A.4. Refer to Table 23.1. For kinetic trials 1, 2, 3, 4, and 7 the preferred procedure for mixing reactants is to add Solution A to Solution B; for kinetic trial 5 the addition can be either Solution A to Solution B or vice versa; for kinetic trial 6 the preferred procedure is to add Solution B to Solution A. Explain.

3. Part A.4. How would the appearance of the solution change for each kinetic trial if
   a. the $Na_2S_2O_3$ solution were omitted from the experiment?
   b. the starch solution were omitted from the experiment?

4. Part C.2. Review the plotted data.
   a. What is the numerical value of the *y*-intercept?
   b. What is the kinetic interpretation of the value for the *y*-intercept?
   c. What does its value equal in Equation 23.8?

5. State the effect that each of the following changes has on the reaction rate in this experiment. (Assume no volume change for any of the concentration changes.)
   a. An increase in the H_2O_2 concentration. Explain.
   b. An increase in the volume of water in Solutions A. Explain.
   c. An increase in the $Na_2S_2O_3$ concentration. Explain.
   d. The substitution of a 0.5% starch solution for one at 0.2%. Explain.

*6. If 0.2 *M* KI replaced the 0.3 *M* KI in this experiment, how would this affect the following?
   a. The rate of the reaction.
   b. The slopes of the graphs used to determine *p* and *q*.
   c. The value of the reaction rate constant.

7. Part E.2. The temperature of the warm water bath is recorded too high. How will this technique error affect the reported activation energy for the reaction . . . too high or too low? Explain.

Experiment 24

LeChâtelier's Principle; Buffers

The chromate ion (left) is yellow and the dichromate ion (right) is orange. An equilibrium between the two ions is affected by changes in pH.

OBJECTIVES

- To study the effects of concentration and temperature changes on the position of equilibrium in a chemical system
- To study the effect of strong acid and strong base addition on the pH of buffered and unbuffered systems
- To observe the common-ion effect on a dynamic equilibrium

TECHNIQUES

The following techniques are used in the Experimental Procedure

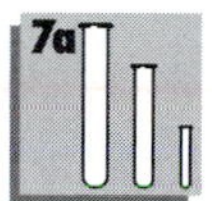

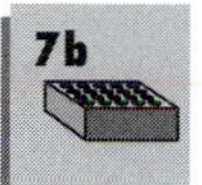

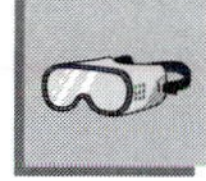

INTRODUCTION

Most chemical reactions do not produce a 100% yield of product, not because of experimental technique or design, but rather because of the chemical characteristics of the reaction. The reactants initially produce the expected products, but after a period of time the concentrations of the reactants and products *stop* changing.

This apparent cessation of the reaction before a 100% yield is obtained implies that the chemical system has reached a state where the reactants combine to form the products at a rate equal to that of the products re-forming the reactants. This condition is a state of **dynamic equilibrium,** characteristic of all reversible reactions.

For the reaction

$$2\ NO_2(g) \rightleftharpoons N_2O_4(g) + 58\ kJ \tag{24.1}$$

chemical equilibrium is established when the rate at which two NO_2 molecules react equals the rate at which one N_2O_4 molecule dissociates (Figure 24.1).

If the concentration of one of the species in the equilibrium system changes, or if the temperature changes, the equilibrium tends to *shift* in a way that compensates for the change. For example, assuming the system represented by Equation 24.1 is in a state of dynamic equilibrium, if more NO_2 is added, the probability of its reaction with other NO_2 molecules increases. As a result, more N_2O_4 forms and the reaction shifts to the *right,* until equilibrium is re-established.

A general statement governing all systems in a state of dynamic equilibrium follows:

> *If an external stress (change in concentration, temperature, etc.) is applied to a system in a state of dynamic equilibrium, the equilibrium shifts in the direction that minimizes the effect of that stress.*

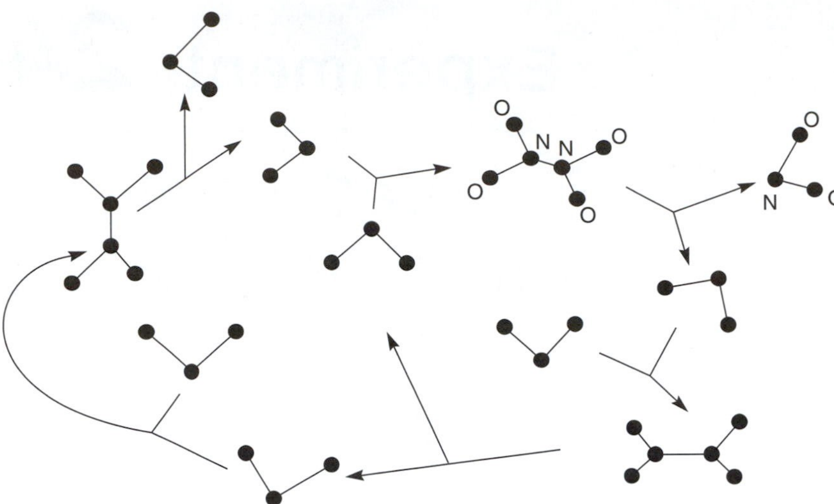

Figure 24.1 A dynamic equilibrium exists between reactant NO_2 molecules and product N_2O_4 molecules.

This is **LeChâtelier's principle,** proposed by Henri Louis LeChâtelier in 1888.

Often the equilibrium concentrations of all species in the system can be determined. From this information an equilibrium constant can be calculated; its magnitude indicates the relative position of the equilibrium. This constant is determined in Experiments 25, 29, and 31.

Two factors affecting equilibrium position are studied in this experiment: changes in concentration and changes in temperature.

Changes in Concentration

Complex ion: a metal ion bonded to a number of Lewis bases. The complex ion is generally identified by its enclosure with brackets, [ ].

Metal–Ammonia Ions. Aqueous solutions of copper ions and nickel ions appear sky blue and green, respectively. The colors of the solutions change, however, in the presence of added ammonia, NH_3. Because the metal–ammonia bond is stronger than the metal–water bond, ammonia substitution occurs and the following equilibria shift *right,* forming the metal–ammonia **complex ions:**

$$[Cu(H_2O)_4]^{2+}(aq) + 4\,NH_3(aq) \rightleftharpoons [Cu(NH_3)_4]^{2+}(aq) + 4\,H_2O(l) \qquad (24.2)$$

$$[Ni(H_2O)_6]^{2+}(aq) + 6\,NH_3(aq) \rightleftharpoons [Ni(NH_3)_6]^{2+}(aq) + 6\,H_2O(l) \qquad (24.3)$$

Addition of strong acid, H^+, affects these equilibria by its reaction with ammonia (a base) on the left side of the equations.

$$NH_3(aq) + H^+(aq) \rightarrow NH_4^+(aq) \qquad (24.4)$$

$[Cu(H_2O)_4]^{2+}$ is a sky-blue color (left), but $[Cu(NH_3)_4]]^{2+}$ is a deep-blue color (right).

The ammonia is removed from the equilibria, and the reactions shift *left* to relieve the stress caused by the removal of the ammonia, re-forming the aqueous Cu^{2+} (sky blue) and Ni^{2+} (green) solutions. For copper ions, this equilibrium shift may be represented as

$$\overset{\leftarrow}{[Cu(H_2O)_4]^{2+}(aq) + 4\,NH_3(aq) \rightleftharpoons [Cu(NH_3)_4]^{2+}(aq) + 4\,H_2O(l)} \qquad (24.5)$$
$$\downarrow \mathbf{4\,\mathit{H}^+(\mathit{aq})}$$
$$\mathbf{4\,NH_4^+(\mathit{aq})}$$

Multiple Equilibria with the Silver Ion

Many salts are only slightly soluble in water. Silver ion, Ag^+, forms a number of these salts. Several equilibria involving the relative solubilities of the silver salts of the carbonate, CO_3^{2-}, chloride, Cl^-, iodide, I^-, and sulfide, S^{2-}, anions are investigated in this experiment.

Silver Carbonate Equilibrium. The first of the silver salt equilibria is a saturated solution of silver carbonate, Ag_2CO_3, in dynamic equilibrium with its silver and carbonate ions in solution.

$$Ag_2CO_3(s) \rightleftharpoons 2\,Ag^+(aq) + CO_3^{2-}(aq) \qquad (24.6)$$

Nitric acid, HNO_3, dissolves silver carbonate: H^+ ions react with (and remove) the CO_3^{2-} ions on the right; the system, in trying to replace the CO_3^{2-} ions, shifts to the *right.* The Ag_2CO_3 dissolves, and carbonic acid, H_2CO_3, forms.

$$\overset{\rightarrow}{Ag_2CO_3(s) \rightleftharpoons 2\,Ag^+(aq) + CO_3^{2-}(aq)} \quad (24.7)$$

$$\downarrow \mathbf{2\,\mathit{H}^+(\mathit{aq})}$$

$$\mathbf{H_2CO_3(\mathit{aq})} \rightarrow H_2O(l) + CO_2(g) \quad (24.8)$$

The carbonic acid, being unstable at room temperature and pressure, decomposes to water and carbon dioxide. The silver ion and nitrate ion (from HNO_3) remain in solution.

Silver Chloride Equilibrium. Chloride ion precipitates silver ion as AgCl. Addition of chloride ion (from HCl) to the above solution, containing Ag^+ and NO_3^-, causes the formation of a silver chloride, AgCl, precipitate, now in dynamic equilibrium with its Ag^+ and Cl^- ions (Figure 24.2).

Figure 24.2 Solid AgCl quickly forms when solutions containing Ag^+ and Cl^- are mixed.

$$Ag^+(aq) + Cl^-(aq) \rightleftharpoons AgCl(s) \quad (24.9)$$

Aqueous ammonia, NH_3, "ties up" (i.e., it forms a complex ion with) silver ion, producing the soluble diamminesilver(I) ion, $[Ag(NH_3)_2]^+$. The addition of NH_3 removes silver ion from the equilibrium in Equation 24.9, shifting its equilibrium position to the *left* and causing AgCl to dissolve.

$$\overset{\leftarrow}{Ag^+(aq) + Cl^-(aq) \rightleftharpoons AgCl(s)} \quad (24.10)$$

$$\downarrow \mathbf{2\,\mathit{NH_3}(\mathit{aq})}$$

$$\mathbf{[Ag(NH_3)_2]^+(\mathit{aq})}$$

Adding acid, H^+, to the solution again frees silver ion to recombine with chloride ion and re-form solid silver chloride. This occurs because H^+ reacts with the NH_3 (see Equation 24.4) in Equation 24.10, restoring the system to that described by Equation 24.9.

$$\overset{\rightarrow}{\mathbf{Ag^+(\mathit{aq})} + Cl^-(aq) \rightleftharpoons AgCl(s)} \quad (24.11)$$

$$\uparrow 2\,NH_3(aq) + 2\,\mathbf{\mathit{H}^+(\mathit{aq})} \rightarrow 2\,NH_4^+(aq)$$

$$[Ag(NH_3)_2]^+(aq)$$

Silver Iodide Equilibrium. Iodide ion, I^- (from KI), added to the $Ag^+(aq) + 2\,NH_3(aq) \leftrightharpoons Ag(NH_3)_2^+\,(aq)$ equilibrium in Equation 24.10 results in the formation of solid silver iodide, AgI.

$$\overset{\leftarrow}{Ag^+(aq) + 2\,NH_3(aq) \rightleftharpoons [Ag(NH_3)_2]^+(aq)} \quad (24.12)$$

$$\downarrow \mathbf{\mathit{I}^-(\mathit{aq})}$$

$$\mathbf{AgI}(s)$$

The iodide ion removes the silver ion, causing a dissociation of the $[Ag(NH_3)_2]^+$ ion and a shift to the *left*.

Silver Sulfide Equilibrium. Silver sulfide, Ag_2S, is less soluble than silver iodide, AgI. Therefore an addition of sulfide ion (from Na_2S) to the $AgI(s) \leftrightharpoons Ag^+(aq) + I^-(aq)$ dynamic equilibrium in Equation 24.12 removes silver ion; AgI dissolves but solid silver sulfide forms.

$$\overset{\rightarrow}{AgI(s) \rightleftharpoons Ag^+\,(aq) + I^-\,(aq)} \quad (24.13)$$

$$\downarrow \mathbf{\tfrac{1}{2}\,S^{2-}(\mathit{aq})}$$

$$\mathbf{\tfrac{1}{2}\,Ag_2S(\mathit{s})}$$

Buffers

In many areas of research, chemists need an aqueous solution that resists a pH change when small amounts of acid or base are added. Biologists often grow cultures that are very susceptible to changes in pH and therefore a buffered medium is required (Figure 24.3).

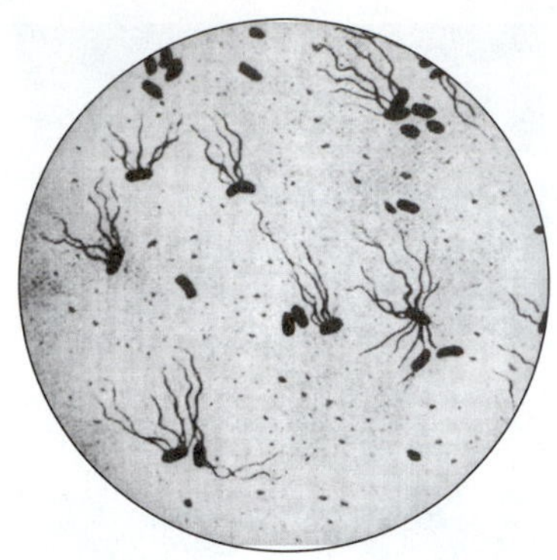

Figure 24.3 Bacteria cultures survive in media that exist over a narrow pH range. Buffers are used to control large changes in pH.

A buffer solution must be able to consume small additions of H_3O^+ and OH^- without undergoing large pH changes. Therefore, it must have present a basic component that can react with the H_3O^+ *and* an acidic component that can react with OH^-. Such a buffer solution consists of a weak acid and its conjugate base (or weak base and its conjugate acid). This experiment shows that an acetic acid–acetate buffer system can minimize large pH changes.

$$CH_3COOH(aq) + H_2O(l) \rightleftharpoons H_3O(aq) + CH_3CO_2^-(aq) \qquad (24.14)$$

The addition of OH^- shifts the buffer equilibrium, according to LeChâtelier's principle, to the *right* because of its reaction with H_3O^+, forming H_2O. The shift right is by an amount that is essentially equal to the moles of OH^- added to the buffer system. Thus, the amount of $CH_3CO_2^-$ increases and the amount of CH_3COOH decreases by an amount equal to the moles of OH^- added.

$$\overset{\rightarrow}{CH_3COOH(aq) + H_2O(l) \rightleftharpoons H_3O^+(aq) + CH_3CO_2^-(aq)} \qquad (24.15)$$
$$\downarrow \boldsymbol{OH^-(aq)}$$
$$\mathbf{2\ H_2O(\mathit{l})}$$

Conversely, the addition of H_3O^+ from a strong acid to the buffer system causes the equilibrium to shift *left:* the H_3O^+ combines with the acetate ion (a base) to form more acetic acid, an amount (moles) equal to the amount of H_3O^+ added to the system.

$$\overset{\leftarrow}{CH_3COOH(aq) + H_2O(l) \rightleftharpoons H_3O^+(aq) + CH_3CO_2^-(aq)} \qquad (24.16)$$
$$\uparrow \boldsymbol{H_3O^+(aq)}$$

As a consequence of the addition of strong acid, the amount of CH_3COOH increases and the amount of $CH_3CO_2^-$ decreases by an amount equal to the moles of strong acid added to the buffer system.

This experiment compares the pH changes of a buffered solution to those of an unbuffered solution when varying amounts of strong acid or base are added to each.

Common-Ion Effect

The effect of adding an ion or ions common to those already present in a system at a state of dynamic equilibrium is called the **common-ion effect.** The effect is observed in this experiment for the following equilibrium:

$$4\ Cl^-(aq) + [Co(H_2O)_6]^{2+}(aq) \rightleftharpoons [CoCl_4]^{2-}(aq) + 6\ H_2O(l) \qquad (24.17)$$

Ligand: a Lewis base that donates a lone pair of electrons to a metal ion, generally a transition metal ion.

Equation 24.17 represents an equilibrium of **ligands** bonded to a metal ion—the equilibrium is shifted because of a change in the concentrations of chloride ion.

Changes in Temperature

Referring again to Equation 24.1,

$$2\ NO_2(g) \rightleftharpoons N_2O_4(g) + 58\ kJ \qquad \text{(repeat of 24.1)}$$

Exothermic: characterized by energy release from the system to the surroundings

The reaction for the formation of colorless N_2O_4 is **exothermic** by 58 kJ. To favor the formation of N_2O_4, the reaction vessel should be kept cool (Figure 24.4); removing heat from the system causes the equilibrium to replace the removed heat and the equilibrium therefore shifts *right.* Added heat shifts the equilibrium in the direction that absorbs heat; for this reaction, a shift to the left occurs with addition of heat.

Coordination sphere: all ligands of the complex ion (collectively with the metal ion they are enclosed in square brackets when writing the formula of the complex ion)

This experiment examines the effect of temperature on the system described by Equation 24.17. This system involves an equilibrium between the **coordination spheres,** the water versus the Cl^- about the cobalt(II) ion; the equilibrium is concentration *and* temperature dependent. The tetrachlorocobaltate(II) ion, $[CoCl_4]^{2-}$ is more stable at higher temperatures.

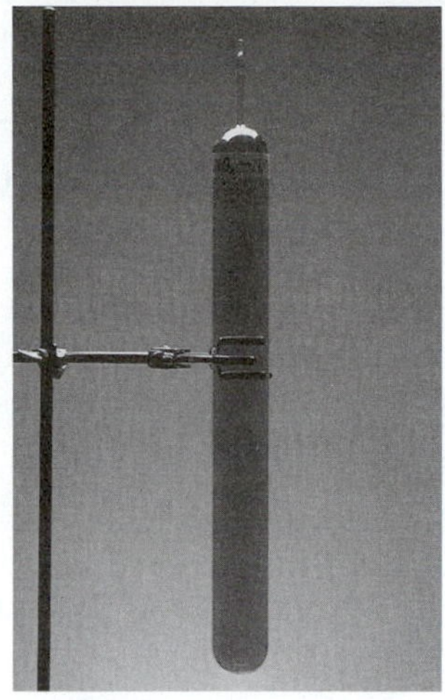

Figure 24.4 NO_2, a red-brown gas (left), is favored at higher temperatures; N_2O_4, a colorless gas (right), is favored at lower temperatures.

Experimental Procedure

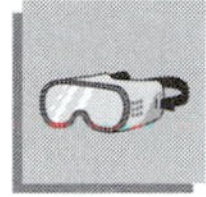

Procedure Overview: A large number of qualitative tests and observations are performed. The effects that concentration changes and temperature changes have on a system at equilibrium are observed and interpreted using LeChâtelier's principle. The functioning of a buffer system and the effect of a common ion on equilibria are observed.

Perform this experiment with a partner. At each circled, superscript (1–21) in the procedure, *stop,* and record your observations on the Report Sheet. Discuss your observations with your lab partner and instructor. Account for the changes in appearance of the solution after each addition in terms of LeChâtelier's principle.

Ask your instructor which parts of the Experimental Procedure are to be completed. Prepare a hot water bath for Part E.

A. Metal-Ammonia Ions

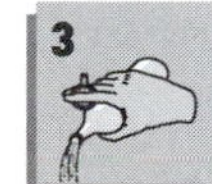

1. **Formation of Metal–Ammonia Ions.** Place 1 mL (<20 drops) of 0.1 *M* $CuSO_4$ (or 0.1 *M* $NiCl_2$) in a small, clean test tube.(1) Add drops of "conc" NH_3 (**caution:** *strong odor, do not inhale*) until a color change occurs and the solution is clear (*not* colorless).(2)
2. **Shift of Equilibrium.** Add drops of 1 *M* HCl until the color again changes.(3)

B. Multiple Equilibria with the Silver Ion

1. **Silver Carbonate Equilibrium.** In a 150-mm test tube (Figure 24.5) add ½ mL (≤10 drops) of 0.01 *M* $AgNO_3$ to ½ mL of 0.1 *M* Na_2CO_3.(4) Add drops of 6 *M* HNO_3 (**caution:** 6 *M* HNO_3 *reacts with the skin!*) to the precipitate until evidence of a chemical change occurs.(5)
2. **Silver Chloride Equilibrium.** To the "clear" solution from Part B.1, add ~ 5 drops of 0.1 *M* HCl.(6) Add drops of conc NH_3 (**caution:** *avoid breathing vapors and avoid skin contact*) until evidence of a chemical change.* (7) Reacidify the solution with 6 *M* HNO_3 and record your observations.(8) What happens if excess conc NH_3 is again added? Try it.(9)
3. **Silver Iodide Equilibrium.** After "trying it," add drops of 0.1 *M* KI.(10)
4. **Silver Sulfide Equilibrium.** To the mixture from Part B.3, add drops of 0.1 *M* Na_2S until evidence of chemical change has occurred.(11)

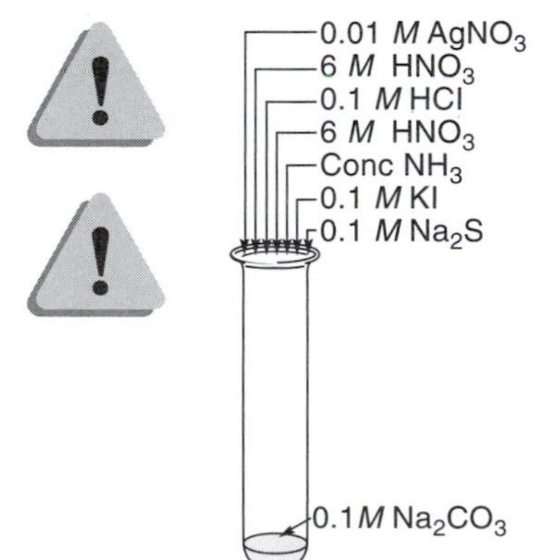

Figure 24.5 Sequence of added reagents for the study of silver ion equilibria.

*At this point, the solution should be "clear and colorless."

Disposal: Dispose of the waste silver salt solutions in the "Waste Silver Salts" container.

CLEANUP: Rinse the test tube twice with tap water and discard in the "Waste Silver Salts" container. Rinse twice with deionized water and discard in the sink.

C. A Buffer System

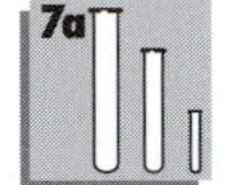

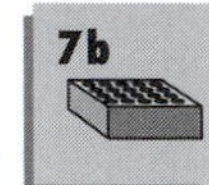

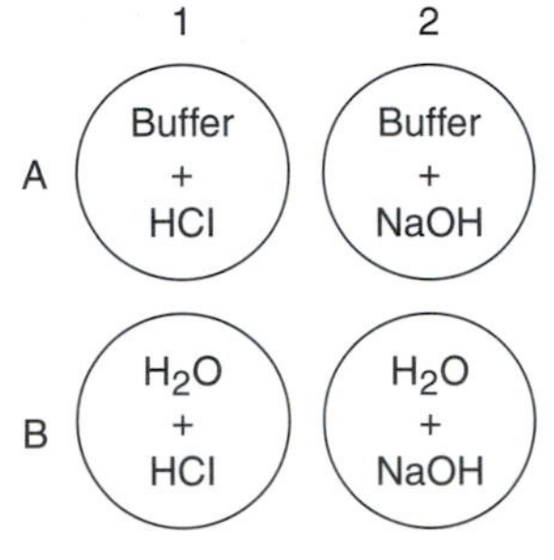

The use of a well plate is recommended for solutions in Parts C and D. Appropriately labeled 75-mm test tubes are equally useful for performing the experiments.

1. **Preparation of Buffered and Unbuffered Systems.** Transfer 10 drops of 0.1 *M* CH_3COOH to wells A1 and A2 of a 24-well plate, add 3 drops of universal indicator,† and note the color.⑫ Compare the color of the solution with the pH color chart for the universal indicator.⑫ Now add 10 drops of 0.1 *M* $NaCH_3CO_2$ to each well.⑬ Place 20 drops of deionized water into wells B1 and B2 and add 3 drops of universal indicator.⑭
2. **Effect of Strong Acid.** Transfer 1–2 drops of 0.10 *M* HCl to wells A1 and B1, estimate the pH, and record each pH *change*.⑮
3. **Effect of Strong Base.** Transfer 1–2 drops of 0.10 *M* NaOH to wells A2 and B2, estimate the pH, and record each pH *change*.⑯
4. **Effect of a Buffer System.** Explain the effect a buffered system (as compared with an unbuffered system) has on pH change when a strong acid or strong base is added to it.⑰

D. $[Co(H_2O)_6]^{2+}$, $[CoCl_4]^{2-}$ Equilibrium (Common-Ion Effect)

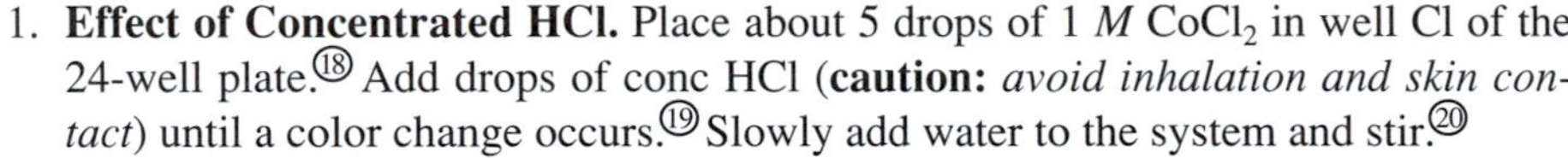

1. **Effect of Concentrated HCl.** Place about 5 drops of 1 *M* $CoCl_2$ in well Cl of the 24-well plate.⑱ Add drops of conc HCl (**caution:** *avoid inhalation and skin contact*) until a color change occurs.⑲ Slowly add water to the system and stir.⑳

E. $[Co(H_2O)_6]^{2+}$, $[CoCl_4]^{2-}$ Equilibrium (Temperature Effect)

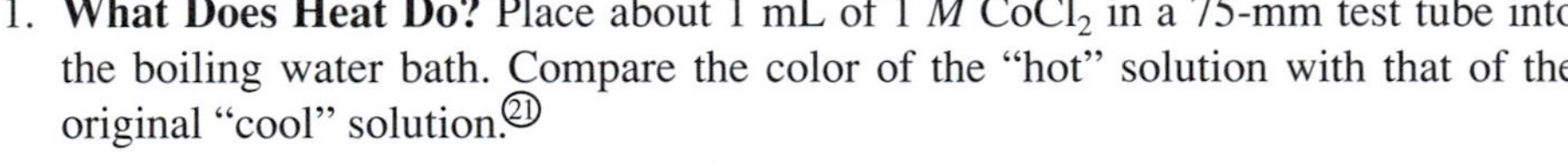

1. **What Does Heat Do?** Place about 1 mL of 1 *M* $CoCl_2$ in a 75-mm test tube into the boiling water bath. Compare the color of the "hot" solution with that of the original "cool" solution.㉑

Disposal for Parts A, C, D, and E: Dispose of the waste solutions in the "Waste Salt Solutions" container.

CLEANUP: Rinse the test tubes and 24-well plate twice with tap water and discard in the "Waste Salt Solutions" container. Do two final rinses with deionized water and discard in the sink.

---

†pH indicator paper may be substituted for the universal indicator to measure the pH of the solutions.

Experiment 24 *Prelaboratory Assignment*

LeChâtelier's Principle; Buffers

Date ________ Lab Sec. ______ Name ________________________________ Desk No. ________

1. Explain the meaning of the phrase, "a system in a state of dynamic equilibrium."

2. a. Describe the dynamic equilibrium that exists in the two water tanks at right.

   b. Explain how LeChâtelier's principle applies when the faucet on the right tank is opened.

   c. Explain how LeChâtelier's principle applies when water is added to the right tank.

3. All of the following chemical equilibria are studied in this experiment. To become familiar with their behavior, indicate the direction, left or right, of the equilibrium shift when the accompanying stress is applied to the system.

   a. $NH_3(aq)$ is added to $Ag^+(aq) + Cl^-(aq) \rightleftharpoons AgCl(s)$ ____________

   b. $HNO_3(aq)$ is added to $Ag_2CO_3(s) \rightleftharpoons Ag^+(aq) + CO_3^{2-}(aq)$ ____________

   c. $KI(aq)$ is added to $Ag^+(aq) + 2\ NH_3(aq) \rightleftharpoons [Ag(NH_3)_2]^+(aq)$ ____________

   d. $Na_2S(aq)$ is added to $AgI(s) \rightleftharpoons Ag^+(aq) + I^-(aq)$ ____________

   e. $KOH(aq)$ is added to $CH_3COOH(aq) + H_2O(l) \rightleftharpoons H_3O^+(aq) + CH_3CO_2^-(aq)$ ____________

   f. $HCl(aq)$ is added to $4\ Cl^-(aq) + Co(H_2O)_6^{2+}(aq) \rightleftharpoons CoCl_4^{2-}(aq) + 6\ H_2O(l)$ ____________

4. The focus of the common-ion effect on equilibria appears in Part D of the experiment. However, the common-ion effect also appears elsewhere in the experiment.
   a. Review the two equilibria appearing in Equation 24.11. Explain how the common-ion effect affects the solubility of AgCl.

   b. Explain, in terms of the common-ion effect, how the addition of $NaCH_3CO_2$ affects the pH of a CH_3COOH solution.

5. A state of dynamic equilibrium, $Ag_2CO_3(s) \rightleftharpoons Ag^+(aq) + CO_3^{2-}(aq)$, exists in solution.

   a. What happens to the equilibrium if more $Ag_2CO_3(s)$ is added to the system?

   b. What happens to the equilibrium if $AgNO_3(aq)$ is added to the system?

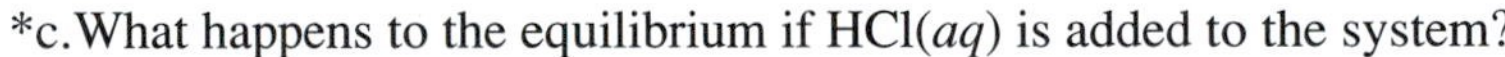

   *c. What happens to the equilibrium if $HCl(aq)$ is added to the system?

6. a. The carbon dioxide of the atmosphere, being a nonmetallic oxide and acid anhydride, has a low solubility in rainwater but produces a slightly acidic solution. Write an equilibrium equation for the dissolution of carbon dioxide in water and identify the acid that is formed.

   b. Calcium carbonate, the major component of limestone, is slightly soluble in water. Write an equilibrium equation showing the slight solubility of calcium carbonate in water.

   c. Write a Brønsted equation for the weak acid formed in part **a,** showing how it produces hydronium ion in aqueous solution.

   d. Using LeChâtelier's principle, explain how the weak acid formed in rainwater causes the dissolution of calcium carbonate. The consequence of this reaction contributes to the degree of hardness in underground and surface water supplies and also accounts for the formation of stalagmites and stalactites.

Experiment 24 *Report Sheet*

LeChâtelier's Principle; Buffers

Date __________ Lab Sec. ______ Name ____ Desk No. __________

A. Metal–Ammonia Ions

$CuSO_4(aq)$ or $NiCl_2(aq)$	$[Cu(NH_3)_4]^{2+}$ or $[Ni(NH_3)_6]^{2+}$	HCl Addition
Color① ______________________	② ______________________	③ ______________________

Account for the effects of $NH_3(aq)$ and $HCl(aq)$ on the $CuSO_4$ or $NiCl_2$ solution.

B. Multiple Equilibria with the Silver Ion

④Observation and net ionic equation for reaction.

⑤Account for the observed chemical change from HNO_3 addition.

⑥Observation from HCl addition and net ionic equation for the reaction.

⑦Effect of conc NH_3. Explain.

⑧What does the HNO_3 do?

⑨What result does excess NH_3 produce?

⑩Effect of added KI. Explain.

⑪Effect of Na_2S and net ionic equation for the reaction.

C. A Buffer System

⑫Write the Brønsted acid equation for $CH_3COOH(aq)$

Color of universal indicator in CH_3COOH ______________________________ pH _____

⑬Color of universal indicator after addition of $NaCH_3CO_2$ ______________________________ pH _____

Effect of $NaCH_3CO_2$ on the equilibrium.

⑭Color of universal indicator in water ______________________________ pH _____

	Buffer System		Water	
	Well A1 (or test tube)	Well A2 (or test tube)	Well B1 (or test tube)	Well B2 (or test tube)
⑮Color after 0.10 *M* HCl addition	______	—	______	—
Approximate pH	______	—	______	—
Approximate ΔpH	______	—	______	—
⑯Color after 0.10 *M* NaOH addition	—	______	—	______
Approximate pH	—	______	—	______
Approximate ΔpH	—	______	—	______

⑰Discuss in detail the magnitude of the changes in pH that are observed in wells A1 and A2 relative to those observed in wells B1 and B2.

D. $[Co(H_2O)_6]^{2+}$, $[CoCl_4]^{2-}$ Equilibrium (Common-Ion Effects)

⑱Color of $CoCl_2(aq)$

⑲Observation from conc HCl addition and net ionic equation for the reaction.

⑳Account for the observation resulting from the addition of water.

E. $[Co(H_2O)_6]^{2+}$, $[CoCl_4]^{2-}$ Equilibrium (Temperature Effects)

(21) Effect of heat. What happens to the equilibrium?

Laboratory Questions

Circle the questions that have been assigned.

1. Part A.1. NH_3 is a weak base; NaOH is a strong base. Predict what would appear in the solution if NaOH were been added to the $CuSO_4$ solution instead of the NH_3. (*Hint:* review the solubility rules.)
2. Part B.1. a. HNO_3, a strong acid, is added to shift the Ag_2CO_3 equilibrium (Equation 24.6) to the right. Explain why the shift occurs.
   b. What would have been observed if HCl (also a strong acid) had been added instead of the HNO_3?
3. Part B.4. Silver bromide, a pale yellow precipitate, is *more* soluble than silver iodide. Predict what would happen if 0.1 *M* NaBr had been added to the solution in Part B.3 instead of the Na_2S solution. Explain.
4. Write an equation that shows the pH dependence on the chromate, CrO_4^{2-}/dichromate, $Cr_2O_7^{2-}$, equilibrium system (see opening photo of the experiment).
5. Part C. HCl(*aq*) is a much stronger acid that CH_3COOH(*aq*). However, when 20 mL of 0.10 *M* HCl(*aq*) is added to 20 mL of a buffer solution that is 0.10 *M* CH_3COOH and 0.10 *M* $CH_3CO_2^-$ only a very small change in pH occurs. Explain.
6. Part E. Consider the following endothermic equilibrium reaction system in aqueous solution:

   $4\,Cl^-(aq) + [Co(H_2O)_6]^{2+}(aq) \leftrightharpoons [CoCl_4]^{2}(aq) + 6\,H_2O(l)$

   If the equilibrium system were stored in an isothermal vessel (with no heat transfer into or out of the vessel from the surroundings), predict what would happen to the temperature reading on a thermometer placed in the solution when hydrochloric acid is added. Explain.

Experiment 25

An Equilibrium Constant

The nearly colorless iron(III) ion (left) forms an intensely colored complex (right) in the presence of the thiocyanate ion.

Objectives

- To use a **spectrophotometer** to determine the equilibrium constant of a chemical system
- To use graphing techniques and data analysis to evaluate data
- To determine the equilibrium constant for a soluble equilibrium

Spectrophotometer: a laboratory instrument that measures the amount of light transmitted through a sample

Techniques

The following techniques are used in the Experimental Procedure

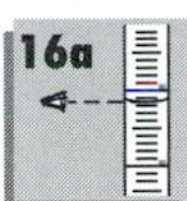

Introduction

A spectrophotometric method of analysis involves the interaction of electromagnetic (EM) radiation with matter. The most common regions of the EM spectrum used for analyses are the ultraviolet, visible, and the infrared regions. We are most familiar with the visible region of the spectrum, having a wavelength range from 400 to 700 nm.

Every chemical substance possesses its own unique set of electronic, vibrational, and rotational **energy states.** When EM radiation falls incident upon an atom or molecule, the radiation absorbed (the absorbed light) is an energy equal to the difference between two energy states in the atom or molecule, placing the atom or molecule in an "excited state." The remainder of the EM radiation (the complementary radiation which is *not* absorbed and called the **transmitted light**) passes through the sample and can be detected by an EM radiation detector. The several energy states of an atom or molecule over a region of EM radiation result in its unique or characteristic **spectrum** (a fingerprint of its energy states).

Energy states: quantized energies can be absorbed or emitted to change the vibrational state, rotational state, or electronic state in a molecule

The higher the concentration of the "absorbing species" in the sample, the more the incident EM radiation is absorbed and the more intense is the complementary EM radiation that is transmitted.

The visible spectra of elements and compounds arise from *electronic* transitions within the atomic and molecular structures. In this experiment, the visible region of the EM spectrum is used to measure the concentration of an absorbing substance in an aqueous solution. The intensity of the transmitted light is measured using an instrument called a **spectrophotometer,** an instrument that measures light intensities with a photosensitive detector at specific (but variable) wavelengths (Figure 25.1). The wavelength where the absorbing substance has a maximum *absorption* of visible radiation is determined and set on the spectrophotometer.

The visible light path through the spectrophotometer from the light source through the sample to the photosensitive detector is shown in Figure 25.2.

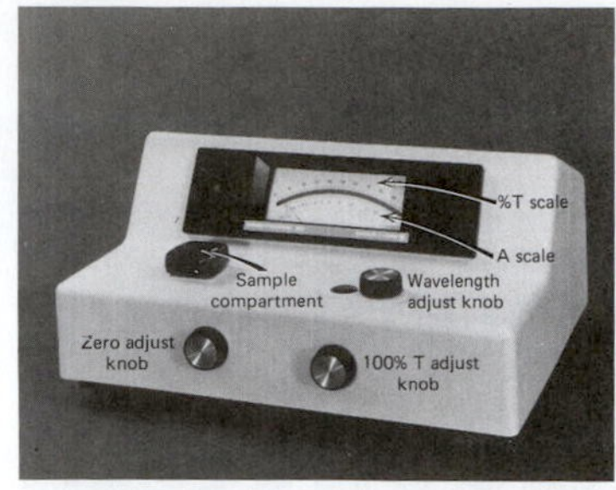

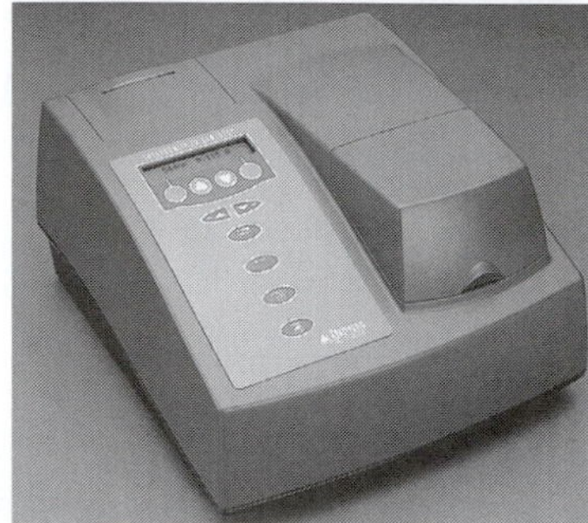

Figure 25.1 Common laboratory visible spectrophotometers.

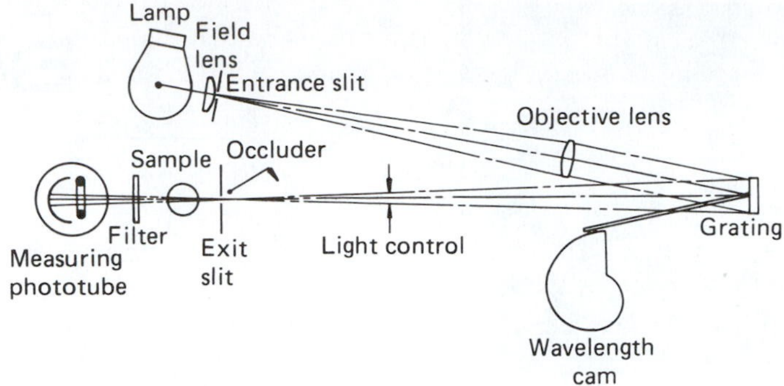

Figure 25.2 The light path through a visible spectrophotometer.

Several factors control the amount of EM radiation (light energy) that a sample absorbs:

- Concentration of the absorbing substance.
- Thickness of the sample containing the absorbing substance (determined by the width of the **cuvet**)
- Probability of light absorption by the absorbing substance (called the **molar absorptivity coefficient** or **extinction coefficient**)

Cuvet: a special piece of glassware to hold solutions for measurement in the spectrophotometer

Incident light: light that enters the sample

The ratio of the intensity of the transmitted light, I_t, to that of the **incident light,** I_0 (Figure 25.3), is called the transmittance, T, of the EM radiation by the sample. This ratio, expressed as percent, is

$$\frac{I_t}{I_0} \times 100 = \%T \qquad (25.1)$$

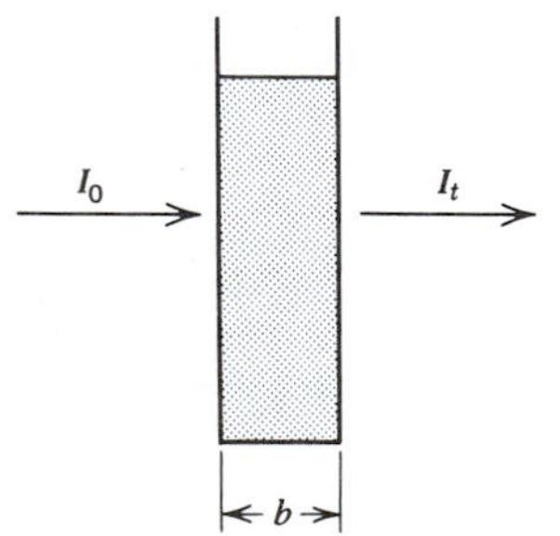

Figure 25.3 Incident light, I_0, and transmitted light, I_t, for a sample of thickness b.

Most spectrophotometers have a $\%T$ (percent transmittance of light) scale. Because it is linear, the $\%T$ scale is easy to read and interpolate. However, chemists often perform calculations based on the amount of light *absorbed* by the sample, rather than the amount of light transmitted, because absorption is directly proportional to the concentration of the absorbing substance. The absorbance, A, of the substance is related to the intensity of the incident and transmitted light (and the percent transmittance) by the equations

$$\mathrm{A} = \log\frac{I_0}{I_t} = \log\frac{1}{T} = \log\frac{100}{\%T} = a \bullet b \bullet c \qquad (25.2)$$

a, the molar absorptivity coefficient, is a constant at any given wavelength for a particular absorbing substance, b is the thickness of the absorbing substance in centimeters, and c is the molar concentration of the absorbing substance.[1]

The absorbance value is directly proportional to the molar concentration of the absorbing substance, *if* the same (or a matched) cuvet and a *set* wavelength are used for all measurements. A plot of absorbance versus concentration data is linear; a calculated slope and absorbance data can be used to determine the molar concentration of the same absorbing species in a solution of unknown concentration (Figure 25.4) from the linear relationship.

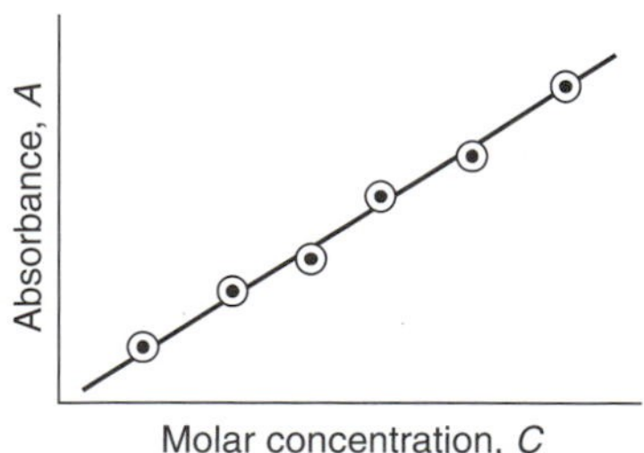

Figure 25.4 A plot of absorbance vs. concentration data.

[1]Since the quantity log (I_0/I_t) is generally referred to as absorbance, Equation 25.2 becomes $A = abc$. This equation is commonly referred to as **Beer's law.**

Measuring an Equilibrium Constant

The magnitude of an equilibrium constant, K_c, expresses the equilibrium position for a chemical system. For the reaction, $aA + bB \rightleftharpoons xX + yY$, the mass action expression, $\frac{[X]^x[Y]^y}{[A]^a[B]^b}$, equals the equilibrium constant, K_c, when a dynamic equilibrium has been established between reactants and products. The brackets in the mass action expression denote the equilibrium molar concentration of the respective substance.

The magnitude of the equilibrium constant indicates the principal species, products or reactants, that exist in the chemical system at equilibrium. For example, a large equilibrium constant indicates that the equilibrium lies to the right with a high concentration of products and correspondingly low concentration of reactants. The value of K_c is constant for a chemical system at a given temperature.

This experiment determines K_c for a chemical system in which all species are soluble. The chemical system involves the equilibrium between iron(III) ion, Fe^{3+}, thiocyanate ion, SCN^-, and thiocyanatoiron(III) ion, $FeNCS^{2+}$:

$$[Fe(H_2O)_6]^{3+}(aq) + SCN^-(aq) \rightleftharpoons [Fe(H_2O)_5NCS]^{2+}(aq) + H_2O(l) \quad (25.3)$$

The "free" thiocyanate ion is commonly written as SCN^-; however its bond to the ferric ion is through the nitrogen atom, thus the formula of the complex is written $FeNCS^{2+}$

Because the concentration of water is essentially constant in dilute aqueous solutions, we omit the waters of hydration and simplify the equation to read

$$Fe^{3+}(aq) + SCN^-(aq) \rightleftharpoons FeNCS^{2+}(aq) \quad (25.4)$$

The mass action expression for the equilibrium system, equal to the equilibrium constant, is

$$K_c = \frac{[FeNCS^{2+}]}{[Fe^{3+}][SCN^-]} \quad (25.5)$$

In Part A you will prepare a set of five **standard solutions** of the $FeNCS^{2+}$ ion. As $FeNCS^{2+}$ is a deep, blood-red complex, its absorption maximum occurs at about 447 nm. The absorbance at 447 nm for each solution is plotted versus the molar concentration of $FeNCS^{2+}$; this establishes a **standardization curve** from which the concentrations of $FeNCS^{2+}$ are determined for the chemical systems in Part B.

Standard solution: a solution with a very well known concentration of solute

Standardization curve: a plot of known data from which further interpretations can be made

In preparing the standard solutions of $FeNCS^{2+}$, the Fe^{3+} concentration is set to *far* exceed the SCN^- concentration. This huge excess of Fe^{3+} pushes the equilibrium (Equation 25.4) *far* to the right, consuming nearly all of the SCN^- placed in the system. As a result the $FeNCS^{2+}$ concentration at equilibrium approximates the original SCN^- concentration. In other words, we assume that the position of the equilibrium is driven so far to the right by the excess Fe^{3+} that all of the SCN^- is **complexed,** forming $FeNCS^{2+}$ (Figure 25.5).

Complexed: the formation of a bond between the Lewis base, SCN^-, and the Lewis acid, Fe^{3+}

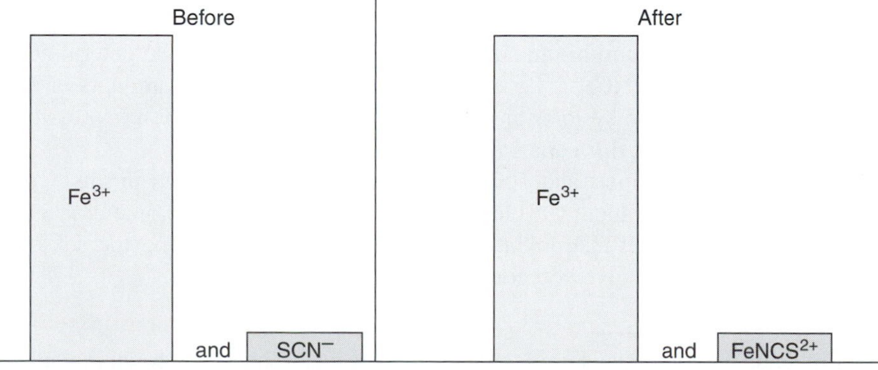

Figure 25.5 A large excess of Fe^{3+} consumes nearly all of the SCN^- to form $FeNCS^{2+}$. The amount of Fe^{3+} remains essentially unchanged in solution.

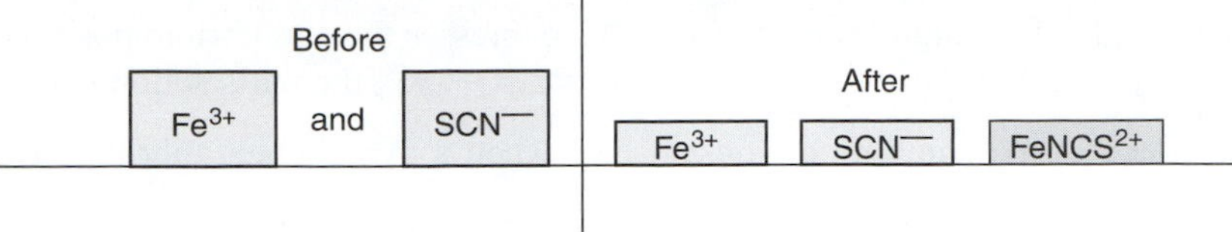

Figure 25.6 Amounts of Fe^{3+} and SCN^- are equally reduced in the formation of $FeNCS^{2+}$.

In Part B, the concentrations of the Fe^{3+} and SCN^- ions in the various test solutions are nearly the same, thus creating equilibrium systems in which there is an appreciable amount of each of the species after equilibrium is established (Figure 25.6).

Calculations of K_c

Accurate volumes of known molar concentrations of Fe^{3+} and SCN^- are mixed for the preparation of the chemical system. The equilibrium molar concentration of $FeNCS^{2+}$ of the system is determined by measuring its absorbance and then using the standardization curve from Part A. Since the total volume of the mixed solution is accurately measured, the *initial* moles of Fe^{3+} and SCN^- and the *equilibrium* moles of $FeNCS^{2+}$ are easily calculated from known molar concentrations.

From Equation 25.4, for every mole of $FeNCS^{2+}$ that exists at equilibrium, an equal number of moles of Fe^{3+} and SCN^- have reacted to reach equilibrium:

$$\text{mol FeNCS}^{2+}{}_{\text{equilibrium}} = \text{mol Fe}^{3+}{}_{\text{reacted}} = \text{mol SCN}^-{}_{\text{reacted}}$$

Therefore, the moles of Fe^{3+} at equilibrium (unreacted) is

$$\text{mol Fe}^{3+}{}_{\text{unreacted}} = \text{mol Fe}^{3+}{}_{\text{initial}} - \text{mol Fe}^{3+}{}_{\text{reacted}}$$

Similarly, the moles of SCN^- at equilibrium (unreacted) is

$$\text{mol SCN}^-{}_{\text{unreacted}} = \text{mol SCN}^-{}_{\text{initial}} - \text{mol SCN}^-{}_{\text{reacted}}$$

Again, since the total volume of the reaction mixture is known, the equilibrium molar concentrations of Fe^{3+} and SCN^- (their *unreacted* concentrations) can be calculated. Knowing the measured equilibrium molar concentration of $FeNCS^{2+}$ from the standardization curve, substitution of the three equilibrium molar concentrations into the mass action expression provides the value of the equilibrium constant, K_c.

The calculations for K_c are involved, but completion of the Prelaboratory Assignment should clarify most of the steps. The Report Sheet is also outlined in such detail as to help with the calculations.

Experimental Procedure

Procedure Overview: One set of solutions having known molar concentrations of $FeNCS^{2+}$ is prepared for a standardization curve, a plot of absorbance versus concentration. A second set of equilibrium solutions is prepared and mixed to determine the molar concentration of $FeNCS^{2+}$. By carefully measuring the initial amounts of reactants placed in the reaction systems and the absorbance, the mass action expression at equilibrium can be solved; this equals K_c.

A large number of pipets and 100-mL volumetric flasks are used in this experiment. Ask your instructor about working with a partner. A spectrophotometer is an expensive, delicate analytical instrument. Operate it with care, following the advice of your instructor, and it will give you good data.

A. A Set of Standard Solutions to Establish a Standardization Curve

The set of standard solutions is used to determine the absorbance of known molar concentrations of $FeNCS^{2+}$. A plot of the data, known as a standardization curve, is used to determine the equilibrium molar concentrations of $FeNCS^{2+}$ in Part B.

Table 25.1 Composition of the Set of Standard $FeNCS^{2+}$ Solutions for Preparing the Standardization Curve

Standard Solution	0.2 *M* $Fe(NO_3)_3$ (in 0.1 *M* HNO_3)	0.001 *M* NaSCN (in 0.1 *M* HNO_3)	0.1 *M* HNO_3
Blank	10.0 mL	0 mL	dilute to 25 mL
1	10.0 mL	1 mL	dilute to 25 mL
2	10.0 mL	2 mL	dilute to 25 mL
3	10.0 mL	3 mL	dilute to 25 mL
4	10.0 mL	4 mL	dilute to 25 mL

Once the standard solutions are prepared, proceed smoothly and methodically through Part A.4. Therefore, read through all of Part A before proceeding.

1. **Prepare a Set of the Standard Solutions.** Prepare the solutions in Table 25.1. Pipet 0, 1, 2, 3, and 4 mL of 0.001 *M* NaSCN into separate, labeled, and clean 25-mL volumetric flasks (or 200-mm test tubes). Pipet 10.0 mL of 0.2 *M* $Fe(NO_3)_3$ into each flask (or test tube) and *quantitatively* dilute to 25 mL (the "mark" on the volumetric flask) with 0.1 *M* HNO_3. Stir/agitate each solution thoroughly to ensure that equilibrium is established. These solutions are used to establish a plot of absorbance versus [$FeNCS^{2+}$], the standardization curve. An additional standard solution, labeled 4A, of 5 mL of 0.001 *M* NaSCN is suggested.

   Record on the Report Sheet the *exact* molar concentrations of the $Fe(NO_3)_3$ and NaSCN reagent solutions.

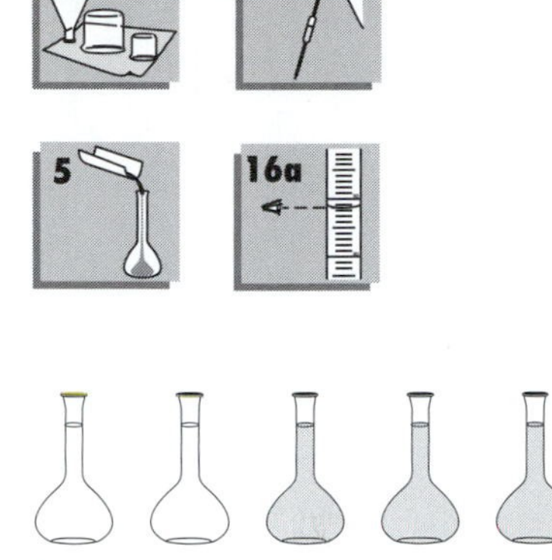

A set of labeled test solutions.

2. **Prepare the Blank Solution.** After the spectrophotometer has been turned on for 10 minutes and the wavelength scale has been set at 447 nm, rinse a cuvet with several portions of the **blank solution.** Dry the outside of the cuvet with a clean Kimwipe, removing water and fingerprints.[2] Handle the lip of the cuvet thereafter.

Blank solution: a solution that contains all light-absorbing species except the one being investigated in the experiment

3. **Calibrate the Spectrophotometer.** Place the cuvet, three-quarters filled with the blank solution, into the sample compartment, align the mark on the cuvet with that on the sample holder, and close the cover. Set the meter on the spectrophotometer to read zero absorbance (or 100%*T*).[3] Remove the cuvet. Consult with your instructor for any further calibration procedures. Once the instrument is set, *do not* perform any additional adjustments for the remainder of the experiment. If you accidentally do, merely repeat the calibration procedure.

4. **Record the Absorbance of the Standard Solutions.** Empty the cuvet and rinse it *thoroughly* with several small portions of Solution 1.[4] Fill it approximately three-fourths full. Again, carefully dry the outside of the cuvet with a clean Kimwipe. Remember, handle only the lip of the cuvet. Place the cuvet into the sample compartment and align the cuvet and sample holder marks; read the absorbance (or percent transmittance if the spectrophotometer has a meter readout) and record. Repeat with Solutions 2, 3, and 4.

   Share the set of standard solutions with other chemists in the laboratory.

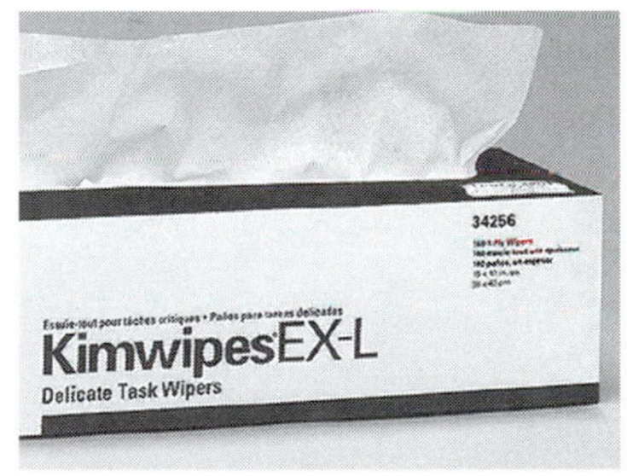

A box of lint-free tissue.

5. **Graph the Data.** Plot absorbance, *A* (ordinate), versus [$FeNCS^{2+}$] (abscissa) for the five solutions on linear graph paper. Draw the *best straight line* through the five (or six) points (see Figure 25.4) to establish the standardization curve. Ask your instructor to approve your graph.

Appendix C

[2]Water and fingerprints (or any foreign material) on the outside of the cuvet reduce the intensity of the light transmitted to the detector.

[3]For spectrophotometers with a meter readout (as opposed to digital), record the percent transmittance and then calculate the absorbance (Equation 25.2). This procedure is more accurate because %*T* is a linear scale (whereas absorbance is logarithmic) and because it is easier to estimate %*T* values more accurately and consistently.

[4]If possible, prepare five matched cuvets, one for each standard solution, and successively measure the absorbance of each, remembering that the first solution is the blank solution.

B. Absorbance for the Set of Test Solutions

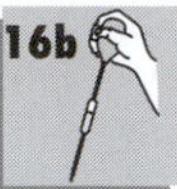

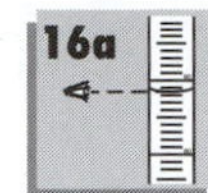

1. **Prepare the Test Solutions.** In *clean* 150-mm test tubes (or 10-mL volumetric flasks)[5] prepare the test solutions in Table 25.2. Use pipets for the volumetric measurements. Be careful not to "mix" pipets to avoid contamination of the reagents prior to the preparation. Also note that the molar concentration of $Fe(NO_3)_3$ for this set of solutions is 0.002 mol/L, *not* the 0.2 mol/L solution used in Part A.

   Record the *exact* molar concentrations of the $Fe(NO_3)_3$ and NaSCN reagent solutions on the Report Sheet.

   Once the test solutions are prepared, proceed smoothly and methodically (you need not hurry!) through Part B.3.

A set of labeled test solutions.

Table 25.2 Composition of the Set of Equilibrium Test Solutions for the Determination of K_c

Test Solution	0.002 *M* $Fe(NO_3)_3$* (in 0.1 *M* HNO_3)	0.002 *M* NaSCN (in 0.1 *M* HNO_3)	0.1 *M* HNO_3
5	5 mL	1 mL	4 mL
6	5 mL	2 mL	3 mL
7	5 mL	3 mL	2 mL
8	5 mL	4 mL	1 mL
9	5 mL	5 mL	—

*If 0.002 *M* $Fe(NO_3)_3$ is not available, dilute 1.0 mL (measure with a 1.0-mL pipet) of the 0.2 *M* $Fe(NO_3)_3$ used in Part A with 0.1 *M* HNO_3 in a 100-mL volumetric flask and then share the remaining solution with other students.

2. **Recalibrate the Spectrophotometer.** Use the blank solution from Part A to check the calibration of the spectrophotometer. See Part A.3.
3. **Determine the Absorbance of the Test Solutions.** Stir or agitate each test solution until equilibrium is reached (approximately 30 seconds). Rinse the cuvet thoroughly with several portions of the test solution and fill it three-fourths full. Clean and dry the outside of the cuvet. Be cautious in handling the cuvets. Record the absorbance of each test solution as was done in Part A.4.

Disposal: Dispose of all waste thiocyanatoiron(III) ion solutions from Parts A and B in the "Waste Salts" container.

CLEANUP: Rinse the volumetric flasks, the pipets, and the cuvets twice with tap water and twice with deionized water. Discard each rinse in the sink.

4. **Use Data to Determine Equilibrium Concentrations.** Using the standardization curve prepared in Part A.5, use the absorbance value for each test solution to determine the equilibrium molar concentration of $FeNCS^{2+}$.
5. **Data Analysis.** Complete the calculations as outlined on the Report Sheet and described in the Introduction. Complete an entire K_c calculation for Test Solution 5 before attempting the calculations for the remaining solutions.

   The equilibrium constant will vary from solution to solution and from chemist to chemist in this experiment, depending on chemical technique and the accumulation and interpretation of the data. Consequently, it is beneficial to work through your own calculations with other colleagues and then "pool" your final, experimental K_c values to determine an accumulated "probable" value and a standard deviation for K_c.

Appendix B

[5]If 10-mL volumetric flasks are used, use pipets to dispense the volumes of the NaSCN and $Fe(NO_3)_3$ solutions and then dilute to the mark with 0.1 *M* HNO_3.

Experiment 25 *Prelaboratory Assignment*

An Equilibrium Constant

Date ________ Lab Sec. ______ Name ______________________________ Desk No. ________

1. a. Identify three factors that affect the amount of light absorbed by a sample.

   b. Which of the three factors remain constant in this experiment? Explain.

2. A 4.00 mL aliquot of 0.001 *M* NaSCN is diluted to 25.0 mL with 0.1 *M* HNO_3 (Part A.1)

   a. What is the molar concentration of SCN^- in the diluted solution?

   b. If all of the SCN^- is complexed with Fe^{3+} to $FeNCS^{2+}$, what is the molar concentration of $FeNCS^{2+}$?

   c. How many *moles* of Fe^{3+} must react to form the $FeNCS^{2+}$ in Part 4.b?

3. For preparing a set of standard solutions of $FeNCS^{2+}$ (Part A.1), the equilibrium molar concentration of $FeNCS^{2+}$ is assumed to equal the initial molar concentration of the SCN^- in the reaction mixture. Why is this assumption valid?

4. The blank solution used to calibrate the spectrophotometer (Part A.3) is 10.0 mL of 0.2 *M* $Fe(NO_3)_3$ diluted to 25.0 mL with 0.1 *M* HNO_3. Why is this solution preferred to simply using de-ionized water for the calibration?

5. Plot the following data as absorbance versus [XX] as a standardization curve:

Absorbance, A	Standard Concentration of XX
0.045	3.0×10^{-4} mol/L
0.097	6.2×10^{-4} mol/L
0.14	9.0×10^{-4} mol/L
0.35	2.2×10^{-3} mol/L
0.51	3.2×10^{-3} mol/L

A "test" solution showed a percent transmittance ($\%T$) reading of 34.2$\%T$. Interpret the calibration curve to determine the molar concentration of XX in the test solution.

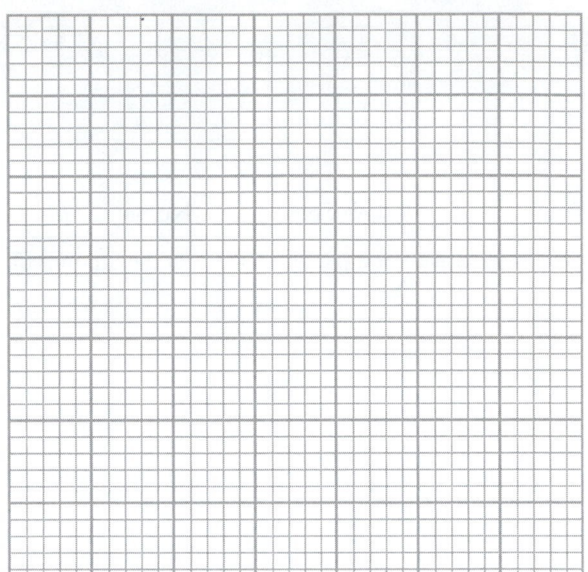

6. A reaction mixture of 4.0 mL of 0.002 M SCN^- and 5.0 mL of 0.002 M Fe^{3+} is diluted to 10.0 mL with deionized water to form the blood-red $FeNCS^{2+}$ complex. The equilibrium molar concentration of the $FeNCS^{2+}$ determined from a standardization curve, is 1.5×10^{-4} mol/L. Calculate, in sequence, each of the following quantities in the aqueous solution to determine the equilibrium constant for the reaction,

$$Fe^{3+}(aq) + SCN^-(aq) \leftrightarrows FeNCS^{2+}(aq)$$

a. *moles* of $FeNCS^{2+}$ that form in reaching equilibrium ______________

b. moles of Fe^{3+} that react to form the $FeNCS^{2+}$ at equilibrium ______________

c. moles of SCN^- that react to form the $FeNCS^{2+}$ at equilibrium ______________

d. moles of Fe^{3+} initially placed in the reaction system ______________

e. moles of SCN^- initially placed in the reaction system ______________

f. moles of Fe^{3+} that remain *un*reacted at equilibrium (**d** − **b**) ______________

g. moles of SCN^- that remain *un*reacted at equilibrium (**e** − **c**) ______________

h. molar concentration of Fe^{3+} (unreacted) at equilibrium ______________

i. molar concentration of SCN^- (unreacted) at equilibrium ______________

j. molar concentration of $FeNCS^{2+}$ at equilibrium 1.5×10^{-4} mol/L

k. $K_c = \dfrac{[FeNCS^{2+}]}{[Fe^{3+}][SCN^-]}$ ______________

Experiment 25 *Report Sheet*

An Equilibrium Constant

Date ________ Lab Sec. ______ Name ________________________________ Desk No. ________

A. A Set of Standard Solutions to Establish a Standardization Curve

Molar concentration of $Fe(NO_3)_3$ ______________

Molar concentration of NaSCN ______________

	Standard Solutions					
	Blank	1*	2	3	4	4A
1. Volume of NaSCN (mL)						
2. Moles of SCN^-						
3. $[SCN^-]$ (25.0 mL)						
4. $[FeNCS^{2+}]$						
5. Percent transmittance, %*T* (for meter readings only)						
6. Absorbance, *A*						

*Calculation for Standard Solution 1.

7. Plot data of *A* versus $[FeNCS^{2+}]$. Instructor's approval ____________________

B. Absorbance for the Set of Test Solutions

Molar concentration of $Fe(NO_3)_3$ ______________

Molar concentration of NaSCN ______________

	Test Solutions				
	5	6	7	8	9
1. Volume of $Fe(NO_3)_3$ (mL)					
2. Moles of Fe^{3+}, initial	*				
3. Volume of NaSCN (mL)					
4. Moles of SCN^-, initial	*				
5. Percent transmittance, %*T* (for meter readings only)					
6. Absorbance, *A*					

*Calculation for Test Solution 5.

Interpreting the Data for Determining K_c

	5	6	7	8	9
1. $[FeNCS^{2+}]$, equilibrium standardization curve					
2. Moles $FeNCS^{2+}$ at equilibrium (10 mL)					

[Fe^{3+}], equilibrium

1. Moles Fe^{3+}, reacted					
2. Moles Fe^{3+}, unreacted					
3. [Fe^{3+}], equilibrium (unreacted)					

*Calculation for Test Solution 5.

[SCN^-], equilibrium	5	6	7	8	9
1. Moles SCN^-, reacted					
2. Moles SCN^-, unreacted					
3. [SCN^-], equilibrium (unreacted)	*				

*Calculation for Test Solution 5.

$K_c = \frac{[FeNCS^{2+}]}{[Fe^{3+}][SCN^-]}$	*				

*Calculation for Test Solution 5.

Average K_c ______________________

Average deviation (Appendix B) ______________________

Standard deviation (Appendix B) ______________________

*Calculation for standard deviation.

Class Data/Group	1	2	3	4	5
Average K_c					

Standard deviation of class data:

Laboratory Questions

Circle the questions that have been assigned.

*1. Part A.2. All spectrophotometers are different. The spectrophotometer is to be set at 447 nm. What experiment could you do, what data would you collect, and how would you analyze the data to ensure that 447 nm is the "best" setting for measuring the absorbance of $FeNCS^{2+}$ in this experiment?

2. Part B.3. Fingerprint smudges are present on the cuvet containing the solution placed into the spectrophotometer for analysis.
   a. How does this technique error affect the absorbance reading for $FeNCS^{2+}$ in the analysis? Explain.
   b. Will the equilibrium concentration of $FeNCS^{2+}$ be recorded as being too high or too low? Explain.
   c. Will the equilibrium concentration of SCN^- be too high, too low, or unaffected by the technique error? Explain.
   d. Will the K_c for the equilibrium be too high, too low or unaffected by the technique error? Explain.

3. Part B.3. For the preparation of Test Solution 7 (Table 25.2), the 2.0 mL of 0.1 *M* HNO_3 is omitted.
   a. How does this technique error affect the absorbance reading for $FeNCS^{2+}$? Explain.
   b. Will the K_c for the equilibrium be too high, too low or unaffected by the technique error? Explain.

*4. The equation, $A = a \bullet b \bullet c$, becomes nonlinear at high concentrations of the absorbing substance. Suppose you prepare a solution with a very high absorbance that is suspect in *not* following the linear relationship. How might you still use the sample for your analysis, rather than discarding the sample and the data?

5. Glass cuvets, used for precision absorbance measurements in a spectrophotometer, are marked so that they always have the *same* orientation in the sample compartment. What error does this minimize and why?

Experiment 26

Antacid Analysis

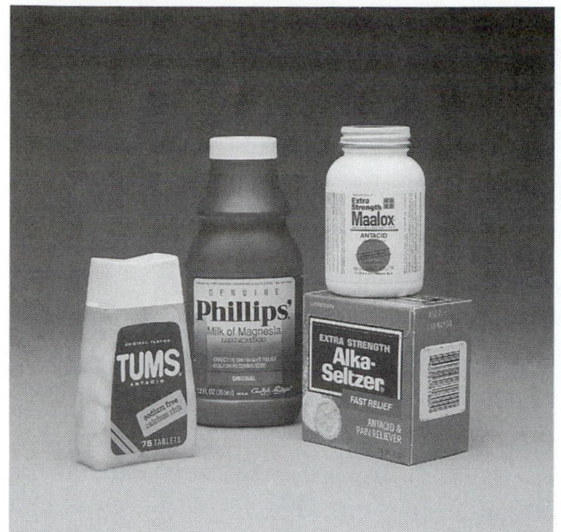

All antacids, as weak bases, reduce the acidity of the stomach.

OBJECTIVE

- To determine the neutralizing effectiveness per gram of a commercial **antacid**

TECHNIQUES

The following techniques are used in the Experimental Procedure

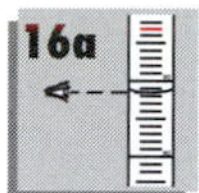

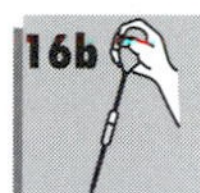

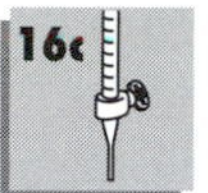

INTRODUCTION

Various commercial antacids claim to be the "most effective" for relieving acid indigestion. All antacids, regardless of their claims or effectiveness, have one purpose—to neutralize the *excess* hydrogen ion in the stomach to relieve acid indigestion.

Antacid: dissolved in water, it forms a basic solution

The acidity (and basicity) of an aqueous solution is often expressed in terms of its **pH.** At 25°C, acidic solutions have a pH less than 7, with the lower values being the more acidic; basic solutions have a pH greater than 7, with the higher values being more basic.

pH: negative logarithm of the molar concentration of hydronium ion, $-\log [H_3O^+]$ (see Experiment 13)

The pH of the "gastric juice" in the stomach ranges from 1.0 to 2.0. This acid, primarily hydrochloric acid, is necessary for the digestion of foods. Acid is continually secreted while eating; consequently, overeating can lead to an excess of stomach acid, leading to acid indigestion and a pH less than 1. An excess of acid can, on occasion, cause an irritation of the stomach lining, particularly the upper intestinal tract, causing "heartburn." An antacid reacts with the hydronium ion to relieve the symptoms. Excessive use of antacids can cause the stomach to have a pH greater than 2, which stimulates the stomach to excrete additional acid, a potentially dangerous condition.

Appendix D

The most common bases used for over-the-counter antacids are the following:

aluminum hydroxide, $Al(OH)_3$	magnesium hydroxide, $Mg(OH)_2$
calcium carbonate, $CaCO_3$	sodium bicarbonate, $NaHCO_3$
magnesium carbonate, $MgCO_3$	potassium bicarbonate, $KHCO_3$

Milk of magnesia (Figure 26.1), an aqueous suspension of magnesium hydroxide, $Mg(OH)_2$, and sodium bicarbonate, $NaHCO_3$, commonly called baking soda, are simple antacids (and thus, bases) that neutralize hydronium ion, H_3O^+:

$$Mg(OH)_2(s) + 2\,H_3O^+(aq) \rightarrow Mg^{2+}(aq) + 4\,H_2O(l) \qquad (26.1)$$

$$NaHCO_3(aq) + H_3O^+(aq) \rightarrow Na^+(aq) + CO_2(g) + 2\,H_2O(l) \qquad (26.2)$$

The release of carbon dioxide gas from the action of sodium bicarbonate on hydronium ion (Equation 26.2) causes one to "belch."

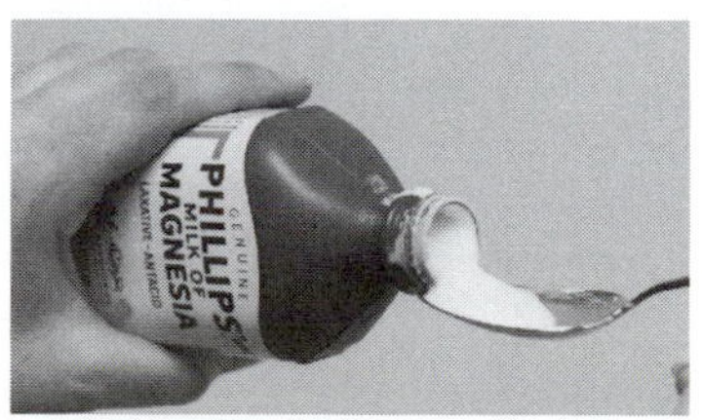

Figure 26.1 Milk of magnesia is an aqueous suspension of slightly soluble magnesium hydroxide.

Table 26.1 Common Antacids

Principal Active Ingredient(s)	Formulation	Commercial Antacid
$CaCO_3$	Tablet	Tums®, Titralac®, Chooz®, Maalox®
$CaCO_3$, $Mg(OH)_2$	Tablet	Rolaids®, Di-Gel®, Mylanta®
$MgCO_3$, $Al(OH)_3$	Tablet	Gaviscon® Extra Strength
$Mg(OH)_2$, $Al(OH)_3$	Tablet	Gelasil®, Tempo®
$NaHCO_3$, citric acid, aspirin	Tablet	Alka-Seltzer®
$Mg(OH)_2$	Tablet	Phillips'® Milk of Magnesia
$Mg(OH)_2$	Liquid	Phillips'® Milk of Magnesia
$Mg(OH)_2$, $Al(OH)_3$	Liquid	Maalox®, Mylanta® Extra Strength
$MgCO_3$, $Al(OH)_3$	Liquid	Gaviscon® Extra Strength

Buffers: substances in an aqueous system that are present for the purpose of resisting changes in acidity or basicity

To decrease the possibility of the stomach becoming too basic from the antacid, **buffers** are often added as part of the formulation of some antacids. The more common, "faster relief" commercial antacids that buffer excess acid in the stomach are those containing calcium carbonate, $CaCO_3$, and/or sodium bicarbonate. A HCO_3^-/CO_3^{2-} buffer system[1] is established in the stomach with these antacids.

$$CO_3^{2-}(aq) + H_3O^+(aq) \rightarrow HCO_3^-(aq) + H_2O(l) \quad (26.3)$$

$$HCO_3^-(aq) + H_3O^+(aq) \rightarrow CO_2(g) + 2\,H_2O(l) \quad (26.4)$$

Rolaids® is an antacid that consists of a combination of $Mg(OH)_2$ and $CaCO_3$ in a mass ratio of 1:5, thus providing the effectiveness of the hydroxide base and the carbonate/bicarbonate buffer. Some of the more common over-the-counter antacids and their major active antacid ingredient(s) are listed in Table 26.1.

Endpoint: the point in the titration when an indicator changes color

In this experiment, the "neutralizing powers" of several antacids are determined using a strong acid–strong base titration. To obtain the quantitative data for the analysis, which requires a well-defined **endpoint** in the titration, the buffer action must be eliminated.

The buffer is eliminated when an excess of hydrochloric acid, HCl, is added to the antacid solution; this addition drives the HCO_3^-/CO_3^{2-} reaction in Equation 26.4 far to the right. The solution is heated to remove carbon dioxide. The *excess* HCl is then titrated with a standardized sodium hydroxide, NaOH, solution.[2] This analytical technique is referred to as a **"back" titration.**

Back titration: an analytical procedure by which the analyte is "swamped" with an excess of a standardized neutralizing agent; the excess neutralizing agent is, in return, neutralized to a final stoichiometric point

The number of moles of base in the antacid of the commercial sample *plus* the number of moles of NaOH used in the titration equals the number of moles of HCl added to the original antacid sample:

$$\text{moles}_{\text{antacid}} + \text{moles}_{\text{NaOH}} = \text{moles}_{\text{HCl}} \quad (26.5)$$

A rearrangement of the equation provides the moles of base in the antacid in the sample:

$$\text{moles}_{\text{antacid}} = \text{moles}_{\text{HCl}} - \text{moles}_{\text{NaOH}} \quad (26.6)$$

The moles of base in the antacid per gram of antacid provide the data required for a comparison of the antacid effectiveness of commercial antacids. For a consumer's point of interest, if purchase prices for the antacids are available, a final cost analysis of various antacids can be made.

[1]A buffer system resists large changes in the acidity of a solution. To analyze for the amount of antacid in this experiment, we want to *remove* this buffering property to determine the total effectiveness of the antacid.

[2]A standardized NaOH solution is one in which the concentration of NaOH has been very carefully determined.

Experimental Procedure

Procedure Overview: The amount of base in an antacid sample is determined. The sample is dissolved, and the buffer components of the antacid are eliminated with the addition of an excess of standardized HCl solution. The unreacted HCl is "back" titrated with a standardized NaOH solution.

At least two analyses should be completed per antacid if two antacids are to be analyzed to compare their neutralizing powers. If only one antacid is to be analyzed, complete three trials.

A. Dissolving the Antacid

1. **Determine the Mass of Antacid for Analysis.** If your antacid is a tablet, pulverize and/or grind the antacid tablet with a mortar and pestle. Measure ($\pm$0.001 g) no more than 0.2 g of the pulverized commercial antacid tablet (or 0.2 g of a liquid antacid) in a 250-mL Erlenmeyer flask with a tared mass.
2. **Prepare the Antacid for Analysis.** Pipet 25.0 mL of a standardized 0.1 *M* HCl solution (stomach acid equivalent) into the flask and swirl.[3] Record the actual molar concentration of the HCl on the Report Sheet. Warm the solution to a very *gentle* boil and maintain the heat for 1 minute to remove dissolved CO_2 (Figure 26.2 or Figure T.13b). Add 4–8 drops of bromophenol blue indicator.[4] If the solution is blue, pipet an additional 10.0 mL of 0.1 *M* HCl into the solution and boil again. Repeat as often as necessary. Record the *total* volume of HCl that is added to the antacid.

B. Analyzing the Antacid Sample

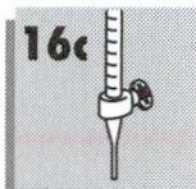

Read Technique 16c closely

Obtain about 75 mL of a standardized 0.1 *M* NaOH solution. The solution may have been previously prepared by the stockroom personnel. If not, prepare a standardized 0.1 *M* NaOH solution, as described in Experiment 9. Consult with your laboratory instructor.

1. **Prepare the Buret for Titration.** Prepare a *clean* buret. Rinse the clean buret with two 3- to 5-mL portions of the standardized NaOH solution. Record the

Figure 26.2 Gently heat the sample to remove CO_2 gas.

[3] If the sample is a tablet, swirl to dissolve. Some of the *inert* ingredients—fillers and binding agents used in the formulation of the antacid tablet—may not dissolve.
[4] Bromophenol blue is yellow at a pH less than 3.0 and blue at a pH greater than 4.6.

actual molar concentration of the NaOH on the Report Sheet. Fill the buret with the NaOH solution; be sure no air bubbles are in the buret tip. Wait for 10–15 seconds, then read and record its initial volume, "using all certain digits (from the labeled calibration marks on the buret) *plus* one uncertain digit (the last digit which is the best estimate between the calibration marks)."

Be constantly aware of the use of significant figures that reflect the precision of your measuring instrument

2. **Titrate the Sample.** Once the antacid solution has cooled, titrate the sample with the NaOH solution to a faint blue endpoint. Watch closely; the endpoint may appear after only a few milliliters of titrant, depending on the concentration of the antacid in the sample. When a single drop (or half-drop) of NaOH solution changes the sample solution from yellow to blue, **stop.** Wait for 10–15 seconds and then read and record the final volume of NaOH solution in the buret.
3. **Repeat the Titration of the Same Antacid.** Refill the buret and repeat the experiment, starting at Part A.1.
4. **Analyze Another Antacid.** Perform the experiment, in duplicate, for another antacid. Record all data on the Report Sheet.

Disposal: Dispose of the test solutions as directed by your instructor.

CLEANUP: Discard the remaining NaOH titrant as directed by your instructor. Flush the buret several times with tap water and dispense through the buret tip, followed by several portions of deionized water. Dispose of all buret washings in the sink.

C. Data Analysis

1. Determine the number of moles of HCl added to the antacid sample.
2. How many moles of NaOH titrant were required to neutralize the *excess* acid?
3. Calculate the number of moles of antacid per gram of sample.
4. (Optional) If the store-bought antacid and its purchase price are available, calculate its cost per gram. Complete a cost analysis—determine the best buy!

Experiment 26 *Prelaboratory Assignment*

Antacid Analysis

Date __________ Lab Sec. ______ Name ____ Desk No. __________

1. Write balanced equations for the reactions of the active ingredients in Gaviscon® Extra Strength with excess acid.

2. Identify the two most common anions present in antacids.

3. If the antacid for analysis (Part A.2) is known to be Phillips'® Milk of Magnesia, the solution does not need to be heated, but if the sample is Extra Strength Maalox®, the solution must be heated. Explain the difference in experimental procedures.

4. One tablet of Regular Strength Maalox® claims to contain 600 mg $CaCO_3$. If 10.0 mL of 0.100 *M* NaOH titrant is to be used to back titrate the excess 0.100 *M* HCl from the analysis of one-third of a Maalox® tablet, how many milliliters of 0.100 *M* HCl must have been initially added to the 200 mg of Maalox® sample?

5. A 0.196-g sample of a CO_3^{2-} antacid is dissolved with 25.0 mL of 0.0946 *M* HCl. The hydrochloric acid that is *not* neutralized by the antacid is titrated to a bromophenol blue endpoint with 6.22 mL of 0.0827 *M* NaOH.
   a. Assuming the active ingredient in the antacid sample is $CaCO_3$, calculate the mass of $CaCO_3$ in the sample.

   b. What is the percent active ingredient in the antacid sample?

6. a. How many moles of stomach acid would be neutralized by one tablet of Regular Strength Maalox® that contains 600 mg of calcium carbonate?

   b. Assuming the volume of the stomach to be 1.0 L, what will be the pH change of the stomach acid resulting from the ingestion of one Regular Strength Maalox®?

7. The label on liquid Gaviscon® states that each tablespoon (15 mL) contains 95 mg of aluminum hydroxide and 412 mg of magnesium carbonate. How many moles of hydronium ion can be neutralized by one tablespoon of Gaviscon®?

Experiment 26 *Report Sheet*

Antacid Analysis

Date __________ Lab Sec. ______ Name ______________________________ Desk No. __________

Commercial Antacid

	Trial 1	*Trial 2*	*Trial 1*	*Trial 2*
1. Mass of flask (g)				
2. Mass of flask + antacid sample (g)				
3. Mass of antacid sample (g)				
4. Volume of HCl added (mL)				
5. Molar concentration of HCl (mol/L)				
6. Molar concentration of NaOH (mol/L)				
7. Buret reading, *initial* (mL)				
8. Buret reading, *final* (mL)				
9. Volume of NaOH (mL)				

Data Analysis

	Trial 1	*Trial 2*	*Trial 1*	*Trial 2*
1. Amount of HCl added, *total* (mol)				
2. Amount of NaOH added (mol)				
3. Amount of base in antacid sample (mol)				
4. $\frac{\text{mol base in antacid}}{\text{mass of antacid sample}}$ (mol/g)				
5. $\frac{\text{average mol base in anatacid}}{\text{mass of antacid sample}}$ (mol/g)				
6. Cost per gram of antacid (cents/g), from bottle (if available)				
7. antacid effectiveness $\left(\frac{\text{mol base in antacid}}{\text{cent}}\right)$				

8. Best buy __________

Calculations

Laboratory Questions

Circle the questions that have been assigned.

1. Part A.1. The antacid tablet for analysis was not finely pulverized before its reaction with hydrochloric acid. How might this technique error affect the reported amount of antacid in the sample? Explain.
2. Part A.2. All of the CO_2 is not removed by gentle boiling after the addition of HCl. Will the reported amount of antacid in the sample be too high, too low, or unaffected? Explain.
3. Part B.1. An air bubble was initially trapped in the buret but was dispensed during the back titration of the excess HCl (Part B.2). As a result of this technique error, will the reported amount of antacid in the sample be too high or too low? Explain.
4. Part B.2. The bromophenol blue indicator changed from yellow to blue as desired in the back titration of the excess HCl. Upon leaving the solution in the flask until the end of the laboratory period however, the solution may return to a yellow color. Explain why this might be possible.
5. Part B.2. The bromophenol blue endpoint is surpassed in the back titration of the excess HCl with the sodium hydroxide titrant. As a result of this technique error, will the reported amount of antacid in the sample be too high or too low? Explain.

*6. A few of the "newer" antacids contain sodium citrate, $Na_3C_6H_5O_7$, as the effective, but more mild antacid ingredient.
   a. Write a balanced equation representing the antacid effect of the citrate ion, $C_6H_5O_7^{3-}$.
   b. Will 500 mg of $Na_3C_6H_5O_7$ (258.1 g/mol) or 500 mg of $CaCO_3$ (100.1 g/mol) neutralize more moles of hydronium ion? Show calculations. Assume that both the $C_6H_5O_7^{3-}$ and the CO_3^{2-} ions become fully protonated.

Experiment 27

Potentiometric Analyses

The electrodes of a pH meter: the saturated calomel electrode (left) and the glass electrode (right).

OBJECTIVES

- To operate a pH meter
- To graphically determine a stoichiometric point
- To determine the molar concentration of a weak acid solution
- To determine the molar mass of a solid weak acid
- To determine the pK_a of a weak acid

TECHNIQUES

The following techniques are used in the Experimental Procedure

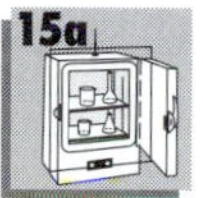

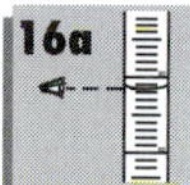

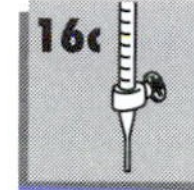

INTRODUCTION

A "probe" connected to an instrument that provides a direct reading of the concentration of a particular substance in an aqueous system is a convenient form of analysis. Such convenience is particularly advantageous when a large number of samples and analyses need to be performed. The probe, or electrode, senses a difference in concentrations between the substance in solution and the substance in the probe itself. The concentration difference causes a voltage (or potential difference), which is recorded by the instrument, called a **potentiometer.**

Potentiometer: an instrument that measures a potential difference—often called a voltmeter

A potentiometric chemical analysis is a powerful, convenient method for determining the concentrations of various ions in solution. To list only a few, the molar concentrations of the cations H^+, Li^+, Na^+, K^+, Ag^+, Ca^{2+}, Cu^{2+}, Pb^{2+}; the anions F^-, Cl^-, Br^-, I^-, CN^-, SO_4^{2-}; and the gases CO_2, NH_3, SO_2, H_2S, NO_x can be measured directly using an electrode specifically designed for its measurement (a specific selective electrode).

pH meter: an instrument that measures the pH of a solution

The pH of a solution is measured with a potentiometer called a **pH meter,** an instrument that measures the potential difference between a reference electrode, generally a **saturated calomel electrode** (SCE), and a silver/silver chloride electrode, called a **glass electrode.** The glass electrode is sensitive to changes in hydrogen ion concentration, $[H^+]$, in solution.

Saturated calomel electrode: a platinum wire that is in contact with solid Hg_2Cl_2, which, in turn, is in contact with Hg metal and a saturated KCl solution:

$$Hg_2Cl_2(s) + 2\ e^- \rightarrow 2\ Hg(l) + 2\ Cl^-(aq)$$

The glass electrode and the standard calomel electrodes are shown schematically in Figure 27.1 and in the opening photo. Newer "combination" electrodes combine the two electrodes into a single probe, shown in Figure 27.2.

Glass electrode: a metal wire that is in contact with solid AgCl, which, in turn, is in contact with Ag metal and a 0.1 M HCl solution, all contained within the special glass:

$$AgCl(s) + e^- \rightarrow Ag(s) + Cl^-(aq)$$

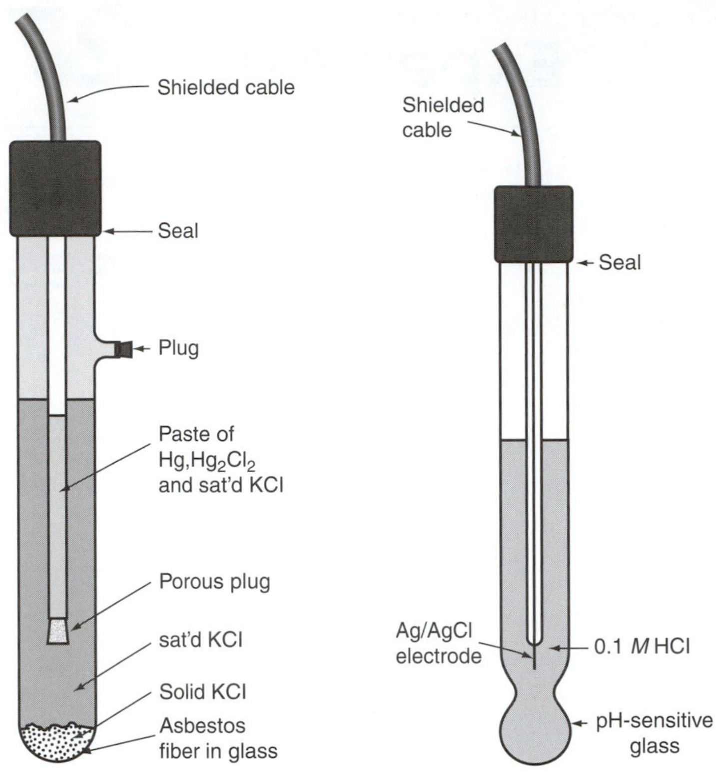

Figure 27.1 Components of the reference or saturated calomel electrode (left) and the glass electrode (right).

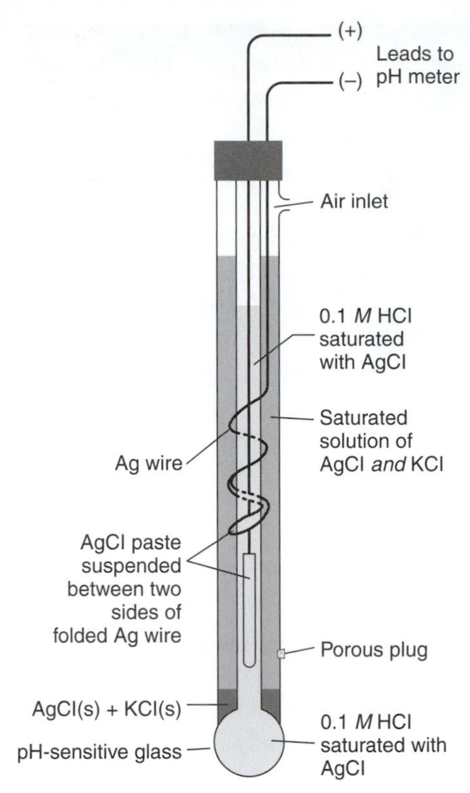

Figure 27.2 A combination electrode combines two glass electrodes—the reference electrode and the glass electrode—into a single probe.

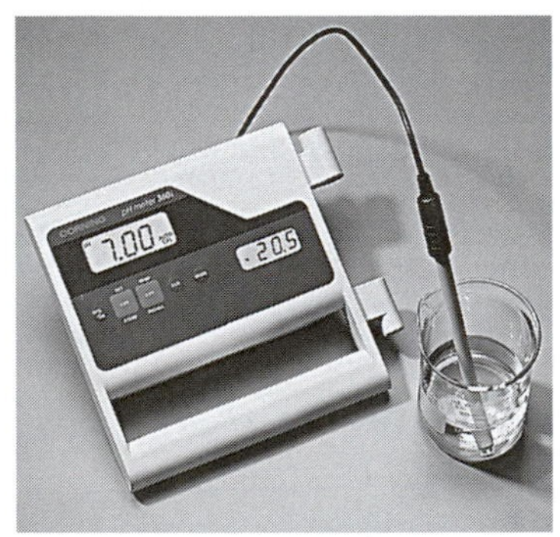

A modern pH meter and a combination electrode.

The measured voltage recorded by the potentiometer, E_{cell}, is a function of the pH of the solution at 25°C by the equation

$$E_{\text{cell}} = E' + 0.0592 \text{ pH} \tag{27.1}$$

E' is a cell constant, an internal parameter that is characteristic of the pH meter and its electrode(s).

Before any pH measurements are made, the pH meter is calibrated (that is, E' is set). The electrodes are placed into a **buffer solution** of known pH, and the E_{cell}, the potential difference between the $[H^+]$ of the glass electrode and the $[H^+]$ of the buffer, is manually adjusted to read the pH of the buffer—this adjustment sets E'.

The readout of the pH meter is E_{cell}, expressed in volts, but since E_{cell} is directly proportional to pH (Equation 27.1), the meter for the readout is expressed directly in pH units.

Molar Concentration of a Weak Acid Solution

Buffer solution: a solution that maintains a relatively constant, reproducible pH

The selection of an indicator for the titration of a strong acid with a strong base is relatively easy in that the color change at the stoichiometric point always occurs at a pH of 7. Usually phenolphthalein can be used, because its color changes at a pH close enough to the stoichiometric point for the analysis; however, when a weak acid is titrated with a strong base, the stoichiometric point is at a pH greater than 7 and a different indicator must be selected.[1] If the weak acid is an unknown acid, then the proper indicator cannot be selected because the pH at the stoichiometric point cannot be pre-

[1]The pH is greater than 7 at the stoichiometric point for the titration of a weak monoprotic acid because of the basicity of the conjugate base, A^-, of the weak acid, HA:

$$A^-(aq) + H_2O(l) \rightarrow HA(aq) + OH^-(aq)$$

determined. The color change of a selected indicator may *not* occur at (or even near) the pH of the stoichiometric point for the titration. To better detect a stoichiometric point for the titration of an unknown weak acid, a pH meter is more reliable.

Molar Mass of an Unknown Acid

Titrimetric analysis: a titration procedure that is chosen for an analysis

Titration curve: a data plot of pH versus volume of titrant

Monoprotic acid: a substance capable of donating a single proton, H^+

Diprotic acid: a substance capable of donating two protons

A **titrimetric analysis** is used to determine the molar concentration of a weak acid solution. A pH meter is used to detect the stoichiometric point of the titration. An acid–base indicator will *not* be used. A standardized sodium hydroxide solution is used as the titrant.[2]

The pH of a weak acid solution increases as the standardized NaOH solution is added. A plot of the pH of the weak acid solution as the strong base is being added, pH versus V_{NaOH}, is called the **titration curve** (Figure 27.3) for the reaction. The inflection point in the sharp vertical portion of the plot (about midway on the vertical rise) is the stoichiometric point.

The moles of NaOH used for the analysis equals the volume of NaOH, dispensed from the buret, times its molar concentration.

$$\text{moles NaOH (mol)} = \text{volume (L)} \times \text{molar concentration (mol/L)} \tag{27.2}$$

For a **monoprotic acid,** HA, one mole of OH^-, neutralizes one mole of acid:

$$HA(aq) + OH^-(aq) \rightarrow H_2O(l) + A^-(aq) \tag{27.3}$$

The molar concentration of the acid is determined by dividing its number of moles in solution by its measured volume in liters:

$$\text{molar concentration Ha (mol/L)} = \frac{\text{mol HA}}{\text{L HA solution}} \tag{27.4}$$

For a **diprotic acid,** H_2X, 2 mol of OH^- neutralizes 1 mol of acid:

$$H_2X(aq) + 2\,OH^-(aq) \rightarrow 2\,H_2O(l) + X^{2-}(aq) \tag{27.5}$$

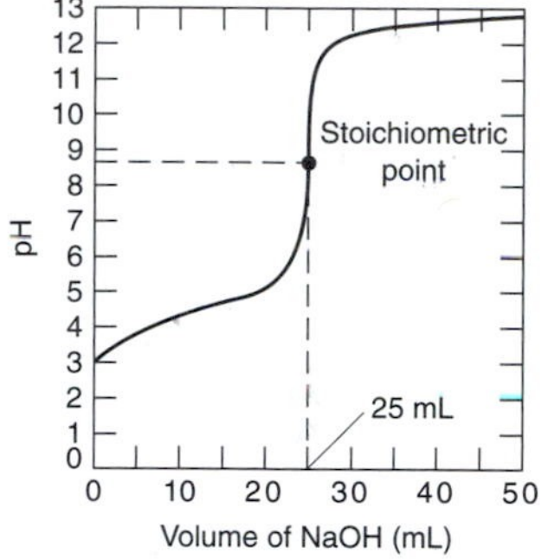

Figure 27.3 A titration curve of a weak acid with a strong base titrant.

Molar Mass of a Weak Acid

In Part B, the molar mass and the pK_a of an unknown *solid* weak acid are determined. The standardized NaOH solution is used to titrate a carefully measured mass of the *dissolved* acid to the stoichiometric point. A plot of pH versus V_{NaOH} is required to define the stoichiometric point.

The moles of acid is determined as described in Equations 27.2 and 27.3.

The molar mass of the acid is calculated from the moles of the solid acid and its measured mass.

$$\text{molar mass (g/mol)} = \frac{\text{mass of solid acid(g)}}{\text{moles of solid acid}} \tag{27.6}$$

pK_a of a Weak Acid

A weak acid, HA, in water undergoes only partial ionization,

$$HA(aq) + H_2O(l) \rightleftharpoons H_3O^+(aq) + A^-(aq) \tag{27.7}$$

At equilibrium conditions, the mass action expression for the weak acid system equals the equilibrium constant, $K_a = \frac{[H^+][A^-]}{[HA]}$. When one-half the moles of the weak acid is neutralized by the NaOH in a titration, an equal number of moles of A^- form such that mol HA = mol A^- or $[HA] = [A^-]$. At this point in the titration, $K_a = [H^+]$. As pH is recorded directly from the pH meter, it is the pK_a of the weak acid that equals the pH when one-half of the weak acid is neutralized; this occurs at the halfway point (when half of the NaOH solution required to reach the stoichiometric point is added to the weak acid solution) in the titration (Figure 27.4).

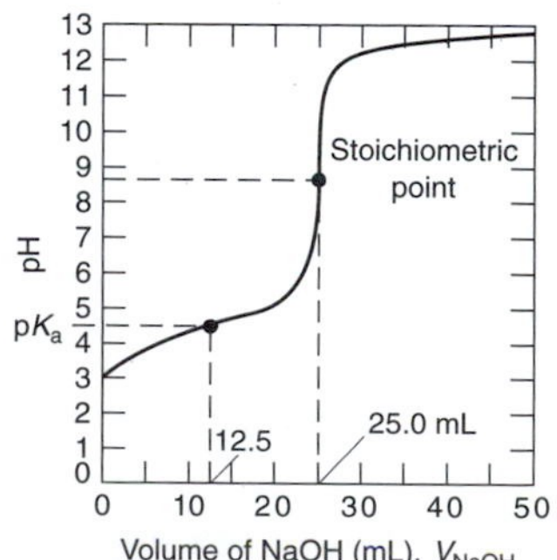

Figure 27.4 pK_a = pH at the halfway point in the titration.

[2]The procedure for preparing a standardized NaOH solution is described in Experiment 9.

The stoichiometric point is again determined from the complete titration curve of pH versus V_{NaOH}. If the weak acid is diprotic and if both stoichiometric points are detected, the pK_{a1} and pK_{a2} can be determined.

Experimental Procedure

Procedure Overview: The pH meter is used in conjunction with a titration apparatus and a standardized sodium hydroxide solution to determine the molar concentration of a weak acid solution and the molar mass and p*K*a of a solid, weak acid. Plots of pH versus volume of NaOH are used to determine the stoichiometric point of each titration.

The number of pH meters in the laboratory is limited. You may need to share one with a partner or with a larger group. Ask your instructor for details of the arrangement. Consult with your instructor for directions on the proper care and use of the pH meter. Also inquire about the standardization of the pH meter.

Because of time and equipment constraints, it may be impossible to do all parts of the experiment in one laboratory period. Time is required not only to collect and graph the data, but also to interpret the data and complete the calculations. Discuss the expectations from the experiment with your instructor.

The pH versus V_{NaOH} curves to be plotted in Parts A.6 and B.3 can be established by using a pH probe that is connected directly to either a calculator or computer with the appropriate software. If this pH sensing/recording apparatus is available in the laboratory, consult with your instructor for its use and adaptation to the experiment. The probe merely replaces the pH electrode in Figure 27.5. However, volume readings from the buret may still need to be recorded.

A. Molar Concentration of a Weak Acid Solution

Obtain about 90 mL of an acid solution with an unknown concentration from your instructor. Your instructor will advise you as to whether your acid is monoprotic or diprotic. Clean three 250-mL beakers.

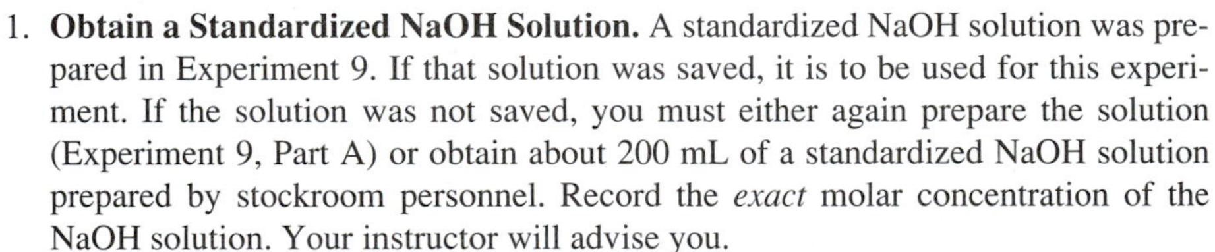

1. **Obtain a Standardized NaOH Solution.** A standardized NaOH solution was prepared in Experiment 9. If that solution was saved, it is to be used for this experiment. If the solution was not saved, you must either again prepare the solution (Experiment 9, Part A) or obtain about 200 mL of a standardized NaOH solution prepared by stockroom personnel. Record the *exact* molar concentration of the NaOH solution. Your instructor will advise you.

2. **Prepare the Buret with the Standardized NaOH Solution.** Properly clean a buret; rinse twice with tap water, twice with deionized water, and finally with three 5-mL portions of the standardized 0.1 *M* NaOH. Drain each rinse through the tip of the buret. Fill the buret with the 0.1 *M* NaOH. After 10–15 seconds, properly read[3] and record the volume of solution, "using all certain digits (from the labeled calibration marks on the buret) *plus* one uncertain digit (the last digit which is the best estimate between the calibration marks)."

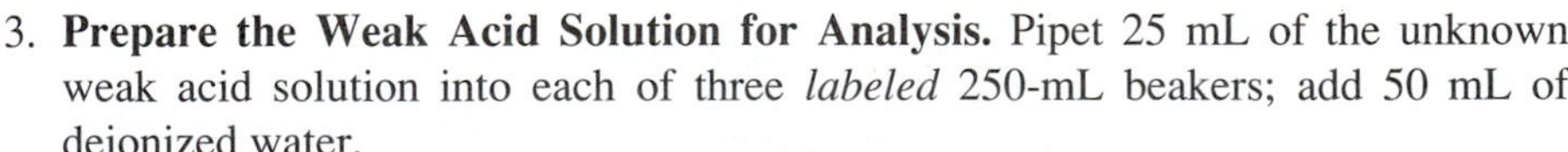

3. **Prepare the Weak Acid Solution for Analysis.** Pipet 25 mL of the unknown weak acid solution into each of three *labeled* 250-mL beakers; add 50 mL of deionized water.

   Set up the titration apparatus as shown in Figure 27.5. Remove the electrode(s) from the deionized water and touch-dry the electrode(s) with lint-free paper (Kimwipes). Immerse the electrode(s) about one-half inch deep into the solution of Beaker 1. Swirl or stir the solution and read and record the initial pH.[4]

[3]Remember to read the bottom of the meniscus with the aid of a black mark drawn on a white card.
[4]A magnetic stirrer and magnetic stirring bar may be used to swirl the solution during the addition of the titrant. Ask your instructor.

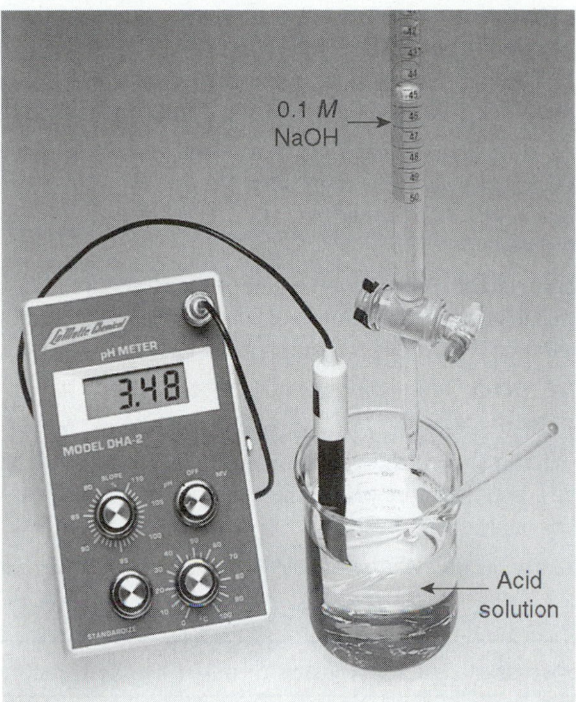

Figure 27.5 The setup for a potentiometric titration. Use a stirring rod or magnetic stirrer to stir the solution.

4. **Titrate the Weak Acid Solution.** Add the NaOH titrant, initially in 1- to 2-mL increments, and swirl or stir the solution. After each addition, allow the pH meter to stabilize; read and record the pH and buret readings on a *self-designed* data sheet. Repeat the additions until the stoichiometric point is near,[5] then slow the addition of the NaOH titrant.[6] When the stoichiometric point is imminent, add drops of NaOH solution. Use a minimum volume of deionized water from a wash bottle to rinse the wall of the beaker or to add half-drop volumes of NaOH. Dilution affects pH readings.

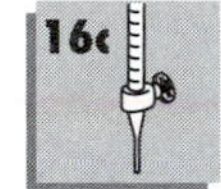

5. **Titrate Beyond the Stoichiometric Point.** After reaching the stoichiometric point, first add drops of NaOH, then 1 mL, and finally 2- to 3-mL **aliquots** until at least 10 mL of NaOH solution has been added beyond the stoichiometric point. Read and record the pH and buret readings after each addition.

Aliquot: an undefined, generally small, volume of a solution

6. **Plot the Data.** Plot the data for the titration curve, pH versus V_{NaOH}. Draw a smooth curve through the data points (do not "follow the dots!"). Properly label your graph and obtain your instructor's approval.

   From the plotted data, determine the pH and the volume of NaOH titrant added to reach the stoichiometric point.

Appendix C

7. **Repeat the Analysis.** Repeat the titration of the samples of weak acid in Beakers 2 and 3. Determine the average molar concentration of the acid.

B. Molar Mass and the pK_a of a Solid Weak Acid

Three samples of the solid weak acid are to be analyzed. Prepare three clean 250-mL beakers for this determination. Obtain an unknown solid acid from your instructor; your instructor will advise you as to whether your unknown acid is monoprotic or diprotic.

[5]The stoichiometric point is near when larger changes in pH occur with smaller additions of the NaOH titrant.

[6]Suggestion: It may save time to quickly titrate a sample to determine an approximate volume to reach the stoichiometric point.

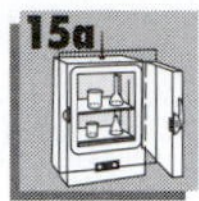

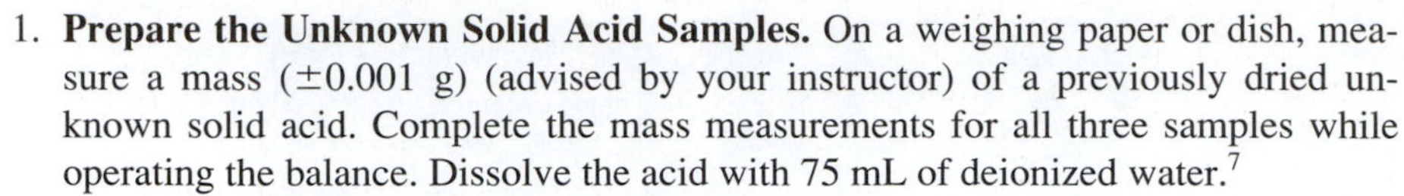

1. **Prepare the Unknown Solid Acid Samples.** On a weighing paper or dish, measure a mass ($\pm$0.001 g) (advised by your instructor) of a previously dried unknown solid acid. Complete the mass measurements for all three samples while operating the balance. Dissolve the acid with 75 mL of deionized water.[7]

2. **Fill the Buret and Titrate.** Refill the buret with the standardized NaOH solution and, after 10–15 seconds, read and record the initial volume and the initial pH. Refer to Parts A.4 and A.5. Titrate each sample to 10 mL beyond the stoichiometric point.

Appendix C

3. **Plot and Interpret the Data.** Plot the data for a titration curve, pH versus V_{NaOH}. From the plot, determine the volume of NaOH used to reach the stoichiometric point of the titration. Obtain your instructor's approval.
4. **Calculate the Molar Mass *and* the pK_a of the Weak Acid.** a. Calculate the molar mass of the weak acid.

   b. Note the coordinates of the stoichiometric point on a plot of the data. Determine the pH (and therefore pK_a of the weak acid) at the point where one-half of the acid was neutralized.
5. **Repeat.** Similarly titrate the other unknown solid acid samples and handle the data accordingly.

Disposal: Dispose of all test solutions as directed by your instructor.

CLEANUP: Discard the sodium hydroxide solution remaining in the buret as directed by your instructor. Rinse the buret twice with tap water and twice with deionized water, discarding each rinse through the buret tip into the sink.

Appendix B

6. **Collect the Data.** Obtain the pK_a for the same sample number from other student chemists in the laboratory. Calculate the standard deviation for the pK_a measurement for the acid.

[7]The solid acid may be relatively insoluble, but with the addition of the NaOH solution from the buret, it will gradually dissolve and react. The addition of 10 mL of ethanol may be necessary to dissolve the acid. Consult with your instructor.

Experiment 27 *Prelaboratory Assignment*

Potentiometric Analyses

Date __________ Lab Sec. ______ Name ____ Desk No. __________

1. As with all laboratory instruments, pH meters must be standardized—calibrated against a solution of known concentration. Briefly explain the procedure for standardizing a pH meter.

2. Briefly explain how the pK_a (and K_a) for a weak acid is determined in this experiment.

3. The pH at the stoichiometric point is 7.00 (at 25°C) for a strong acid (analyte)–strong base (titrant) titration, but it is greater than 7.00 for a weak acid–strong base titration. Explain.

4. A pH meter is often used to measure the pH of an existing solution. An old bottle of hydrochloric acid was found in the stockroom with an unknown concentration (the label fell off), and it is desirable to dispose of the solution. A pH meter was used to determine that the pH of the hydrochloric acid solution is 1.80.
   a. How many moles of hydronium ion are present in 25.0 mL of the sample?

   b. How many milliliters of 0.0110 M NaOH would be required to neutralize a 50.0-mL sample of the hydrochloric acid solution—increase its pH to 7.00—so that it can be discarded?

5. Data in the following table were obtained for the titration of 20.00 mL of a weak acid with a 0.200 *M* KOH solution. Plot (at right) pH (ordinate) versus V_{base} (abscissa).

V_{KOH} added (mL)	pH
0.00	2.59
1.00	3.27
2.00	3.61
4.00	4.02
5.00	4.19
7.00	4.56
9.00	5.15
9.50	5.49
9.80	5.89
10.00	8.50
10.20	11.12
11.00	11.81
13.00	12.26
15.00	12.46
24.00	12.80

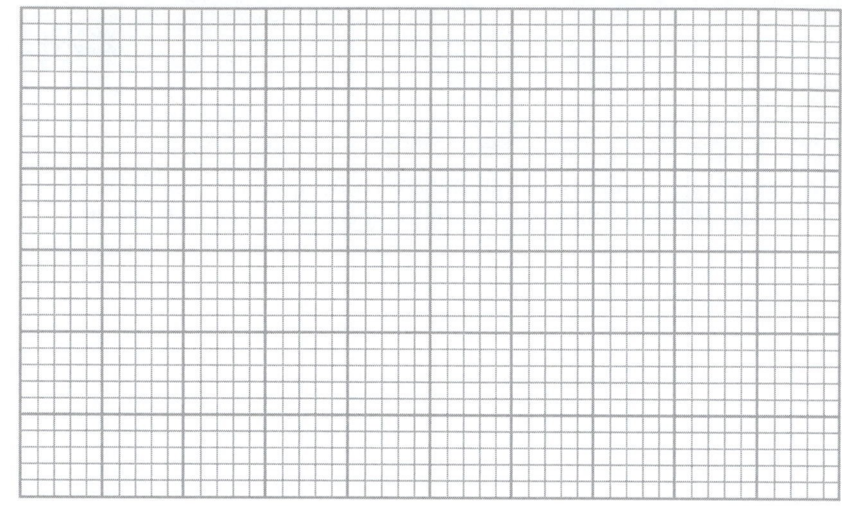

a. What volume of the KOH solution is required to reach the stoichiometric point?

b. What is the pH at the stoichiometric point?

c. What is the pK_a and K_a of the weak acid?

d. What is the molar concentration of the weak acid?

Experiment 27 *Report Sheet*

Potentiometric Analyses

Date __________ Lab Sec. ______ Name ____ Desk No. __________

A. Molar Concentration of a Weak Acid Solution

Sample No. _______________ Monoprotic or diprotic acid? _______________

	Trial 1	*Trial 2*	*Trial 3*
1. Volume acid (mL)			
2. Molar concentration of NaOH (mol/L)			
3. Instructor's approval of pH versus V_{NaOH} graph			
4. Volume NaOH to stoichiometric point (mL)			
5. Amount of acid (mol)			
6. Molar concentration of acid (mol/L)			
7. Average molar concentration of acid			

B. Molar Mass and the pK_a of a Solid Weak Acid

Sample No. _______________ Monoprotic or diprotic acid? _______________ Suggested mass _______________

	Trial 1	*Trial 2*	*Trial 3*
1. Mass of dry, solid acid (g)			
2. Molar concentration of NaOH (mol/L)			
3. Instructor's approval of pH versus V_{NaOH} graph			
4. Volume of NaOH to stoichiometric point (mL)			
5. Amount of acid (mol)			
6. Molar mass of acid (g/mol)	*		
7. Average molar mass of acid			
8. Volume of NaOH halfway to stoichiometric point (mL)			
9. pK_a of weak acid	*		
10. Average pK_a			

*Show calculations for Trial 1 on next page.

Calculations for Trial 1.

Class Data/Group	1	2	3	4	5	6
pK_a for Sample No. ______						

Standard deviation for the pK_a of an acid from the class data.

Laboratory Questions

Circle the questions that have been assigned.

1. The pH meter was calibrated with a buffer solution that reads 1.0 pH unit lower than what the label indicates.
   a. Part B. How will this miscalibration affect the determination of the molar concentration of the weak acid? Explain.
   b. Part C. Is the determined pK_a of the weak acid too high, too low, or unaffected by the miscalibration? Explain.
2. a. Part A.4. The pH reading is taken before the pH meter stabilizes. As a result the pH reading may be too low. Explain.
   b. Part A.4. Explain why it is good technique to slow the addition of NaOH titrant near the stoichiometric point.
   c. Part A.5. While not absolutely necessary, why is it good technique to add NaOH titrant beyond the stoichiometric point?
3. Part B.4. The NaOH titrant is added in large increments so that the stoichiometric point can be quickly reached.
   a. Will the molar mass of the weak acid be reported too high, too low, or be unaffected by this technique error? Explain.
   b. Will the pK_a of the weak acid (most likely) be reported too high, too low, or be unaffected by this technique error? Explain.
4. Ideally how many stoichiometric points would be observed on a pH titration curve for a diprotic acid? Sketch the appearance of its titration curve and explain.
5. For polyprotic acids the successive conjugate bases increase in strength. Explain why the "last" stoichiometric point for a polyprotic acid (e.g., H_3PO_4) may be ill-defined or undefined on a pH titration curve?

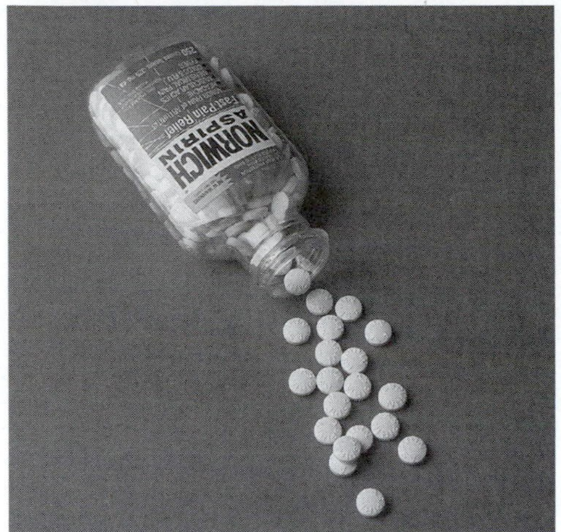

Experiment 28

Aspirin Synthesis and Analysis

Aspirin is a leading commercial pain reliever, synthesized in a pure and stable form by Felix Hoffman in 1897.

OBJECTIVES

- To synthesize aspirin
- To determine the purity of the synthesized aspirin or a commercial aspirin tablet

TECHNIQUES

The following techniques are used in the Experimental Procedure

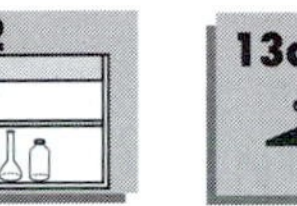

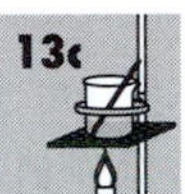

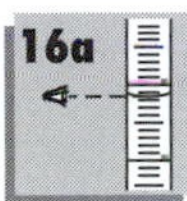

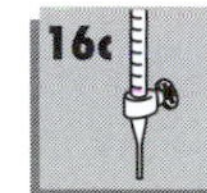

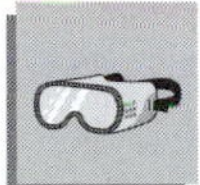

INTRODUCTION

Pure aspirin, chemically called acetylsalicylic acid, is both an organic **ester** and an **organi acid.** It is used extensively as a painkiller (analgesic) and as a fever-reducing drug (antipyretic). When ingested, acetylsalicylic acid remains intact in the acidic stomach, but in the basic medium of the upper intestinal tract, it **hydrolyzes** forming the salicylate and acetate ions.

Ester: a $-C(=O)-O-$ *grouping of atoms in a molecule*

Organic acid: a grouping of atoms in a molecule $-C(=O)-O-H$

Hydrolyzes: reacts with water

$$\text{aspirin (acetylsalicylic acid)} + 2\,OH^- \longrightarrow \text{salicylate ion} + \text{acetate ion} + H_2O \qquad (28.1)$$

aspirin (acetylsalicylic acid) — salicylate ion — acetate ion

The analgesic action of aspirin is undoubtedly due to the salicylate ion; however, its additional physiological effects and biochemical reactions are still not thoroughly understood. It is known that **salicylic acid** has the same therapeutic effects as aspirin; however, due in part to the fact that it is an acid, salicylic acid causes a more severe upset stomach than does aspirin.

salicylic acid

Aspirin (molar mass of 180.2 g/mol) is prepared by reacting salicylic acid (molar mass of 138.1 g/mol) with acetic anhydride (molar mass of 102.1 g/mol). Aspirin, like many other organic acids, is a weak **monoprotic acid.**

Monoprotic acid: a molecule that provides one proton for neutralization

$$\text{salicylic acid} + \text{acetic anhydride} \longrightarrow \text{aspirin (acetylsalicylic acid)} + \text{acetic acid} \quad (28.2)$$

salicylic acid   acetic anhydride   aspirin (acetylsalicylic acid)   acetic acid

Qualitatively, the purity of an aspirin sample can be determined from its melting point. The melting point of a substance is essentially independent of atmospheric pressure, but it is always lowered by the presence of impurities (a colligative property of pure substances; see Experiment 20). The degree of lowering of the melting point depends on the nature and the concentration of the impurities.

Quantitatively, the purity of an aspirin sample can be determined by a simple acid–base titration. The acetylsalicylic acid reacts with hydroxide ion, from a standardized sodium hydroxide solution, accordingly.

$$C_6H_4(OCOCH_3)COOH + OH^- \longrightarrow C_6H_4(OCOCH_3)COO^- + H_2O \quad (28.3)$$

Phenolphthalein: an acid-base indicator that is colorless at a pH less than 8.2 and pink at a pH greater than 10.0

A standardized NaOH solution titrates the acetylsalicylic acid to the **phenolphthalein** endpoint, where

$$\text{volume of NaOH (L)} \times \text{molar concentration of NaOH (mol/L)} = \text{mol NaOH} \quad (28.4)$$

According to Equation 28.3, one mole of OH^- reacts with one mole of acetylsalicylic acid; thus, the moles and mass of acetylsalicylic acid in the prepared sample are calculated. Knowing the calculated mass of the acid and the measured mass of the aspirin sample, the percent purity of the aspirin sample can be calculated:

$$\text{mol acetylsalicyclic acid} \times \frac{180.2\text{ g}}{\text{mol}} = \text{g acetylsalicyclic acid} \quad (28.5)$$

$$\%\text{ purity} = \frac{\text{g acetylsalicyclic acid}}{\text{g aspirin sample}} \times 100 \quad (28.6)$$

In Part C, the analysis for the percent acetylsalicylic acid in an aspirin sample can be performed on the aspirin prepared in Part A *or* on a commercial aspirin tablet.

Experimental Procedure

Procedure Overview: Crystalline aspirin is synthesized and then purified by the procedure of recrystallization. The melting point and the percent purity of the aspirin are measured, the latter by titration with a standardized NaOH solution.

A. Preparation of Aspirin

It is safest to prepare the aspirin in a fume hood. Set up a boiling water bath in a 400-mL beaker. Prepare about 100 mL of deionized ice water. Also, set up an ice bath.

1. **Mix the Starting Materials and Heat.** Measure about 2 g (±0.01 g) of salicylic acid (**caution:** *this is a skin irritant*) in a *dry* 125-mL Erlenmeyer flask. Cover the crystals with 4–5 mL of acetic anhydride. (**Caution:** *Acetic anhydride is a severe eye irritant—avoid skin and eye contact.*) Swirl the flask to wet the salicylic acid crystals. Add 5 drops of conc H_2SO_4 to the mixture and gently heat the flask in a boiling water bath (Figure 28.1) for 5–10 minutes. (**Caution:** H_2SO_4 *causes severe skin burns.*)

2. **Cool to Crystallize the Aspirin.** Remove the flask from the hot water bath and, to the reaction mixture, add 10 mL of deionized ice water to decompose any excess acetic anhydride. Chill the solution in an ice bath until crystals of aspirin no longer form, stirring occasionally to decompose residual acetic anhydride. *If* an "oil" appears instead of a solid, reheat the flask in the hot water bath until the oil disappears and again cool.
3. **Separate the Solid Aspirin from the Solution.** Set up a vacuum filtration apparatus and "turn it on." Seal the filter paper with water in the Büchner funnel. *Decant* the liquid onto the filter paper; minimize any transfer of the solid aspirin. Some aspirin, however, may be inadvertently transferred to the filter; that's O.K.
4. **Wash and Transfer the Aspirin to the Filter Paper and Wash.** Add 15 mL of ice water to the flask, swirl, chill briefly, and decant onto the filter. Repeat until the transfer of the crystals to the vacuum filter is complete; maintain the vacuum to dry the crystals as best possible. Wash the aspirin crystals on the filter paper with 10 mL of ice water. Keep all of the filtrate until the aspirin has been transferred from the flask.

   If aspirin forms in the filtrate, transfer the filtrate and aspirin to a beaker, chill in an ice bath, and vacuum filter as before, using a new piece of filter paper.

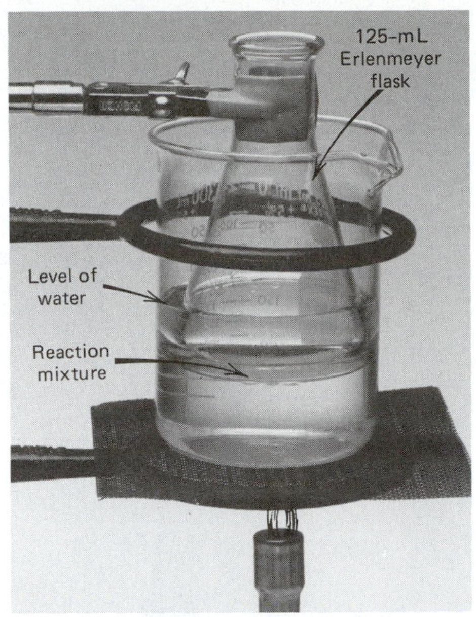

Figure 28.1 Boiling water bath for the dissolution of the acetylsalicylic acid crystals.

Disposal: Dispose of the "final" filtrate as directed by your laboratory instructor.

5. **Recrystallize the Aspirin.** Transfer the crystals from the filter paper(s) to a 100-mL beaker. Add repetitive small volumes of ethanol (e.g., 3-mL volumes) to the aspirin until the crystals *just* dissolve ($\leq$20 mL is required). Warm the mixture in a 60°C water bath (**caution:** *no flame,* use a hot plate or a hot water bath). Pour 50 mL of ~60°C water into the solution. If a solid forms, continue warming until the solid dissolves.

   Cover the beaker with a watchglass, remove it from the heat, and set it aside to cool slowly to room temperature. Then set the beaker in an ice bath. Beautiful needlelike crystals of acetylsalicylic acid form.

6. **How Much Did You Prepare?** Vacuum filter the crystals on filter paper, the mass of which has been previously measured ($\pm$0.01 g). Wash the crystals with two 10-mL volumes of *ice* water. Place the filter paper and aspirin sample on a watchglass and allow them to air-dry. The time for air-drying the sample may require that it be left in your lab drawer until the next laboratory period.

   Determine the mass of the dry filter paper and sample. Dispose of the filtrate as directed by your laboratory instructor.

7. **Correct for Residual Solubility.** The solubility of acetylsalicylic acid is 0.25 g per 100 mL of water. Correcting for this inherent loss of product due to the wash water in Part A.6, calculate the percent yield.

8. **What Do You Do with It?** Don't use it for a headache! Place the sample in a properly labeled test tube, stopper, and submit it along with your Report Sheet to your laboratory instructor at the conclusion of the experiment.

B. Melting Point of the Aspirin Sample

The melting point of the aspirin sample can be determined with either a commercial melting point apparatus or with the apparatus shown in Figure 28.2 and described in Part B.1. Consult with your instructor.

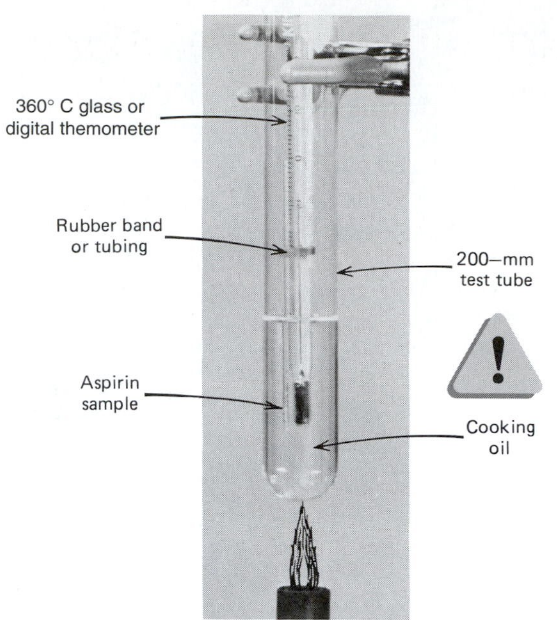

Figure 28.2 Melting point apparatus for aspirin.

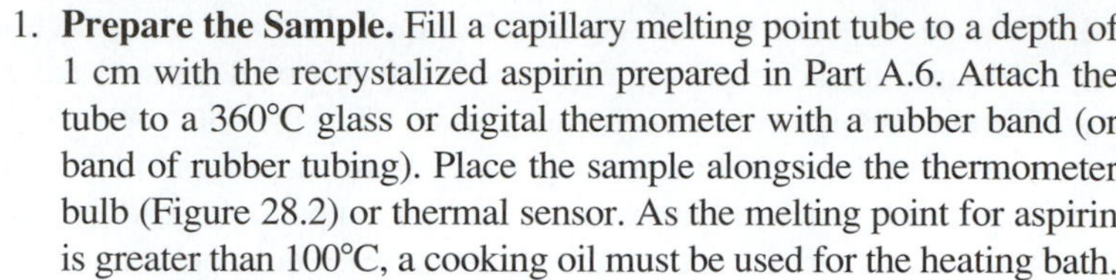

1. **Prepare the Sample.** Fill a capillary melting point tube to a depth of 1 cm with the recrystalized aspirin prepared in Part A.6. Attach the tube to a 360°C glass or digital thermometer with a rubber band (or band of rubber tubing). Place the sample alongside the thermometer bulb (Figure 28.2) or thermal sensor. As the melting point for aspirin is greater than 100°C, a cooking oil must be used for the heating bath.
2. **Determine the Melting Point.** Slowly and gently heat the oil bath at a rate of ~5°C per minute until the aspirin melts. (**Caution:** *The oil bath is at a temperature greater than 100°C—do not touch!*) Cool the bath and aspirin to just below this approximate melting point until the aspirin in the tube solidifies; at a slower 1°C per minute rate, heat again until it melts; this is the melting point of your prepared aspirin.
3. **A Purity Check of the Sample.** If the melting point of your prepared aspirin sample is less than 130°C, repeat Part A.5 to recrystallize the sample for the purpose of increasing its purity. After the recrystallization, repeat Parts B.1 and B.2.
4. **Repeat the Melting Point Measurement.** Again, cool the bath and aspirin to just below the melting point until the aspirin in the tube solidifies; at a 1°C per minute rate, heat again until it melts.

Disposal: Ask your instructor about proper disposal of the oil. Be sure the oil is cool when handling it. Dispose of the capillary tube in the "Waste Glass" container.

C. Percent Acetylsalicyclic Acid in the Aspirin Sample

Three trials are to be completed in the analysis of the aspirin. Prepare three clean 125- or 250-mL Erlenmeyer flasks and determine the mass of three aspirin samples while occupying the balance. Obtain a 50-mL buret.

1. **Prepare the Aspirin Sample for Analysis.** *Assuming* 100% purity of your aspirin sample, calculate the mass of aspirin that requires 20 mL of 0.1 *M* NaOH to reach the stoichiometric point. Show the calculation on the Report Sheet. On weighing paper measure the calculated mass (±0.001 g) of the aspirin you have just prepared (or a crushed commercial aspirin tablet) and transfer it to the flask. Add 10 mL of 95% ethanol, followed by about 50 mL of deionized water, and swirl to dissolve the aspirin. Add 2 drops of phenolphthalein indicator.

2. **Prepare the Buret for Titration.** Prepare a clean buret, rinse, and fill it with a standardized 0.1 *M* NaOH solution.[1] Be sure that no air bubbles are present in the buret tip. After 10–15 seconds, read and record the volume, and the actual molar concentration of the NaOH solution.

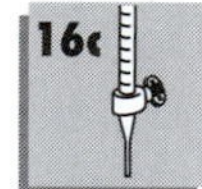

3. **Titrate the Sample.** Slowly add the NaOH solution from the buret to the dissolved aspirin sample until the endpoint is reached. The endpoint in the titration should be within one-half drop of a faint pink color. The color should persist for 30 seconds. Read and record the final volume of NaOH in the buret.

Disposal: Discard the test solution in the "Waste Acids" container or as advised by your instructor.

CLEANUP: Discard the NaOH titrant into a properly labeled bottle; rinse the buret with several 5-mL volumes of tap water, followed by two 5-mL volumes of deionized water.

[1]You may need to prepare the 0.1 *M* NaOH solution using the procedure in Experiment 9, or the stockroom personnel may have it already prepared.

Experiment 28 *Prelaboratory Assignment*

Aspirin Synthesis and Analysis

Date __________ Lab Sec. ______ Name ____ Desk No. __________

1. A 0.331-g sample of aspirin prepared in the laboratory is dissolved in 95% ethanol, diluted with water, and titrated to the phenolphthalein endpoint with 16.7 mL of 0.107 *M* NaOH.
   a. How many moles of acetylsalicylic acid are present in the sample?

   b. Calculate the percent purity of acetylsalicylic acid in the aspirin sample?

2. Determine the number of grams of acetylsalicylic acid that will react with 20.0 mL of 0.100 *M* NaOH (see Part C.1).

3. In the experiment 2.00 g of salicylic acid reacts with an excess amount of acetic anhydride. Calculate the theoretical yield of acetylsalicylic acid for this synthesis. (This calculation is similar to the one for the Report Sheet, Part A.2).

4. The aspirin that is synthesized in the experiment is purified by a recrystallization procedure. In general terms describe what is involved in a recrystallization procedure.

5. Search a chemistry handbook or the internet for the melting point of acetylsalicylic acid. Reference the location of your source.

6. Technique 11e. Describe the procedure for "seating" the filter paper in the funnel for a vacuum filtration.

7. Identify the five **cautions** cited in the Experimental Procedure for this experiment.

Experiment 28 *Report Sheet*

Aspirin Synthesis and Analysis

Date _________ Lab Sec. ______ Name ______________________________ Desk No. _________

A. Preparation of Aspirin

1. Mass of salicylic acid (g) _______________
2. Theoretical yield of aspirin (g) _______________
3. Experimental yield of aspirin (g) _______________
4. Experimental yield, corrected for solubility (g) _______________
5. Percent yield (%) _______________

B. Melting Point of the Aspirin Sample

1. Melting point measurements (°C) _______________ _______________ _______________

C. Percent Acetylsalicylic Acid in the Aspirin Sample

Calculation for the mass of aspirin for the titrimetric analysis.

	Trial 1	*Trial 2*	*Trial 3*
1. Mass of weighing paper plus aspirin sample (g)			
2. Mass of weighing paper (g)			
3. Mass of aspirin sample (g)			
4. Molar concentration of the NaOH solution (mol/L)			
5. Buret reading, *initial* (mL)			

	Trial 1	*Trial 2*	*Trial 3*
6. Buret reading, *final* (mL)			
7. Volume of NaOH used (mL)			
8. Amount of NaOH added (mol)			
9. Amount of acetylsalicylic acid (mol)			
10. Mass of acetylsalicylic acid (g)			
11. Percent purity of aspirin sample (%)			
12. Average percent purity of aspirin sample (%)			

Class Data/Group	1	2	3	4	5	6
Average percent acid in sample						

Determine the average deviation of the percent acetylsalicylic acid in aspirin from the class data. See Appendix B.

Laboratory Questions

Circle the questions that have been assigned.

1. Part A.1. "Anhydride" means without water. What happens to the yield of acetylsalicylic acid if the reaction were to be conducted in a *wet* 125-mL Erlenmeyer flask?
2. Part A.1. According to LeChâtelier's principle, explain why it is necessary to add the conc H_2SO_4 during the preparation of the acetylsalicylic acid. Also see Equation 28.1.
3. Part A.2. Deionized ice water is added to decompose excess acetic anhydride. What is the product for the hydrolysis of acetic anhydride? Write a balanced equation.
4. Part A.4. Some of the aspirin passed through the filter to form in the filtrate. How does the aspirin in the filtrate differ from that collected on the filter paper?
5. Part A.5. The product crystals are dissolved in a minimum volume of ethanol. Is acetylsalicylic acid more soluble in ethanol or water? Explain.
6. Part B.2. a. Explain why the melting point of the synthesized acetylsalicylic acid is likely to be *less than* that for pure acetylsalicylic acid?
   b. Would the product isolated after Part A.4 have a higher or lower melting point than that isolated after Part A.6? Explain.
7. Part C.3. If the endpoint for the titration of the acetylsalicylic acid is surpassed, will the percent purity of the aspirin sample be reported too high, too low, or unaffected? Explain.

Experiment 29

Molar Solubility, Common-Ion Effect

Silver oxide forms a brown mudlike precipitate from a mixture of silver nitrate and sodium hydroxide solutions.

OBJECTIVES

- To determine the molar solubility and the solubility constant of calcium hydroxide
- To study the effect of a common ion on the molar solubility of calcium hydroxide

TECHNIQUES

The following techniques are used in the Experimental Procedure

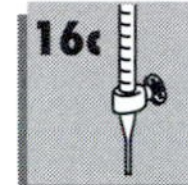

INTRODUCTION

Slightly soluble salt: a qualitative term that reflects the very low solubility of a salt

Dynamic equilibrium: the rate of the forward reaction equals the rate of the reverse reaction

Salts that have a very limited solubility in water are called **slightly soluble** (or "insoluble") **salts.** A saturated solution of a slightly soluble salt is a result of a dynamic equilibrium between the solid salt and its ions in solution; however, because the salt is only slightly soluble, the concentrations of the ions in solution are low. For example, in a saturated silver sulfate, Ag_2SO_4, solution, the **dynamic equilibrium** between solid Ag_2SO_4 and the Ag^+ and SO_4^{2-} ions in solution lies far to the *left* because of the low solubility of silver sulfate.

$$Ag_2SO_4(s) \rightleftharpoons 2\ Ag^+(aq) + SO_4^{2-}(aq) \qquad (29.1)$$

The mass action expression for this system is

$$[Ag^+]^2[SO_4^{2-}] \qquad (29.2)$$

As Ag_2SO_4 is a solid, its concentration is constant and, therefore, does not appear in the mass action expression. At equilibrium, the mass action expression equals K_{sp}, called the **solubility product** or, more simply, the equilibrium constant for this slightly soluble salt.

Molar solubility: the number of moles of salt that dissolve per liter of (aqueous) solution

The **molar solubility** of Ag_2SO_4, determined experimentally, is 1.4×10^{-2} mol/L. This means that in 1.0 L of a saturated Ag_2SO_4 solution, only 1.4×10^{-2} mol of silver sulfate dissolves, forming 2.8×10^{-2} mol of Ag^+ and 1.4×10^{-2} mol of SO_4^{2-}. The solubility product of silver sulfate equals the product of the molar concentrations of the ions, each raised to the power of its coefficient in the balanced equation:

$$K_{sp} = [Ag^+]^2[SO_4^{2-}] = [2.8 \times 10^{-2}]^2[1.4 \times 10^{-2}] = 1.1 \times 10^{-5} \qquad (29.3)$$

What happens to the molar solubility of a salt when an ion, common to the salt, is added to the saturated solution? According to LeChâtelier's principle (Experiment 24),

Figure 29.1 The solid $Ca(OH)_2$ in a saturated $Ca(OH)_2$ solution is slow to settle.

the equilibrium for the salt shifts to compensate for the added ions; that is, it shifts *left* to favor the formation of more of the solid salt. This effect, caused by the addition of an ion common to an existing equilibrium, is called the **common-ion effect.** As a result of the common-ion addition and the corresponding shift in the equilibrium, fewer moles of the salt dissolve in solution, lowering the molar solubility of the salt.

In this experiment, you will determine the molar solubility and the solubility product for calcium hydroxide, $Ca(OH)_2$. A saturated $Ca(OH)_2$ solution[1] (Figure 29.1) is prepared; after an equilibrium is established between the solid $Ca(OH)_2$ and the Ca^{2+} and OH^- ions in solution, the supernatant solution is analyzed. The hydroxide ion, OH^-, in the supernatant solution is titrated with a standardized HCl solution to determine its molar concentration.

According to the equation

$$Ca(OH)_2(s) \rightleftharpoons Ca^{2+}(aq) + 2\,OH^-(aq) \tag{29.4}$$

for each mole of $Ca(OH)_2$ that dissolves, 1 mol of Ca^{2+} and 2 mol of OH^- are present in solution. Thus, by determining the molar concentration of hydroxide ion, the $[Ca^{2+}]$, the K_{sp}, and the molar solubility of $Ca(OH)_2$ can be calculated.

$$\begin{gathered}[Ca^{2+}] = \tfrac{1}{2}[OH^-] \\ K_{sp} = [Ca^{2+}][OH^-]^2 = \tfrac{1}{2}[OH^-]^3 \\ \text{molar solubility of } Ca(OH)_2 = [Ca^{2+}] = \tfrac{1}{2}[OH^-]\end{gathered} \tag{29.5}$$

Likewise, the same procedure is used to determine the molar solubility of $Ca(OH)_2$ in the presence of added calcium ion, an ion common to the slightly soluble salt equilibrium.

Experimental Procedure

Procedure Overview: The supernatant from a saturated calcium hydroxide solution is titrated with a standardized hydrochloric acid solution to the methyl orange endpoint. An analysis of the data results in the determination of the molar solubility and solubility product of calcium hydroxide. The procedure is repeated on the supernatant from a saturated calcium hydroxide solution containing added calcium ion.

[1]A saturated solution of calcium hydroxide is called **limewater.**

A. Molar Solubility and Solubility Product of Calcium Hydroxide

Three analyses are to be completed. To hasten the analyses in Parts A.5 and A.7, prepare three, *clean* labeled 125- or 250-mL Erlenmeyer flasks. Obtain no more than 50 mL of standardized 0.05 *M* HCl for use in Part A.4.

Ask your laboratory instructor about the status of the saturated $Ca(OH)_2$ solution: if you are to prepare the solution, then omit Part A.3; if the stockroom personnel has prepared the solution, then omit Parts A.1 and A.2.

1. **Prepare the Stock Calcium Hydroxide Solution.** Prepare a saturated $Ca(OH)_2$ solution 1 week before the experiment by adding approximately 3 g of $Ca(OH)_2$ to 120 mL of boiled, deionized water in a 125-mL Erlenmeyer flask. Stir the solution and stopper the flask.

2. **Transfer the Saturated Calcium Hydroxide Solution.** Allow the solid $Ca(OH)_2$ to remain settled on the bottom of the flask (in Part A.1). *Carefully* [try not to disturb the finely divided $Ca(OH)_2$ solid] decant about 90 mL of the saturated $Ca(OH)_2$ solution into a second 125-mL flask.
3. **Obtain a Saturated Calcium Hydroxide Solution (Alternate).** Submit a clean, dry 150-mL beaker to your laboratory instructor (or stockroom) for the purpose of obtaining ~90 mL of supernatant from a saturated $Ca(OH)_2$ solution for analysis.

4. **Set Up the Titration Apparatus.** Prepare a clean, 50-mL buret for titration. Rinse the clean buret and tip with three 5-mL portions of the standardized 0.05 *M* HCl solution and discard. Fill the buret with standardized 0.05 *M* HCl, remove the air bubbles in the buret tip, and, after 10–15 seconds, read and record the initial volume, "using all certain digits (from the labeled calibration marks on the buret) *plus* one uncertain digit (the last digit which is the best estimate between the calibration marks)." Record the *actual* concentration of the 0.05 *M* HCl on the Report Sheet. Place a sheet of white paper beneath the receiving flask.

5. **Prepare a Sample for Analysis.** Rinse a 25-mL pipet twice with 1- to 2-mL portions of the saturated $Ca(OH)_2$ solution and discard. Pipet 25 mL of the saturated $Ca(OH)_2$ solution into a 125-mL flask and add 2 drops of methyl orange indicator.

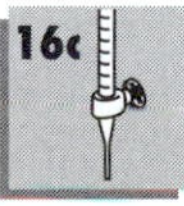

6. **Titrate.** Titrate the $Ca(OH)_2$ solution with the standardized HCl solution to the methyl orange endpoint, where the color changes from yellow to a faint red. Remember the addition of HCl should stop within *one-half drop* of the endpoint. After 10–15 seconds of the persistent endpoint, read and record the final volume of standard HCl in the buret.
7. **Repeat.** Titrate two additional samples of the saturated $Ca(OH)_2$ solution until 1% reproducibility is achieved.

8. **Do the Calculations.** Complete your calculations as outlined on the Report Sheet. The reported values for the K_{sp} of $Ca(OH)_2$ will vary from chemist to chemist.

B. Molar Solubility of Calcium Hydroxide in the Presence of a Common Ion

Three analyses are to be completed. Clean and label three 125- or 250-mL Erlenmeyer flasks.

Again, as in Part A, ask your instructor about the procedure by which you are to obtain the saturated $Ca(OH)_2$ solution with the added $CaCl_2$ for analysis. If you are to prepare the solution, then complete Parts B.1 and A.2; if the stockroom personnel prepared the "spiked" saturated $Ca(OH)_2$ solution, then repeat Part A.3.

In either case, complete Part B.2 in its entirety for the analysis of the saturated $Ca(OH)_2$ solution with the added $CaCl_2$.

1. **Prepare the Stock Solution.** Mix 3 g of $Ca(OH)_2$ and 1 g of $CaCl_2{\cdot}2H_2O$ with 120 mL of boiled, deionized water in a 125-mL flask 1 week before the experiment. Stir and stopper the flask.

2. **Prepare a Buret for Analysis, Prepare the Sample, and Titrate.** Repeat Parts A.4–A.8.

Disposal: Discard all of the reaction mixtures as advised by your instructor.

CLEANUP: Discard the HCl solution in the buret as advised by your instructor. Rinse the buret twice with tap water and twice with deionized water.

Experiment 29 *Prelaboratory Assignment*

Molar Solubility, Common-Ion Effect

Date __________ Lab Sec. ______ Name ______________________________ Desk No. __________

1. The molar solubility of PbI_2 is 1.5×10^{-3} mol/L (see photo).
   a. What is the molar concentration of iodide ion in a saturated PbI_2 solution?

   b. Determine the solubility constant, K_{sp}, for lead(II) iodide.

Lead iodide precipitate.

2. a. Referring to Question 1, qualitatively how is the molar solubility affected by the addition of lead(II) nitrate to the solution? Explain.

   b. What is the molar solubility of lead(II) iodide in a solution containing 0.10 M $Pb(NO_3)_2$?

3. A saturated solution of strontium hydroxide is prepared and the excess solid strontium hydroxide is allowed to settle. A 25.0-mL aliquot of the saturated solution is withdrawn and transferred to an Erlenmeyer flask, and two drops of methyl orange indicator are added. A 0.073 *M* HCl solution (titrant) is dispensed from a buret into the solution (analyte). The solution turns from yellow to a very faint red after the addition of 29.5 mL.
   a. How many moles of hydroxide ion are neutralized in the analysis?

   b. What is the molar concentration of the hydroxide ion in the saturated solution?

   c. What is the molar solubility of strontium hydroxide?

   d. What is the solubility product, K_{sp}, for strontium hydroxide?

   e. For the determination of the K_{sp} for strontium hydroxide, the stoichiometric point (as determined by the endpoint of methyl orange indicator) should not be surpassed. Suppose the stoichiometric point in an experiment is surpassed (i.e., surpassing the faint red of the indicator). Will the reported solubility product of strontium hydroxide be too high, too low, or unaffected? Explain.

*4. Phenolphthalein has a color change over the pH range of 8.2 to 10.0; methyl orange has a color change over the pH range of 3.2 to 4.4. Although the phenolphthalein indicator is commonly used for neutralization reactions, why instead is the methyl orange indicator recommended for this experiment?

Experiment 29 *Report Sheet*

Molar Solubility, Common-Ion Effect

Date ________ Lab Sec. ______ Name ______________________________ Desk No. ________

A. Molar Solubility and Solubility Product of Calcium Hydroxide

	Trial 1	*Trial 2*	*Trial 3*
1. Concentration of standardized HCl solution (mol/L)		________	
2. Buret reading, *initial* (mL)	________	________	________
3. Buret reading, *final* (mL)	________	________	________
4. Volume of HCl added (mL)	________	________	________
5. Amount of HCl added (mol)	________	________	________
6. Amount of OH^- in satd solution (mol)	________	________	________
7. Volume of satd $Ca(OH)_2$ solution (mL)	25.0	25.0	25.0
8. $[OH^-]$, equilibrium (mol/L)	________	________	________
9. $[Ca^{2+}]$, equilibrium (mol/L)	________	________	________
10. Molar solubility of $Ca(OH)_2$	________*	________	________
11. Average molar solubility of $Ca(OH)_2$		________	
12. K_{sp} of $Ca(OH)_2$	________*	________	________
13. Average K_{sp}		________	

*Calculations for Trial 1.

B. Molar Solubility of Calcium Hydroxide in the Presence of a Common Ion

	Trial 1	*Trial 2*	*Trial 3*
1. Concentration of standardized HCl solution (mol/L)			
2. Buret reading, *initial* (mL)			
3. Buret reading, *final* (mL)			
4. Volume of HCl added (mL)			
5. Amount of HCl added (mol)			
6. Amount of OH^- in satd solution (mol)			
7. Volume of satd $Ca(OH)_2$ with added $CaCl_2$ solution (mL)	25.0	25.0	25.0
8. $[OH^-]$, equilibrium (mol/L)			
9. Molar solubility of $Ca(OH)_2$ with added $CaCl_2$	*		
10. Average molar solubility of $Ca(OH)_2$ with added $CaCl_2$			

*Calculations for Trial 1.

Laboratory Questions

Circle the questions that are to be answered.

1. Part A.2. Suppose some of the solid calcium hydroxide is inadvertently transferred along with the supernatant liquid for analysis.
   a. Will more, less, or the same amount of hydrochloric acid titrant be used for the analysis in Part A.6? Explain.
   b. How will this inadvertent transfer affect the calculated solubility product for calcium hydroxide? Explain.
   c. How will this inadvertent transfer affect the calculated molar solubility for calcium hydroxide? Explain.
2. Part A.4–6. A dirty buret was used for the analysis causing droplets of the standardized HCl solution to cling to the buret wall. If HCl droplets cling to the buret wall between the initial and final buret readings for an analysis, will the molar solubility of $Ca(OH)_2$ be reported too high or too low? Explain.
3. Refer to the Prelaboratory Assignment, Question 4. If phenolphthalein had been used in this experiment instead of the methyl orange, would the reported solubility product for calcium hydroxide be too high, too low, or unaffected? Explain.
4. Part A.6. Does adding boiled, deionized water to the titrating flask to wash the wall of the Erlenmeyer flask and the buret tip affect the K_{sp} value of the $Ca(OH)_2$? Explain.

*5. How will tap water instead of boiled, deionized water affect the K_{sp} value of $Ca(OH)_2$ in Part A? *Hint:* How will the minerals in the water affect the solubility of $Ca(OH)_2$?

Experiment 30

Hard Water Analysis

Deposits of hardening ions (generally calcium carbonate deposits) can reduce the flow of water in plumbing.

OBJECTIVES

- To learn the cause and effects of hard water
- To determine the hardness of a water sample

TECHNIQUES

The following techniques are used in the Experimental Procedure

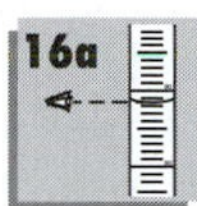

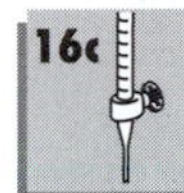

INTRODUCTION

Hardening ions present in natural waters are the result of slightly acidic rainwater flowing over mineral deposits of varying compositions; the acidic rainwater[1] reacts with the *very* slightly soluble carbonate salts of calcium and magnesium and with various iron-containing rocks. A partial dissolution of these salts releases the ions into the water supply, which may be **surface water** or ground water.

$$CO_2(aq) + H_2O(l) + CaCO_3(s) \rightarrow Ca^{2+}(aq) + 2\ HCO_3^-(aq) \qquad (30.1)$$

Surface water: water that is collected from a watershed, e.g., lakes, rivers, and streams

Hardening ions, such as Ca^{2+}, Mg^{2+}, and Fe^{3+}, form insoluble compounds with soaps and causes many detergents to be less effective. Soaps, which are sodium salts of fatty acids such as sodium stearate, $C_{17}H_{35}CO_2^-Na^+$, are very effective cleansing agents so long as they remain soluble; the presence of the hardening ions however causes the formation of a gray, insoluble soap scum such as $(C_{17}H_{35}CO_2)_2Ca$.

$$2\ C_{17}H_{35}CO_2^-Na^+(aq) + Ca^{2+}(aq) \rightarrow (C_{17}H_{35}CO_2)_2Ca(s) + 2\ Na^+(aq) \quad (30.2)$$

This gray precipitate appears as a "bathtub ring" and it also clings to clothes, causing white clothes to appear gray. Dishes and glasses may have spots, shower stalls and lavatories may have a sticky film, clothes may feel rough and scratchy, hair may be dull and unmanageable, and your skin may be irritated and sticky because of hard water.

Hard water is also responsible for the appearance and undesirable formation of "boiler scale" on tea kettles and pots used for heating water. The boiler scale is a poor conductor of heat and thus reduces the efficiency of transferring heat. Boiler scale also builds on the inside of hot water pipes causing a decrease in the flow of water (see opening photo); in extreme cases, this buildup causes the pipe to break.

[1] CO_2 dissolved in rainwater makes rainwater slightly acidic.

$$CO_2(g) + 2\ H_2O(l) \rightarrow H_3O^+(aq) + HCO_3^-(aq)$$

Figure 30.1 Stalactite and stalagmite formations are present in regions having large deposits of limestone, a major contributor of hardening ions.

Table 30.1 Hardness Classification of Water*

Hardness (*ppm $CaCO_3$*)	Classification
<17.1 ppm	soft water
17.1 ppm–60 ppm	slightly hard water
60 ppm–120 ppm	moderately hard water
120 ppm–180 ppm	hard water
>180 ppm	very hard water

*U.S. Department of Interior and the Water Quality Association

Boiler scale consists primarily of the carbonate salts of the hardening ions and is formed according to

$$Ca^{2+}(aq) + 2\,HCO_3^{-}(aq) \xrightarrow{\Delta} CaCO_3(s) + CO_2(g) + H_2O(l) \qquad (30.3)$$

Notice that this reaction is just the reverse of the reaction for the formation of hard water (Equation 30.1). The same two reactions are also key to the formation of stalactites and stalagmites for caves located in regions with large limestone deposits (Figure 30.1).

Because of the relatively large natural abundance of limestone deposits and other calcium minerals, such as gypsum, $CaSO_4{\bullet}2H_2O$, it is not surprising that Ca^{2+} ion, in conjunction with Mg^{2+}, is a major component of the dissolved solids in hard water.

Hard water, however, is not a health hazard. In fact, the presence of Ca^{2+} and Mg^{2+} in hard water can be considered dietary supplements to the point of providing their daily recommended allowance (RDA). Some research studies (though disputed) have also indicated a positive correlation between water hardness and decreased heart disease.

ppm: parts per million by mass, e.g., 15 ppm $CaCO_3$ is 15 g $CaCO_3$ in 10^6 g solution or 15 mg $CaCO_3$ in 1.0 L of solution

The concentration of the hardening ions in a water sample is commonly expressed as though the hardness is due exclusively to $CaCO_3$. Hardness is commonly expressed as mg $CaCO_3$/L, which is also **p p m** $CaCO_3$.[2] A general classification of hard waters is listed in Table 30.1

Theory of Analysis

Complex ion: generally a cation of a metal ion to which is bonded a number of molecules or anions

Titrant: the solution placed in the buret in a titrimetric analysis

Analyte: the solution containing the substance being analyzed, generally in the receiving flask in a titration setup

Na_2H_2Y

In this experiment a titration technique is used to measure the combined Ca^{2+} and Mg^{2+} concentrations in a water sample. The titrant is the disodium salt of ethylenediaminetetraacetic acid (abbreviated Na_2H_2Y).[3]

In aqueous solution Na_2H_2Y dissociates into Na^+ and H_2Y^{2-} ions. The H_2Y^{2-} ion reacts with the hardening ions, Ca^{2+} and Mg^{2+}, to form very stable **complex ions,** especially in a solution buffered at a pH of about 10. An ammonia–ammonium ion buffer is often used for this pH adjustment in the analysis.

As H_2Y^{2-} **titrant** is added to the **analyte,** it complexes with the "free" Ca^{2+} and Mg^{2+} of the water sample to form the respective complex ions:

$$Ca^{2+}(aq) + H_2Y^{2-}(aq) \rightarrow [CaY]^{2-}(aq) + 2\,H^+(aq) \qquad (30.4a)$$

$$Mg^{2+}(aq) + H_2Y^{2-}(aq) \rightarrow [MgY]^{2-}(aq) + 2\,H^+(aq) \qquad (30.4b)$$

From the balanced equations, it is apparent that once the molar concentration of the Na_2H_2Y solution is known, the moles of hardening ions in a water sample can be calculated, a 1:1 stoichiometric ratio.

$$\begin{aligned}\text{volume } H_2Y^{2-} \times \text{molar concentration of } H_2Y^{2-} &= \text{moles } H_2Y^{2-}\\ &= \text{moles hardening ions} \qquad (30.5)\end{aligned}$$

[2]ppm means "parts per million"—1 mg of $CaCO_3$ in 1 000 000 mg (or 1 kg) solution is 1 ppm $CaCO_3$. Assuming the density of the solution is 1 g/mL (or 1 kg/L), then 1 000 000 mg solution = 1 L solution. Therefore, 1 mg/L is an expression of ppm.

[3]**E**thylene**d**iamine**t**etra**a**cetic acid is often simply referred to as EDTA with an abbreviated formula of H_4Y.

The hardening ions, for reporting purposes, are assumed to be exclusively Ca^{2+} from the dissolving of $CaCO_3$. Since one mole of Ca^{2+} forms from one mole of $CaCO_3$, the hardness of the water sample expressed as mg $CaCO_3$ per liter of sample is

$$\text{moles hardening ions} = \text{moles } Ca^{2+} = \text{moles of } CaCO_3 \qquad (30.6)$$

$$\text{ppm } CaCO_3 \left(\frac{\text{mg } CaCO_3}{\text{L sample}}\right) = \frac{\text{mol } CaCO_3}{\text{L sample}} \times \frac{100.1 \text{ g } CaCO_3}{\text{mol}} \times \frac{\text{mg}}{10^{-3}\text{g}} \qquad (30.7)$$

The Indicator for the Analysis

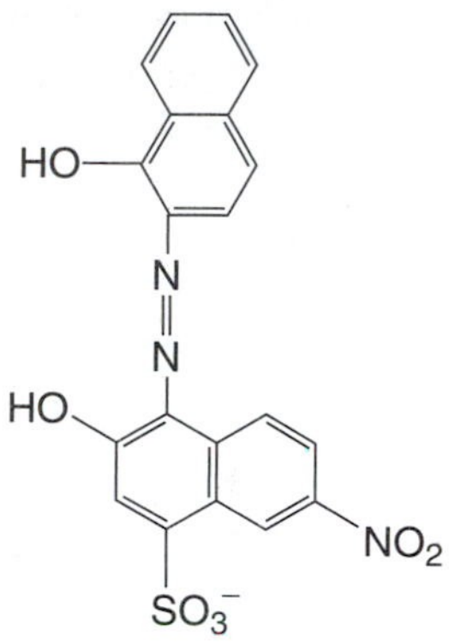

A special indicator is used to detect the endpoint in the titration. Called Eriochrome Black T (EBT), it also forms complex ions with the Ca^{2+} and Mg^{2+} ions, but binds more strongly to Mg^{2+} ions. Because only a small amount of EBT is added, only a small quantity of Mg^{2+} is complexed; no Ca^{2+} ion complexes to EBT—therefore, most of the hardening ions remain "free" in solution. The EBT indicator is sky-blue in solution but forms a wine-red complex with $[Mg\text{-}EBT]^{2+}$.

HO, N=N, HO, NO_2, SO_3^-

Eriochrome Black T

$$\underset{\text{sky-blue}}{Mg^{2+}(aq) + EBT(aq)} \rightleftharpoons \underset{\text{wine-red}}{[Mg\text{-}EBT]^{2+}(aq)} \qquad (30.8)$$

Therefore even before any H_2Y^{2-} titrant is added for the analysis, the analyte is wine-red because of the $[Mg\text{-}EBT]^{2+}$ complex ion.

Once the H_2Y^{2-} titrant complexes all of the "free" Ca^{2+} and Mg^{2+} from the water sample, it then removes the trace amount of Mg^{2+} from the wine-red $[Mg\text{-}EBT]^{2+}$ complex; the solution changes from wine-red back to the sky-blue color of the EBT indicator, and the endpoint is reached—all of the hardening ions have been complexes with H_2Y^{2-}.

$$\underset{\text{wine-red}}{[Mg^{2+}\text{-}EBT]^{2+}(aq)} + H_2Y^{2-}(aq) \rightarrow MgY^{2-}(aq) + 2\,H^+(aq) + \underset{\text{sky-blue}}{EBT(aq)} \qquad (30.9)$$

Therefore, Mg^{2+} *must* be present in the water sample for the EBT indicator to "work." Oftentimes, a small amount of Mg^{2+} as MgY^{2-} is initially added to the analyte to ensure/enhance the formation of the wine-red color with EBT by the equation,

$$MgY^{2-}(aq) + 2\,H^+(aq) + EBT(aq) \rightarrow \underset{\text{wine-red}}{[Mg{-}EBT]^{2+}(aq)} + H_2Y^{2-}(aq) \qquad (30.10)$$

This reaction is then reversed at the stoichiometric point.

$$[Mg\text{-}EBT]^{2+}(aq) + H_2Y^{2-}(aq) \rightarrow MgY^{2-}(aq) + 2\,H^+(aq) + \underset{\text{blue}}{EBT(aq)} \qquad (30.11)$$

Because H_2Y^{2-} is freed initially but then consumed at the endpoint (previous two equations), no additional H_2Y^{2-} titrant is required for the analysis of hardness in the water sample because of this added Mg^{2+}.

A Standard $Na_2H_2Y_2$ Solution

The concentration of a $Na_2H_2Y_2$ solution is determined by its reaction with a known amount of calcium ion in a (primary) standard Ca^{2+} solution (Equation 30.4a). The measured aliquot of the standard Ca^{2+} solution is buffered to a pH of 10 and titrated with the $Na_2H_2Y_2$ solution to the Eriochrome Black T *blue* endpoint (Equation 30.9). To achieve the endpoint a small amount of Mg^{2+} in the form of MgY^{2-} is added to the standard Ca^{2+} solution (Equations 30.10 and 30.11).

Note that the standardization of the $Na_2H_2Y_2$ solution with a standard Ca^{2+} solution in Part A is reversed in Part B where the (now) standardized $Na_2H_2Y_2$ solution is used to determine an unknown Ca^{2+} solution.

Experimental Procedure

Procedure Overview: A (primary) standard solution of Ca^{2+} is used to standardize a prepared ~0.01 *M* $Na_2H_2Y_2$ solution. The (secondary) standardized $Na_2H_2Y_2$ solution is subsequently used to titrate a water sample to the Eriochrome Black T indicator endpoint.

Three trials are to be completed for the standardization of the ~0.01 *M* $Na_2H_2Y_2$ solution *and* for determining the hardness of a water sample. To save time, initially prepare three Erlenmeyer flasks for Part A.3.

A. A Standard 0.01 *M* Disodium Ethylenediaminetetraacetate, Na_2H_2Y, Solution

1. **Measure of the Mass of the $Na_2H_2Y_2$ Solution.** Calculate the mass of $Na_2H_2Y_2{\bullet}2H_2O$ (molar mass = 372.24 g/mol) required to prepare 250 mL of a 0.01 *M* $Na_2H_2Y_2$ solution. Show this calculation on the Report Sheet. Tare this mass on weighing paper, transfer it to a 250-mL volumetric flask containing 100 mL of deionized water, swirl to dissolve, and dilute to the "mark" (slight heating may be required).
2. **Prepare a buret for titration.** Rinse the buret with the Na_2H_2Y solution and then fill. Record the volume of the titrant "using all certain digits (from the labeled calibration marks on the buret) plus one uncertain digit (the last digit which is the best estimate between the calibration mark)."

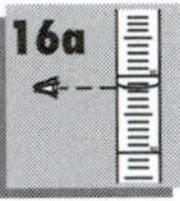

3. **Prepare the Standard Ca^{2+} Solution.** Obtain ~80 mL of a standard Ca^{2+} solution and record its exact molar concentration. Pipet 25.0 mL of the standard Ca^{2+} solution into a 125 mL Erlenmeyer flask, add 1 mL of buffer (pH = 10) solution, and 2 drops of EBT indicator (containing a small amount of MgY^{2-}).[4]

4. **Titrate the Standard Ca^{2+} Solution.** Titrate the standard Ca^{2+} solution with the Na_2H_2Y titrant; swirl continuously. Near the endpoint, slow the rate of addition to drops; the last few drops should be added at 3–5 s intervals. The solution changes from wine-red to purple to blue—no tinge of the wine-red color should remain; the solution is *blue* at the endpoint. Record the final volume in the buret.

5. **Repeat the Titration of the Standard Ca^{2+} Solution.** Repeat the titrations on the remaining two samples. Calculate the molar concentration of the Na_2H_2Y solution. Save the standard Ca^{2+} solution for Part B.

B. Analysis of Water Sample

Complete three trials for your analysis. The first trial is an indication of the hardness of your water sample. You may want to adjust the volume of water for the analysis of the second and third trials.

1. **Obtain the Water Sample for Analysis.** Obtain about 100 mL of a water sample from your instructor. You may use your own water sample or simply the tap water in the laboratory.

   If the water sample is from a lake, stream, or ocean, you will need to gravity filter the sample before the analysis.

   If your sample is acidic, add 1 *M* NH_3 until it is basic to litmus.

2. **Prepare the Water Sample for Analysis.** Pipet 25.0 mL of your (filtered, if necessary) water sample[5] into a 125 mL Erlenmeyer flask, add 1 mL of the buffer (pH = 10) solution, and 2 drops of EBT indicator.

3. **Titrate the Water Sample.** Titrate the water sample with the standardized $Na_2H_2Y_2$ until the *blue* endpoint appears (as described in Part A.4). Repeat (twice) the analysis of the water sample to determine its hardness.

> *Disposal:* Dispose of the analyzed solutions in the "Waste EDTA" container.

[4] The indicator, calgamite, can be substituted for the Eriochrome Black T indicator.

[5] If your water is known to have a high hardness, decrease the volume of the water proportionally until it takes about 15 mL of Na_2H_2Y titrant for your second and third trials. Similarly if your water sample is known to have a low hardness, increase the volume of the water proportionally.

Experiment 30 *Prelaboratory Assignment*

Hard Water Analysis

Date ________ Lab Sec. ______ Name ______________________________ Desk No. ________

1. What ions are responsible for water hardness?

2. What is boiler scale? What substance(s) causes boiler scale?

3. Water hardness is not only reported in ppm $CaCO_3$ but also in grains/gallon, where 1 grain/gallon equal 17.1 ppm $CaCO_3$. Express the "moderately hard water" range (Table 30.1) in units of grains/gallon.

4. a. Which hardening ion, Ca^{2+} or Mg^{2+}, binds more tightly to (forms a stronger complex ion with) the Eriochrome Black T indicator used for today's analysis?

   b. What is the color change at the endpoint?

5. A 50.0 mL water sample requires 5.14 mL of 0.0100 *M* Na_2H_2Y to reach the Eriochrome Black T endpoint.
   a. Calculate the moles of hardening ions in the water sample.

   b. Assuming the hardness is due exclusively to $CaCO_3$, express the hardness concentration in mg $CaCO_3$/L sample.

   c. What is this hardness concentration expressed in ppm $CaCO_3$?

   d. Classify the hardness of this water according to Table 30.1.

6. The hardness of a water sample is known to be 500 ppm $CaCO_3$. In an analysis of a 50 mL water sample, what volume of 0.0100 *M* Na_2H_2Y is needed to reach the Eriochrome Black T endpoint?

Experiment 30 *Report Sheet*

Hard Water Analysis

Date __________ Lab Sec. ______ Name ______________________________ Desk No. __________

A. A Standard 0.01 *M* Disodium Ethylenediaminetetraacetate, Na_2H_2Y, Solution

Calculation for mass of $Na_2H_2Y_2 \cdot 2H_2O$ required to prepare 250 mL of a 0.01 *M* $Na_2H_2Y_2$ solution.

	Trial 1	*Trial 2*	*Trial 3*
1. Volume of standard Ca^{2+} solution (mL)	25.0	25.0	25.0
2. Concentration of standard Ca^{2+} solution			
3. Mol Ca^{2+} = mol Na_2H_2Y (mol)			
4. Buret reading, *initial* (mL)			
5. Buret reading, *final* (mL)			
6. Volume of Na_2H_2Y titrant (mL)			
7. Molar concentration of Na_2H_2Y solution (mol/L)			
8. Average molar concentration of Na_2H_2Y solution (mol/L)			

B. Analysis of Water Sample

	Trial 1	*Trial 2*	*Trial 3*
1. Volume of water sample (mL)			
2. Buret reading, *initial* (mL)			

3. Buret reading, *final* (mL) ________ ________ ________

4. Volume of Na_2H_2Y titrant (mL) ________ ________ ________

5. Mol Na_2H_2Y = mol hardening ions, Ca^{2+} and Mg^{2+} (mol) ________ ________ ________

6. Mass of equivalent $CaCO_3$ (g) ________ ________ ________

7. ppm $CaCO_3$ (mg $CaCO_3$/L sample) ________ ________ ________

8. Average ppm $CaCO_3$ ________________

Laboratory Questions

Circle the questions that have been assigned.

1. Part A.3. State the purpose for the 1 mL of buffer (pH = 10) being added to the standard Ca^{2+} solution?

*2. Part A.3. The buffer solution is omitted from the titration procedure, the Eriochrome Black T indicator and a small amount of Mg^2 are added, and the standard Ca^{2+} solution is acidic.
   a. What is the color of the solution? Explain.
   b. The $Na_2H_2Y_2$ solution is dispensed from the buret. What color changes are observed? Explain.

3. Part A.4. Deionized water from the wash bottle is used to wash the side of the Erlenmeyer flask. How does this affect the reported molar concentration of the $Na_2H_2Y_2$ solution . . . too high, too low, or unaffected? Explain.

4. Part A.4. The dispensing of the $Na_2H_2Y_2$ solution from the buret is discontinued when the solution turns purple. Because of this technique error, will the reported molar concentration of the $Na_2H_2Y_2$ solution be too high, too low, or unaffected? Explain.

5. Part B.3. The dispensing of the $Na_2H_2Y_2$ solution from the buret is discontinued when the solution turns purple. Because of this technique error, will the reported hardness of the water sample be too high, too low, or unaffected? Explain.

6. Part A.4 and Part B.3. The dispensing of the $Na_2H_2Y_2$ solution from the buret is discontinued when the solution turns purple. However in Part B.3, the standardized $Na_2H_2Y_2$ solution is then used to titrate a water sample to the (correct) *blue* endpoint. Will the reported hardness of the water sample be too high, too low, or unaffected? Explain.

*7. Washing soda, $Na_2CO_3{\bullet}10H_2O$, is often used to "soften" hard water, i.e., to remove hardening ions. Assuming hardness is due to Ca^{2+}, the $CO_3^{\,2-}$ ion precipitates the Ca^{2+}.

$$Ca^{2+}(aq) + CO_3^{\,2-}(aq) \rightarrow CaCO_3(s)$$

How many grams and pounds of washing soda are needed to remove the hardness from 500 gallons of water having a hardness of 150 ppm $CaCO_3$ (see Appendix A for conversion factors)?

Experiment 31

The Thermodynamics of the Dissolution of Borax

Borax is a commonly added to (clothes) wash water to increase the pH for more effective cleansing.

OBJECTIVES

- To determine the solubility product of borax as a function of temperature
- To determine the standard free energy, standard enthalpy, and standard entropy changes for the dissolution of borax in an aqueous solution

TECHNIQUES

The following techniques are used in the Experimental Procedure

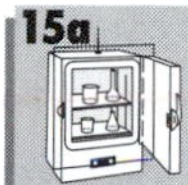

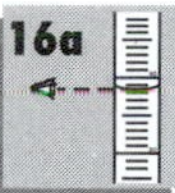

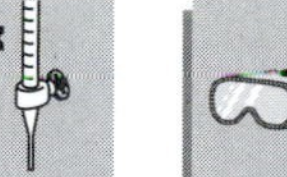

INTRODUCTION

Large deposits of borax are found in the arid regions of the southwestern United States, most notably in the Mojave Desert (east central) region of California. Borax is obtained as tincal, $Na_2B_4O_5(OH)_4{\bullet}8H_2O$, and kernite, $Na_2B_4O_7{\bullet}4H_2O$, from an open pit mine near Boron, California, and as tincal from brines from Searles Lake near Trona, California. Borax, used as a washing powder for laundry formulations, is commonly sold as 20-Mule Team Borax®. Historically, borax was mined in Death Valley, California, in the late nineteenth century. To transport the borax from this harsh environment, teams of 20 mules were used to pull a heavy wagon loaded with borax (and a water wagon) across the desert and over the mountains to railroad depots for shipment to other parts of the world. Borax is used as a cleansing agent, in the manufacture of glazing paper and varnishes, and as a flux in soldering and brazing; however, its largest current use is in the manufacture of borosilicate glass.

The free energy change of a chemical process is proportional to its equilibrium constant according to the equation

$$\Delta G^\circ = -RT \ln K \tag{31.1}$$

where R, the gas constant, is 8.314×10^{-3} kJ/mol•K and T is the temperature in kelvins. The equilibrium constant, K, is expressed for the equilibrium system when the reactants and products are in their **standard states.** For a slightly soluble salt in an

Standard state: the state of a substance at one atmosphere (and generally 25°C)

aqueous system, the precipitate and the ions in solution correspond to the standard states of the reactants and products, respectively.

The "standard state" equilibrium for the slightly soluble silver chromate salt is

$$Ag_2CrO_4(s) \rightleftharpoons 2\,Ag^+(aq) + CrO_4^{2-}(aq) \tag{31.2}$$

The solubility product, K_{sp}, is set equal to the product of the molar concentrations of the ions, each raised to the power of their respective coefficients in the balanced equation—this is the mass action expression for the system:

$$K_{sp} = [Ag^+]^2\,[CrO_4^{2-}] \tag{31.3}$$

Free energy change, $\Delta G°$: negative value, spontaneous process; positive value, nonspontaneous process

and the **free energy change** for the equilibrium is

$$\Delta G° = -RT \ln K_{sp} = -RT \ln [Ag^+]^2\,[CrO_4^{2-}] \tag{31.4}$$

Enthalpy change, $\Delta H°$: negative value, exothermic process; positive value, endothermic process

Additionally, the free energy change of a chemical process is a function of the **enthalpy change** and the **entropy change** of the process:

Entropy change, $\Delta S°$: negative value, decrease in randomness of the process; positive value, increase in randomness of the process

$$\Delta G° = \Delta H° - T\Delta S° \tag{31.5}$$

When the two free energy expressions are set equal for a slightly soluble salt, such as silver chromate, then

$$-RT \ln K_{sp} = \Delta H° - T\Delta S° \tag{31.6}$$

Rearranging and solving for $\ln K_{sp}$,

$$\ln K_{sp} = -\frac{\Delta H°}{R}\left(\frac{1}{T}\right) + \frac{\Delta S°}{R} \text{ (analogous to the equation } y = mx + b) \tag{31.7}$$

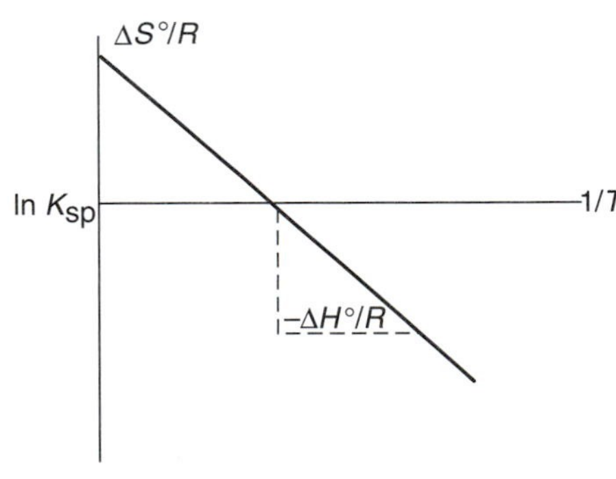

This equation can prove valuable in determining the thermodynamic properties of a chemical system, such as that of a slightly soluble salt. A linear relationship exists when the values of $\ln K_{sp}$ obtained at various temperatures are plotted as a function of the reciprocal temperature. The (negative) slope of the line equals $-\Delta H°/R$, and the y-intercept (where $x = 0$) equals $\Delta S°/R$. Since R is a constant, the $\Delta H°$ and the $\Delta S°$ for the equilibrium system can easily be calculated.

Since the values of $\ln K_{sp}$ may be positive or negative for slightly soluble salts and T^{-1} values are always positive, the data plot of $\ln K_{sp}$ versus $1/T$ appears in the first and fourth quadrants of the Cartesian coordinate system.

The Borax System

Borax is often given the name sodium tetraborate decahydrate and the formula $Na_2B_4O_7{\bullet}10H_2O$. However, according to its chemical behavior, a more defining formula for borax is $Na_2B_4O_5(OH)_4{\bullet}8H_2O$, also called tincal.

In this experiment the thermodynamic properties, $\Delta G°$, $\Delta H°$, and $\Delta S°$, are determined for the aqueous solubility of borax (tincal), $Na_2B_4O_5(OH)_4{\bullet}8H_2O$. Borax dissolves and dissociates in water according to the equation

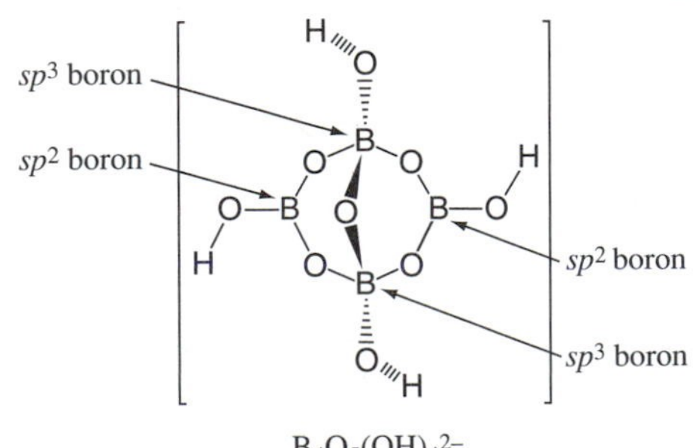

$$Na_2B_4O_5(OH)_4{\bullet}8H_2O(s) \rightleftharpoons 2\,Na^+(aq) + B_4O_5(OH)_4^{2-}(aq) + 8\,H_2O(l) \tag{31.8}$$

The mass action expression, set equal to the solubility product at equilibrium, for the solubility of borax is

$$K_{sp} = [Na^+]^2\,[B_4O_5(OH)_4^{2-}] \tag{31.9}$$

The $B_4O_5(OH)_4^{2-}$ anion, because it is the conjugate base of the weak acid boric acid, is capable of accepting two protons from a strong acid in an aqueous solution:

H—O—B(—O—H)—O—H

H_3BO_3

$$B_4O_5(OH)_4^{2-}(aq) + 2\,H^+(aq) + 3\,H_2O(l) \rightleftharpoons 4\,H_3BO_3(aq) \qquad (31.10)$$

Therefore, the molar concentration of the $B_4O_5(OH)_4^{2-}$ anion in a saturated borax solution can be measured with a titrimetric analysis of the saturated borax solution using a standardized hydrochloric acid solution as the titrant.

$$\text{mol } B_4O_5(OH)_4^{2-} = \text{volume (L) HCl} \times \frac{\text{mol HCl}}{\text{volume (L) HCl}} \times \frac{1 \text{ mol } B_4O_5(OH)_4^{2-}}{2 \text{ mol HCl}} \qquad (31.11)$$

$$[B_4O_5(OH)_4^{2-}] = \frac{\text{mol } B_4O_5(OH)_4^{2-}}{\text{volume (L) sample}} \qquad (31.12)$$

This analysis is also a measure of the molar solubility of borax in water at a given temperature—according to the stoichiometry, one mole of the $B_4O_5(OH)_4^{2-}$ anion forms for every mole of borax that dissolves.

$$[B_4O_5(OH)_4^{2-}] = \text{molar solubility of borax} \qquad (31.13)$$

Temperature changes do affect the molar solubility of most salts, and borax is no exception. For example, the solubility of borax is 2.01 g/100 mL at 0°C and is 170 g/100 mL at 100°C.[1]

As a consequence of the titration and according to the stoichiometry of the dissolution of the borax, the molar concentration of the sodium ion in the saturated solution is twice that of the experimentally determined $B_4O_5(OH)_4^{2-}$ anion concentration.

$$[Na^+] = 2 \times [B_4O_5(OH)_4^{2-}] \qquad (31.14)$$

The solubility product for borax at a measured temperature is, therefore,

$$K_{sp} = [Na^+]^2[B_4O_5(OH)_4^{2-}] = [2 \times [B_4O_5(OH)_4^{2-}]]^2[B_4O_5(OH)_4^{2-}]$$
$$= 4[B_4O_5(OH)_4^{2-}]^3 = 4[\text{molar solubility of borax}]^3 \qquad (31.15)$$

To obtain the thermodynamic properties for the dissolution of borax, values for the molar solubility and the solubility product for borax are determined over a range of temperatures.

Standardized HCl Solution

A standardized HCl solution is prepared using anhydrous sodium carbonate as the primary standard. Sodium carbonate samples of known mass are transferred to Erlenmeyer flasks, dissolved in deionized water, and titrated to a methyl orange endpoint (pH range 3.1 to 4.4) with the prepared hydrochloric acid solution.

$$CO_3^{2-}(aq) + 2\,H_3O^+(aq) \rightarrow 3\,H_2O(l) + CO_2(g) \qquad (31.16)$$

The flask is heated to near boiling close to the stoichiometric point of the analysis to remove the carbon dioxide gas produced in the reaction.

The objectives of this experiment are fourfold: (1) determine the molar concentration of the $B_4O_5(OH)_4^{2-}$ anion and the molar solubility of borax at *at least* five different temperatures using the titration technique with a standardized hydrochloric acid solution; (2) calculate the solubility product of borax at each temperature; (3) plot the natural logarithm of the solubility product versus the reciprocal temperature of the measurements; (4) extract from the plotted data the thermodynamic properties of $\Delta H°$ and $\Delta S°$ for the dissolution of borax, from which its $\Delta G°$ can be calculated.

[1]The solubility of borax at room temperature is about 6.3 g/100 mL.

EXPERIMENTAL PROCEDURE

Procedure Overview: This experiment is to be completed in cooperation with other chemists/chemist groups in the laboratory. In Part A, a standardized solution of hydrochloric acid is to be prepared. In Part B, four warm water baths are to be set up (see Part B.2 and Figure 31.2) at the beginning of the laboratory, each at a different temperature, but at a maximum of 60°C. *The water baths then are to be shared.* Consult with your colleagues (and your laboratory instructor) in coordinating the setup of the apparatus. In conjunction with the four warm water baths for Part B, about 150 mL of warm (~60°C) deionized water is to be prepared for Part C.1.

A. Standardized HCl Solution

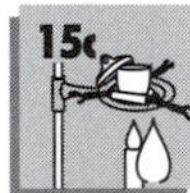

1. **Dry the Primary Standard.** Dry 2–3 g of anhydrous sodium carbonate, Na_2CO_3, for several hours in a drying oven set at about 110°C. Cool the sample in a desiccator.
2. **Prepare the HCl Solution.** Prepare 200 mL of ~0.25 *M* HCl starting with concentrated (12 *M*) HCl. (**Caution:** *Concentrated HCl causes severe skin burns.*)
3. **Prepare the Primary Standard.** Calculate the mass of sodium carbonate that neutralizes 15–20 mL of 0.25 *M* HCl at the stoichiometric point. Show the calculation on the Report Sheet. Measure this mass (±0.001 g) on a tared weighing paper or dish and transfer to a 125-mL Erlenmeyer flask. Prepare *at least* three samples of sodium carbonate for the analysis of the HCl solution.

Keep the dry sample in a desiccator.

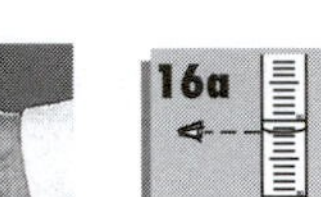

4. **Prepare the Buret.** Clean a buret and rinse with several 3- to 5-mL portions of the ~0.25 *M* HCl solution. Use a clean funnel to fill the buret with the ~0.25 *M* HCl solution. After 10–15 seconds use the proper technique to read and record the volume of HCl solution in the buret, "using all certain digits (from the labeled calibration marks on the buret) *plus* one uncertain digit (the last digit which is the best estimate between the calibration marks)."

5. **Titrate the Primary Standard.** To each solid sodium carbonate sample add ~50 mL of deionized water and several drops of methyl orange indicator.[2] Place a sheet of white paper beneath the Erlenmeyer flask. Dispense the HCl solution from the buret, swirling the Erlenmeyer flask during the addition.

   Very near or at the apparent endpoint of the indicator, heat the flask over a direct flame to near boiling to drive off the carbon dioxide gas. Carefully (dropwise) add additional HCl titrant until the endpoint is reached and the color persists for 30 seconds (a color change caused by the addition of one "additional" drop of the HCl solution from the buret). Stop the addition of the HCl titrant. After 10–15 seconds read and record the volume of HCl in the buret.
6. **Repeat the Analysis.** Refill the buret and complete the standardization procedure with the remaining sodium carbonate samples.
7. **Do the Calculations.** Calculate the experimental molar concentration of the prepared HCl solution from the sodium carbonate samples. The molar concentrations of the HCl solution from the trials should agree within ±1%; if not, complete additional trials as necessary.

Disposal: Dispose of the analyzed samples in the "Waste Salts" container.

[2]Methyl orange indicator changes color over the pH range of 3.1–4.4; its color appears yellow at a pH greater than 4.4, but orange at a pH less than 3.1.

B. Preparation of Borax Solutions

Five or six borax samples can be prepared for the experiment. The sixth sample (a saturated solution in an ice bath) will have a *very* low solubility of borax. Consult with your instructor.

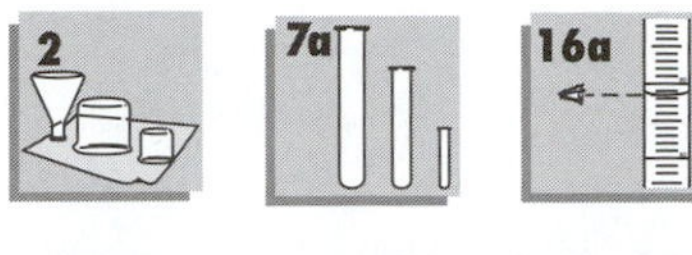

1. **Calibrate Test Tubes.** Pipet 5 mL of deionized water into five (or six) *clean* medium-sized (~150 mm) test tubes (Figure 31.1). Mark the bottom of the meniscus (use a marking pen or tape). Discard the water and allow the test tubes to oven-dry. Label the test tubes.
2. **Prepare Stock Solution of Borax.** Using a 125- or 250-mL Erlenmeyer flask, dissolve 30–35 g of borax in 100 mL of deionized water. Stopper the flask and agitate the mixture for several minutes to prepare a saturated solution.
3. **Prepare the Test Samples of Borax.** a. *Second set of test tubes.* Label a second set of clean, medium-sized test tubes for use in the setup shown in Figure 31.2.

   b. *Half-fill the test tubes with stock solution.* Again, thoroughly agitate the borax stock solution and then, with the stock solution, half-fill this second set of medium-sized test tubes. Place the test tubes in the respective baths shown Figure 31.2. The temperatures of the baths need not be exactly those indicated, but the measured temperature is important. The "first" bath should *not* exceed 60°C.[3]
4. **Prepare Saturated Solutions of Borax.** a. *Saturate the solutions.* Occasionally agitate the test tubes in the baths (for 10–15 minutes), assuring the formation of a saturated solution—solid borax should always be present—add more solid borax if necessary.

   b. *Allow sample to settle.* Allow the borax to settle until the solution is clear (this will require several minutes, be patient!) and has reached thermal equilibrium. Allow 10–15 minutes for thermal equilibrium to be established.
5. **Prepare Borax Solutions for Analysis.** Quickly, but carefully, transfer 5 mL of the clear solutions to the correspondingly labeled, calibrated test tubes from Part B.1 and Figure 31.1 (do *not* transfer any of the solid borax!).

   Record the *exact* temperature of the respective water baths.

   Remove the heat from the water baths.

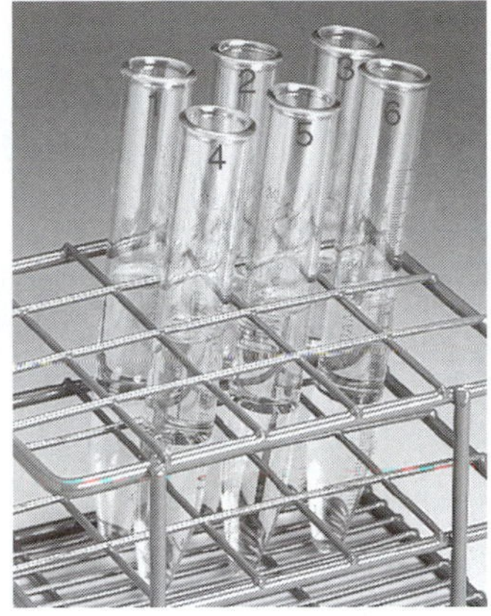

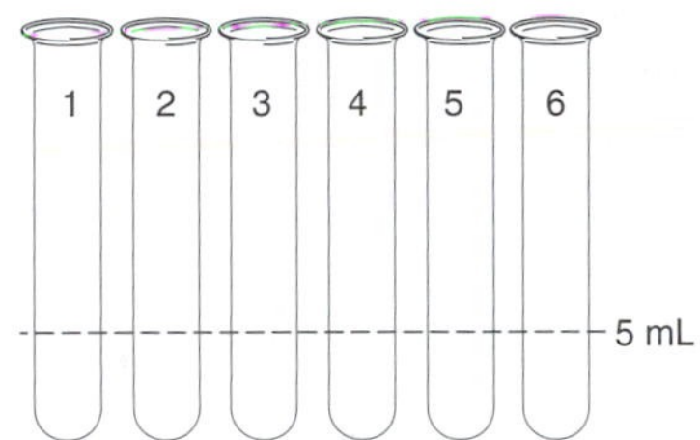

Figure 31.1 Calibrate six, clean 150-mm test tubes at the 5-mL mark. (Part B.1)

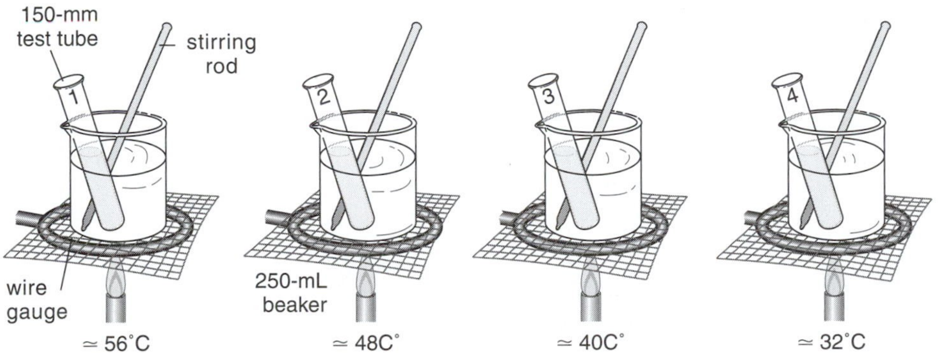

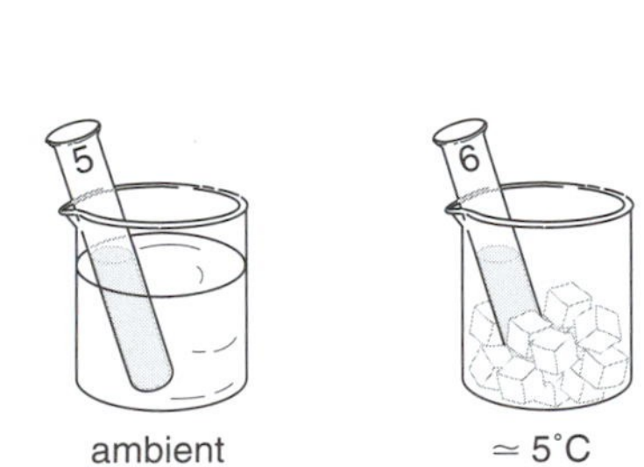

Figure 31.2 Six (different temperature) baths for preparing saturated borax solutions. (Part B.3)

---

[3]Borax (tincal) is stable in aqueous solutions at temperatures less than 61°C; at higher temperatures, some dehydration of the $B_4O_5(OH)_4^{2-}$ anion occurs.

C. Analysis of Borax Test Solutions

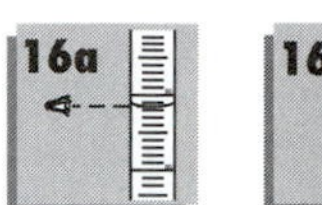

1. **Transfer the Samples.** a. *Prepare for sample transfer.* Set up and label a set of five (or six) clean, 125- or 250-mL Erlenmeyer flasks. As the samples (Part B.4) cool to room temperature in the test tubes, some borax may crystallize. Return those samples to a *warm* (~55°C) water bath until all of the solid dissolves.

   b. *Transfer the samples.* After the sample is clear, transfer it to the correspondingly labeled Erlenmeyer flask, rinse the test tube with two or three ~5-mL portions of *warm* deionized water, and combine the washings with the sample.
2. **Titrate the Samples.** Dilute each sample to about 25 mL with warm deionized water. Add 2–3 drops of bromocresol green.[4] Titrate each of the five samples to a yellow endpoint with the standardized HCl solution prepared in Part A. Remember to record the buret readings before and after each analysis of a sample.

D. Data Analysis

> *Disposal:* Dispose of the analyte in the "Waste Salts" container and the titrant in the "Waste Acids" container.

Six (or five) repeated calculations are required in order to establish the data plot for the determination of $\Delta H°$ and $\Delta S°$ for the dissolution of borax. The lengthy task of completing the calculations and for minimizing errors in the calculations can be reduced with the use of an Excel (or similar) spreadsheet. The data from the calculations can then be plotted using the embedded graphing capabilities of Excel.

Appendix C

The calculations and the analyses that are required are:

1. Calculate the molar solubility of borax at each of the measured temperatures.
2. Calculate the solubility product of borax at each of the measured temperatures.
3. Plot the natural logarithm of the solubility product versus the reciprocal temperature (K^{-1}) for each sample and draw the "best straight line" through the data points.
4. Determine the slope of the linear plot and calculate the standard enthalpy of solution for borax.
5. Determine the y-intercept (at $x = 0$) of the linear plot and calculate the standard entropy of solution for borax.

[4] The pH range for the color change of bromocresol green is 4.0 to 5.6. The indicator appears yellow at a pH less than 4.0 and blue at a pH greater than 5.6.

Experiment 31 *Prelaboratory Assignment*

The Thermodynamics of the Dissolution of Borax

Date __________ Lab Sec. ______ Name ____ Desk No. __________

1. The standard free energy change for the decomposition of two moles of hydrogen peroxide at 25°C is −224 kJ.

$$2\ H_2O_2(l) \rightarrow 2\ H_2O(l) + O_2(g) \quad \Delta G° = -224\ \text{kJ}$$

   a. Calculate the equilibrium constant for the reaction.

   b. What is the chemical significance of the value of the equilibrium constant?

   c. The standard enthalpy change, $\Delta H°$, for the decomposition of hydrogen peroxide is +196.1 kJ. Determine the standard entropy change, $\Delta S°$, for the reaction at 25°C.

   d. i) What is the significance of the standard free energy change?

   ii) What is the significance of the standard enthalpy change?

   iii) What is the significance of the standard entropy change?

2. Describe the preparation of 200 mL of 0.25 *M* HCl, starting with conc HCl (12 *M*). See the Experimental Procedure, Part A.2.

3. The $B_4O_5(OH)_4^{2-}$ ion, present in 5.0-mL of a saturated $Na_2B_4O_5(OH)_4$ solution at a measured temperature, is titrated to the bromocresol endpoint with 3.25 mL of 0.291 *M* HCl.
   a. How many moles of $B_4O_5(OH)_4^{2-}$ are present in the sample?

   b. What is the molar concentration of $B_4O_5(OH)_4^{2-}$ in the sample?

   c. Calculate the K_{sp} for $Na_2B_4O_5(OH)_4$ from this data.

   d. What is the free energy change for the dissolution of $Na_2B_4O_5(OH)_4$ at this temperature?

4. Consider the Experimental Procedure, Part C.
   a. How many saturated solutions of borax are to be titrated?

   b. Each saturated solution of borax is prepared at a different temperature. How many trials per saturated solution are to be completed?

5. Consider the Experimental Procedure, Part D. A graph is to be constructed from the data.
   a. What are the coordinates of the graph?

   b. How many data points are to be on the graph?

Experiment 31 *Report Sheet*

The Thermodynamics of the Dissolution of Borax

Date ________ Lab Sec. ______ Name ______________________________ Desk No. ________

A. Standardized HCl Solution

Calculation for the mass of sodium carbonate sample in Part A.3.

	Trial 1	*Trial 2*	*Trial 3*	*Trial 4*
1. Tared mass of Na_2CO_3 (g)				
2. Amount of Na_2CO_3 (mol)				
3. Amount of HCl (mol)				
4. Buret reading, *initial* (mL)				
5. Buret reading, *final* (mL)				
6. Volume of HCl added (mL)				
7. Molar concentration of HCl (mol/L)				

8. Average molar concentration of HCl (mol/L) ____________

B. Preparation of Borax Solution

Sample number	1	2	3	4	5	6
1. Volume of sample (mL)	5	5	5	5	5	5
2. Temperature of sample (°C)						

C. Analysis of Borax Test Solutions

1. Buret reading, *initial* (mL)						
2. Buret reading, *final* (mL)						
3. Volume of HCl added (mL)						

D. Data Analysis

	1	2	3	4	5	6
1. Temperature (K)						
2. $1/T$ (K^{-1})						
3. Amount of HCl used (mol)						
4. Amount of $B_4O_5(OH)_4^{2-}$ (mol)						
5. $[B_4O_5(OH)_4^{2-}]$ (mol/L)						
6. Molar solubility of borax (mol/L)						
7. Solubility product, K_{sp}						
8. $\ln K_{sp}$						

Instructor's approval of plotted data ______________

From the data plot determine:

$-\Delta H°/R$ ______________

$\Delta S°/R$ ______________

$\Delta H°$ (kJ/mol) ______________

$\Delta S°$ (J/mol•K) ______________

$\Delta G°$ (kJ), at 298 K ______________

Laboratory Questions

Circle the questions that have been assigned.

1. Part A.1. No desiccator is available. The sodium carbonate is cooled to room temperature, but the humidity in the room is high. How will this affect the reported molar concentration of the hydrochloric acid solution in Part A.7 . . . too high, too low, or unaffected?
2. Part A.5. The endpoint in the titration is "overshot."
   a. Is the reported molar concentration of the hydrochloric acid solution too high or too low? Explain.
   b. As a result of this poor titration technique, is the reported number of moles of $B_4O_5(OH)_4^{2-}$ in each of the analyses (Part C.2) too high or too low? Explain.
3. Part B.2. The solid borax reagent is contaminated with a water-soluble substance that does not react with hydrochloric acid. How will this contamination affect the reported solubility product of borax? Explain.
4. Part B.4. For the borax solution at 48°C, no solid borax is present in the test tube. Five milliliters is transferred to the corresponding calibrated test tube and subsequently titrated with the standardized hydrochloric acid solution. How will this oversight in technique affect the reported K_{sp} of borax at 48°C . . . too high, too low, or unaffected? Explain
5. Part B.5. A "little more" than 5 mL of a saturated solution is transferred to the corresponding calibrated test tube and subsequently titrated with the standardized hydrochloric acid solution. How will this "generosity" affect the reported molar solubility of borax for that sample . . . too high, too low, or unaffected? Explain.
6. Part C.2. The saturated solution of borax is diluted with "about" 25 mL of deionized water. How does this dilution affect the reported number of moles of $B_4O_5(OH)_4^{2-}$ in the saturated solution . . . too high, too low, or unaffected? Explain.

Experiment 32

Galvanic Cells, the Nernst Equation

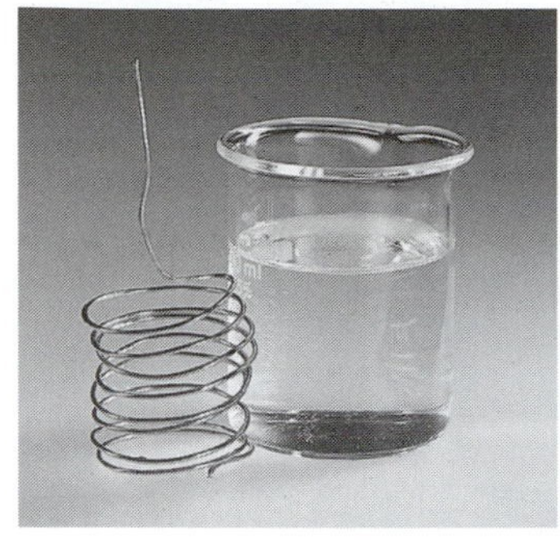

Copper metal spontaneously oxidizes to copper(II) ion in a solution containing silver ion. Silver metal crystals form on the surface of the copper metal.

OBJECTIVES

- To measure the relative reduction potentials for a number of redox couples
- To develop an understanding of the movement of electrons, anions, and cations in a galvanic cell
- To study factors affecting cell potentials
- To estimate the concentration of ions in solution using the Nernst equation

TECHNIQUES

The following techniques are used in the Experimental Procedure

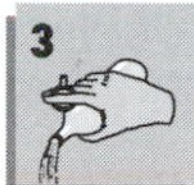

INTRODUCTION

When iron corrodes, a change in the oxidation number of the iron atoms occurs; when gasoline burns, a change in the oxidation number of the carbon atoms occurs. A change in the oxidation number of an atom in a chemical reaction is the result of an exchange of electrons between the reactants. A chemical reaction that involves the transfer of electrons from one substance to another is an **oxidation–reduction (redox) reaction** (see Experiment 14).

Experimentally, when copper wire is placed into a silver ion solution (see opening photo), copper atoms *spontaneously* lose electrons (copper atoms are oxidized) to the silver ions (which are reduced). Silver ions migrate to the copper atoms to pick up electrons and form silver atoms at the copper metal/solution **interface;** the copper ions that form then move into the solution away from the interface. The overall reaction that occurs at the interface is

Interface: the boundary between two phases; in this case, the boundary that separates the solid metal from the aqueous solution

$$Cu(s) + 2\,Ag^+(aq) \rightarrow 2\,Ag(s) + Cu^{2+}(aq) \quad (32.1)$$

This redox reaction can be divided into an oxidation and a reduction half-reaction. Each half-reaction, called a **redox couple,** consists of the reduced state and the oxidized state of the substance.

Redox couple: an oxidized and reduced form of an ion/substance appearing in a reduction or oxidation half-reaction, generally associated with galvanic cells

$$Cu(s) \rightarrow Cu^{2+}(aq) + 2\,e^- \qquad \text{oxidation half-reaction (redox couple)} \quad (32.2)$$

$$2\,Ag^+(aq) + 2\,e^- \rightarrow 2\,Ag(s) \qquad \text{reduction half-reaction (redox couple)} \quad (32.3)$$

A **galvanic cell** is designed to take advantage of this *spontaneous transfer of electrons.* Instead of electrons being transferred at the interface between the copper metal and the silver ions in solution, a galvanic cell separates the copper metal from the sil-

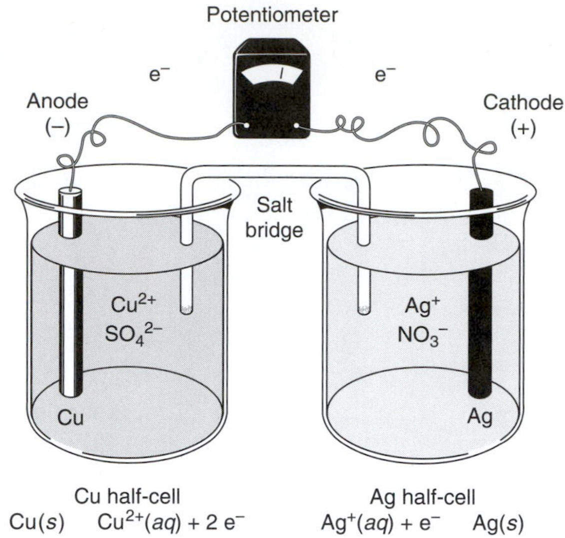

Figure 32.1 Schematic diagram of a galvanic cell.

ver ions and forces the electrons to pass externally through a wire, an external circuit. Figure 32.1 is a schematic diagram of a galvanic cell setup for these two redox couples.

Half-cell: a part of the galvanic cell that hosts a redox couple

External circuit: the movement of charge as electrons through a wire connecting the two half-cells, forming one half of the electrical circuit in a galvanic cell

Salt bridge: paper moistened with a salt solution, or an inverted tube containing a salt solution, that bridges two half-cells to complete the solution part of an electrical circuit

Internal circuit: the movement of charge as ions through solution from one half-cell to the other, forming one half of the electrical circuit in a galvanic cell

The two redox couples are placed in separate compartments, called **half-cells.** Each half-cell consists of an electrode, usually the metal (reduced state) of the redox couple, and a solution containing the corresponding cation (oxidized state) of the redox couple. The electrodes of the half-cells are connected by a wire; this is where the electrons flow, providing current for the **external circuit.**

A **salt bridge,** which connects the two half-cells, completes the construction of the galvanic cell (and the circuit). The salt bridge permits limited movement of ions from one half-cell to the other, the **internal circuit,** so that when the cell operates, electrical neutrality is maintained in each half-cell. For example, when copper metal oxidizes to copper(II) ions in the Cu^{2+}/Cu half-cell, either anions must enter or copper ions must leave the half-cell to maintain neutrality. Similarly, when silver ions reduce to form silver metal in its half-cell, either anions must leave or cations must enter its half-cell to maintain neutrality.

The electrode at which reduction occurs is called the **cathode;** the electrode at which oxidation occurs is called the **anode.** Because oxidation releases electrons to the electrode to provide a current in the external circuit, the anode is designated the *negative* electrode in a galvanic cell. The reduction process draws electrons from the circuit and supplies them to the ions in solution; the cathode is the *positive* electrode. This sign designation allows us to distinguish the anode from the cathode in a galvanic cell.

Different metals, such as copper and silver, have different tendencies to oxidize; similarly, their ions have different tendencies to undergo reduction. The **cell potential** of a galvanic cell is due to the difference in tendencies of the two metals to oxidize (lose electrons) or of their ions to reduce (gain electrons). Commonly, a measured **reduction potential,** the tendency for a substance to gain electrons, is the value used to identify the relative ease of reduction for a half-reaction.

A **potentiometer,** placed in the external circuit between the two electrodes, measures the cell potential, E_{cell}, a value that represents the *difference* between the tendencies of the metal ions in their respective half-cells to undergo reduction (i.e., the difference between the reduction potentials of the two redox couples).

For the copper and silver redox couples, we can represent their reduction potentials as $E_{Cu^{2+},Cu}$ and $E_{Ag^+,Ag}$, respectively. The cell potential being the difference of the two reduction potentials is therefore

$$E_{cell} = E_{Ag^+,Ag} - E_{Cu^{2+},Cu} \tag{32.4}$$

Experimentally, silver ion has a greater tendency than does copper ion to be in the reduced (metallic) state; therefore, Ag^+ has a greater (more positive) reduction potential. As the cell potential, E_{cell}, is measured as a positive value, $E_{Ag^+,Ag}$ is placed before $E_{Cu^{2+},Cu}$ in Equation 32.4.

Silver jewelry is longer lasting than copper jewelry; therefore silver has a higher tendency to be in the reduced state, a higher reduction potential

The measured cell potential corresponds to the **standard cell potential** when the concentrations of all ions are 1 mol/L and the temperature of the solutions is 25°C.

The *standard* reduction potential for the Ag^+(1 *M*)/Ag redox couple, $E°_{Ag^+,Ag}$, is +0.80 V, and the *standard* reduction potential for the Cu^{2+}(1 *M*)/Cu redox couple, $E°_{Cu^{2+},Cu}$, is +0.34 *V*. Theoretically, a potentiometer should show the difference between these two potentials, or, at standard conditions,

$$E°_{cell} = E°_{Ag^+,Ag} - E°_{Cu^{2+},Cu} = +0.80\text{ V} - (+0.34\text{ V}) = +0.46\text{V} \tag{32.5}$$

Deviation from the theoretical value may be the result of surface activity at the electrodes or activity of the ions in solution.

In Part A of this experiment, several cells are "built" from a selection of redox couples and data are collected. From an analysis of the data, the relative reduction potentials for the redox couples are determined and placed in an order of decreasing reduction potentials.

In Part B, the formations of the complex, $[Cu(NH_3)_4]^{2+}$, and the precipitate, CuS, are used to change the concentration of $Cu^{2+}(aq)$ in the Cu^{2+}/Cu redox couple. The observed changes in the cell potentials are interpreted.

The Nernst equation is applicable to redox systems that are *not* at standard conditions, most often when the concentrations of the ions in solution are *not* 1 mol/L. At 25°C, the measured cell potential, E_{cell}, is related to $E°_{cell}$ and ionic concentrations by

$$E_{cell} = E°_{cell} - \frac{0.0592}{n} \log Q \tag{32.6}$$

where *n* represents the moles of electrons exchanged according to the cell equation. For the copper/silver cell, $n = 2$; two electrons are lost per copper atom and two electrons are gained per two silver ions (see Equations 32.1–32.3). For dilute ionic concentrations, the *reaction quotient, Q,* equals the **mass action expression** for the cell reaction. For the copper/silver cell (see Equation 32.1),

Mass action expression: the product of the molar concentrations of the products divided by the product of the molar concentrations of the reactants, each concentration raised to the power of its coefficient in the balanced cell equation

$$Q = \frac{[Cu^{2+}]}{[Ag^+]^2}$$

In Part C of this experiment, we study, in depth, the effect that changes in concentration of an ion have on the potential of the cell. The cell potentials for a number of zinc/copper redox couples are measured in which the copper ion concentrations are varied but the zinc ion concentration is maintained constant.

$$Zn(s) + Cu^{2+}(aq) \rightarrow Cu(s) + Zn^{2+}(aq)$$

The Nernst equation for this reaction is

$$E_{cell} = E°_{cell} - \frac{0.0592}{2} \log \frac{[Zn^{2+}]}{[Cu^{2+}]} \tag{32.7}$$

Rearrangement of this equation yields an equation for a straight line:

$$\underset{y\ =}{E_{cell}} = \underbrace{E°_{cell} - \frac{0.0592}{2} \log [Zn^{2+}]}_{b} + \underset{m}{\frac{0.0592}{2}} \underset{x}{\log [Cu^{2+}]} \tag{32.8}$$

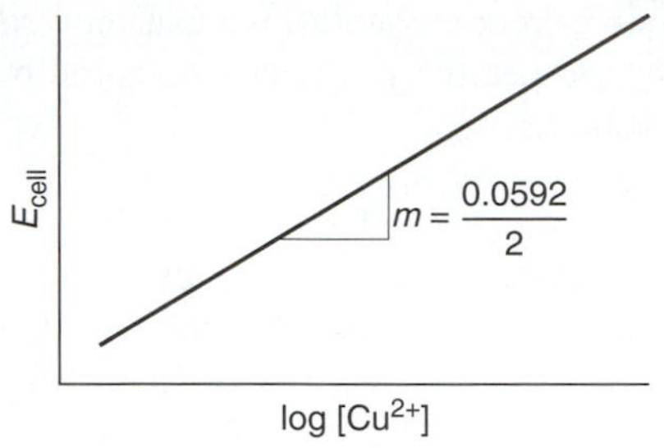

Figure 32.2 The variation of $E°_{cell}$ versus the log [Cu^{2+}].

A plot of E_{cell} versus log [Cu^{2+}] for solutions of known copper ion concentrations has a slope of 0.0592/2 and an intercept b that not only includes the constants in Equation 32.8, but also the inherent characteristics of the cell and potentiometer (Figure 32.2).

The E_{cell} of a solution with an *unknown* copper ion concentration is then measured; from the linear plot, its concentration is determined.

Experimental Procedure

Procedure Overview: The cell potentials for a number of galvanic cells are measured, and the redox couples are placed in order of decreasing reduction potentials. The effects of changes in ion concentrations on cell potentials are observed and analyzed.

Perform the experiment with a partner. At each circled superscript (1–12) in the procedure, *stop,* and record your observation on the Report Sheet. Discuss your observations with your lab partner and your instructor.

A. Reduction Potentials of Several Redox Couples

1. **Collect the Electrodes, Solutions, and Equipment.** Obtain four, small (~50 mL) beakers and fill them three-fourths full of the 0.1 *M* solutions as shown in Figure 32.3. Share these solutions with other chemists/groups of chemists in the laboratory.

   Polish strips of copper, zinc, magnesium, and iron metal with steel wool, rinse briefly with dilute (~1 *M*) HNO_3 **(caution!)**, and rinse with deionized water. These polished metals, used as electrodes, should be bent to extend over the lip of their respective beakers. Check out a potentiometer (or a voltmeter) with two electrical wires (preferably a red and black wire) attached to "alligator" clips.

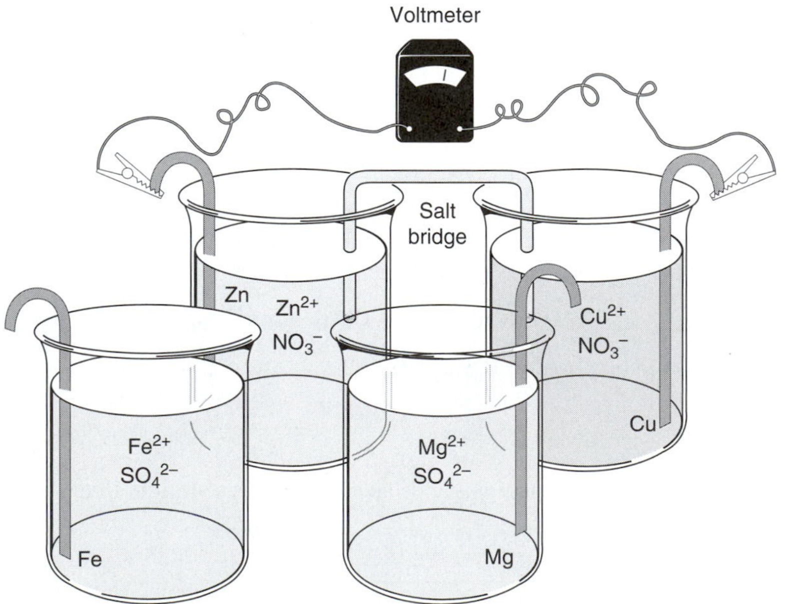

Figure 32.3 Setup for measuring the cell potentials of six galvanic cells.

2. **Set Up the Copper/Zinc Cell.** Place a Cu strip (electrode) in the $Cu(NO_3)_2$ solution and a Zn strip (electrode) in the $Zn(NO_3)_2$ solution. Roll and flatten a piece of filter paper; wet the filter paper with a 0.1 *M* KNO_3 solution. Fold and insert the ends of the filter paper into the solutions in the two beakers; this is the salt bridge. Connect one electrode to the negative terminal of the potentiometer and the other to the positive terminal.[1]
3. **Determine the Copper/Zinc Cell Potential.** If the needle on the potentiometer swings the "wrong way" (i.e., if it reads a negative potential), reverse the connections to the electrodes. Read and record the (positive) cell potential. Identify the metal strips that serve as the cathode (positive terminal) and the anode. Write an equation for the half-reaction occurring at each electrode. Combine the two half-reactions to write the equation for the cell reaction.①

Chemists often use the "red, right, plus" rule in connecting the red wire to the right-side positive electrode (cathode) of the galvanic cell

4. **Repeat for the Remaining Cells.** Determine the cell potentials for all possible galvanic cells that can be constructed from the four redox couples. Refer to the Report Sheet for the various galvanic cells. Prepare a "new" salt bridge for each galvanic cell.②
5. **Determine the Relative Reduction Potentials.** Assuming the reduction potential of the Zn^{2+}(0.1 *M*)/Zn redox couple is −0.79 V, determine the reduction potentials of all other redox couples.[2]③
6. **Determine the Reduction Potential of the Unknown Redox Couple.** Place a 0.1 *M* solution and electrode obtained from your instructor in a small beaker. Determine the reduction potential, relative to the Zn^{2+}(0.1 *M*)/Zn redox couple, for your unknown redox couple.④

B. Effect of Concentration Changes on Cell Potential

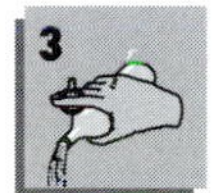

1. **Effect of Different Molar Concentrations.** Set up the galvanic cell shown in Figure 32.4, using 1 *M* $CuSO_4$ and 0.001 *M* $CuSO_4$ solutions. Immerse a polished copper electrode in each solution. Prepare a salt bridge to connect the two redox couples. Measure the cell potential. Determine the anode and the cathode. Write an equation for the reaction occurring at each electrode.⑤
2. **Effect of Complex Formation.** Add 2–5 mL of 6 *M* NH_3 to the 0.001 *M* $CuSO_4$ solution, until any precipitate redissolves.[3] (**Caution:** *Do not inhale* NH_3.) Observe and record any changes in the half-cell and the cell potential.⑥
3. **Effect of Precipitate Formation.** Add 2–5 mL of 0.2 *M* Na_2S to the 0.001 *M* $CuSO_4$ solution, now containing the added NH_3. What is observed in the half-cell and what happens to the cell potential? Record your observations.⑦

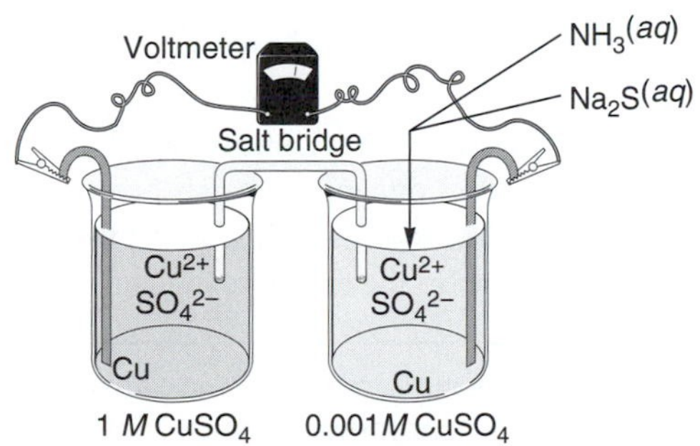

Figure 32.4 Setup for measuring the cell potential of a Cu^{2+} concentration cell.

[1] You have now combined two half-cells to form a galvanic cell.

[2] *Note:* These are *not* standard reduction potentials because 1 *M* concentrations of cations at 25°C are not used.

[3] Copper ion forms a complex with ammonia: $Cu^{2+}(aq) + 4\ NH_3(aq) \rightarrow [Cu(NH_3)_4]^{2+}(aq)$

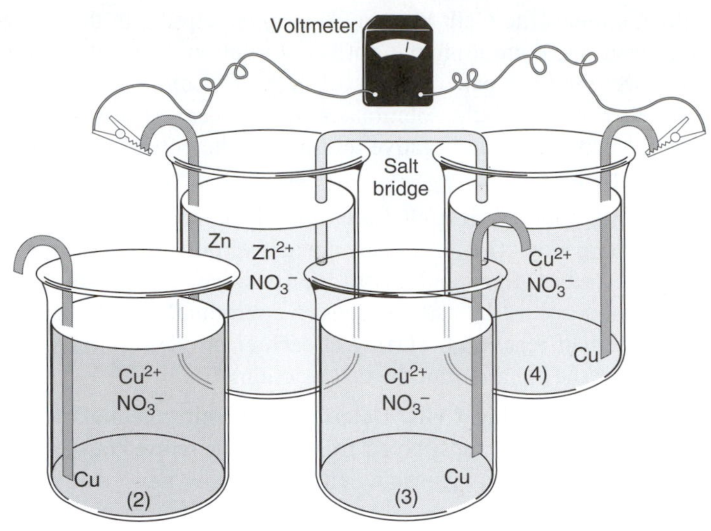

Figure 32.5 Setup to measure the effect that diluted solutions have on cell potentials.

C. The Nernst Equation and an Unknown Concentration

If an expanded scale potentiometer (0–2 or 0–1.5 V) is not available, *only* calculate the theoretical potentials for each solution in Parts C.1-C.3.[4] Complete Parts C.4 and C.5 as suggested.

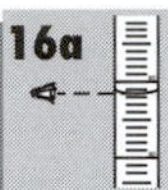

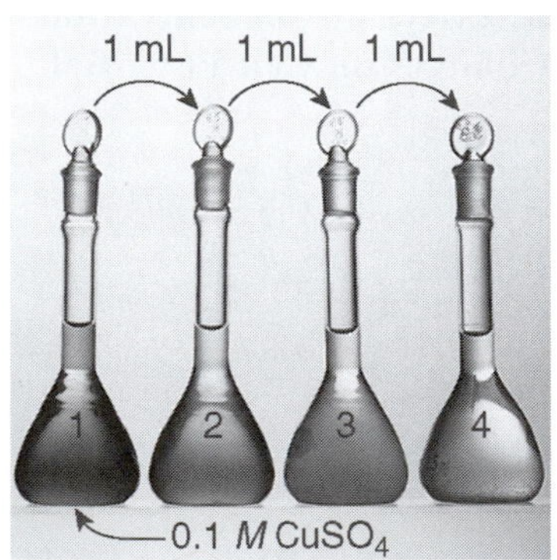

Successive quantitative dilution, starting with 0.1 *M* $CuSO_4$.

Appendix C

1. **Prepare the Diluted Solutions.** Prepare Solutions 1 through 4 as shown in the margin figure using a 1-mL pipet and 100-mL volumetric flasks.[5] Be sure to rinse the pipet with the more concentrated solution before making the transfer. Use deionized water for dilution "to the mark" in the volumetric flasks. Calculate the molar concentration of each solution and record.⑧
2. **Measure and Calculate the Cell Potential for Solution 4.** Set up the experiment as shown in Figure 32.5, using small (~50 mL) beakers.

   The Zn^{2+}/Zn redox couple is the reference half-cell for this part of the experiment. Connect the two half-cells with a "new" salt bridge. Connect the electrodes to the potentiometer and record the potential difference, E_{cell}.⑨ Calculate the *theoretical* cell potential. (Use a table of standard reduction potentials and the Nernst equation.)⑩
3. **Measure and Calculate the Cell Potentials for Solutions 3 and 2.** Repeat Part C.2 with Solutions 3 and 2, respectively. A freshly prepared salt bridge is required for each cell.
4. **Plot the Data.** Plot $E_{cell, expt}$ *and* $E_{cell, calc}$ (ordinate) versus log $[Cu^{2+}]$ (abscissa) on the *same* piece of linear graph paper for the *four* concentrations of $Cu(NO_3)_2$ (see data from Part A.3 for the potential of Solution 1). Have your instructor approve your graph.⑪
5. **Determine the Concentration of the Unknown.** Obtain a $Cu(NO_3)_2$ solution with an "unknown" copper ion concentration from your instructor and set up a galvanic cell. Determine E_{cell} as in Part C.2. Using the graph, determine the unknown copper ion concentration in the solution.⑫

Disposal: Dispose of the waste zinc, copper, magnesium, and iron solutions in the "Waste Metal Solutions" container. Return the metals to appropriately marked containers.

CLEANUP: Rinse the beakers twice with tap water and twice with deionized water. Discard the rinses in the "Waste Metal Solutions" container.

[4]Part C will be a "dry lab" only if the expanded scale potentiometer is not available.
[5]Share these prepared solutions with other chemists/groups of chemists in the laboratory.

Experiment 32 *Prelaboratory Assignment*

Galvanic Cells, the Nernst Equation

Date __________ Lab Sec. ______ Name ______________________________ Desk No. __________

1. In a galvanic cell
   a. oxidation occurs at the (name of electrode) ____________________

   b. the cathode is the (sign) electrode ____________________

   c. cations flow toward the (name of electrode) ____________________

   d. electrons flow from the (name of electrode) to (name of electrode) ____________________

2. a. What is the purpose of a salt bridge? Explain.

   b. How is the salt bridge prepared in this experiment?

3. Consider a galvanic cell consisting of the Bi^{3+}/Bi ($E° = +0.20$ V) and the Ag^{+}/Ag ($E° = +0.80$ V) redox couples.
   a. Write the half-equation that occurs at the cathode:

   b. Write the half-reaction that occurs at the anode:

   c. Write the equation for the cell reaction.

   d. What is the standard cell potential, $E°_{cell}$, for the galvanic cell?

4. a. Use the Nernst equation to determine the cell potential, E_{cell}, of the galvanic cell consisting of the following redox couples:

Ag^+ (0.010 M) + $e^- \rightleftharpoons$ Ag $E° = +0.80$ V

Bi^{3+} (0.010 M) + 3 $e^- \rightleftharpoons$ Bi $E° = +0.20$ V

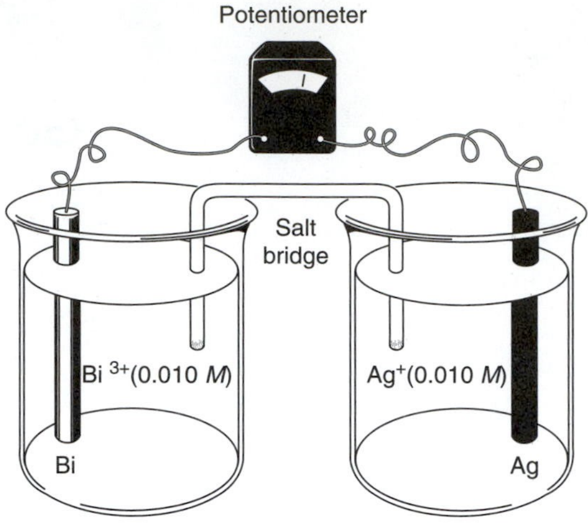

b. If the E_{cell} measures 0.50 V, what is a more accurate molar concentration of Bi^{3+}, assuming the molar concentration of Ag^+ is accurate?

5. Refer to Equation 32.8. Assuming that "b" equals 0.76 V,
a. what is the [Cu^{2+}] if the measured cell potential is 0.70 V?

b. what should be the cell potential if the [Cu^{2+}] is 1.0×10^{-3} mol/L?

*6. The extent of corrosion in the steel reinforcing (rebar) of concrete is measured by the galvanic cell shown in the diagram of the instrument. The half-cell of the probe (the reference electrode) is usually a AgCl/Ag redox couple:

AgCl + $e^- \rightarrow$ Ag + Cl^- (1.0 M) $E° = +0.23$ V

Corrosion is said to be "severe" if the cell potential is measured at greater than 0.41 V. Under these conditions, what is the iron(II) concentration on the rebar?

Fe^{2+} + 2 $e^- \rightarrow$ Fe $E° = -0.44$ V

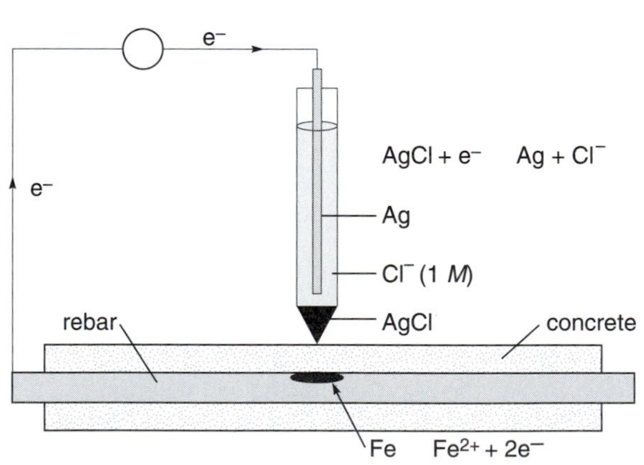

Experiment 32 *Report Sheet*

Galvanic Cells, the Nernst Equation

Date __________ Lab Sec. ______ Name ______________________________________ Desk No. __________

A. Reduction Potentials of Several Redox Couples

Fill in the following table with your observations and interpretations from the galvanic cells.

Galvanic Cell	Measured E_{cell}	Anode	Equation for Anode Reaction	Cathode	Equation for Cathode Reaction
①Cu–Zn					
②Cu–Mg					
Cu–Fe					
Zn–Mg					
Fe–Mg					
Zn–Fe					

Write balanced equations for the six cell reactions.

Compare the *sum* of the Zn–Mg and Cu–Mg cell potentials with the Zn–Cu cell potential. Explain.

Compare the *sum* of the Zn–Fe and Zn–Mg cell potentials with the Fe–Mg cell potential. Explain.

5. Arrange the four redox couples in order of decreasing (measured) reduction potentials. List the reduction potential for each redox couple relative to that of the Zn^{2+}(0.1 *M*)/Zn couple, which is −0.79 V. Use a table of standard reduction potentials and the Nernst equation to *calculate* the reduction potentials for each of these redox couples.

Redox Couple	Reduction Potential (measured)	Reduction Potential (calculated)	% Error
③			

*Calculation of reduction potential.

6. ④Reduction potential of the unknown redox couple: ____________________

B. Effect of Concentration Changes on Cell Potential

1. ⑤Cell potential of "concentration cell": ____________________

   Anode reaction: ____________________

   Cathode reaction: ____________________

   Explain *why* a potential is recorded.

2. ⑥Cell potential from complex formation: ____________________

   Observation of solution in half-cell.

   Explain *why* the potential changes with the addition of $NH_3(aq)$.

3. ⑦Cell potential from precipitate formation: ____________________

   Observation of solution in half-cell.

   Explain *why* the potential changes with the addition of Na_2S.

C. The Nernst Equation and an Unknown Concentration

1. Complete the following table with the concentrations of the $Cu(NO_3)_2$ solutions and the measured cell potentials. Use a table of standard reduction potentials and the Nernst equation to *calculate* the E_{cell}.

Solution Number	⑧Concentration of $Cu(NO_3)_2$	log $[Cu^{2+}]$	⑨E_{cell} (measured)	⑩E_{cell} (calculated)
1	0.1 mol/L	−1	______	______
2	______	______	______	______
3	______	______	______	______
4	______	______	______	______

2. ⑪Instructor's approval of graph: ______________________________

   Account for any significant difference between the measured and calculated E_{cell} values.

3. ⑫E_{cell} for the solution of unknown concentration: ____________________

   Molar concentration of Cu^{2+} in the unknown: ____________________

Laboratory Questions

Circle the questions that have been assigned.

1. Part A.3. The filter paper salt bridge is *not* wetted with the 0.1 *M* KNO_3 solution. How will this affect the measured potential of the cell? Explain.
2. Part A.3. A positive potential is recorded when the copper electrode is the positive electrode. Is the copper electrode the cathode or the anode of the cell? Explain.
3. Part B.2. How would the cell potential have been affected if the $NH_3(aq)$ had been added to the 1 *M* $CuSO_4$ solution instead of the 0.0010 *M* $CuSO_4$ solution of the cell? Explain.
4. Part B.3. The cell potential increased (compared to Part B.2) with the addition of the Na_2S solution to the 0.001 *M* $CuSO_4$ solution. Explain.
5. Part C. As the concentration of the copper(II) increased from Solution 4 to Solution 1, what happened to the measured cell potentials? Explain.
6. Part C. Suppose the 0.1 *M* Zn^{2+} solution had been diluted (instead of the Cu^{2+} solution), what would have happened to the measured cell potentials? Explain.

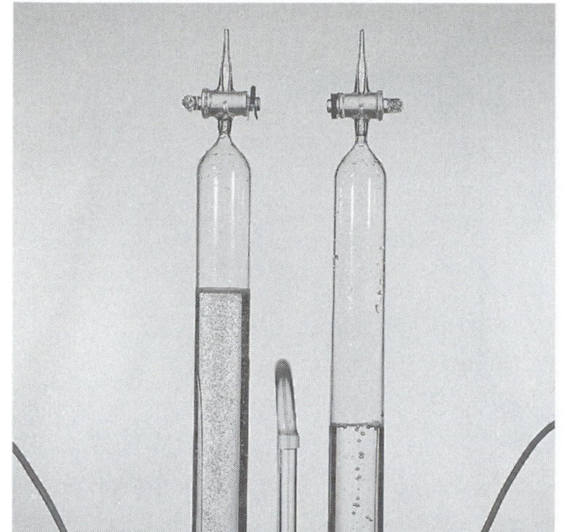

A 1 : 2 mole (and volume) ratio of oxygen (left) to hydrogen (right) is produced from the electrolysis of water.

Experiment 33

Electrolytic Cells, Avogadro's Number

OBJECTIVES

- To identify the reactions occurring at the anode and cathode during the electrolysis of various aqueous salt solutions
- To determine Avogadro's number and the Faraday constant

TECHNIQUES

The following techniques are used in the Experimental Procedure

INTRODUCTION

Electrolysis: Use of electrical energy to cause a chemical reaction to occur

Electrolytic cell: an apparatus used for an electrolysis reaction

Electrolysis processes are very important in achieving high standards of living. The industrial production of metals such as aluminum and magnesium, and nonmetals such as chlorine and fluorine, occurs in electrolytic cells. The highly refined copper metal required for electrical wiring is obtained through an electrolytic process.

In an **electrolytic cell** the input of an electric current causes an otherwise nonspontaneous oxidation–reduction reaction, a nonspontaneous transfer of electrons, to occur. For example, sodium metal, a very active metal, and chlorine gas are prepared industrially by the electrolysis of molten sodium chloride. Electrical energy is supplied to a molten NaCl system (Figure 33.1) by a direct current (dc) power source (set at an appropriate voltage) across the electrodes of an electrolytic cell.

The electrical energy causes the reduction of the sodium ion, Na^+, at the cathode and oxidation of the chloride ion, Cl^-, at the anode. Because cations migrate to the cathode and anions migrate to the anode, the cathode is the negative electrode (opposite charges attract) and the anode is designated the positive electrode.

$$\text{cathode } (-) \text{ reaction:} \quad Na^+ (l) + e^- \rightarrow Na(l)$$
$$\text{anode } (+) \text{ reaction:} \quad 2\,Cl^-(l) \rightarrow Cl_2(g) + 2\,e^-$$

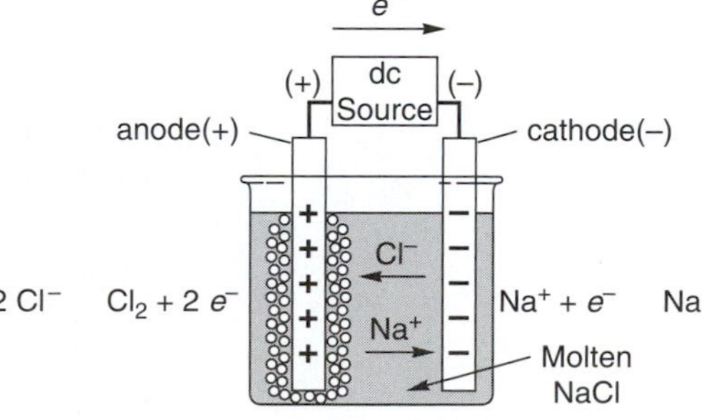

Figure 33.1 Schematic diagram of the electrolysis of molten sodium chloride.

Electrolysis reactions also occur in aqueous solutions. For example, in the electrolysis of an aqueous copper(II) bromide, $CuBr_2$, solution, copper(II) ions, Cu^{2+}, are reduced at the cathode and bromide ions, Br^-, are oxidized at the anode (Figure 33.2).

$$\text{cathode } (-) \text{ reaction:} \quad Cu^{2+}(aq) + 2\,e^- \rightarrow Cu(s)$$
$$\text{anode } (+) \text{ reaction:} \quad 2\,Br^-(aq) \rightarrow Br_2(l) + 2\,e^-$$
$$\text{cell reaction:} \quad Cu^{2+}(aq) + 2\,Br^-(aq) \rightarrow Cu(s) + Br_2(l)$$

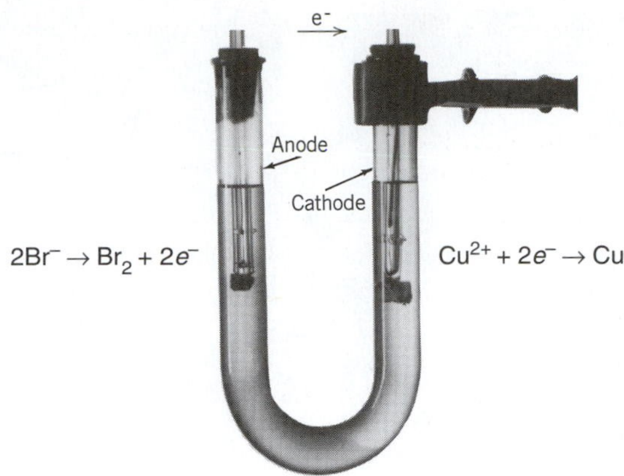

Figure 33.2 Electrolysis of a copper(II) bromide solution; the anode is on the left and the cathode is on the right.

In an aqueous solution, however, the reduction of water at the cathode (the negative electrode) and the oxidation of water at the anode (the positive electrode) are also possible reactions.

cathode (−) reaction for water: $2\ H_2O(l) + 2\ e^- \rightarrow H_2(g) + 2\ OH^-(aq)$ (33.1)

anode (+) reaction for water: $2\ H_2O(l) \rightarrow O_2(g) + 4\ H^+(aq) + 4\ e^-$ (33.2)

If the reduction of water occurs at the cathode, hydrogen gas is evolved and the solution near the cathode becomes basic as a result of the production of hydroxide ion. If oxidation of water occurs at the anode, oxygen gas is evolved and the solution near the anode becomes acidic. The acidity (or basicity) near the respective electrodes can be detected with litmus paper or another appropriate acid–base indicator.

When two or more competing reduction reactions are possible at the cathode, the reaction that occurs most easily (the one with the *higher* reduction potential) is the one that usually occurs. Conversely, for two or more competing oxidation reactions at the anode, the reaction that takes place most easily (the one with the *higher* oxidation potential or the *lower* reduction potential) is the one that usually occurs.

In the electrolysis of the aqueous copper(II) bromide solution, Cu^{2+} has a higher reduction potential than H_2O and is therefore preferentially reduced at the cathode; Br^- has a greater tendency to be oxidized than water, and so Br^- is oxidized.

Very few metals exist as metals in the natural state (gold being an exception), but rather most exist as cations in the earth's crust. Therefore, metal cations from the various ores must be reduced for the production of metals for commercial or industrial use. Some metals (such as iron and zinc) are reduced using a simple redox reaction, but others (such as aluminum and magnesium) require electrolytic processes. The molten salts (to avoid water interferences) of the latter (more active) metals are electrolyzed with the metals being electroplated at the cathode. By careful control of the applied potential in an electrolytic cell, metal cations of a mixture can be selectively electroplated, thus making electrolysis a process of refining the metal to a very high degree of purity. Copper metal is electrolytically refined for use in electrical wiring.

In Part A of this experiment, a number of aqueous salt solutions, using different electrodes, are electrolyzed. The anode and cathode are identified and the products that are formed at each electrode are identified.

In Part B, a quantitative investigation of the electrolytic reduction of copper(II), $Cu^{2+}(aq)$, is used to determine Avogadro's number and the Faraday constant.

$$Cu^{2}(aq) + 2\ e^- \rightarrow Cu(s) \qquad (33.3)$$

1 faraday = 1 mol e^- = 96,485 coulombs

Two moles of electrons (or 2 **faradays**) reduce one mole of Cu^{2+}; therefore, a mass measurement of the copper cathode before and after the electrolysis determines the moles of copper that electroplate. This in turn is used to calculate the moles of electrons that pass through the cell.

$$\text{moles of electrons} = \text{mass Cu} \times \frac{\text{mol Cu}}{63.54\text{ g}} \times \frac{2\text{ mol }e^-}{\text{mol Cu}} \tag{33.4}$$

The actual *number* of electrons that pass through the cell is calculated from the electrical current, measured in amperes (= coulombs/second), that passes through the cell for a recorded time period (seconds). The total charge (**coulombs,** C) that passes through the cell is

Coulomb: SI base unit for electrical charge

$$\text{number of coulombs} = \frac{\text{coulombs}}{\text{second}} \times \text{seconds} \tag{33.5}$$

As the charge of one electron equals 1.60×10^{-19} C, the number of electrons that pass through the cell can be calculated.

$$\text{number of electrons} = \text{number of coulombs} \times \frac{\text{electron}}{1.60 \times 10^{-19}\text{ C}} \tag{33.6}$$

Therefore, since the number of electrons (Equation 33.6) and the moles of electrons (Equation 33.4) can be separately determined, Avogadro's number is calculated as

$$\text{Avogadro's number} = \frac{\text{number of electrons}}{\text{mole of electrons}} \tag{33.7}$$

In addition, the number of coulombs (Equation 33.5) per mole of electrons (Equation 33.4) equals the Faraday constant. With the available data, the Faraday constant can also be calculated:

$$\text{Faraday constant} = \frac{\text{number of coulombs}}{\text{mole of electrons}} \tag{33.8}$$

Experimental Procedure

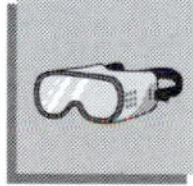

Procedure Overview: The products that result from the electrolysis of various salt solutions are observed and identified; these are qualitative measurements. An experimental setup is designed to measure quantitatively the flow of current and consequent changes in mass of the electrodes in an electrolytic cell; from these data, experimental "constants" are calculated.

The electrolysis apparatus may be designed differently than the one described in this experiment. Ask your instructor.

A. Electrolysis of Aqueous Salt Solutions

1. **Set Up the Electrolysis Apparatus.** Connect two wire leads (different colors) attached to alligator clips to a direct current (dc) power supply.[1] Mount the glass U-tube on a ring stand (see Figure 33.3). Connect the alligator clips to the corresponding electrodes, listed in Table 33.1.
2. **Electrolyze the Solutions.** Fill the U-tube three-fourths full with a solution from Table 33.1 and electrolyze for 5 minutes. Insert the corresponding electrodes into the solution. During the electrolysis, watch for any evidence of a reaction in the anode and cathode chambers.

- Does the pH of the solution change at each electrode? Test each chamber with litmus.[2] Compare the color with a litmus test on the original solution.
- Is a gas evolved at either or both electrodes?
- Look closely at each electrode. Is a metal depositing on the electrode or is the metal electrode slowly disappearing?

[1]The dc power supply can be a 9-V transistor battery.

[2]Several drops of universal indicator can be added to the solution in both chambers to detect pH changes.

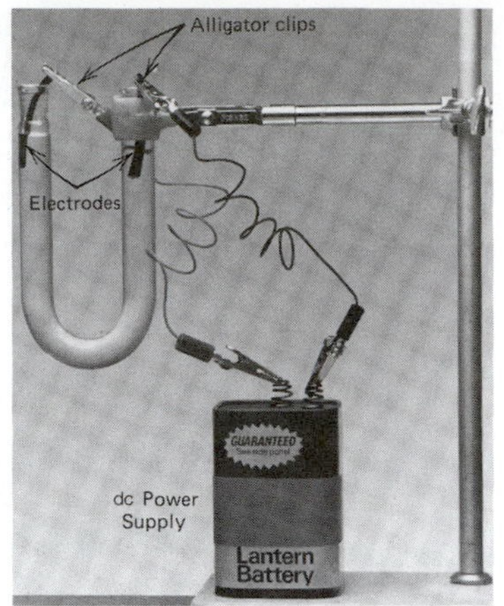

Figure 33.3 Electrolysis apparatus.

Table 33.1 Electrolytic Cells for Study

Solution*	Electrodes
2 g NaCl/100 mL	Carbon (graphite)
2 g NaBr/100 mL	Carbon (graphite)
2 g KI/100 mL	Carbon (graphite)
0.1 *M* $CuSO_4$	Carbon (graphite)
0.1 *M* $CuSO_4$	Polished copper metal strips

*Try other solutions as suggested by your laboratory instructor. Also, which electrodes will be used?

3. **Account for Your Observations.** Write the equations for the reactions occurring at the anode and cathode and for each cell reaction.

Disposal: Discard the salt solutions into the "Waste Salts" container.

CLEANUP: Rinse the U-tube twice with tap water and twice with deionized water before preparing the next solution for electrolysis. Discard each rinse in the sink.

B. Determination of Avogadro's Number and the Faraday Constant Using Faraday's Law

1. **Set Up the Apparatus in Figure 33.4.** The U-tube from Part A can again be used. The dc power supply must provide 3–5 volts (two or three flashlight batteries in series or a lantern battery); the ammeter must read from 0.2 to 1.0 A. Polish two copper metal strips with steel wool, dip them briefly in 1 *M* HNO_3 (**caution:** *do not allow skin contact*), and rinse with deionized water. Dry and measure the mass (±0.001 g) of these two "electrodes." Label the two copper strips because the mass of each will be determined after electrolysis and a mass difference of each electrode will be calculated. Connect the copper metal strip with the lesser mass to the positive terminal (anode) and the other copper strip to the negative terminal (cathode) of the dc power supply.

2. **Electrolyze the $CuSO_4$ solution.** Add 100 mL of 1.0 *M* $CuSO_4$ (in 0.1 *M* H_2SO_4) to the 150-mL beaker (or fill the U-tube) and adjust the variable resistance to its maximum value.[3] Turn on the power supply and start timing the electrolysis. During the electrolysis do not move the electrodes; this changes current flow. Adjust the current with the variable resistor to about 0.5 A and, periodically during the course of the electrolysis, readjust the current to 0.5 A.[4] Record the data.

3. **Dry and Measure the Mass.** Discontinue the electrolysis after 15 minutes. Record the exact time (minutes and seconds) of the electrolysis process. Carefully remove the electrodes (be careful not to loosen the electroplated copper metal from the cathode); rinse each electrode by carefully dipping it into a 400-mL beaker of deionized water. Air-dry, determine the mass of each electrode, and record.

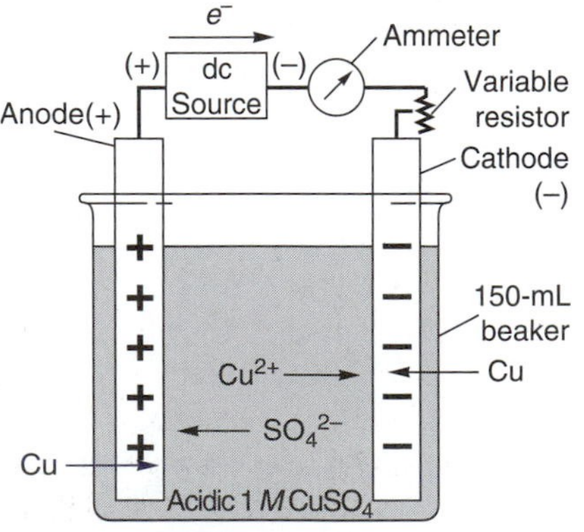

Figure 33.4 Setup for determining Avogadro's number and the Faraday constant.

Disposal: Discard the copper(II) sulfate solution into the "Waste Salts" container.

CLEANUP: Rinse the beaker twice with tap water and twice with deionized water. Discard each rinse as directed by your instructor.

[3]If a variable resistor is unavailable, record the current at 1-minute intervals and then calculate an *average* current over the entire electrolysis period.

[4]If the current is greater or less than 0.5 A, vary the time of electrolysis proportionally.

Experiment 33 *Prelaboratory Assignment*

Electrolytic Cells, Avogadro's Number

Date __________ Lab Sec. ______ Name ____ Desk No. __________

1. The standard reduction potential for the Cu^{2+}/Cu redox couple is +0.34 V; that for H_2O/H_2, OH^- at a pH of 7 is −0.41 V. For the electrolysis of a neutral 1.0 *M* $CuSO_4$ solution, write the equation for the half reaction occurring at the cathode at standard conditions.

2. In electrolytic cell
   a. reduction occurs at the (name of electrode) ____________________

   b. the anode is the (sign) electrode ____________________

   c. anions flow toward the (name of electrode) ____________________

   d. electrons flow from the (name of electrode) to (name of electrode) ____________________

3. a. Identify a chemical test(s) to determine if water is oxidized at the anode in an electrolytic cell.

   b. Similarly, identify a chemical test(s) to determine if water is reduced at the cathode in an electrolytic cell.

4. A current of 2.0 A passes through an electrolytic cell for 20 minutes.
   a. How many coulombs have passed through the cell?

   b. How many electrons and moles of electrons have passed through the cell?

   c. How many moles of oxygen gas *could* be produced for the oxidation of water at the anode?

5. Very pure copper metal is produced by the electrolytic refining of blister (impure) copper. In the cell at right, label the anode, the cathode, and the polarity (+, −) of each.

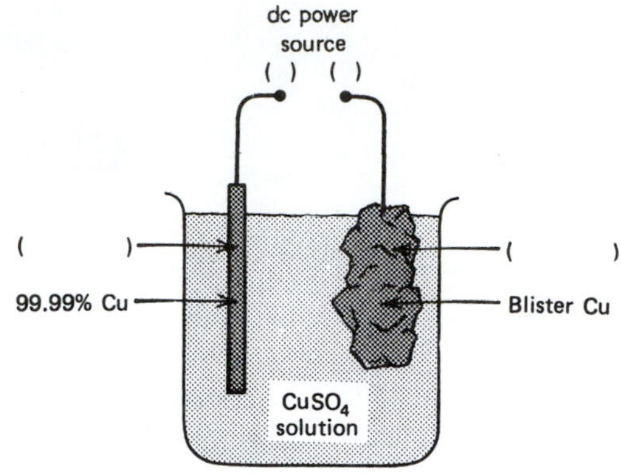

6. a. When a solution of $MgCl_2$ is electrolyzed, blue litmus turns red and a gas is evolved in the anode chamber. Write the equation for the half-reaction occurring at the anode.

b. When a solution of nickel(II) sulfate adjusted to a pH of 7 is electrolyzed, the green color of the solution becomes less intense in the cathodic chamber and gas bubbles are detected in the anodic chamber. Write the equations for the half-reactions occurring in the cathodic and the anodic chambers.

Experiment 33 *Report Sheet*

Electrolytic Cells, Avogadro's Number

Date __________ Lab Sec. ______ Name ____ Desk No. __________

A. Electrolysis of Aqueous Salt Solutions

Solution	Electrodes	Litmus Test	Gas Evolved?	Balanced Equations for Reactions
NaCl	C(*gr*)			Anode ____________________ Cathode ____________________ Cell ____________________
NaBr	C(*gr*)			Anode ____________________ Cathode ____________________ Cell ____________________
KI	C(*gr*)			Anode ____________________ Cathode ____________________ Cell ____________________
$CuSO_4$	C(*gr*)			Anode ____________________ Cathode ____________________ Cell ____________________
$CuSO_4$	Cu(*s*)			Anode ____________________ Cathode ____________________ Cell ____________________

B. Determination of Avogadro's Number and the Faraday Constant Using Faraday's Law

	Trial 1	*Trial 2*
1. *Initial* tared mass of copper anode (g)		
2. *Initial* tared mass of copper cathode (g)		
3. Time of electrolysis (s)		
4. Current (or average current) (A)		
5. *Final* tared mass of copper anode (g)		
6. *Final* tared mass of copper cathode (g)		
7. Mass of copper reduced at cathode (g)		
8. Amount of copper reduced (mol)		
9. Amount of electrons transferred (mol e^-)		
10. Coulombs passed through cell (C)		
11. Electrons passed through cell (e^-)		
12. Avogadro's number (e^-/mol e^-)		
13. Average value of Avogadro's number		
14. Literature value of Avogadro's number		
15. Percent error		
16. Faraday constant (C/mol e^-)		
17. Average Faraday constant (C/mol e^-)		
18. Literature value of Faraday constant		
19. Percent error		

Laboratory Questions

Circle the questions that have been assigned.

1. Part A.2. If nickel electrodes are used instead of the graphite electrodes, the reaction occurring at the anode may be different but the reaction occurring at the cathode would remain unchanged. Explain.
2. Part B. Repeat the calculation of Avogadro's number, using the mass loss of the anode *instead* of the mass gain of the cathode. Account for any difference in the calculated values.
3. Part B.2. If the current is recorded as being greater than it actually is, would Avogadro's number be calculated too high, too low, or be unaffected? Explain.
4. Part B.2. If the time for the electrolysis would have been extended to 30 minutes (instead of the suggested 15 minutes), would Avogadro's number be calculated too high, too low, or be unaffected? Explain.
5. For the electrolysis of an aqueous solution of $MgCl_2$, hydrogen gas and hydroxide ions are produced at the cathode. However in the electrolysis of molten $MgCl_2$, magnesium metal is produced at the cathode. Account for the difference in products produced at the cathode.

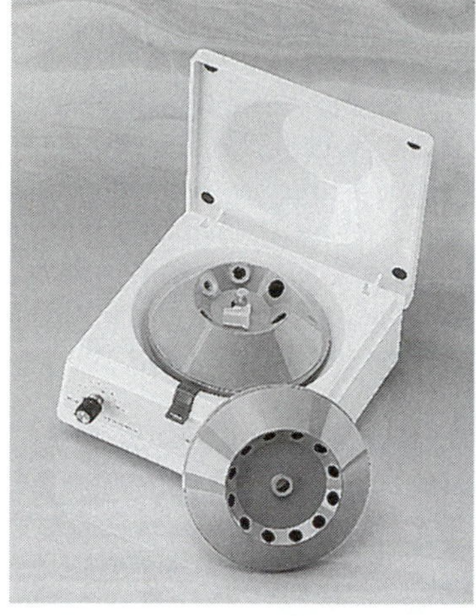

Dry Lab 4

Preface to Qualitative Analysis

A centrifuge compacts a precipitate by centrifugal force.

Some rocks have a reddish tint; others are nearly black. Table salt is white, but not all salts are white. A quick, yet simple, identification of the ions of a salt or salt mixture is often convenient. Gold prospectors were quick to identify the presence of gold and/or silver. It is the characteristic physical and chemical properties of an ion that will allow us in the next series of experiments to identify its presence in a sample. For example, the Ag^+ ion is identified as being present in a solution by its precipitation as the chloride, $AgCl(s)$. Although other cations precipitate as the chloride, silver chloride is the only one that is soluble in an ammoniacal solution.[1] Similarly, from the solubility rules (Appendix G) that you learned earlier in the course, you know that $BaSO_4$ is a white precipitate, whereas most other sulfate salts are soluble (*except* Pb^{2+}, Sr^{2+}, and $Hg_2{}^{2+}$).

With enough knowledge of the chemistry of the various ions, a unique separation and identification procedure for each ion can be developed. Some procedures are quick, one-step, tests; others are more exhausting. The experimental procedure, however, must systematically eliminate all other ions that may interfere with the specific ion test.

Many ions, such as Cl^-, Br^-, and I^-, exhibit similar chemical behavior. Anions can be sorted into groups according to similar chemical properties; the anions within each group can then be further separated and characteristically identified. Cations are similarly systematically separated and identified. A procedure that follows this pattern of analysis is called **qualitative analysis,** one with which we will become quite familiar in the next several experiments.

Qualitative analysis: a systematic procedure by which the presence (or absence) of a substance (usually a cation or anion) can be determined

The separation and identification of the ions in a mixture require the application of many chemical principles, many of which we will cite as we proceed. An *understanding* of the chemistry of precipitate formation, ionic equilibrium, acids and bases, pH, oxidation and reduction reactions, and complex formation is necessary for their successful separation and identification. To help you to understand these principles and test procedures, each experiment presents some pertinent chemical equations, but you are asked to write equations for other reactions that occur in the separation and identification of the ions.

To complete the procedures for the separation and identification of ions, you will need to practice good laboratory techniques and, in addition, develop several new techniques. The most critical techniques in qualitative analysis are the maintenance of clean glassware and the prevention of contamination of the testing reagents.

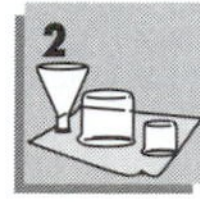

[1]This was one test procedure that prospectors for silver used in the early prospecting days.

Measuring and Mixing Test Solutions

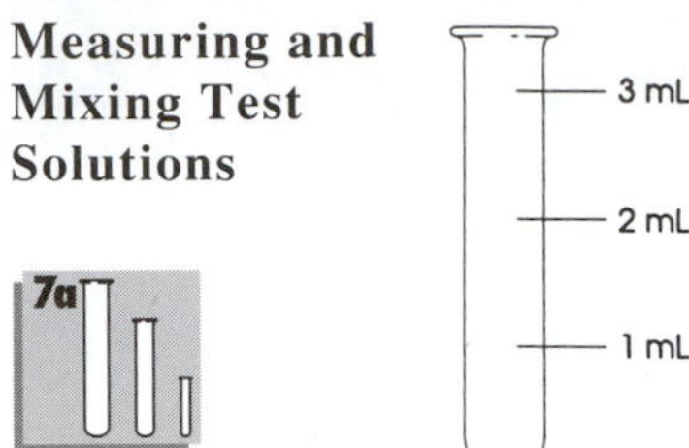

Figure D4.1 Use the 10-mL graduated cylinder to transfer 1 mL, 2 mL, and 3 mL of water to a small test tube. Mark the test tube at each mark for later reference.

Most of the testing for ions will be performed in small (~3 mL) test tubes or centrifuge tubes that fit the centrifuges used in your laboratory. Reagents will be added with dropper bottles or dropping pipets (~15–20 drops/mL; you should do a preliminary check with your dropping pipet to determine the drops/mL). If the procedure dictates the addition of 1 mL, do *not* use a graduated cylinder to transfer the 1 mL, but instead use the dropping pipet or estimate the addition of 1 mL in the (~3-mL) test tube (Figure D4.1). Do *not* mix the different dropping pipets with the various test reagents you will be using and do *not* contaminate a reagent by inserting your pipet or dropping pipet into it. Instead, if the procedure calls for a larger volume, first dispense a small amount of reagent into one of your *small* beakers or test tubes.

When mixing solutions in a test tube, agitate by tapping the side of the test tube. Break up a precipitate with a stirring rod or cork the test tube and invert, but *never* use your thumb (Figure D4.2)!

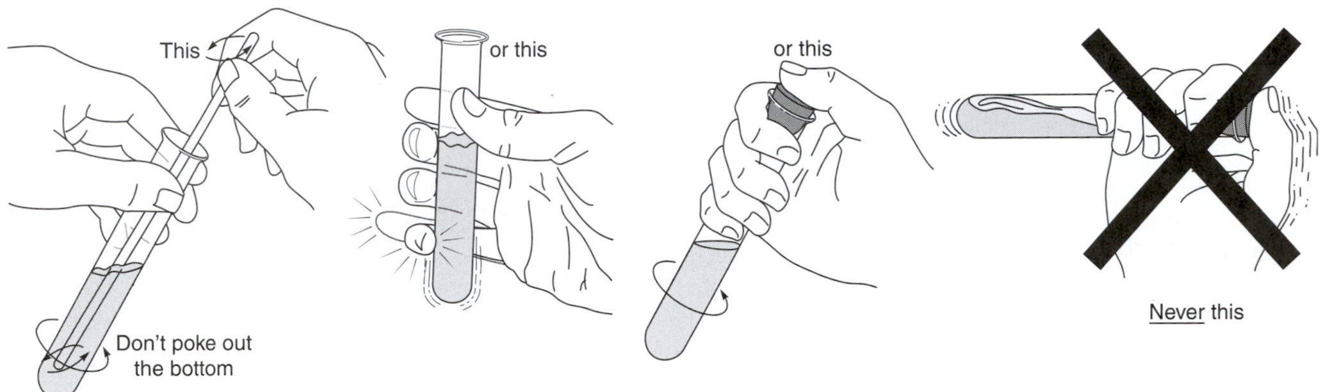

Figure D4.2 Technique for mixing solutions in a test tube.

Testing for Complete Precipitation

Precipitating reagent: a solution containing an ion(s) that, when added to a second solution, causes a precipitate to form

Oftentimes it is advisable to test the supernatant to determine if complete precipitation of an ion has occurred. After the mixture has been centrifuged, add a drop of the **precipitating reagent** to the supernatant. If a precipitate forms, add several more drops, disperse the mixture with a stirring rod or by gentle agitation, and centrifuge (Figure D4.3). Repeat the test for complete precipitation.

Washing a Precipitate

A precipitate must often be washed to remove occluded impurities. Add deionized water or wash liquid to the precipitate, disperse the solid thoroughly with a stirring rod or by gentle agitation, centrifuge, and decant. Usually the wash liquid can be discarded. Two washings are usually satisfactory. Failure to properly wash precipitates often leads to errors in the analysis (and arguments with your laboratory instructor!) because of the presence of occluded contaminating ions.

As you will be using the centrifuge frequently in the next several experiments, *be sure* to read carefully Technique 11F in the Laboratory Techniques section of this manual.

Heating and Cooling Solutions

Many procedures call for the reference solution to be heated. Heating a mixture either accelerates the rate of a chemical reaction or causes the formation of larger crystals of precipitate, allowing its separation to be more complete. ***Never** heat the small test tubes directly with a flame*. Heat one or several test tubes in a hot water bath; a 150-mL beaker containing 100 mL of deionized water is satisfactory. The test tube can be placed

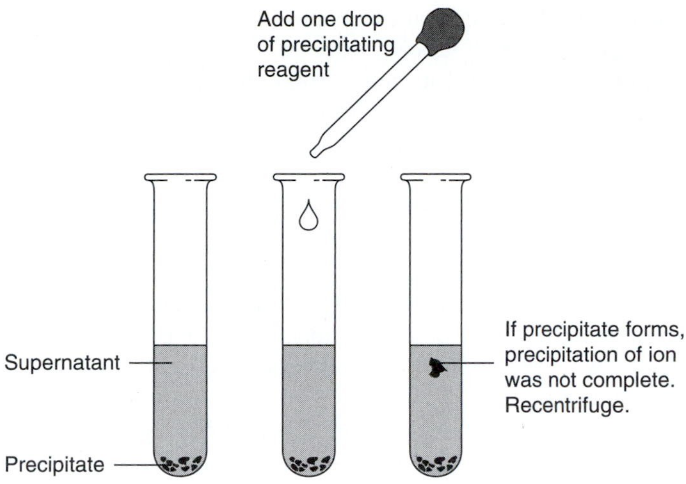

Figure D4.3 A test for the complete precipitation of an ion.

directly into the bath, supported against the wall of the beaker (Technique 13D). Read the Experimental Procedure before lab; if a hot water bath is needed, start heating the water at the beginning of the laboratory period and keep it warm with a **"cool" flame.**

A solution or mixture can be cooled by placing the test tube under cold running tap water or by submerging the test tube in a beaker of ice water.

Cool flame: a nonluminous flame supplied with a reduced supply of fuel. The rule of thumb for creating a cool flame for heating a liquid in a test tube is, if you can feel the heat of the flame with the hand that is holding the test tube clamp, the flame is too hot!

Constructing Flow Diagrams

A flow diagram is often used to organize the sequence of test procedures for the separation and identification of the large number of ions in a mixture. A flow diagram uses several standard notations:

- Brackets, [ ], indicate the use of a test reagent written in molecular form.
- A longer single horizontal line, ______, indicates a separation of a precipitate from a solution, most often with a centrifuge.
- A double horizontal line, ═══, indicates the soluble ions in the solution.
- Two short vertical lines, ‖ , indicate the presence of a precipitate; these lines are drawn to the *left* of the single horizontal line.
- One short vertical line,—, indicates a supernatant and is drawn to the *right* of the single horizontal line.
- Two branching diagonal lines, ∧, indicate a separation of the existing solution into two portions.
- A rectangular box, □, placed around a compound or the result of a test confirms the presence of the ion.

The flow diagram for the anions is presented in Experiment 34. Study it closely and become familiar with the symbols and notation as you read the Introduction and Experimental Procedure.

Partially completed flow diagrams for the various groups of cations are presented in Experiments 35 through 37. The last section of Experiment 37 challenges you to identify any number of the various cations in a general unknown. Ask your instructor about that assignment.

How to Effectively Do "Qual"

The following suggestions are offered before and during the following "qual" experiments:

- Use good laboratory techniques during the analyses. Review the suggested techniques that appear as icons in the Experimental Procedure prior to beginning the analysis.

- *Always* read the Experimental Procedure in detail. Is extra equipment necessary? Is a hot water bath needed? Maintain a water bath during the laboratory period if one is needed. What cautions are to be taken?
- Understand the principles of the separation and identification of the ions. Is this an acid–base separation, redox reaction, or complex formation? Why is this reagent added at this time?
- Closely follow, simultaneously, the principles used in each test, the flow diagram, the Experimental Procedure, and the Report Sheet during the analysis.
- Mark with a "magic marker" 1-, 2-, and 3-mL intervals on the small test tube used for testing your sample to quickly estimate volumes (see Figure D4.1).
- Keep a number of *clean* dropping pipets, stirring rods, and small test tubes available; always rinse each test tube several times with deionized water immediately after use.[2]
- Estimate the drops/mL of one or more of your dropping pipets.
- Keep a wash bottle filled with deionized water available at all times.
- Maintain a "file" of confirmatory tests of the ions in the test tubes that result from the analysis on your *reference* solution; in that way, comparisons for tests on the unknown test solution can be quick.

Caution: *In the next several experiments you will be handling a large number of chemicals (acids, bases, oxidizing and reducing agents, and, perhaps, even some toxic chemicals), some of which are more concentrated than others and must be handled with care and respect!*

Carefully, *handle all chemicals.* **Do not** *intentionally inhale the vapors of any chemical unless you are specifically told to do so;* **avoid** *skin contact with any chemicals—wash the skin immediately in the laboratory sink, eye wash fountain, or safety shower;* **clean up** *any spilled chemical—if you are uncertain of the proper cleanup procedure, flood with water, and consult your laboratory instructor;* **be aware** *of the techniques and procedures of neighboring chemists—discuss potential hazards with them.*
And finally, ***dispose of the waste chemicals*** *in the appropriately labeled waste containers. Consult your laboratory instructor to ensure proper disposal.*

[2]Failure to maintain clean glassware during the analysis causes more spurious data and reported errors in interpretation than any other single factor in qualitative analysis.

Dry Lab 4 *Prelaboratory Assignment*

Preface to Qualitative Analysis

Date __________ Lab Sec. ______ Name ____ Desk No. __________

1. a. The approximate volume of a standard 75-mm test tube is ___________________.

   b. Small volumes of reagents are usually added using a ___________________.

   c. A ___________________ is commonly used to break up a precipitate.

   d. A ___________________ is an instrument used to separate and compact a precipitate in a test tube.

   e. The clear solution above a precipitate is called the ___________________.

   f. The number of drops of water equivalent to 1 mL is about ___________________.

   g. A solution should be centrifuged for (how long?) ___________________.

   h. On a flow diagram, a double set of vertical lines, ‖ , means ___________________.

   i. On a flow diagram, a single horizontal line, ——, means ___________________.

   j. On a flow diagram, ▭ means ___________________.

   k. On a flow diagram, a double horizontal line, ═══, indicates a ___________________.

   l. Agitate the solution in a test tube by ____.

2. How is a centrifuge balanced?

3. Describe the technique for washing a precipitate.

4. Describe the procedure used to test for the completeness of precipitation of an ion.

5. Describe the procedure for mixing/agitating a solution in a test tube.

Experiment 34

Common Anions

Calcium ion and carbonate ion combine to form a calcium carbonate precipitate, a preliminary test for the presence of carbonate ion in a solution.

Objectives

- To determine some of the chemical and physical properties of anions
- To separate and characteristically identify the presence of a particular anion in a solution containing a number of anions

Techniques

The following techniques are used in the Experimental Procedure

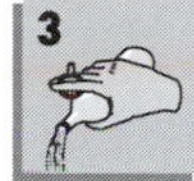

Introduction

Common anions in aqueous solution are either single atom anions (Cl^-, Br^-, I^-) or polyatomic anions usually containing oxygen (OH^-, SO_4^{2-}, CO_3^{2-}, PO_4^{3-}). In nature the most common anions are the chlorides, silicates, carbonates, phosphates, sulfates, sulfides, nitrates, aluminates, and combinations thereof.

Specific anion tests are subject to interference from other anions and cations. Therefore, to characteristically identify an anion in a mixture, preliminary elimination of the interferences is necessary.

Most common anions may be classified into three groups according to their chemical properties. The anion groups and the anions that are characteristically identified in this experiment are:

Anions I: The insoluble salts of barium ion in a basic solution. The anions in this group include SO_4^{2-}, PO_4^{3-}, and CO_3^{2-}.

Anions II: The insoluble salts of silver ion in a dilute nitric acid solution. The anions in this group include Cl^-, I^-, and S^{2-}.

Anions III: The soluble anions. An anion in this group is NO_3^-.

The separation and identification of the anions are outlined in the **flow diagram** on the next page. Follow the diagram as you read through the Introduction and follow the Experimental Procedure.

3A	4A	5A	6A	7A	Noble Gases
	CO_3^{2-}	NO_3^-			
		PO_4^{3-}	SO_4^{2-} S^{2-}	Cl^-	
				I^-	

Common anions detected in this experiment.

Flow diagram: a diagram that summarizes a procedure for following a rigid sequence of steps

Flow Diagram for Anion "Qual" Scheme

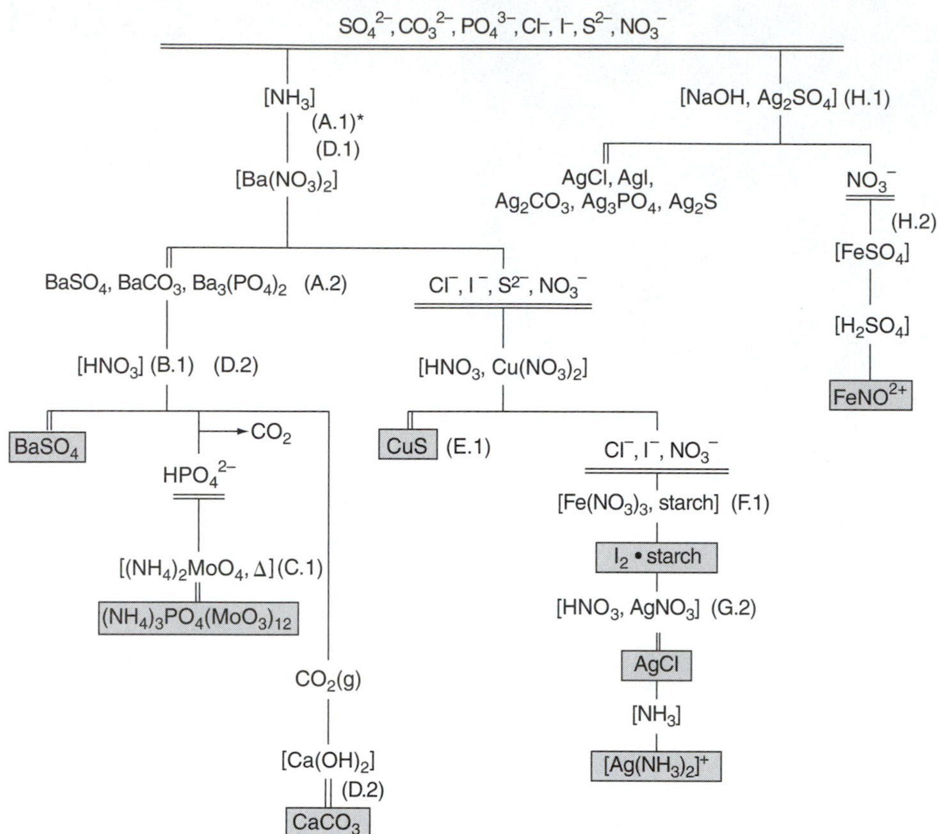

* Numbers in parentheses refer to parts of the Experimental Procedure

Sulfate Ion

All sulfate salts are *soluble* except those of strontium, barium, mercury(I), and lead (Appendix G). The addition of barium ion to an acidified solution containing sulfate ions produces insoluble barium sulfate, a finely divided, *white precipitate,* which is insoluble in acidic solutions.

$$Ba^{2+}(aq) + SO_4^{2-}(aq) \rightarrow \boxed{BaSO_4(s)} \qquad (34.1)$$

This confirms the presence of the sulfate ion in the test solution.

Phosphate Ion

The phosphate ion, PO_4^{3-}, is a strong Brønsted base (proton acceptor)

All phosphate salts are *insoluble* except those of the Group 1A cations and ammonium ion (Appendix G). The phosphate salt (*and* the sulfate salt) of barium forms a *white precipitate* in a basic solution. Unlike barium sulfate, however, barium phosphate dissolves in an acidic solution.

$$3\ Ba^{2+}(aq) + 2\ PO_4^{3-}(aq) \rightarrow Ba_3(PO_4)_2(s) \qquad (34.2)$$
$$\downarrow \mathbf{2\ H^+(aq)}$$
$$3\ Ba^{2+}(aq) + 2\ HPO_4^{2-}(aq)$$

Ammonium molybdate added to an acidified solution of the hydrogen phosphate ion precipitates *yellow* ammonium phosphomolybdate, confirming the presence of phosphate ion in the test solution.

$$HPO_4^{2-}(aq) + 12\ (NH_4)_2MoO_4(aq) + 23\ H^+(aq) \rightarrow$$
$$\boxed{(NH_4)_3PO_4(MoO_3)_{12}(s)} + 21\ NH_4^+(aq) + 12\ H_2O(l) \qquad (34.3)$$

The rate of precipitate formation depends on the concentration of the phosphate ion in solution.

The reactions of the arsenate ion (AsO_4^{3-}) are identical to those of the phosphate ion and therefore would, if present, interfere with the test.

Carbonate Ion

All carbonate salts are *insoluble* except those of the Group 1A cations and ammonium ion (Appendix G). Acidification of a solution containing carbonate ion produces carbon dioxide gas (Figure 34.1).

$$CO_3^{2-}(aq) + 2\,H^+(aq) \rightarrow H_2O(l) + CO_2(g) \quad (34.4)$$

When the evolved CO_2 comes into contact with a basic solution containing calcium ion, the carbonate ion re-forms and reacts with the calcium ion, forming a *white precipitate* of calcium carbonate:

$$CO_2(g) + 2\,OH^-(aq) \rightarrow CO_3^{2-}(aq) + H_2O(l) \quad (34.5)$$

$$\downarrow \boldsymbol{Ca^{2+}(aq)}$$

$$\boxed{CaCO_3(s)} \quad (34.6)$$

The precipitate confirms the presence of carbonate ion in the test solution. The sulfite ion, SO_3^{2-}, if present, would interfere with the test; under similar conditions, it produces sulfur dioxide gas and insoluble calcium sulfite.

Figure 34.1 Acidifying a solution containing carbonate ion produces carbon dioxide gas.

CO_2 is an acidic anhydride and, thus, readily reacts with a base

Sulfide Ion

The sulfide ion exists only in basic solutions. Most sulfide salts are insoluble, including CuS. When Cu^{2+} is added to a solution containing sulfide ion, a *black precipitate* of copper(II) sulfide, CuS, forms, confirming the presence of sulfide ion in the test solution:

$$Cu^{2+}(aq) + S^{2-}(aq) \rightarrow \boxed{CuS(s)} \quad (34.7)$$

Halide Ions

The salts of the chloride, bromide, and iodide ions are soluble with the exception of the Ag^+, Pb^{2+}, and Hg_2^{2+} halides. A simple reaction with silver ion would cause a mixture of the silver halides to precipitate, and therefore no separation or identification could be made. Instead, differences in the ease of oxidation of the halide ions are used for their identification (see Experiment 11, Part D). The iodide ion is most easily oxidized, followed by the bromide and chloride ions. A weak oxidizing agent oxidizes only the iodide ion. In this experiment iron(III) ion oxidizes iodide ion to yellow-brown molecular iodine.

$$2\,Fe^{3+}(aq) + 2\,I^-(aq) \rightarrow 2\,Fe^{2+}(aq) + I_2(aq) \quad (34.8)$$

The molecular iodine then reacts with starch to form a *deep blue complex*, I_2•starch, confirming the presence of iodide ion in the sample.

$$I_2\,(aq) + \text{starch}(aq) \rightarrow I_2\text{•starch}\,(aq,\ \textit{deep blue}) \quad (34.9)$$

The bromide ion can be detected by being oxidized to molecular bromine with a stronger oxidizing agent than iron(III) ion, an oxidizing agent such as $KMnO_4$.

The chloride ion is then precipitated as a *white precipitate* of silver chloride.[1]

$$Cl^-(aq) + Ag^+(aq) \rightarrow \boxed{AgCl(s)} \quad (34.10)$$

[1] A faint cloudiness with the addition of Ag^+ is inconclusive as the chloride ion is one of those "universal impurities" in aqueous solutions.

To further confirm the presence of chloride ion in the test solution, aqueous ammonia is added to dissolve the silver chloride which again precipitates with the addition of nitric acid:

$$AgCl(s) + 2\ NH_3(aq) \rightleftharpoons [Ag(NH_3)_2]^+(aq) + Cl^-(aq) \qquad (34.11)$$

Nitrate Ion

As all nitrate salts are soluble, no precipitate can be used for identification of the nitrate ion. The nitrate ion is identified by the "brown ring" test. The nitrate ion is reduced to nitric oxide by iron(II) ions in the presence of concentrated sulfuric acid:

$$NO_3^-(aq) + 3\ Fe^{2+}(aq) + 4\ H^+(aq) \xrightarrow{\text{conc } H_2SO_4} 3\ Fe^{3+}(aq) + NO(aq) + 2\ H_2O(l) \qquad (34.12)$$

The nitric oxide, NO, combines with excess iron(II) ions, forming the *brown* $FeNO^{2+}$ ion at the interface (where acidity is high) of the aqueous layer and a concentrated sulfuric acid layer that underlies the aqueous layer.

$$Fe^{2+}(aq) + NO(aq) \rightarrow \boxed{FeNO^{2+}(aq)} \qquad (34.13)$$

$FeNO^{2+}$ is more stable at low temperatures. This test has many sources of interference: (1) sulfuric acid oxidizes bromide and iodide ions to bromine and iodine, and (2) sulfites, sulfides, and other reducing agents interfere with the reduction of NO_3^- to NO. A preparatory step of adding sodium hydroxide and silver sulfate removes these interfering anions, leaving only the nitrate ion in solution.

Experimental Procedure

Procedure Overview: Two solutions are tested with various reagents in this analysis: (1) a reference solution containing all seven of the anions for this analysis and (2) a test solution containing any number of anions. Separations and observations are made and recorded. Equations that describe the observations are also recorded. Comparative observations of the two solutions result in the identification of the anions in the test solution. All tests are qualitative; only identification of the anion(s) is required.

To become familiar with the identification of these anions, take a sample that contains all of the anions (the reference solution) and analyze it according to the procedure. At each circled, superscript (e.g.,①), *stop* and record on the Report Sheet. After each anion is confirmed, **save** it in the test tube so that its appearance can be compared to that of your test solution.

To analyze for anions in your test solution, place the test solution alongside the reference solution during the analysis. As you progress through the procedure, perform the same test on both solutions and make comparative observations. Check (√) the findings on the Report Sheet. Do not discard any solutions (but keep all solutions labeled) until the experiment is complete.

The test solution may be a water sample from some location in the environment, e.g., a lake, a stream, or a drinking water supply. Ask your instructor about this option.

Before proceeding, review the techniques outlined in Dry Lab 4. The review of these procedures may expedite your analysis with less frustration.

Contamination by trace amounts of metal ions in test tubes and other glassware leads to "unexplainable" results in qualitative analysis. Thoroughly clean all glassware with soap and tap water; rinse twice with tap water and twice with deionized water before use.

Disposal: Dispose of all test solutions and precipitates in the appropriate waste container.

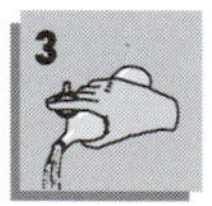

***Caution:** A number of concentrated and 6 M acids and bases are used in the analysis of these anions. Handle each of these solutions with care. Read the Laboratory Safety section for instructions in handling acids and bases.*

The expression "small test tube" that is mentioned throughout the Experimental Procedure refers to a 75-mm test tube (~3 mL volume) *or* a centrifuge tube of the size that fits into your laboratory centrifuge. Consult with your laboratory instructor.

A. Separation of Sulfate, Carbonate, and Phosphate Anions

The Experimental Procedure is written for a single solution. If you are simultaneously identifying anions in *both* a reference solution *and* a test solution, adjust the procedure accordingly. If the test solution is a sample with an environmental origin, gravity filter 10–15 mL before beginning the Experimental Procedure.

1. **Precipitate the SO_4^{2-}, CO_3^{2-}, and PO_4^{3-}.** Place about 1.5 mL of the reference solution in a small test tube. Test the solution with litmus. If acidic, add drops of 3 *M* NH_3 until the solution is basic to litmus; then add 3–4 more drops. Add 10–12 drops of 0.1 *M* $Ba(NO_3)_2$ until the precipitation of the anions is complete.①
2. **Separate the Solution from Precipitate.** Centrifuge the solution. Decant the supernatant② into a small test tube and save for Part E. Wash the precipitate *twice* with about 1 mL of deionized water (see Dry Lab 4). Discard the washings as directed by your instructor. Save the precipitate for Parts B and C.

B. Test for Sulfate Ion

1. **Confirmatory Test.** Acidify the precipitate from Part A.2 with 6 *M* HNO_3 **(caution!)** and centrifuge. The *white precipitate* that does *not* dissolve is barium sulfate; this confirms the presence of the sulfate ion③ in the solution.

C. Test for Phosphate Ion

1. **Confirmatory Test.** Transfer the supernatant from Part B.1 to a small test tube. Add 1 mL of 0.5 *M* $(NH_4)_2MoO_4$. Shake and warm slightly in a warm water (~60°C) bath and let stand for 10–15 minutes. A *slow* formation of a *yellow precipitate* confirms the presence of the phosphate ion④ in the solution.[2]

D. Test for Carbonate Ion

1. **Precipitate the CO_3^{2-}.** Repeat Part A. Centrifuge the mixture; save the precipitate, but discard the supernatant. Dip a glass rod into a saturated $Ca(OH)_2$ solution.
2. **Confirmatory Test.** Add 5–10 drops of 6 *M* HNO_3 to the precipitate and immediately insert the glass rod into the test tube (Figure 34.2). *Do not* let the glass rod touch the test tube wall or the solution. The formation of a *milky solution* on the glass rod confirms the presence of carbonate ion⑤ in the test solution.

E. Test for Sulfide Ion

1. **Confirmatory Test.** To the supernatant from Part A.2, add 2–4 drops of 6 *M* HNO_3 until the solution is acid to litmus and then drops of 1 *M* $Cu(NO_3)_2$ until precipitation is complete. Be patient, allow ~2 minutes to form.⑥ Centrifuge; save the supernatant for Part F. The *black precipitate* confirms the presence of sulfide ion in the solution.

F. Test for Iodide Ion

1. **Confirmatory Test.** To about 1 mL of the supernatant from Part E.1 add 5 drops of 0.2 *M* $Fe(NO_3)_3$. Agitate the solution. The formation of molecular iodine is slow—allow 2–3 minutes. Add 2 drops of 1% starch solution. The *deep blue I_2•starch complex* confirms the presence of iodide ion in the sample.⑦

[2]A *white* precipitate may form if the solution is heated too long or if the solution is not acidic enough. The precipitate is MoO_3, *not* a pale form of the phosphomolybdate precipitate.

Stirring rod
Small test tube
Previously dipped in saturated $Ca(OH)_2$
5–10 drops 6 *M* HNO_3
$CaCO_3$ precipitate

Figure 34.2 Position a stirring rod dipped into a saturated $Ca(OH)_2$ solution just above the solid/HNO_3 mixture.

G. Test for Chloride Ion

1. Add 2–3 drops of 6 *M* HNO_3 to the solution from Part F.1.
2. **Confirmatory Test.** Add drops of 0.01 *M* $AgNO_3$ to the aqueous solution (sample) and centrifuge. A *white precipitate* indicates the likely presence of Cl^-.⑧ Discard the supernatant. The addition of 6 *M* NH_3 quickly dissolves the precipitate if Cl^- is present.⑨ Reacidification of the solution with 6 *M* HNO_3 re-forms the silver chloride precipitate.

H. Test for Nitrate Ion

1. **Precipitate Anions I and II.** Place 1½ mL of the original sample into a small test tube. Add drops of 3 *M* NaOH until the solution is basic to litmus. Add drops of a saturated (0.04 *M*) Ag_2SO_4 solution until precipitation appears complete. Centrifuge and save the supernatant for Part H.2.⑩ Test for complete precipitation in the supernatant and, if necessary, centrifuge again.
2. **Confirmatory Test.** Decant 0.5 mL of the supernatant into a small test tube and acidify (to litmus) with 3 *M* H_2SO_4. Add 0.5 mL of a saturated iron(II) sulfate, $FeSO_4$, solution and agitate. Cool the solution in an ice bath. Holding the test tube at a 45° angle (Figure 34.3), add, with a dropping pipet, *slowly and cautiously,* down its side, about 0.5 mL of conc H_2SO_4. (**Caution:** *Concentrated* H_2SO_4 *causes severe skin burns.*[3]) Do *not* draw conc H_2SO_4 into the bulb of the dropping pipet. Do *not* agitate the solution. The more dense conc H_2SO_4 should underlie the aqueous layer. Use extreme care to avoid mixing the conc H_2SO_4 with the solution. Allow the mixture to stand for several minutes. A *brown ring* at the interface between the solution and the conc H_2SO_4 confirms the presence of the nitrate ion⑪ in the test solution.

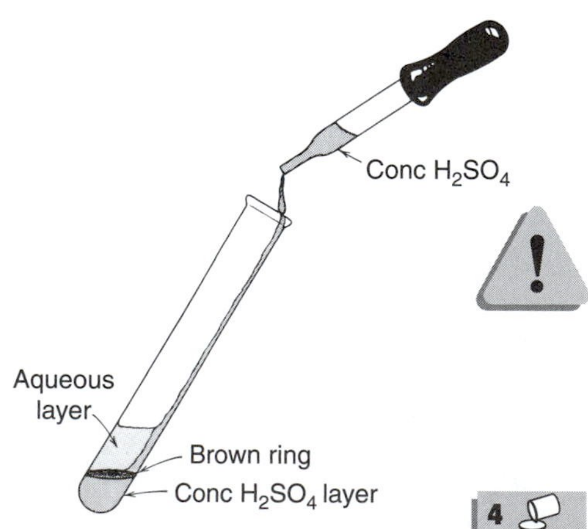

Figure 34.3 Test for the nitrate ion.

Disposal: Dispose of the concentrated acid solution in the "Waste Acids" container.

[3]The conc H_2SO_4 has a greater density than water and therefore underlies the aqueous layer.

Experiment 34 *Prelaboratory Assignment*

Common Anions

Date __________ Lab Sec. ______ Name ____ Desk No. __________

1. a. Identify the anion that produces a gas in its reaction with an acid. Write the balanced equation.

   b. Identify the anion that can be identified from the formation of a black precipitate. Write the balanced equation.

   c. Identify the anion in which all of its salts are soluble. Write the balanced equation that expresses a positive identification of its presence in solution.

2. Which group of anions is separated from other anions as insoluble barium salts?

3. On a flow diagram,
   a. what is the meaning of a single horizontal line?

   b. what is the meaning of a pair of short vertical lines?

   c. what is the meaning of a pair of horizontal lines?

   d. what is the meaning of the brackets, [ ], around a reagent?

4. Three anions in this experiment are identified by the complexes they form. Which anions are so identified?

5. Only two anions are confirmed present by the formation of a white precipitate. Which two anions are so confirmed? Write the formula of the precipitates.

6. Identify a single reagent that distinguishes the carbonate ion from the chloride ion. Write the balanced equation(s) that makes the distinction.

7. Identify the single reagent that separates the sulfate ion from the phosphate ion. Explain.

Experiment 34 *Report Sheet*

Common Anions

Date __________ Lab Sec. ______ Name ______________________________ Desk No. __________

Procedure Number and Ion	Test Reagent or Technique	Evidence of Chemical Change*	Chemical(s) Responsible for Observation	Equation(s) for Observed Reaction	Check (√) if Observed in Unknown
①					
②					
③ SO_4^{2-}					☐
④ PO_4^{3-}					☐
⑤ CO_3^{2-}					☐
⑥ S^{2-}					☐
⑦ I^-					☐
⑧ Cl^-					☐
⑨					
⑩					
⑪ NO_3^-					☐

*See Experiment 12, Introduction.

Anions present in unknown test solution no. _____:______________________________

Instructor's approval: ______________________________

Laboratory Questions

Circle the questions that have been assigned.

1. Part A.1. The *test* solution is made basic and drops of 0.1 *M* $Ba(NO_3)_2$ are added but no precipitate forms. To what part of the Experimental Procedure do you proceed? Explain.
2. Part A.1. The *reference* solution is made acidic instead of basic. How would this change the composition of the precipitate and the test in Part D.1? Explain.
3. Part B.1. The *test* solution is acidified with 6 *M* HNO_3 but no precipitate is present? What can be concluded from the observation?
4. Part E.1. 6 *M* HNO_3 could not be found on the reagent shelf. Instead 6 *M* HCl is added to the *test* solution.
   a. What affect does this have on the test for sulfide ion? Explain.
   b. What affect does this have on subsequent tests of the supernatant from Part E.1? Explain.
5. Part G.2. 6 *M* NH_3, a basic solution, cannot be found on the reagent shelf, but 6 *M* NaOH, also a base, is available. What would be observed if the 6 *M* NaOH is substituted for the 6 *M* NH_3 in testing the *reference* solution? Explain.
6. Part G.2. Both iodide ion and chloride ion are in the *test* solution. Describe the appearance of the contents of the test tube after drops of 0.01 *M* $AgNO_3$ have been added and the solution has been centrifuged.
7. Part H.2. There is no other "brown ring" test commonly known in chemistry. What is the substance producing the brown ring?

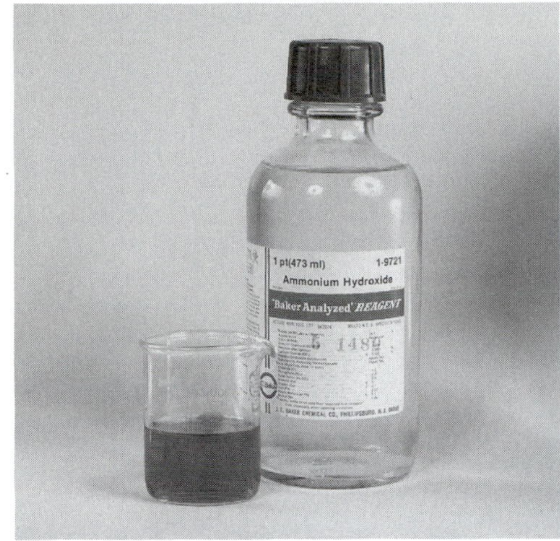

Experiment 35

Qual I. Na^+, K^+, NH_4^+, Ag^+, Cu^{2+}, Bi^{3+}

Ammonia forms a complex with copper(II) ion, intensifying the color of the copper(II) solution.

OBJECTIVES

- To observe and utilize the chemical and physical properties of Na^+, K^+, NH_4^+, Ag^+, Cu^{2+}, and Bi^{3+}
- To separate and identify the presence of one or more of the cations, Na^+, K^+, NH_4^+, Ag^+, Cu^{2+}, Bi^{3+}, in an aqueous solution

TECHNIQUES

The following techniques are used in the Experimental Procedure

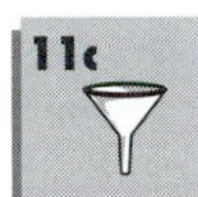

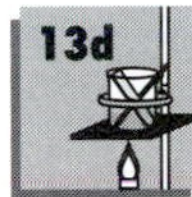

INTRODUCTION

In identifying a particular cation in a mixture of cations, it would be ideal if we could detect each cation in the mixture by merely adding a specific reagent that would produce a characteristic color or precipitate; however, no such array of reagents has been developed. Instead, the cations must first be separated into groups having similar chemical properties; each group is then further divided until each ion is isolated. From there, the cation responds to a specific reagent to produce the characteristic color or precipitate for identification. The qualitative analysis of a mixture of cations requires such a plan of investigation.

The analysis of Qual I cations requires good laboratory technique for their separation and confirmation, as well as careful preparation and understanding of the procedure. If in following the prescribed steps for separation and identification, you only rely on a "cookbook" procedure, expect unexplainable results. Please read carefully the remainder of this Introduction and the Experimental Procedure, review your laboratory techniques, and complete the Prelaboratory Assignment before beginning the analysis; it will save you time and minimize frustration.

This experiment is the first in a series (Experiment 36 for Qual II cations and Experiment 37 for Qual III cations) in which a set of reagents and techniques are used to identify the presence of a particular cation among a larger selection of cations.[1] We will approach this study as an experimental chemist: we will conduct some tests, write down our observations, and then write a balanced equation that agrees with our data.

[1]The mercurous ion, Hg_2^{2+}, and the lead(II) ion, Pb^{2+}, are often included in Qual I cation analysis. Because of their potential environmental effects, both ions are omitted here.

As an introduction to the chemistry of the cations in this experiment, keep in mind the following:

- Most Na^+, K^+, and NH_4^+ salts are soluble.
- The chloride salt of Ag^+ is "insoluble."
- The sulfide salts of Cu^{2+} and Bi^{3+} are also considered "insoluble."

Thus, these ions can be separated from a large number of other cations.[2]

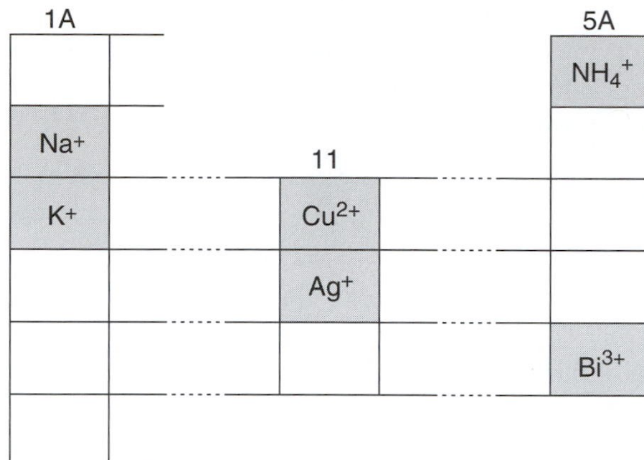

Cations of the Qual I group.

Sodium and potassium ions form a very limited number of salts with low solubilities and only carefully controlled conditions provide confirmatory tests; however, both ions provide very sensitive, distinctive flame tests.[3] Traditionally qual tests for these cations are completed after all other cations have been precipitated, but because sodium ion is a "universal" impurity, and because it is used for a number of separation tests in Qual II (Experiment 36) and Qual III (Experiment 37), the tests will be performed here, after all other possible interfering ions have been removed.

All ammonium salts are water soluble. The ammonium ion is a weak acid; its conjugate base is ammonia, which is a gas. When the pH of an aqueous solution of ammonium ion is increased, ammonia gas is evolved and is easily detected with red litmus.

The sulfide salts of Cu^{2+} and Bi^{3+} are insoluble in solutions with pH values as low as 0.5 (e.g., 0.3 M H_3O^+) and higher. If the pH is greater than 0.5, not only will the Cu^{2+} and Bi^{3+} sulfide salts precipitate but also the Qual II metal cations (Experiment 36). Therefore, pH control is extremely important in the separation of Cu^{2+} and Bi^{3+} from the Qual II cations.

Once the pH of the test solution is set at about 0.5, the sulfide ion is generated *in situ* in the form of $H_2S(aq)$. The H_2S is *slowly* generated from the thermal hydrolysis of thioacetamide, CH_3CSNH_2.

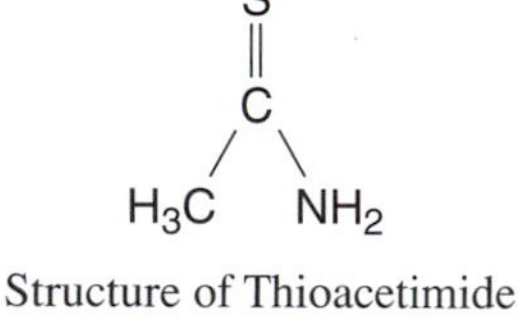

Structure of Thioacetimide

$$CH_3CSNH_2(aq) + 2\,H_2O(l) \xrightarrow{\Delta} CH_3CO_2^-(aq) + NH_4^+(aq) + H_2S(aq) \quad (35.1)$$

The slow generation of H_2S in solution minimizes the escape of this foul-smelling, highly toxic gas into the laboratory. When H_2S is produced slowly, more compact sulfide precipitates form, making them easier to separate by centrifugation.

$H_2S(aq)$ is a weak, diprotic acid ionizing slightly to produce the sulfide ion that is necessary for Cu^{2+} and Bi^{3+} precipitation:

$$H_2S(aq) + 2\,H_2O(l) \rightleftharpoons 2\,H_3O^+(aq) + S^{2-}(aq) \quad (35.2)$$

---

[2]Review the solubility rules for salts in Appendix G or in your text.

[3]The sodium flame test is so sensitive that the sodium ion washed (with several drops of deionized water) from a fingerprint can be detected.

For a Cu^{2+} or Bi^{3+} sulfide salt to precipitate, the product of the molar concentrations of the cation and sulfide ion in solution (the mass action expression for the salt) must exceed its K_{sp} value. Using LeChâtelier's principle, a high H_3O^+ concentration (pH $\approx$ 0.5) shifts this equilibrium to the *left,* reducing the sulfide concentration. Hence, pH or H_3O^+ controls the S^{2-} concentration.

As the Cu^{2+} and Bi^{3+} cations precipitate at a low pH, only small amounts of sulfide ion are needed for their precipitation; these cations have low molar solubilities and also small K_{sp} values, at least smaller than those for the Qual II cations.

Sodium Ion

A flame test to detect the sodium ion produces a characteristic, fluffy yellow flame. The color is so intense that it masks the color produced by other ions and is comparatively persistent.

Potassium Ion

A flame test is also used to detect the presence of potassium ion in a sample. The flame is viewed through cobalt blue glass. The glass absorbs the intense yellow color of sodium, but transmits the lavender color of the potassium flame. The lavender flame is of short duration. A comparative test of a solution known to contain only potassium ion is often necessary for the determination.

Ammonium Ion

To test for ammonium ion, advantage is taken of the following equilibrium:

$$NH_3(aq) + H_2O(l) \rightleftharpoons NH_4^+(aq) + OH^-(aq) \quad (35.3)$$

When hydroxide ion (a common ion in the equilibrium) is added to a solution containing ammonium ion, this equilibrium shifts *left.* Heating drives *ammonia gas* from the system; its presence is detected using moist litmus paper (its odor is also frequently detected).

$$\boxed{NH_3(g)} + H_2O(l) \overset{\leftarrow}{\rightleftharpoons} NH_4^+(aq) + OH^-(aq) \quad (35.4)$$
$$\uparrow \boldsymbol{OH^-(aq)}$$

The ☐ *is used to note the confirmation of the presence of an ion in the test solution*

Silver Ion

When hydrochloric acid is added to a solution containing Ag^+ ions, AgCl precipitates. A large concentration of chloride ion, however, must be avoided to prevent the formation of the soluble silver chloride complex anion:

$$AgCl(s) + Cl^-(aq) \rightleftharpoons [AgCl_2]^-(aq) \quad (35.5)$$

In an ammoniacal solution, AgCl dissolves to form the $[Ag(NH_3)_2]^+$ complex:

$$AgCl(s) + 2\,NH_3(aq) \rightleftharpoons [Ag(NH_3)_2]^+(aq) + Cl^-(aq) \quad (35.6)$$

The $[Ag(NH_3)_2]^+$ ion is unstable in an acidic solution; H_3O^+ reacts with NH_3, forming NH_4^+. The equilibrium shifts to the left and the AgCl re-forms as a *white precipitate:*

$$\boxed{AgCl(s)} + 2\,NH_3(aq) \overset{\leftarrow}{\rightleftharpoons} [Ag(NH_3)_2]^+(aq) + Cl^-(aq) \quad (35.7)$$
$$\downarrow \boldsymbol{2\,H_3O^+(aq)}$$
$$2\,NH_4^+(aq) + 2\,H_2O(l)$$

$[H_3N—Ag—NH_3]^+$

Structure of $[Ag(NH_3)_2]^+$

Separation of Copper (II) and Bismuth (III) Ions from Qual II and Qual III Cations

If Qual II and Qual III cations are a part of the separation and identification scheme, thioacetamide is added to an acidified solution of the cations to generate the sulfide ion. Copper(II) sulfide and bismuth(III) sulfide precipitate, and the supernatant solution contains the cations of Qual II and Qual III.

The CuS and Bi_2S_3 precipitates (both black in color) dissolve in hot HNO_3. The hot HNO_3 oxidizes the sulfide ion in the precipitates to (free) elemental sulfur, forming $NO(g)$ as its reduction product. For CuS,

$$3\,CuS(s) + 8\,H^+(aq) + 2\,NO_3^-(aq) \xrightarrow{\Delta} 3\,Cu^{2+}(aq) + 2\,NO(g) + 3\,S(s) + 4\,H_2O(l) \quad (35.8)$$

Copper Ion

Structure of $[Cu(NH_3)_4]^{2+}$

Structure of $[Fe(CN)_6]^{4-}$

The addition of aqueous NH_3 to a solution containing the Bi^{3+} and Cu^{2+} cations precipitates white bismuth hydroxide, but complexes the Cu^{2+} as a soluble *deep blue* $[Cu(NH_3)_4]^{2+}$ *complex*. This confirms the presence of Cu^{2+} in the solution.

$$Bi^{3+}(aq) + 3\,NH_3(aq) + 3\,H_2O(l) \rightarrow Bi(OH)_3(s) + 3\,NH_4^+(aq) \quad (35.9)$$

$$Cu^{2+}(aq) + 4\,NH_3(aq) \rightarrow \boxed{[Cu(NH_3)_4]^{2+}(aq)} \quad (35.10)$$

A second confirmatory test for the presence of Cu^{2+} is the addition of potassium hexacyanoferrate(II), $K_4[Fe(CN)_6]$, which produces a *red-brown precipitate* of $Cu_2[Fe(CN)_6]$.

$$2[Cu(NH_3)_4]^{2+}(aq) + [Fe(CN)_6]^{4-}(aq) \rightarrow \boxed{Cu_2[Fe(CN)_6](s)} + 8\,NH_3(aq) \quad (35.11)$$

The Cu^{2+} ion in solution (right) has a less intense color than does the deep blue $[Cu(NH_3)_4]^{2+}$ complex.

Bismuth Ion

The *white* bismuth hydroxide (Equation 35.9) is dissolved and confirmed with the addition of *freshly prepared* sodium stannite, $Na_2Sn(OH)_4$, in a basic solution. The Bi^{3+} ion is reduced to *black* bismuth *metal,* confirming the presence of bismuth in the sample.

$$2\,Bi(OH)_3(s) + 3\,[Sn(OH)_4]^{2-}(aq) \rightarrow \boxed{2\,Bi(s)} + 3\,[Sn(OH)_6]^{2-}(aq) \quad (35.12)$$

As a guide to an understanding of the separation and identification of these six cations, carefully read the Experimental Procedure and then complete the flow diagram on the Prelaboratory Assignment. Review Dry Lab 4 for an understanding of the symbolism.

Experimental Procedure

Procedure Overview: Two solutions are tested with various reagents in this analysis: (1) a reference solution containing all six of the cations of Qual I and (2) a test solution containing any number of Qual I cations. Separations and observations are made and recorded. Equations that describe the observations are also recorded. Comparative observations of the two solutions result in the identification of the cations in the test solution. All tests are qualitative; only identification of the cation(s) is required.

To become familiar with the separation and identification of Qual I cations, take a sample that contains the six cations (reference solution) and analyze it according to the Experimental Procedure. At each circled, superscript (e.g.,①), *stop* and record data on the Report Sheet. After the presence of a cation is confirmed, **save** the characteristic appearance of the cation in the test tube so that it can be compared with observations made in the analysis of your test solution.

To analyze for cations in your test solution, place the test solution alongside the reference solution during the analysis. As you progress through the procedure, perform the same test on both solutions and make comparative observations. Check (√) the findings on the Report Sheet. Do not discard any solutions (but keep all solutions labeled) until the experiment is complete.

The test solution may be a water sample from some location in the environment, e.g., a lake, a stream, or a drinking water supply. Ask your instructor about this option.

Before proceeding, review the techniques outlined in Dry Lab 4. The review of these procedures may expedite your analysis with less frustration.

Contamination by trace amounts of metal ions in test tubes and other glassware leads to "unexplainable" results in qualitative analysis. Thoroughly clean all glassware with soap and tap water; rinse twice with tap water and twice with deionized water before use.

Caution: *A number of 6 M acids and bases are used in the analysis of these cations. Handle each of these solutions with care. Read the Laboratory Safety section for instructions in handling acids and bases.*

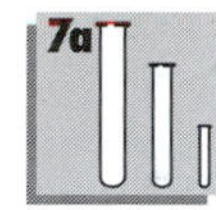

The expression "small test tube" that is mentioned throughout the Experimental Procedure refers to a 75-mm test tube (~3 mL volume) *or* a centrifuge tube of the size that fits into your laboratory centrifuge. Consult with your laboratory instructor.

A. Test for Sodium Ion

The Experimental Procedure is written for a single solution. If you are simultaneously identifying cations in *both* a reference solution *and* a test solution, adjust the procedure accordingly. If the test solution is a sample with an environmental origin, gravity filter 10–15 mL before beginning the Experimental Procedure.

1. **Remove the Interfering Ions.** Place no more than 2 mL of the reference solution in an evaporating dish. Add solid CaO or $Ca(OH)_2$, (**Caution:** *avoid skin contact*) with stirring, until the solution is basic to litmus; add a slight excess of the solid and then a pinch of solid $(NH_4)_2CO_3$. Heat the solution to a moist residue (Figure 35.1)

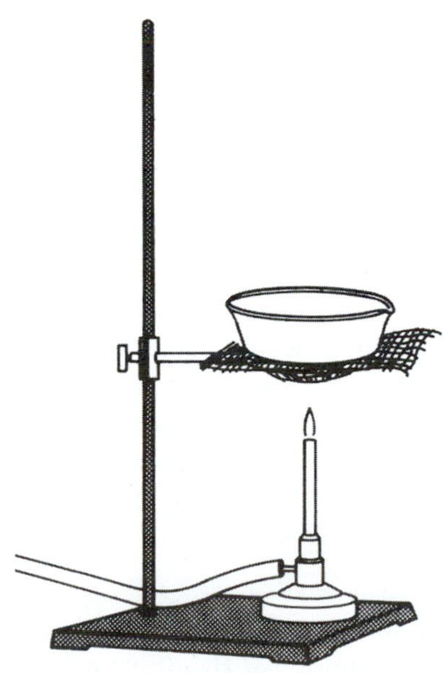

Figure 35.1 Heat the test solution to a moist residue.

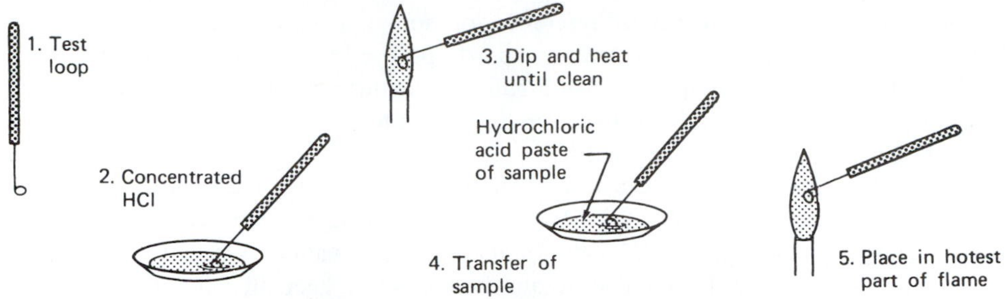

Figure 35.2 Flame test procedure.

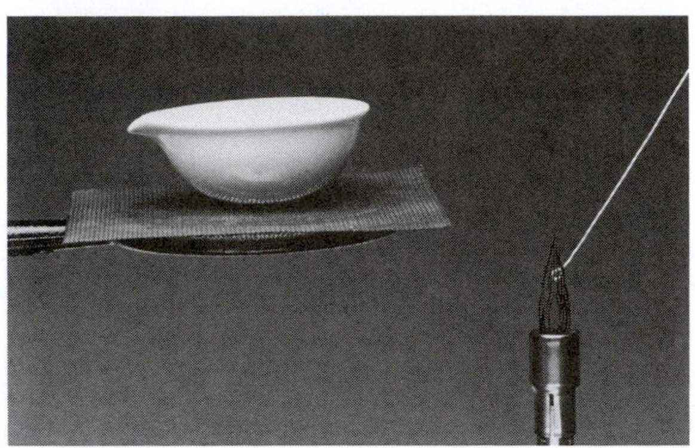

Figure 35.3 Flame test for the presence of sodium ion.

and allow the evaporating dish to cool. Add up to 1 mL of deionized water, stir, and decant into a small beaker or well plate.

2. **Confirmatory Test.** The flame test for sodium ion is reliable but also requires some technique. Clean the flame test wire by dipping it in 6 *M* HCl **(caution!)** and heating it in the hottest part of a Bunsen flame until the flame is colorless (Figure 35.2). Repeat as necessary. Dip the flame test wire into the solution and place it in the flame (Figure 35.3). A *brilliant yellow* persistent flame indicates the presence of sodium.① Conduct the sodium flame test on a 0.2 *M* NaCl solution for comparison.

B. Test for Potassium Ion

1. **Confirmatory Test.** Repeat Part A.2. A fleeting lavender flame confirms the presence of potassium. If sodium is present, view the flame through cobalt blue glass. Several trials are necessary as the test is judgmental. Conduct the potassium flame test on a 0.2 *M* KCl solution for comparison.②

C. Test for Ammonium Ion

1. **Prepare the Sample.** Transfer 5 mL of the *original* reference solution to a 100-mL beaker, support it on a wire gauze, and heat until a moist residue forms (do not evaporate to dryness!). Moisten the residue with 1–2 mL of deionized water. Moisten a piece of red litmus paper and with water attach it to the convex side of a small watchglass (Figure 35.4). Cover the beaker with the convex side *down*.

2. **Confirmatory Test.** Add 1–2 mL of 6 *M* NaOH to the reference solution and very gently warm the mixture—*do not boil.* (**Caution:** *Be careful not to let the* NaOH *contact the litmus paper.*) A change in litmus from *red* to *blue* confirms ammonia.③ The nose is also a good detector, but it is not always as sensitive as the litmus test.

D. Test for Silver Ion

1. **Precipitate the Silver Ion.** Place 2 mL of the *original* reference solution in a small test tube, add 4–5 drops of 6 *M* HCl (Figure 35.5) until no further evidence

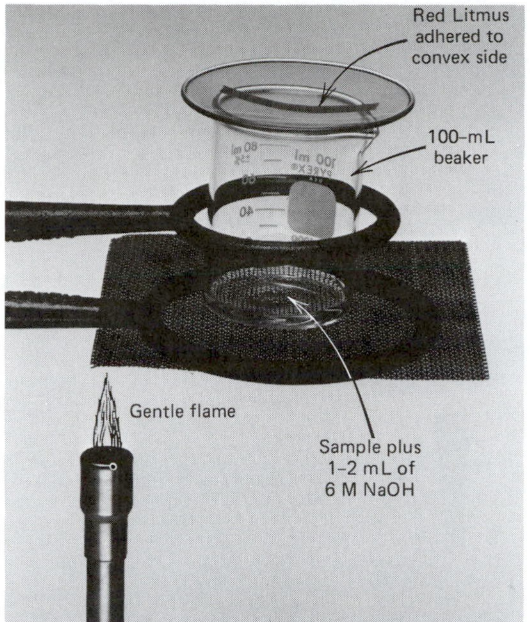

Figure 35.4 Absorbing NH_3 vapors on red litmus.

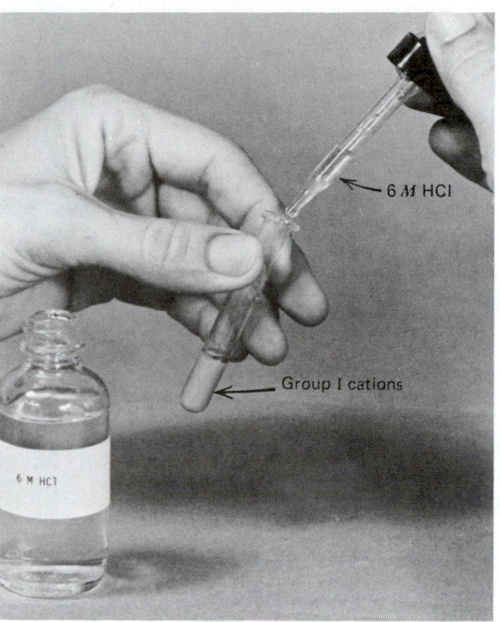

Figure 35.5 Precipitation of silver ion from solution.

of precipitation occurs.④ (**Caution:** *Handle acid with care.*) Stir with a stirring rod and centrifuge the mixture. Test for complete precipitation by adding 1–2 additional drops of 6 *M* HCl. Ask your instructor about the safe operation of the centrifuge. Decant and save the supernatant for Part E. Wash the precipitate[4] with 5 drops of deionized water, centrifuge, and combine the supernatants.

2. **Confirmatory Test.** Dissolve the precipitate with 6 *M* NH_3 **(caution!)**.⑤ Acidify the solution (check with litmus) with drops of 6 *M* HNO_3. (**Caution:** *Avoid skin contact; clean up spills immediately.*) A *white precipitate* confirms the presence of silver ion in the reference solution.⑥

E. Separation of Copper(II) and Bismuth(III) Ions from Qual II and Qual III Cations

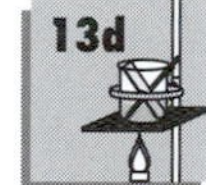

If cations from Qual II and Qual III are not present in your unknown solution, omit Part E and proceed to Part F.

1. **Precipitate the Cu^{2+} and Bi^{3+} Sulfides.** To the supernatant reference solution from Part D.1 add 10–15 drops of 1 *M* CH_3CSNH_2, heat in a hot water bath (~95°C, Figure 35.6) for several minutes, cool, and centrifuge. Test the solution for complete precipitation by repeating the thioacetamide addition to the supernatant. The precipitate contains the CuS and⑦ Bi_2S_3 salts[7], the supernatant contains cations that do not precipitate under these conditions (on the advice of your laboratory instructor, save for Experiments 36 and 37). Decant the supernatant.
2. **Wash the Precipitate.** Wash the precipitate *twice* with 0.1 *M* HCl, stir, and set aside each washing.
3. **Dissolve the Cu^{2+} and Bi^{3+} Sulfides.** Add 15 drops of 6 *M* HNO_3 **(caution!)** to the precipitate from Part E.2 and *very carefully* heat with a direct *cool* flame or a hot water bath until the precipitates dissolve. (**Caution:** *Read again Technique 13A for heating test tubes with a cool, direct flame.*) Cool and centrifuge. Save the supernatant⑧ and discard any free sulfur.

Danger! *The contents of the test solution can be ejected if the procedure is not performed as prescribed*

[4] Review the procedure for washing a precipitate in Dry Lab 4.

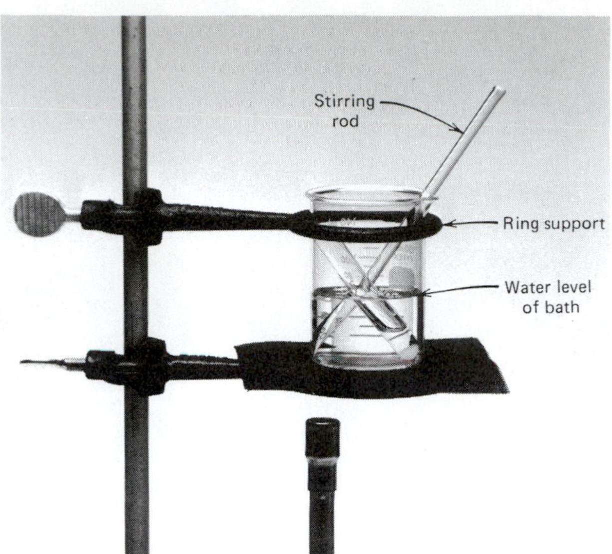

Figure 35.6 A hot water bath for sulfide precipitation of cations.

F. Test for Copper Ion

1. **Confirmatory Test.** Add drops of conc NH_3 (**caution:** *avoid inhalation or skin contact!*) to the solution from Part E.3 (or Part D.1, if no Qual II or Qual III cations are present). The *deep-blue* solution confirms the presence of Cu^{2+} in the sample.⑨ Centrifuge, decant the solution, and save the precipitate⑩ for Part G.
2. **A Second Confirmatory Test.** Acidify the supernatant from Part F.1 with 6 *M* CH_3COOH (**caution!**). Add 3 drops of 0.2 *M* $K_4[Fe(CN)_6]$. A *red-brown precipitate* reconfirms the presence of Cu^{2+} ion.⑪

G. Test for Bismuth Ion

1. **Prepare the Sodium Stannite Solution.** Prepare a fresh $Na_2Sn(OH)_4$ solution by placing 2 drops of 1 *M* $SnCl_2$ in a small test tube, followed by drops of 6 *M* NaOH (**caution!**) until the $Sn(OH)_2$ precipitate just dissolves. Agitate or stir the solution.
2. **Confirmatory Test.** Add several drops of the $Na_2Sn(OH)_4$ solution to the precipitate from Part F.1. The immediate formation of a *black precipitate*⑫ confirms the presence of the Bi^{3+} ion in the sample.

Disposal: Dispose of all test solutions and precipitates in the "Waste Metal Salts" container.

CLEANUP: Rinse each test tube twice with tap water. Discard each rinse in the "Waste Metal Salts" container. Thoroughly clean each test tube with soap and tap water; rinse twice with tap water and twice with deionized water.

Experiment 35 *Prelaboratory Assignment*

Qual I. Na^+, K^+, NH_4^+, Ag^+, Cu^{2+}, Bi^{3+}

Date __________ Lab Sec. ______ Name ____ Desk No. __________

1. Identify the Qual I cation(s) that is (are) confirmed present in a test solution by
   a. the formation of a precipitate

   b. the color of a soluble complex ion

   c. the characteristic color of a flame test

   d. the evolution of a gas

2. Identify the Qual I cation(s) that is (are) confirmed present in a test solution as a result of
   a. a Brønsted acid–base reaction

   b. an oxidation–reduction reaction

3. Identify the reagent that separates
   a. Ag^+ from Cu^{2+}. Explain the chemistry of the separation.

   b. Cu^{2+} from Bi^{3+}. Explain the chemistry of the separation.

4. Write the balanced redox equation for the dissolution of Bi_2S_3, similar to Equation 35.8.

5. Part E.3. Describe the danger encountered in this step of the experiment.

6. Part A.1. Why is CaO or $Ca(OH)_2$ added to the original test solution? *Hints:* Ca^{2+} is a non-interfering cation. Review the solubility rules for hydroxide compounds.

7. Complete the following flow diagram for the Qual I cations. Refer to Dry Lab 4 for a review of the symbolism on a flow diagram.

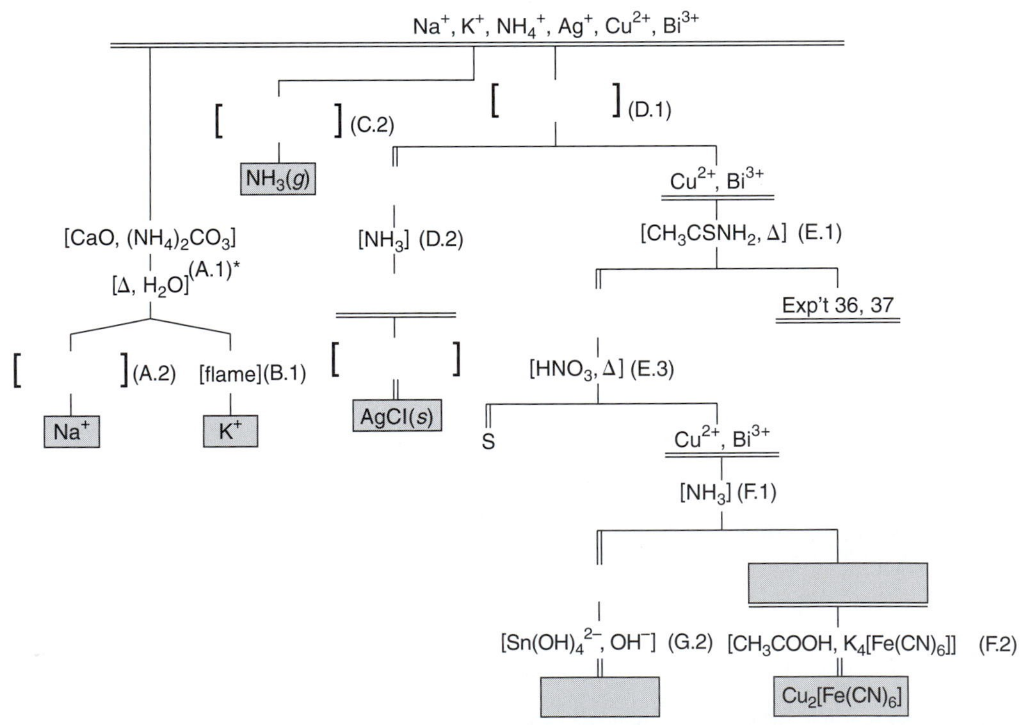

*Numbers in parentheses refer to parts of the Experimental Procedure.

Experiment 35 *Report Sheet*

Qual I. Na^+, K^+, NH_4^+, Ag^+, Cu^{2+}, Bi^{3+}

Date ________ Lab Sec. ______ Name ______________________________ Desk No. ________

Procedure Number and Ion	Test Reagent or Technique	Evidence of Chemical Change*	Chemical(s) Responsible for Observation	Equation(s) for Observed Reaction	Check (√) if Observed in Unknown
① Na^+	flame				☐
② K^+	flame				☐
③ NH_4^+					☐
④ Ag^+					
⑤					
⑥					☐
⑦					
⑧					
⑨ Cu^{2+}					☐
⑩					
⑪					☐
⑫ Bi^{3+}					☐

*See Experiment 12, Introduction.

Cations present in unknown test solution no. _____: ______________________________

Instructor's approval: ______________________________

Laboratory Questions

Circle the questions that have been assigned.

1. Explain why a positive flame test for sodium is *not* an absolute confirmation of sodium ion in a test sample.
2. Part A.1. The CaO or $Ca(OH)_2$ addition is omitted in the procedure. How does this affect the appearance of the flame test in Part A.2?
3. Part C.2. Instead of 6 *M* NaOH being added to the test solution, 6 *M* HCl is added. How will this affect the test for the presence of ammonium ion in the test solution? Explain.
4. Part D.1. Instead of 6 *M* HCl being added to the test solution, 6 *M* HNO_3 is added (both are strong acids). How will this affect the test for the presence of silver ion in the test solution? Explain.
5. Part D.2. Instead of 6 *M* NH_3 being added to the test solution, 6 *M* NaOH is added (both are bases). How will this affect the test for the presence of silver ion in the test solution? Explain.
6. Part E.1. Is the solution highly acidic, weakly acidic, or basic? Explain.
7. Part F.1. Instead of conc NH_3 being added to the test solution, 6 *M* NaOH is added (both are bases). What will be the composition of the precipitate that forms? Explain.
8. Part G.2. Write balanced half-reactions for the conversions of (see Equation 35.12)
   a. $Sn(OH)_4^{2-}$ to $Sn(OH)_6^{2-}$
   b. $Bi(OH)_3$ to Bi

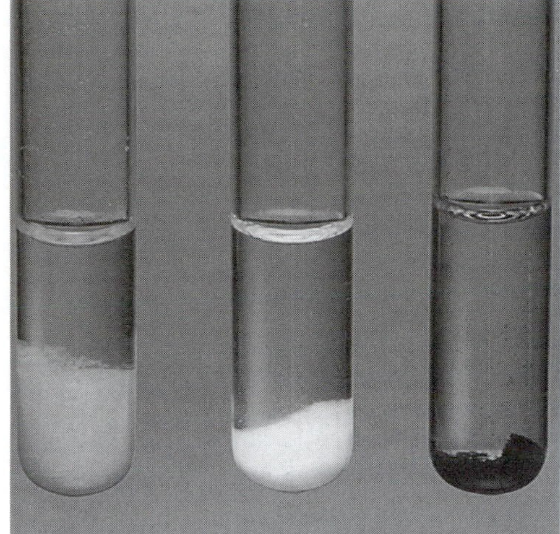

Experiment 36

Qual II. Mn^{2+}, Ni^{2+}, $Fe^{3+}(Fe^{2+})$, Al^{3+}, Zn^{2+}

Sulfides of the manganese(II), zinc(II), and iron(II) ions (left to right) precipitate from a neutral solution saturated with hydrogen sulfide.

OBJECTIVES

- To observe and utilize the chemical and physical properties of Mn^{2+}, Ni^{2+}, Fe^{3+} (and Fe^{2+}), Al^{3+}, Zn^{2+}
- To separate and identify the presence of one or more of the cations, Mn^{2+}, Ni^{2+}, $Fe^{3+}(Fe^{2+})$, Al^{3+}, Zn^{2+}, in an aqueous solution

TECHNIQUES

The following techniques are used in the Experimental Procedure

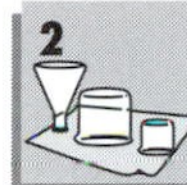

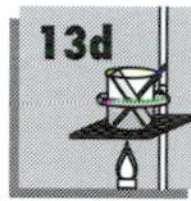

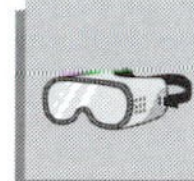

INTRODUCTION

The Qual II cations are for the most part more common and with the exception of Cu^{2+} are more colorful than are the Qual I cations. The Qual II cations are soluble as chlorides and as sulfides in an acidic solution (Qual I), but they form insoluble salts in sulfide-containing solutions at a pH near neutrality or higher. The ions Ni^{2+}, $Fe^{3+}(Fe^{2+})$, Zn^{2+}, and Mn^{2+} precipitate as sulfides and Al^{3+} as the hydroxide at a higher pH.

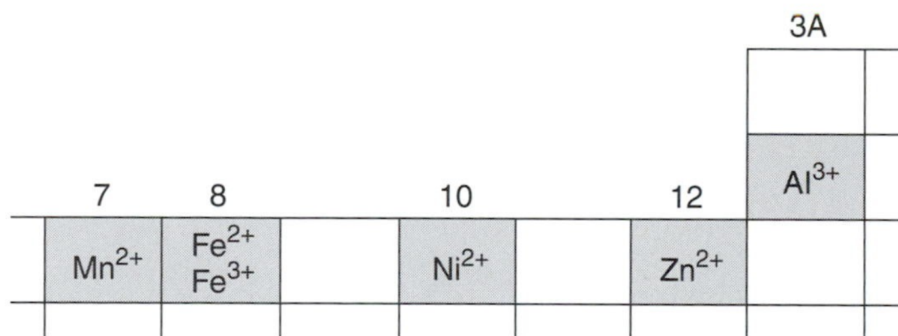

Cations of the Qual II Group

The addition of a base to a solution containing H_2S shifts the following equilibrium to the *right:*

$$\begin{array}{c} \rightarrow \\ H_2S(aq) + 2\,H_2O(l) \rightleftharpoons 2\,H_3O^+(aq) + S^{2-}(aq) \\ \downarrow \boldsymbol{2\,OH^-(aq)} \\ 4\,H_2O(l) \end{array} \quad (36.1)$$

As a result the sulfide concentration increases to a point where the product of the molar concentrations of the Qual II cation and the sulfide ion exceeds the K_{sp} of the sulfide salt, and the precipitate forms. The pH of the solution determines the sulfide ion concentration. In this basic solution, the Al^{3+} precipitates as $Al(OH)_3$ (Figure 36.1).

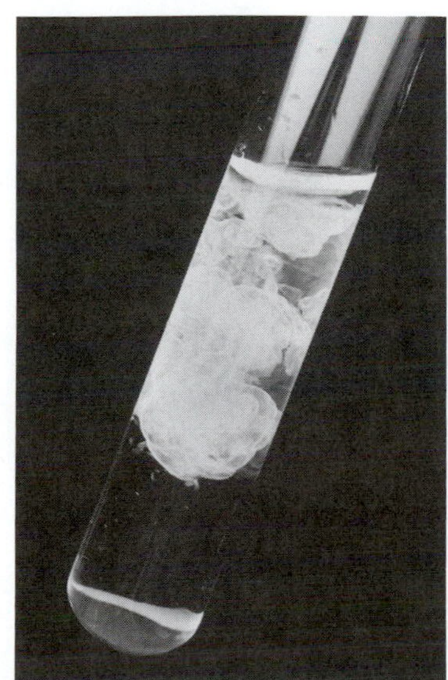

Figure 36.1 Formation of the aluminum hydroxide precipitate.

Thioacetamide, CH_3CSNH_2, provides the sulfide ion in the form of H_2S;[1] ammonia, a weak base, is used to adjust the pH of the equilibrium system. A weak base is used to adjust pH instead of a strong base because the addition of a strong base, such as NaOH, may cause precipitation of some Qual III cations (Experiment 37), which would then interfere with some Qual II tests.

Preparation of the Cations for Analysis—Separation from Qual III Cations

Hydrogen sulfide is also a mild reducing agent, reducing iron(III) ion to iron(II) ion in an aqueous solution, with elemental sulfur forming the oxidized product:

$$2\,Fe^{3+}(aq) + H_2S(aq) + 2\,H_2O(l) \rightarrow 2\,Fe^{2+}(aq) + S(s) + 2\,H_3O^+(aq) \quad (36.2)$$

Therefore, when a test solution containing all of the Qual II cations is treated with thioacetamide in a solution made basic with NH_3 and heated, precipitates of MnS, NiS, FeS, ZnS, and $Al(OH)_3$ form. The Qual III cations remain in solution.

Gelatinous: jellylike due to adsorbed and occluded water molecules

NiS and FeS are black precipitates, ZnS is white, MnS is pink, and $Al(OH)_3$ is a blue-white, **gelatinous** precipitate. After the addition of thioacetamide to a solution of Qual II cations, the precipitate may appear black because of the masking effect of the nickel(II) and iron(II) sulfides, if present.

The MnS, FeS, ZnS, and $Al(OH)_3$ precipitates dissolve with a strong acid; however, hot conc HNO_3 is necessary to dissolve the NiS. The HNO_3 oxidizes the sulfide ion of the NiS, forming free, elemental sulfur and NO as products; in addition, the HNO_3 oxidizes Fe^{2+} to Fe^{3+} in the solution.

$$3\,NiS(s) + 8\,H^+(aq) + 2\,NO_3^-(aq) \rightarrow 3\,Ni^{2+}(aq) + 2\,NO(g) + 3\,S(s) + 4\,H_2O(l) \quad (36.3)$$

$$3\,Fe^{2+}(aq) + 4\,H^+(aq) + NO_3^-(aq) \rightarrow 3\,Fe^{3+}(aq) + NO(g) + 2\,H_2O(l) \quad (36.4)$$

Amphoteric (also amphiprotic): the chemical property of a substance as having both acidlike and baselike properties

Therefore, HNO_3 is used to dissolve the precipitates forming the soluble Qual II cations, sulfur, and NO gas as the products.

Separation of Mn^{2+}, Ni^{2+}, and Fe^{3+} from Zn^{2+} and Al^{3+}

A strong base, OH^-, added to an aqueous solution of the five cations precipitates Mn^{2+}, Ni^{2+}, and Fe^{3+} as gelatinous hydroxides, but the Zn^{2+} and Al^{3+} hydroxides, being **amphoteric,** redissolve in excess base, forming the aluminate, $[Al(OH)_4]^-$, and zincate, $[Zn(OH)_4]^{2-}$, ions.

Structures of $[Al(OH)_4]^-$ and $[Zn(OH)_4]^{2-}$

$$Al^{3+}(aq) + 4\,OH^-(aq) \rightleftharpoons [Al(OH)_4]^-(aq) \quad (36.5)$$

$$Zn^{2+}(aq) + 4\,OH^-(aq) \rightleftharpoons [Zn(OH)_4]^{2-}(aq) \quad (36.6)$$

The gelatinous hydroxides of Mn^{2+}, Ni^{2+}, and Fe^{3+} dissolve with the addition of nitric acid, after the soluble aluminate and zincate ions are separated.

Manganese Ion

Structure of MnO_4^-

When a reference solution containing only the Mn^{2+}, Ni^{2+}, and Fe^{3+} ions is treated with sodium bismuthate, $NaBiO_3$, a strong oxidizing agent, the BiO_3^- ion oxidizes only the Mn^{2+} ion to the soluble *purple* permanganate anion, MnO_4^-. The color change confirms the presence of Mn^{2+} in the original test solution.

$$2\,Mn^{2+}(aq) + 5\,BiO_3^-(aq) + 14\,H^+(aq) \rightarrow \boxed{2\,MnO_4^-(aq)} + 5\,Bi^{3+}(aq) + 7\,H_2O(l) \quad (36.7)$$

[1]See Experiment 35, Introduction, Equation 35.1 and related discussion.

Iron Ion as Iron(III)

When a reference solution containing the Mn^{2+}, Ni^{2+}, and Fe^{3+} ions is treated with an excess of ammonia, NH_3, *brown* $Fe(OH)_3$ precipitates and the soluble *blue* hexaammine *complex*, $[Ni(NH_3)_6]^{2+}$, forms in solution.

$$Fe^{3+}(aq) + 3\,NH_3(aq) + 3\,H_2O(l) \rightarrow Fe(OH)_3(s) + 3\,NH_4^+(aq) \quad (36.8)$$

$$Ni^{2+}(aq) + 6\,NH_3(aq) \rightleftharpoons [Ni(NH_3)_6]^{2+}(aq) \quad (36.9)$$

Acid dissolves the $Fe(OH)_3$ precipitate; addition of thiocyanate ion, SCN^-, forms the *blood-red* thiocyanatoiron(III) complex, $[FeNCS]^{2+}$, a confirmation of the presence of Fe^{3+} in the test solution (see Experiment 25).

Structure of $[Ni(NH_3)_6]^{2+}$

$$Fe^{3+}(aq) + SCN^-(aq) \rightarrow \boxed{[FeNCS]^{2+}(aq)} \quad (36.10)$$

Other forms of the Fe^{3+}–SCN^- complex are $[Fe(NCS)_2]^+$, $[Fe(NCS)_4]^-$, and $[Fe(NCS)_6]^{3-}$ depending on the SCN^- concentration.

Nickel Ion

A supernatant from the test for iron(III) ion is used to test for the presence of nickel ion. The confirmation of Ni^{2+} ion in the reference solution is the appearance of a bright *pink* (or brick-red) *precipitate* formed with the addition of dimethylglyoxime, H_2DMG,[2] to a solution of the hexaamminenickel(II) complex ion.

$$[Ni(NH_3)_6]^{2+}(aq) + 2\,H_2DMG(aq) \rightarrow \boxed{Ni(HDMG)_2(s)} + 2\,NH_4^+(aq) + 4\,NH_3(aq) \quad (36.11)$$

Structure of $Ni(HDMG)_2(s)$

Aluminum Ion

A solution containing $[Al(OH)_4]^-$ and $[Zn(OH)_4]^{2-}$ ions, acidified with HNO_3, re-forms the Al^{3+} and Zn^{2+} ions. The subsequent addition of ammonia reprecipitates Al^{3+} as the gelatinous hydroxide, but Zn^{2+} forms the soluble tetraammine complex, $[Zn(NH_3)_4]^{2+}$:

$$[Al(OH)_4]^-(aq) + 4\,H^+(aq) \rightarrow Al^{3+}(aq) + 4\,H_2O(l) \quad (36.12)$$

$$\downarrow \mathbf{3\,NH_3(aq)}$$

$$Al(OH)_3(s) + 3\,NH_4^+(aq) + H_2O(l) \quad (36.13)$$

$$[Zn(OH)_4]^{2-}(aq) + 4\,H^+(aq) \rightarrow Zn^{2+}(aq) + 4\,H_2O(l) \quad (36.14)$$

$$\downarrow \mathbf{4\,NH_3(aq)}$$

$$[Zn(NH_3)_4]^{2+}(aq) \quad (36.15)$$

$Al(OH)_3$, an opaque, blue-white, gelatinous precipitate, is not easy to detect. To confirm the presence of Al^{3+} ion, the $Al(OH)_3$ precipitate is dissolved with HNO_3, aluminon reagent[3] is added, and the $Al(OH)_3$ is reprecipitated with the addition of NH_3. In the process of reprecipitating, the aluminon reagent, a red dye, adsorbs onto the surface of the gelatinous $Al(OH)_3$ precipitate, giving it a *pink/red* appearance, confirming the presence of Al^{3+} in the test solution.

$$Al^{3+}(aq) + 3\,NH_3(aq) + 3\,H_2O(l) + \text{aluminon}(aq) \rightarrow \boxed{Al(OH)_3 \bullet \text{aluminon}(s)} + 3\,NH_4^+(aq) \quad (36.16)$$

[2] Dimethylglyoxime, abbreviated as H_2DMG for convenience in this experiment, is an organic chelating agent, specific for the precipitation of the nickel ion.
[3] The aluminon reagent is the ammonium salt of aurin tricarboxylic acid, a red dye.

Zinc Ion

When potassium hexacyanoferrate(II), $K_4[Fe(CN)_6]$, is added to an acidified solution of the $[Zn(NH_3)_4]^{2+}$ ion, a *light green precipitate* of $K_2Zn_3[Fe(CN)_6]_2$ forms, confirming the presence of Zn^{2+} in the test solution.

$$3\,[Zn(NH_3)_4]^{2+}(aq) + 4\,H^+(aq) \rightarrow 3\,Zn^{2+}(aq) + 4\,NH_4^+(aq) \qquad (36.17)$$

$$\downarrow \mathbf{2\,K_4[Fe(CN)_6](aq)}$$

$$\boxed{K_2Zn_3[Fe(CN)_6]_2(s)} + 6\,K^+(aq) \qquad (36.18)$$

As a guide to an understanding of the separation and identification of these five cations, carefully read the Experimental Procedure and then complete the flow diagram on the Prelaboratory Assignment. Review Dry Lab 4 for an understanding of the symbolism in the flow diagram.

Experimental Procedure

Procedure Overview: Two solutions are tested with various reagents in this analysis: (1) a reference solution containing the Mn^{2+}, Ni^{2+}, Fe^{3+} (or Fe^{2+}), Al^{3+}, and Zn^{2+} ions of Qual II and (2) a test solution containing any number of Qual II cations. Separations and observations are made and recorded. Equations that describe the observations are also recorded. Comparative observations of the two solutions result in the identification of the cations in the test solution. All tests are qualitative; only identification of the cation(s) is required.

To become familiar with the identification of these cations, take a sample that contains the five cations (reference solution) and analyze it according to the procedure. At each circled, superscript (e.g., ①), *stop* and record on the Report Sheet. After each cation is confirmed, **save** it in the test tube so that its appearance can be compared with that of your test solution.

To analyze for cations in your test solution, place the test solution alongside the reference solution during the analysis. As you progress through the procedure, perform the same test on both solutions and make comparative observations. Check (√) the findings on the Report Sheet. Do not discard any solutions (but keep all solutions labeled) until the experiment is complete.

The test solution may be a water sample from some location in the environment, e.g., a lake, a stream, or a drinking water supply. Ask your instructor about this option.

Before proceeding, review the techniques outlined in Dry Lab 4. The review of these procedures may expedite your analysis with less frustration.

Contamination by trace amounts of metal ions in test tubes and other glassware leads to "unexplainable" results in qualitative analysis. Thoroughly clean all glassware with soap and tap water; rinse twice with tap water and twice with deionized water before use. **Caution:** *A number of concentrated and 6 M acids and bases are used in the analysis of these cations. Handle each of these solutions with care. Read the Laboratory Safety section for instructions in handling acids and bases.*

The expression "small test tube" that is mentioned throughout the Experimental Procedure refers to a 75-mm test tube (~3 mL volume) *or* a centrifuge tube of the size that fits into your laboratory centrifuge. Consult with your laboratory instructor.

A. Separation of Qual II from Qual III Cations

The Experimental Procedure is written for a single solution. If you are simultaneously identifying cations in *both* a reference solution *and* a test solution, adjust the procedure accordingly. If the test solution is a sample with an environmental origin, gravity filter 10–15 mL before beginning the Experimental Procedure.

If your sample contains only *Qual II cations, you will* not *need to precipitate the cations in a sulfide solution at the relatively high pH, but instead you may proceed directly to Part B.*

1. **Precipitate the Qual II Cations.** a. Transfer no more than 2 mL of a reference solution that may contain Qual III cations for analysis (Experiment 37) to a small test tube. The reference solution may be the supernatant solution from

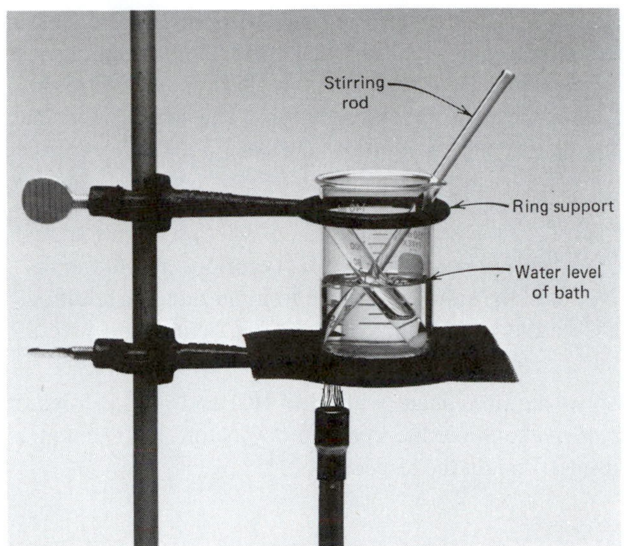

Figure 36.2 A hot water bath for sulfide precipitation of cations.

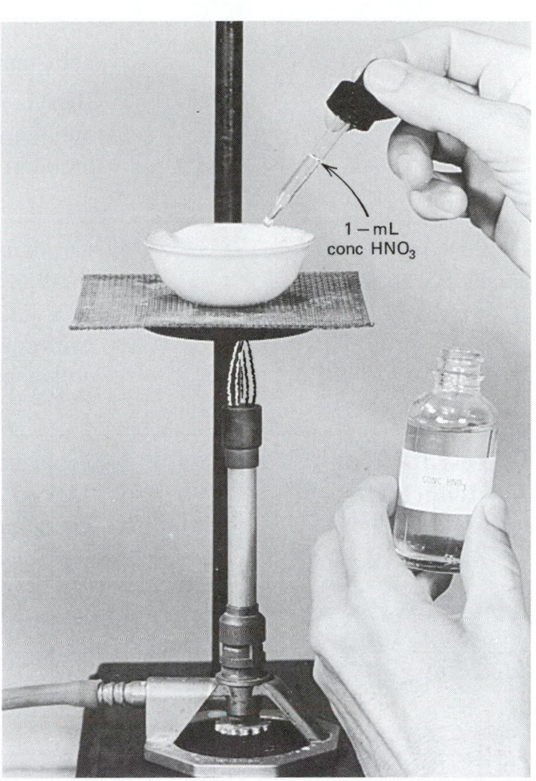

Figure 36.3 Addition of conc HNO_3 **(Caution!)** to the moist residue.

Experiment 35, Part E.1, or it may be an original reference solution of Qual II and Qual III cations.

b. Add 10 drops of 4 *M* NH_4Cl. Add 6 *M* NH_3 **(caution!)** until the solution is basic to litmus and then add 5 additional drops. Saturate the solution with H_2S by adding 6 drops of 1 *M* CH_3CSNH_2. Heat the solution in a hot water bath (~95°C, see Figure 36.2) for several minutes, cool, and centrifuge. Test for complete precipitation by adding several more drops of CH_3CSNH_2. Save the precipitate① for Part A.2 and the supernatant for analysis in Experiment 37.

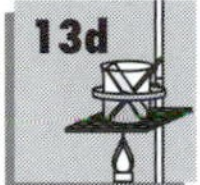

2. **Dissolve the Precipitates.** a. Wash the precipitate[4] twice with 1 mL of water and discard the washings. Add 9 drops of 6 *M* HCl and 3 drops of 6 *M* HNO_3 and again heat in the hot water bath until the precipitates dissolve. (**Caution:** *Be careful in handling acids.*) Cool and centrifuge; transfer the supernatant② to an evaporating dish and discard any free, elemental sulfur.

b. Heat the supernatant gently until a moist residue remains (do *not* heat to dryness). Add 1–2 mL of conc HNO_3 (Figure 36.3). Do not allow the conc HNO_3 to enter the rubber bulb! (**Caution:** *Concentrated HNO_3 is a strong oxidizing agent—do not allow contact with the skin or clothing!*) Reheat with a gentle flame until a moist residue again forms.

c. Dissolve the residue in 1–2 mL of water and transfer the solution to a small test tube.

B. Separation of Mn^{2+}, Ni^{2+}, and Fe^{3+} from Zn^{2+} and Al^{3+}

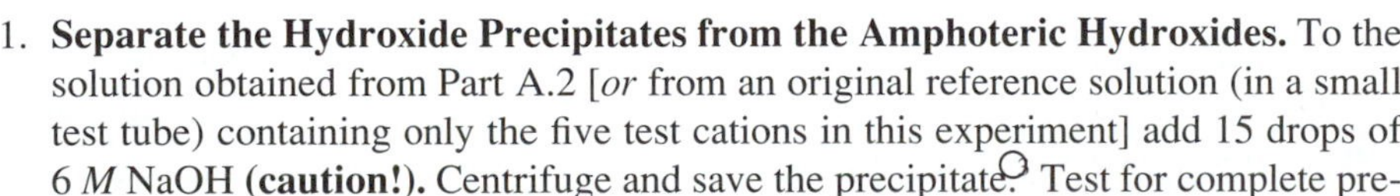

1. **Separate the Hydroxide Precipitates from the Amphoteric Hydroxides.** To the solution obtained from Part A.2 [*or* from an original reference solution (in a small test tube) containing only the five test cations in this experiment] add 15 drops of 6 *M* NaOH **(caution!).** Centrifuge and save the precipitate③ Test for complete pre-

[4]See Dry Lab 4 for the proper technique in washing precipitates.

cipitation by adding several drops of 6 *M* NaOH to the supernatant. Decant the supernatant④ into a small test tube and save for Part F.

2. **Dissolve the Hydroxide Precipitates.** Dissolve the precipitate⑤ with a minimum number of drops of conc HNO_3. (**Caution:** *Be careful!!*) If necessary, heat the solution in the hot water bath for several minutes.

C. Test for Manganese Ion

1. **Confirmatory Test.** Decant about one-half of the solution from Part B.2 into a small test tube; save the other half of the solution for Part D. Add a slight excess of solid $NaBiO_3$ to the solution and centrifuge.⑥ A soluble *deep purple* MnO_4^- confirms the presence of manganese in the test solution.[5] The MnO_4^- ion may only appear pink, depending on its concentration in solution.

D. Test for Iron Ion Iron(III)

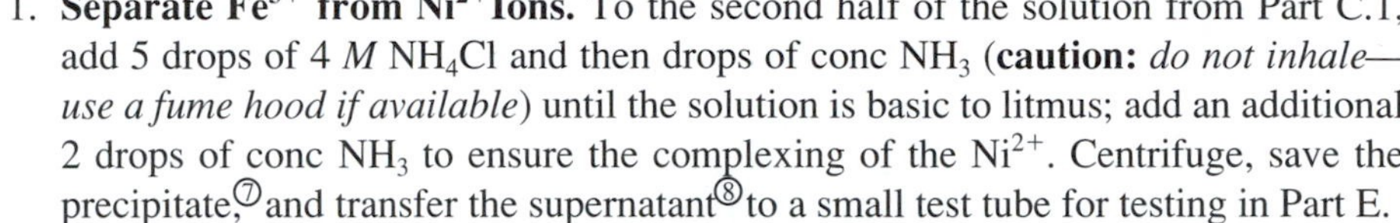

1. **Separate Fe^{3+} from Ni^{2+} Ions.** To the second half of the solution from Part C.1, add 5 drops of 4 *M* NH_4Cl and then drops of conc NH_3 (**caution:** *do not inhale—use a fume hood if available*) until the solution is basic to litmus; add an additional 2 drops of conc NH_3 to ensure the complexing of the Ni^{2+}. Centrifuge, save the precipitate,⑦ and transfer the supernatant⑧ to a small test tube for testing in Part E.

2. **Confirmatory Test.** Dissolve the precipitate with 6 *M* HCl and add 5 drops of 0.1 *M* NH_4SCN.⑨ The *blood-red* solution due to the thiocyanatoiron(III) complex confirms the presence of iron(III) ion in the test solution.

E. Test for Nickel Ion

1. **Confirmatory Test.** To the supernatant solution from Part D.1, add 3 drops of dimethylglyoxime solution.⑩ Appearance of a *pink* (brick-red) *precipitate* confirms the presence of nickel ion in the test solution.

F. Test for Aluminum Ion

1. **Separate Al^{3+} from Zn^{2+}.** Acidify the supernatant from Part B.1 to litmus with 6 *M* HNO_3. Add drops of 6 *M* NH_3 until the solution is now basic to litmus; then add 5 more drops. Heat the solution in the hot water bath for several minutes to **digest the** gelatinous **precipitate.**⑪ Centrifuge and decant the supernatant⑫ into a small test tube and save for the Zn^{2+} analysis in Part G.

To digest the precipitate: to make the precipitate more compact

2. **Confirmatory Test.** Wash the precipitate *twice* with 1 mL of hot, deionized water and discard each washing. Centrifugation is necessary after each washing. Add 6 *M* HNO_3 until the precipitate *just* dissolves. Add 2 drops of the aluminon reagent, stir, and add drops of 6 *M* NH_3 until the solution is again basic and a precipitate re-forms.⑬ Centrifuge the solution; if the $Al(OH)_3$ *precipitate* is now *pink or red* and the solution is colorless, Al^{3+} is present in the sample.

G. Test for Zinc Ion

1. **Confirmatory Test.** To the supernatant from Part F.1, add 6 *M* HCl until the solution is acid to litmus; then add 3 drops of 0.2 *M* $K_4[Fe(CN)_6]$ and stir. A very light green precipitate⑭ confirms the presence of Zn^{2+} in the sample. The precipitate is slow to form and difficult to see. Centrifugation may be necessary.

> *Disposal:* Dispose of all test solutions and precipitates in the "Waste Metal Salts" container.

CLEANUP: Rinse each test tube twice with tap water. Discard each rinse in the "Waste Metal Salts" container. Thoroughly clean each test tube with soap and tap water; rinse twice with tap water and twice with deionized water.

---

[5]If the deep purple color forms and then fades, Cl^- is present and is being oxidized by the MnO_4^-; add more $NaBiO_3$.

Experiment 36 *Prelaboratory Assignment*

Qual II. Mn^{2+}, Ni^{2+}, $Fe^{3+}(Fe^{2+})$, Al^{3+}, Zn^{2+}

Date __________ Lab Sec. ______ Name ____ Desk No. __________

1. Identify the Qual II cation(s) that is (are) confirmed present in a test solution by
   a. the formation of a precipitate

   b. the color of a soluble complex ion

2. Identify the Qual II cation that is confirmed present in a test solution as a result of an oxidation–reduction reaction.

3. Identify the reagent that separates
   a. Mn^{2+} from Al^{3+}. Explain the chemistry of the separation.

   b. Fe^{3+} from Ni^{2+}. Explain the chemistry of the separation.

   c. Al^{3+} from Zn^{2+}. Explain the chemistry of the separation.

4. Aluminum hydroxide is amphoteric but ferric hydroxide is not. Write equations to show the difference in the chemical property of the two ions.

5. Part A.1. The reference solution will form a black precipitate. If the test solution does *not* form a black precipitate, what can be concluded?

6. Complete the Qual III flow diagram at right.

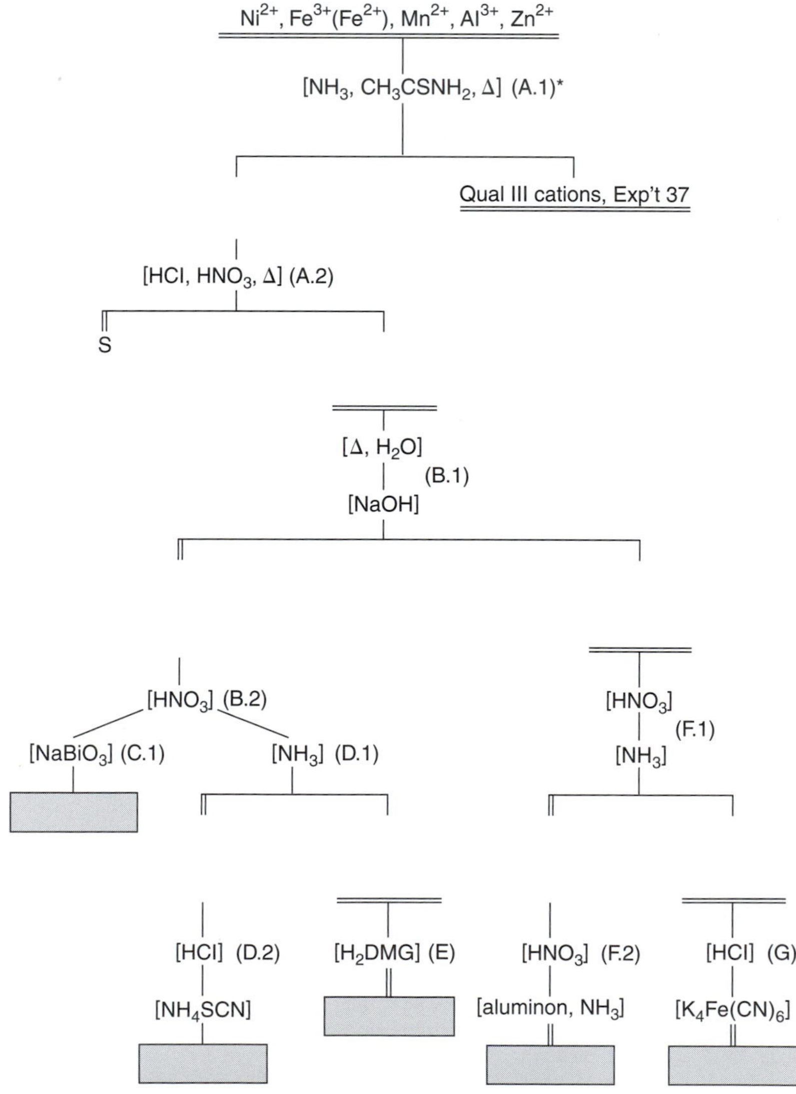

*Numbers in parentheses refer to parts of the Experimental Procedure.

Experiment 36 *Report Sheet*

Qual II. Mn^{2+}, Ni^{2+}, $Fe^{3+}(Fe^{2+})$, Al^{3+}, Zn^{2+}

Date __________ Lab Sec. ______ Name ______________________________ Desk No. __________

Procedure Number and Ion	Test Reagent or Technique	Evidence of Chemical Change*	Chemical(s) Responsible for Observation	Equation(s) for Observed Reaction	Check (√) if Observed in Unknown
①					
②					
③					
④					
⑤					
⑥ Mn^{2+}					☐
⑦ $Fe^{3+(2+)}$					
⑧					
⑨					☐
⑩ Ni^{2+}					☐
⑪ Al^{3+}					
⑫					
⑬					☐
⑭ Zn^{2+}					☐

*See Experiment 12, Introduction.

Cations present in unknown test solution no. _____: ______________________________

Instructor's approval: ______________________________

Laboratory Questions

Circle the questions that have been assigned.

1. Part A.1. What is the reagent that reduces Fe^{3+} to Fe^{2+}?
2. Part A.2b. The 6 *M* HNO_3 is omitted from the procedure because only 6 *M* HCl is available. What is the likely consequence of this omission?
3. Part B.1. Instead of 6 *M* NaOH being added to the test solution, 6 *M* NH_3 is added (both are bases). How will this affect the separation of the ions in the test solution? Explain.
4. Part D.1. Instead of conc NH_3 being added to the test solution, 6 *M* NaOH is added (both are bases). How will this affect the separation of the Fe^{3+} from the Ni^{2+} ions in the test solution? Explain.
5. Part F.2. Aluminon is a dye. Explain "how" aluminon is used in the detection of the aluminum ion in the test solution.
6. Part G.1. Why is the test solution acidified with 6 *M* HCl before the addition of $K_4[Fe(CN)_6]$? Explain.

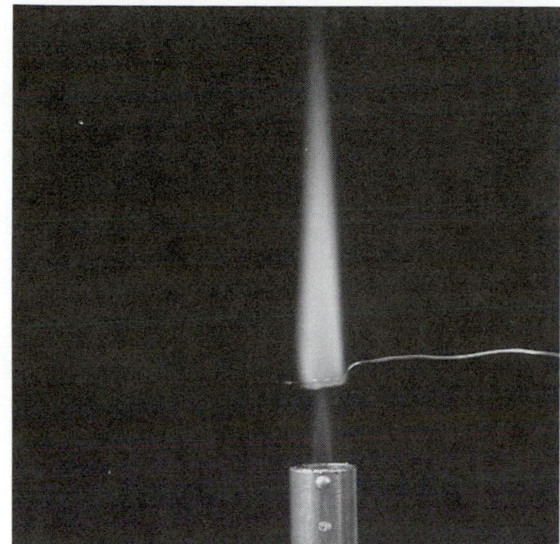

Experiment 37

Qual III. Mg^{2+}, Ca^{2+}, Ba^{2+}; General Unknown Examination

Flame tests provide a tentative identification for many metal ions.

Objectives

- To observe and utilize the chemical and physical properties of Mg^{2+}, Ca^{2+}, Ba^{2+}
- To separate and identify the presence of one or more of the cations, Mg^{2+}, Ca^{2+}, Ba^{2+}, in an aqueous solution
- To identify the presence of 2 to 5 cations from a total of 14 cations

Techniques

The following techniques are used in the Experimental Procedure

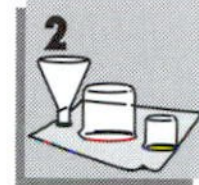

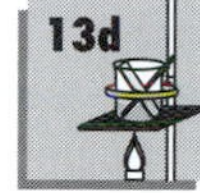

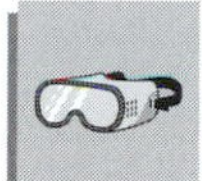

Introduction

Most salts of the Qual III cations are soluble; certainly the chloride (Experiment 35) and sulfide (Experiments 35 and 36) salts of Mg^{2+}, Ca^{2+}, and Ba^{2+} are soluble. As a result these cations appear as a supernatant from the Qual II analysis.

The separation and identification of the alkaline earth metal ions, Mg^{2+}, Ca^{2+}, and Ba^{2+}, require good laboratory technique. The anions SO_4^{2-}, $C_2O_4^{2-}$, and HPO_4^{2-} are used for the analyses. Some attention to pH adjustments is necessary.

Preparation of Qual III Cations

The supernatant from Qual II, Experiment 36, Part A.1, is reduced in volume by evaporating to dryness. This also removes any remaining sulfide ion from the solution.

Barium Ion

The sulfate salt of barium is insoluble. Barium sulfate forms a *white precipitate:*

$$Ba^{2+}(aq) + SO_4^{2-}(aq) \rightarrow \boxed{BaSO_4(s)} \quad (37.1)$$

A flame test of the wet precipitate produces a *yellow-green flame.*

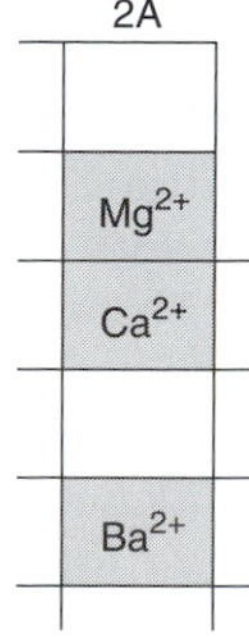

Cations of the Qual III group.

Calcium Ion

A *white precipitate* of calcium oxalate forms in an ammoniacal solution, confirming the presence of calcium ion in the test solution.

$$Ca^{2+}(aq) + C_2O_4^{2-}(aq) \rightarrow \boxed{CaC_2O_4(s)} \quad (37.2)$$

The dissolution of calcium oxalate with hydrochloric acid, followed by a flame test, produces a *yellow-red flame,* characteristic of the calcium ion.

Magnesium Ion

The addition of monohydrogen phosphate ion, HPO_4^{2-}, to an ammoniacal solution containing magnesium ion causes, after heating, the formation of a *white precipitate.*

$$Mg^{2+}(aq) + HPO_4^{2-}(aq) + NH_3(aq) \xrightarrow{\Delta} \boxed{MgNH_4PO_4(s)} \quad (37.3)$$

Magnesium ion does not produce a characteristic flame test.

General Unknown Examination (optional)

Qualitative tests for 14 cations were performed in Experiments 35 through 37. A maximum of 5 cations are contained in the general unknown solution. At the conclusion of the Qual III analysis, your instructor will designate which cations *may be* in your unknown.

As a guide to an understanding of the separation and identification of the Qual III cations, carefully read the Experimental Procedure and then construct a flow diagram on the Prelaboratory Assignment. Review Dry Lab 4 for an understanding of the symbolism.

Experimental Procedure

Procedure Overview: Two solutions are tested with various reagents in this analysis: (1) a reference solution containing the Mg^{2+}, Ca^{2+}, and Ba^{2+} ions of Qual III and (2) a test solution containing any number of Qual III cations. Separations and observations are made and recorded. Equations that describe the observations are also recorded. Comparative observations of the two solutions result in the identification of the cations in the test solution. All tests are qualitative; only identification of the cation(s) is required.

At the conclusion of the Qual III analysis of the unknown solution, a general unknown, containing cations from Qual I, II, and III, may be issued and analyzed, using the procedures outlined in Experiments 35, 36, and 37.

To become familiar with the identification of the Qual III cations, take a sample that contains the three cations (reference solution) and analyze it according to the procedure. At each circled, superscript (e.g., ①), *stop* and record on the Report Sheet. After each cation is confirmed, **save** it in the test tube so that its appearance can be compared with that for your test solution.

To analyze for cations in your test solution, place the test solution alongside the reference solution during the analysis. As you progress through the procedure, perform the same test on both solutions and make comparative observations. Check (√) the findings on the Report Sheet. Do not discard any solutions (but keep all solutions labeled) until the experiment is complete.

The test solution may be a water sample from some location in the environment, e.g., a lake, a stream, or a drinking water supply. Ask your instructor about this option.

Before proceeding, review the techniques outlined in Dry Lab 4. The review of these procedures may expedite your analysis with less frustration.

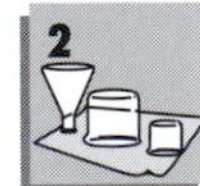

Contamination by trace amounts of metal ions in test tubes and other glassware leads to "unexplainable" results in qualitative analysis. Thoroughly clean all glassware with soap and tap water; rinse twice with tap water and twice with deionized water before use.

Caution: *A number of concentrated and 6 M acids and bases are used in the analysis of these cations. Handle each of these solutions with care. Read the Laboratory Safety section for instructions in handling acids and bases.*

The expression "small test tube" that is mentioned throughout the Experimental Procedure refers to a 75-mm test tube (~3 mL volume) *or* a centrifuge tube of the size that fits into your laboratory centrifuge. Consult with your laboratory instructor.

A. Preparation of Qual III Cations for Analysis

The Experimental Procedure is written for a single solution. If you are simultaneously identifying cations in *both* a reference solution *and* a test solution, adjust the procedure accordingly. If the test solution is a sample with an environmental origin, gravity filter 10–15 mL before beginning the Experimental Procedure.

If your sample contains only the cations in this experiment, you will not *need to remove any sulfide ion from solution. Proceed directly to Part B.*

1. **Remove the Excess Sulfide Ion.** a. Transfer the supernatant from Qual II (Experiment 36, Experimental Procedure, Part A.1) to an evaporating dish, supported on a wire gauze, and evaporate to a moist residue (Figure 37.1). Allow the contents to cool.

   b. Dissolve the moist residue with about 1 mL of 6 *M* HCl and transfer to a small test tube.

B. Test for Barium Ion

1. **Confirmatory Test.** To the reference solution from Part A.1 [or to $1\frac{1}{2}$ mL of the reference solution (in a small test tube) containing only the three Qual III cations] add drops of 6 *M* H_2SO_4 until any precipitation appears complete. A *white precipitate* confirms the presence of Ba^{2+} in the sample.① Centrifuge, decant, and save the supernatant for Part C.
2. **Perform a Flame Test.** Add a few drops of conc HCl to the moist precipitate in the test tube and perform a flame test.[1] (**Caution:** *Do not allow skin contact; avoid inhalation; clean up any spills.*) A weak *yellow-green flame* is characteristic of barium ion.②

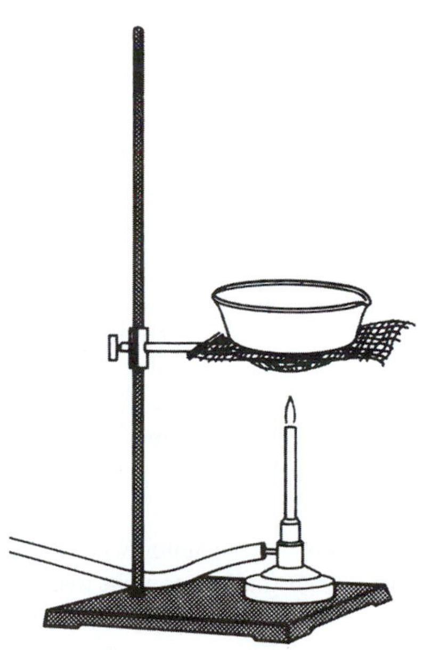

Figure 37.1 Heat the test solution to a moist residue.

[1]Review the technique for performing a flame test, described in Experiment 35.

C. Test for Calcium Ion

1. **Confirmatory Test.** To the supernatant from Part B.1, add 6 *M* NH_3 until the solution is basic to litmus. Add 2–3 drops of 1 *M* $K_2C_2O_4$. A *white precipitate* confirms the presence of Ca^{2+} in the sample.③ If no precipitate appears immediately, warm the solution in a water bath. Centrifuge, decant, and save the supernatant for Part D.
2. **Perform a Flame Test.** Wet the precipitate to a moist paste with 6 *M* HCl and perform a flame test. A fleeting *yellow-red flame* is characteristic of calcium ion.④

D. Test for Magnesium Ion

1. **Confirmatory Test.** Add 1–2 drops of 6 *M* NH_3 to the supernatant from Part C.1. Add 2–3 drops of 1 *M* Na_2HPO_4, heat in a hot water (~90°C) bath, and allow to stand. The precipitate⑤ may be slow in forming; be patient.

Disposal: Dispose of all test solutions and precipitates in the "Waste Metal Salts" container.

CLEANUP: Rinse each test tube twice with tap water. Discard each rinse in the "Waste Metal Salts" container. Thoroughly clean each test tube with soap and tap water; rinse twice with tap water and twice with deionized water.

E. General Unknown Examination

1. **Assignment of Cations.** Consult with your instructor to determine to which of the five cations from Qual I, II, and III your unknown is limited. Record these at the top of your flow diagram (see next paragraph).
2. **Design the Flow Diagram.** Prior to the examination and on your own paper, construct a flow diagram for only the cations that your instructor assigns. Use the flow diagrams from Experiments 34–37 as guides in its construction.

   On the flow diagram, list only the reagents (or reaction conditions) that are necessary to separate and confirm the presence (or absence) of your list of cations. Omit all unnecessary steps, but be sure to include vital steps. Do *not* include details such as drops, volume, decant, and centrifuge. The reasons for these additions and techniques should be understood from the "chemistry" of the separation and from the proper construction of the flow diagram (Dry Lab 4). You *must know the chemistry of the cations* involved in your unknown.

   Your instructor will approve your proposed flow diagram *prior* to your receiving the general unknown. Time may be a factor in your analysis, so be prepared.
3. **Analysis of the General Unknown.** Critically observe the unknown solid sample for color(s). Dissolve the solid in ~2 mL of deionized water, add several drops of 6 *M* HCl, and again observe. Can any predictions (or guesses) be made? Depending on the cations in your unknown, begin with the appropriate part in the Experimental Procedures for analysis of the cations appearing in Experiments 35, 36, and 37; omit all parts of the procedures that do not relate to the separation and identification of the cations in *your* unknown.
4. **Report.** At least two analyses of your unknown with the same results should be completed before reporting your findings to your instructor.

Experiment 37 *Prelaboratory Assignment*

Qual III. Mg^{2+}, Ca^{2+}, Ba^{2+}; General Unknown Examination

Date __________ Lab Sec. ______ Name ____ Desk No. __________

1. Identify the Qual III cation(s) that is (are) confirmed present in a test solution by
   a. the formation of a precipitate

   b. the characteristic color of a flame test

2. If a flame test is conducted on the original reference solution of the Qual III cations, what would be concluded?

3. Identify the reagent that separates
   a. Mg^{2+} from Ca^{2+}. Explain the chemistry of the separation.

   b. Ba^{2+} from Mg^{2+}. Explain the chemistry of the separation.

4. a. The precipitation of the calcium ion with the oxalate ion must be done in a basic solution. Why might the oxalate ion *not* precipitate the calcium if the solution were acidic? *Hint:* base your explanation using Brønsted acid-base theory.

b. Similarly the precipitation of the magnesium ion with the monohydrogenphosphate ion must occur in a basic solution. Explain.

5. Construct a flow diagram for the three cations in this experiment. Read the Introduction and Experimental Procedure to help you in its construction.

Experiment 37 *Report Sheet*

Qual III. Mg^{2+}, Ca^{2+}, Ba^{2+}; General Unknown Examination

Date __________ Lab Sec. ______ Name ____ Desk No. __________

Procedure Number and Ion	Test Reagent or Technique	Evidence of Chemical Change*	Chemical(s) Responsible for Observation	Equation(s) for Observed Reaction	Check (√) if Observed in Unknown
① Ba^{2+}	________	________	________	________	☐
②	flame	________	________	________	☐
③ Ca^{2+}	________	________	________	________	☐
④	flame	________	________	________	☐
⑤ Mg^{2+}	________	________	________	________	☐

*See Experiment 12, Introduction.

Cations present in unknown test solution no. _____: ____

Instructor's approval: ____

Laboratory Questions

Circle the questions that have been assigned.

1. Part B.1. A precipitate does not appear with the addition of H_2SO_4 to the test solution. What is the next step in the Experimental Procedure? Explain.
2. Part B.1. Instead of 6 *M* H_2SO_4 being added to the test solution, 6 *M* HCl is added (both are acids). What will be observed as a result? Explain.
3. Part C.1. Instead of 6 *M* NH_3 being added to the test solution, 6 *M* NaOH is added (both are bases) before the addition of the $K_2C_2O_4$. What would be the appearance of the solution? Explain.
4. Part D.1. Instead of 6 *M* NH_3 being added to the test solution, 6 *M* NaOH is added (both are bases). What would be the appearance of the solution? Explain.

General Unknown Examination

Submit your flow diagram for your five cations that were assigned by the instructor. At the bottom of your flow diagram, record the cations that you found present in your unknown; submit the flow diagram and your findings to your laboratory instructor.

Instructor's approval of flow diagram: ____

Cations reported in the general unknown test solution no. _____: ________________________________

Laboratory Questions for General Unknown

Circle the questions that have been assigned.

1. Identify a reagent (as suggested in Experiments 35–37) that will separate
   a. Ag^+ from Mn^{2+}. Explain the chemistry of the separation.
   b. Fe^{3+} from Mg^{2+}. Explain the chemistry of the separation.
   c. Cu^{2+} from Ba^{2+}. Explain the chemistry of the separation.
   d. Bi^{3+} from Al^{3+}. Explain the chemistry of the separation.
2. Devise a flow diagram, as concise as possible, for identifying the presence of Cu^{2+}, Zn^{2+}, and Mn^{2+} in a test solution.
3. Devise a flow diagram, as concise as possible, for identifying the presence of Na^+, Ca^{2+}, and Ba^{2+} in a test solution.

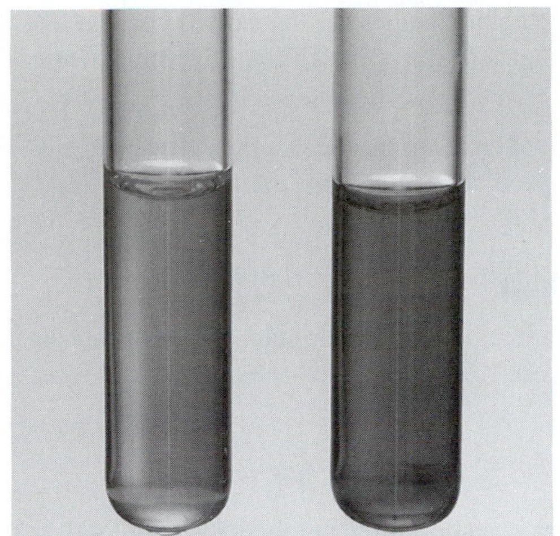

Experiment 38

Transition Metal Chemistry

The appearance of a nickel(II) ion complex depends on the ligands; at left is $[Ni(H_2O)_6]^{2+}$ and at right is $[Ni(NH_3)_6]^{2+}$.

OBJECTIVES

- To observe the various colors associated with transition metal ions
- To determine the relative strengths of ligands
- To compare the stability of complexes
- To synthesize a coordination compound

TECHNIQUES

The following techniques are used in the Experimental Procedure

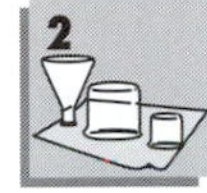
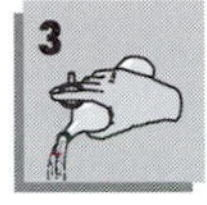

INTRODUCTION

One of the most intriguing features of the transition metal ions is their vast array of colors. The blues, greens, and reds that we associate with chemicals are oftentimes due to the presence of transition metal ions. We observed these colors in Experiment 12, where it was noted that some colors are characteristic of certain hydrated transition metal ions; for example, hydrated Cu^{2+} salts are blue, Ni^{2+} salts are green, and Fe^{3+} salts are rust-colored.

A second interesting feature of the transition metal ions is that subtle, and on occasion very significant, color changes occur when molecules or ions other than water bond to the metal ion to form a **complex.** These molecules or ions including water, called **ligands,** are Lewis bases (electron pair donors) which bond directly to the metal ion, producing a change in the electronic energy levels of the metal ion. As a result, the energy (and also the wavelength) of light absorbed by the electrons in the transition metal ion and, consequently, the energy (and wavelength) of light transmitted also change.[1] The solution has a new color.

Complex: an ion or molecule formed between a metal ion (a Lewis acid) and anions or molecules (Lewis bases)

Ligand: a Lewis base (an electron-pair donor) that combines with a metal ion to form a complex

The complex has a number of ligands bonded to the transition metal ion which form a **coordination sphere.** The complex along with its neutralizing ion is called a **coordination compound.** For example, $K_4[Fe(CN)_6]$ is a coordination compound: the six CN^- ions are the ligands, $Fe(CN)_6$ is the coordination sphere, and $[Fe(CN)_6]^{4-}$ is the complex.

Coordination sphere: the metal ion and all ligands of the complex

Coordination compound: a neutral compound which has a cationic and/or an anionic complex ion

[1]See Dry Lab 3 for a discussion of the theory of the origin of color in substances.

Figure 38.1 A photographic negative and a positive print made from it.

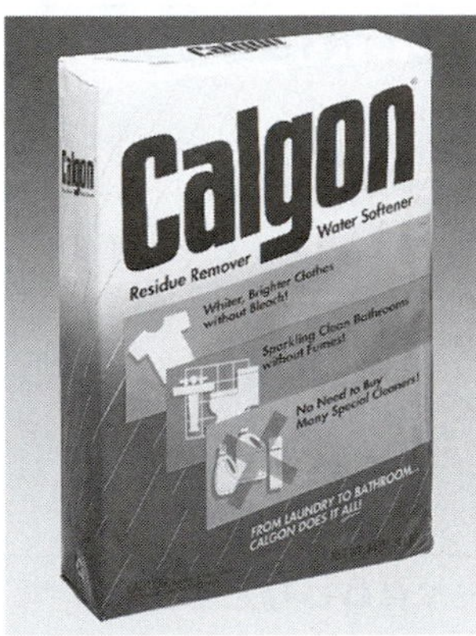

Sodium tripolyphosphate, the major component of Calgon®, complexes and solubilizes $Ca^{2+}(aq)$—thus the undesirable "calcium is gone" from the wash water!

Figure 38.2 The complex shows three ethylenediamine ligands, each as a bidentate ligand. Ethylenediamine is a chelating agent.

More than the intrigue of the color and color changes are the varied uses that these complexes have in chemistry. A few examples of the applicability of complex formation follow:

- For photographic film development (Figure 38.1), sodium thiosulfate, $Na_2S_2O_3$, called **hypo,** removes the unsensitized silver ion from the film in the form of the soluble silver complex, $[Ag(S_2O_3)_2]^{3-}$:

$$AgBr(s) + 2\ S_2O_3^{2-}(aq) \rightarrow [Ag(S_2O_3)_2]^{3-}(aq) + Br^-(aq)$$

- For the removal of calcium ion hardness from water, soluble polyphosphates, such as sodium tripolyphosphate, $Na_5P_3O_{10}$, are added to detergents to form a soluble calcium complex:

$$Ca^{2+}(aq) + P_3O_{10}^{5-}(aq) \rightarrow [CaP_3O_{10}]^{3-}(aq)$$

- For mercury and lead poisoning by ingestion, mercaptol (British antilewisite), $C_3H_8OS_2$, is swallowed for purposes of forming the complex with the metal ion and rendering the "free" ion ineffective:

$$Hg^{2+}(aq)\ [\text{or } Pb^{2+}\ (aq)] + 2\ \underset{SH}{\underset{|}{CH_2}}—\underset{SH}{\underset{|}{CH}}—\underset{OH}{\underset{|}{CH_2}}(aq) \rightarrow [Hg(C_3H_8OS_2)_2]^{2+}(aq)$$

- In the qualitative identification of transition metal ions in earlier experiments, many of the transition metal ions were confirmed present as a result of complex formation: copper ion as $[Cu(NH_3)_4]^{2+}$, nickel ion as $Ni(HDMG)_2$, iron(III) ion as $[FeNCS]^{2+}$, and zinc ion as $[Zn_3[Fe(CN)_6]_2]^{2-}$.
- Trace amounts of metal ions, such as zinc, aluminum, and iron, tend to catalyze the air oxidation (the spoilage) of various foods. To retard spoilage and extend the shelf life, some food companies add a small amount of calcium disodium ethylenediaminetetraacetate, $CaNa_2EDTA$, to their product. The $EDTA^{4-}$ (Table 38.1) complexes (sequesters) the metal ions and nullifies their catalytic activity.

The bond strength between the transition metal ion and its ligands varies, depending on the electron pair donor (Lewis base) strength of the ligand and the electron pair acceptor (Lewis acid) strength of the transition metal ion. Ligands may be neutral (e.g., H_2O, NH_3, $H_2NCH_2CH_2NH_2$) or anionic (e.g., CN^-, SCN^-, Cl^-).

A single ligand may form one bond to the metal ion (a **monodentate** ligand), two bonds to the metal ion (a **bidentate** ligand), three bonds to the metal ion (a **tridentate** ligand), and so on. Ligands that form two or more bonds to a transition metal ion are also called **chelating agents** and **sequestering agents** (Figure 38.2). In Table 38.1 note that a nonbonding electron pair (a Lewis base) is positioned on each atom of the ligand that serves as a bonding site to the transition metal ion.

The complex formed between a chelating agent (a polydentate ligand) and a metal ion is generally *more* stable than that formed between a monodentate ligand and a metal ion. The explanation is that the several ligand–metal bonds between a polydentate ligand and a metal ion are more difficult to break than a single bond between a monodentate ligand and a metal ion. The stability of complexes having polydentate ligands will be compared with the stability of those having only monodentate ligands in this experiment.

The *number of bonds* between a metal ion and its ligands is called the **coordination number** of the complex. If four monodentate ligands or if two bidentate ligands bond to a metal ion, the coordination number for the complex is 4. Six water molecules, or six cyanide ions, or three ethylenediamine (a bidentate ligand) molecules (see Figure 38.2), or two diethylenetriamine (a tridentate ligand) molecules bonded to a given metal ion all form a complex with a coordination number of 6

Table 38.1 Common Ligands

Monodentate Ligands

H_2O:, :NH_3, :CN^-, :SCN^-, :$S_2O_3^{2-}$, :F^-, :Cl^-, :Br^-, :I^- :OH^-

Bidentate Ligands (chelating agents)

Ethylenediamine (en)

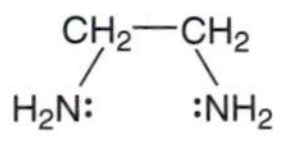

Oxalate ion

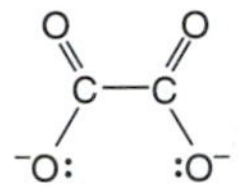

Mercaptol

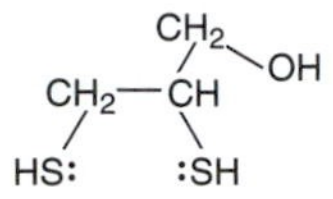

Tartrate ion

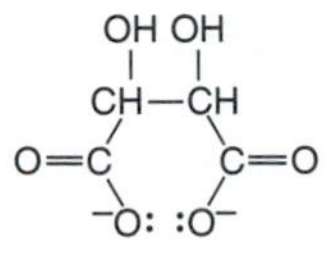

Polydentate Ligands (chelating agents)

Diethylenetriamine

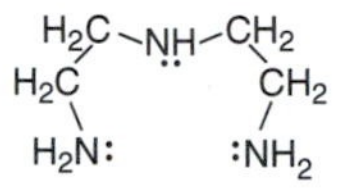

Tripolyphosphate ion, $P_3O_{10}^{5-}$

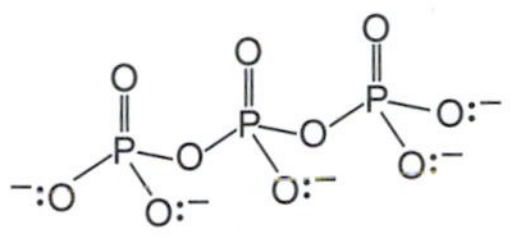

Ethylenediaminetetraacetate ion, $EDTA^{4-}$

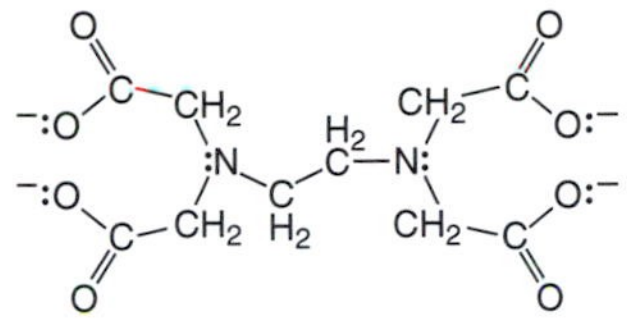

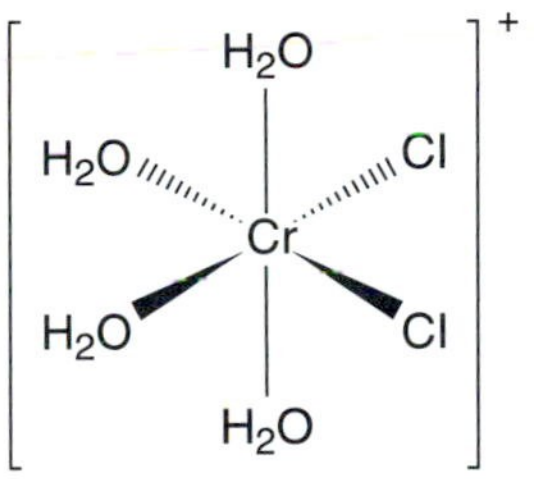

Figure 38.3 The complex has a coordination number of six, each ligand is monodentate.

Table 38.2 Common Coordination Numbers of Some Transition Metal Ions

Coordination Number	Transition Metal Ions
2	Ag^+, Au^+
4	Hg^{2+}, Cu^{2+}, Ni^{2+}, Co^{2+}, Zn^{2+}, Pd^{2+}, Pt^{2+}, Au^{3+}
6	Co^{2+}, Co^{3+}, Fe^{2+}, Fe^{3+}, Cr^{3+}, Ni^{2+}, Cu^{2+}

(Figure 38.3). Coordination numbers of 2, 4, and 6 are most common among the transition metal ions.

Table 38.2 lists the common coordination numbers for some transition metal ions.

In Parts A, B, C, and D of this experiment we will observe the formation of complexes between Cu^{2+}, Ni^{2+}, and Co^{2+} ions and the ligands Cl^-, H_2O, NH_3, ethylenediamine (en), and the thiocyanate ion, SCN^-. In specified examples, we will determine the stability of the complex.

Stablity of a Complex

The stability of a complex can be determined by mixing it with an anion known to form a precipitate with the cation. The anion most commonly used to measure the stability of a complex is the hydroxide ion. For example, consider the copper(II) ion and the generic mondentate ligand, X^-, forming the complex $[CuX_4]^{2-}$:

$$Cu^{2+}(aq) + 4\,X^-(aq) \rightleftharpoons [CuX_4]^{2-}(aq)$$

When hydroxide ion is added to this system (in a state of dynamic equilibrium), the "free" copper(II) ion has a choice of now combining with the anion that forms the stronger bond—in other words, a competition for the copper(II) ion between the two anions exists in solution. If the ligand forms the stronger bond to the copper(II) ion, the complex remains in solution and no change is observed; if, on the other hand, the hydroxide forms a stronger bond to the copper(II) ion, $Cu(OH)_2$ precipitates from solution.

$$Cu^{2+}(aq) + 2\ OH^-(aq) \rightleftharpoons Cu(OH)_2(s)$$

Therefore, a measure of the stability of the complex is determined.

To write the formulas of the complexes in this experiment, we will assume that the coordination number of Cu^{2+} is always 4 and that of Ni^{2+} and Co^{2+} is always 6.

Parts E, F, and G outline the syntheses of several coordination compounds:

- tetraamminecopper(II) sulfate monohydrate, $[Cu(NH_3)_4]SO_4 \cdot H_2O$
- hexaamminenickel(II) chloride, $[Ni(NH_3)_6]Cl_2$
- *tris*(ethylenediamine)nickel(II) chloride dihydrate, $[Ni(en)_3]Cl_2 \cdot 2H_2O$

Tetraamminecopper(II) Sulfate Monohydrate

Ammonia is added to an aqueous solution of copper(II) sulfate pentahydrate, $[Cu(H_2O)_4]SO_4 \cdot H_2O$. Ammonia displaces the four water ligands from the coordination sphere:

$$[Cu(H_2O)_4]SO_4 \cdot H_2O + 4\ NH_3 \xrightarrow{\text{aqueous}} [Cu(NH_3)_4]SO_4 \cdot H_2O + 4\ H_2O \qquad (38.1)$$

The solution is cooled and 95% ethanol is added to reduce the solubility of the *blue* tetraamminecopper(II) sulfate salt.

Hexaamminenickel(II) Chloride

For the preparation of hexaamminenickel(II) chloride, six ammonia molecules displace the six water ligands in $[Ni(H_2O)_6]Cl_2$.

$$[Ni(H_2O)_6]Cl_2 + 6\ NH_3 \xrightarrow{\text{aqueous}} [Ni(NH_3)_6]Cl_2 + 6\ H_2O \qquad (38.2)$$

The $[Ni(NH_3)_6]Cl_2$ coordination compound is cooled and precipitated with 95% ethanol; the lower polarity of the 95% ethanol causes the *lavender* salt to become less soluble.

Tris(ethylenediamine)nickel (II) Chloride Dihydrate

Three ethylenediamine (en) molecules displace the six water ligands of $[Ni(H_2O)_6]Cl_2$ for the preparation of the *violet tris*(ethylenediamine)nickel(II) chloride dihydrate.

$$[Ni(H_2O)_6]Cl_2 + 3\ en \xrightarrow{\text{aqueous}} [Ni(en)_3]Cl_2 \cdot 2H_2O + 4\ H_2O \qquad (38.3)$$

Experimental Procedure

Procedure Overview: Several complexes of Cu^{2+}, Ni^{2+}, and Co^{2+} are formed and studied. The observations of color change that result from the addition of a ligand are used to understand the relative stability of the various complexes that form. One or more coordination compounds are synthesized and isolated.

Notes on observation. Some colors may be difficult to distinguish. If a precipitate initially forms when a solution of the ligand is added, add more of the solution. *Always* compare the test solution with the original aqueous solution. On occasion you may need to discard some of the test solution if too much of the original solution was used for testing at the outset. If a color change occurs (*not* a change in color intensity) then a new complex has formed. Also, after each addition of the ligand-containing solution, tap the test tube to agitate the mixture (Figure 38.4) and view the solution at various angles to note the color (change) (Figure 38.5).

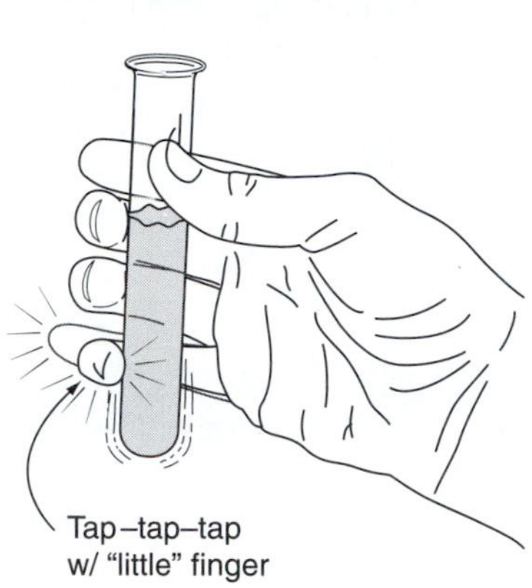

Figure 38.4 Shake the contents of the test tube with the "little" finger.

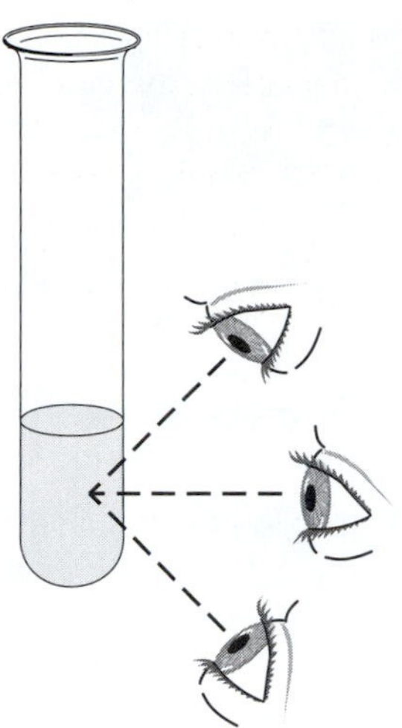

Figure 38.5 View solution from all angles.

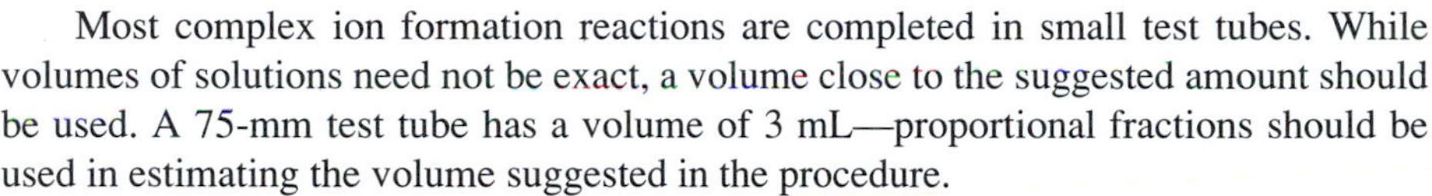

Caution: *Several of the ligand-containing solutions should be handled with care. The conc HCl, conc NH_3, and ethylenediamine reagents produce characteristic odors that are strong skin and respiratory irritants. Use "drops" as suggested and avoid inhalation and skin contact.*

Most complex ion formation reactions are completed in small test tubes. While volumes of solutions need not be exact, a volume close to the suggested amount should be used. A 75-mm test tube has a volume of 3 mL—proportional fractions should be used in estimating the volume suggested in the procedure.

From three to six small test tubes are used for Parts A, B, C, and D of the experiment. Plan to keep them clean, rinsing thoroughly after each use. However, do *not* discard any test solutions until the entire "Part" of the Experimental Procedure has been completed.

For Parts A–D, perform the experiment with a partner. At each circled superscript(1–9) in the procedure, *stop,* and record your observation on the Report Sheet. Discuss your observations with your lab partner and your instructor. Consult with your laboratory instructor to determine which of the coordination compounds outlined in Parts E, F, and G you are to synthesize.

A. Chloro Complexes of the Copper(II), Nickel(II), and Cobalt(II) Ions

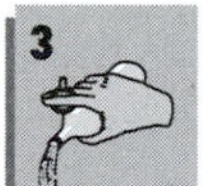

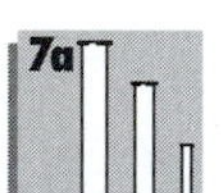

1. **Form the Complexes.** Place about 0.5 mL of 0.1 *M* $CuSO_4$, 0.1 *M* $Ni(NO_3)_2$, and 0.1 *M* $CoCl_2$ into each of three separate small test tubes.① Add 1 mL (20 drops) of conc HCl to each. (**Caution:** *Do not allow conc HCl to contact skin or clothing. Flush immediately the affected area with water.*) Tap the test tube to agitate. Record your observations on the Report Sheet.②
2. **Dilute the Complexes.** Slowly add 1–2 mL of water to each test tube. Compare the colors of the solutions to about 2.5 mL of the original solutions. Does the original color return? Record.③

B. Complexes of the Copper(II) Ion

1. **Form the Complexes.** Place about 0.5 mL of 0.1 *M* $CuSO_4$ in each of five test tubes and transfer them to the fume hood. Add 5 drops of conc NH_3 (**caution:** *avoid breathing vapors*) to the first test tube, 5 drops of ethylenediamine (**caution:** *avoid breathing its vapors, too*) to the second, 5 drops of 0.1 *M* KSCN to the third test tube, and nothing more to a fourth test tube (Figure 38.6).

   The fifth test tube can be used to form a complex with a ligand selected by your instructor—the thiosulfate anion, $S_2O_3^{2-}$, the oxalate anion, $C_2O_4^{2-}$, the

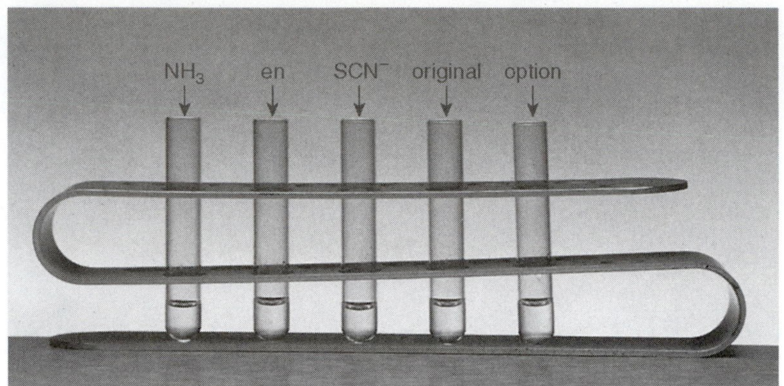

Figure 38.6 Set up of five test tubes for ligand addition to a metal ion.

tartrate anion, $(CHOH)_2(CO_2)_2^{2-}$, and the ethylenediaminetetraacetate anion, $EDTA^{4-}$, readily form complexes and are suggested.

If a precipitate forms in any of the solutions, *add an excess* of the ligand-containing solution. Compare the appearance of the test solutions with the 1 mL of original $CuSO_4$ solution in the fourth test tube.④

2. **Test for Stability.** Add 3–5 drops of 1 *M* NaOH to each of the five solutions. Account for your observations.⑤

C. Complexes of the Nickel(II) Ion

1. **Form the Complexes.** Repeat Part B.1, substituting 0.1 *M* $NiCl_2$ for 0.1 *M* $CuSO_4$. Record your observations.⑥
2. **Test for Stability.** Add 3–5 drops of 1 *M* NaOH to each test solution. Account for your observations.⑦

D. Complexes of the Cobalt(II) Ion

1. **Form the Complexes.** Repeat Part B.1, substituting 0.1 *M* $Co(NO_3)_2$ for 0.1 *M* $CuSO_4$. Record your observations.⑧
2. **Test for Stability.** Add 3–5 drops of 1 *M* NaOH to each test solution. Account for your observations.⑨

Disposal: Dispose of the waste solutions from Parts A, B, C, and D in the "Waste Metal Ion Solutions" container.

CLEANUP: Rinse the test tubes twice with tap water and twice with deionized water and discard in the "Waste Metal Ion Solution" container. Additional rinses can be discarded in the sink, followed by a generous amount of tap water.

E. Synthesis of Tetraamminecopper(II) Sulfate Monohydrate

1. **Dissolve the Starting Material.** Measure 6 g (±0.01 g) of copper(II) sulfate pentahydrate, $[Cu(H_2O)_4]SO_4{\bullet}H_2O$, in a clean, dry 125-mL Erlenmeyer flask. Dissolve the sample in 15 mL of deionized water. Heating may be necessary. Transfer the beaker to the fume hood.
2. **Precipitate the Complex Ion.** Add conc NH_3 (**caution:** *do not inhale*) until the precipitate that initially forms has dissolved. Cool the deep blue solution in an ice bath. Cool 20 mL of 95% ethanol (**caution:** *flammable, extinguish all flames*) to ice bath temperature and then slowly add it to the solution. The blue, solid complex should form.
3. **Isolate the Product.** Premeasure the mass (±0.01 g) of a piece of filter paper and fit it in a Büchner funnel. Vacuum filter the solution (Figure 38.7); wash the solid with two 5-mL portions of cold 95% ethanol (*not* water!) and 5 mL of acetone.

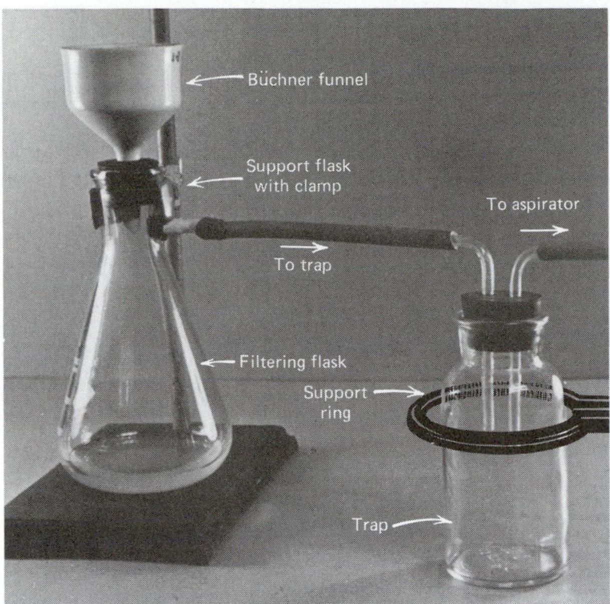

Figure 38.7 A vacuum filtration setup for filtering the coordination compound.

Place the filter and sample on a watchglass and allow the sample to air-dry. Determine the mass of the filter paper and product. Calculate the percent yield.

4. **Obtain the Instructor's Approval.** Transfer your product to a *clean, dry* test tube, stopper and (properly) label the test tube, and submit it to your instructor for approval.

F. Synthesis of Hexaamminenickel (II) Chloride

1. **Form the Complex Ion.** In a 125-mL Erlenmeyer flask, dissolve 6 g ($\pm$0.01 g) of nickel(II) chloride hexahydrate, $[Ni(H_2O)_6]Cl_2$, in (no more than) 10 mL of warm ($\sim$50°C) deionized water. In a fume hood, slowly add 20 mL of conc NH_3 **(caution!).**
2. **Precipitate the Complex Ion.** Cool the mixture in an ice bath. Cool 15 mL of 95% ethanol (**caution:** *flammable, extinguish all flames*) to ice bath temperature and then slowly add it to the solution. Allow the mixture to settle for complete precipitation of the lavender product. The supernatant should be nearly colorless.

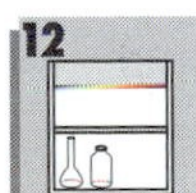

3. **Isolate the Product.** Premeasure the mass ($\pm$0.01 g) of a piece of filter paper, fitted for a Büchner funnel. Vacuum filter the product; wash with two 5-mL volumes of cold 95% ethanol and 5 mL of acetone. Air-dry the product and filter paper on a watchglass. Determine the mass of the filter paper and product. Calculate the percent yield.

4. **Obtain the Instructor's Approval.** Transfer your product to a *clean, dry* test tube, stopper and (properly) label the test tube, and submit it to your instructor for approval.

G. Synthesis of *Tris*(ethylenediamine)nickel (II) Chloride Dihydrate

1. **Form the Complex Ion.** In a 125-mL Erlenmeyer flask, dissolve 6 g ($\pm$0.01 g) of nickel(II) chloride hexahydrate, $[Ni(H_2O)_6]Cl_2$, in 10 mL of warm ($\sim$50°C) deionized water. Cool the mixture in an ice bath. In a fume hood, slowly add 10 mL ethylenediamine. (**Caution:** *Avoid skin contact and inhalation.*)
2. **Precipitate and Isolate the Product.** Complete as in Parts F.2 and F.3.
3. **Obtain the Instructor's Approval.** Transfer your product to a *clean, dry* test tube, stopper and (properly) label the test tube, and submit it to your instructor for approval.

> *Disposal:* Dispose of the waste solutions from Parts E, F, and G in the "Waste Metal Ion Solutions" container.

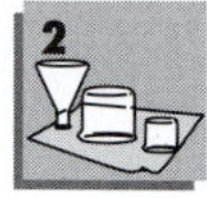

CLEANUP: Rinse all glassware twice with tap water and discard in the "Waste Metal Ion Solutions" container. Rinse the glassware twice with deionized water and discard in the sink; follow with a generous supply of tap water.

Transition Metal Chemistry

Date ________ Lab Sec. ______ Name ______________________________ Desk No. ________

1. Consider the coordination compound, $[Co(NH_3)_3(H_2O)_3]_2(SO_4)_3$. Use the definitions in the Introduction to identify the following with the formula and charge (if applicable).
   a. the ligand(s)

   b. the complex

   c. the coordination sphere

   d. the coordination number of cobalt

2. Write the formula of the complex ion formed between
   a. the ligand, Cl^-, and the copper(II) ion with a coordination number of four

   b. the ligand, NH_3, and the nickel(II) ion with a coordination number of six

   c. the ligand, ethylenediamine (abbreviated "en"), and the chromium(III) ion with a coordination number of six

3. The cyanide ion, CN^-, is a much *stronger* ligand than the chloride ion, Cl^-. Explain.

4. Complex ions with the ligand ethylenediamine are more stable than those with the ligand NH_3, assuming the same metal ion. Explain.

5. When 6 *M* NaOH is slowly added to a solution containing chromium(III) ion, a precipitate forms. However when an excess of 6 *M* NaOH is added, the precipitate dissolves forming a complex ion with a coordination number of four.
   a. Write the formula of the precipitate.

   b. Write the formula of the complex ion.

6. A 6.0-g sample of $[Ni(H_2O)_6]Cl_2$ is dissolved in aqueous solution. If an excess of ethylenediamine (en) is added to the solution, solid $[Ni(en)_3]Cl_2{\bullet}2H_2O$ forms. What is the theoretical yield of product from the reaction?

Experiment 38 *Report Sheet*

Transition Metal Chemistry

Date ________ Lab Sec. ______ Name ________________________________ Desk No. ________

A. Chloro Complexes of the Copper(II), Nickel(II), and Cobalt(II) Ions

Solution	① Color/H_2O	② Color/HCl	Formula of Complex	③ Effect of H_2O
0.1 *M* $CuSO_4$				
0.1 *M* $Ni(NO_3)_2$				
0.1 *M* $CoCl_2$				

For each metal ion state whether the aqua complex or the chloro complex is more stable:

Cu^{2+} ____________; Ni^{2+} ____________; Co^{2+} ____________

B. Complexes of the Copper(II) Ion

Ligand	④ Color	Formula of Complex	⑤ Effect of OH^-
NH_3			
Ethylenediamine			
SCN^-			
H_2O			
________			

C. Complexes of the Nickel(II) Ion

Ligand	⑥ Color	Formula of Complex	⑦ Effect of OH^-
NH_3			
Ethylenediamine			
SCN^-			
H_2O			
________			

D. Complexes of the Cobalt(II) Ion

Ligand	⑧ Color	Formula of Complex	⑨ Effect of OH^-
NH_3			
Ethylenediamine			
SCN^-			
H_2O			
________			

Review of Data

Of the complexes formed with copper(II), nickel(II), and colbalt(II), which metal ion appears to form the most stable complexes. Explain. Also see Laboratory Question 4.

Synthesis of a Coordination Compound

Name and formula of coordination compound ______________________________

1. Mass of starting material (g) _______________
2. Mass of filter paper (g) _______________
3. Mass of filter paper + product (g) _______________
4. Mass of product (g) _______________
5. Instructor's approval ______________________________
6. Theoretical yield of product* (g) _______________
7. Percent yield* (%) _______________

*Show calculations.

Laboratory Questions

Circle the questions that have been assigned

1. Part A.2. Is the chloride ion or water a stronger ligand? Explain.
2. Part B.1. Is water or ammonia a stronger ligand. Explain.
3. Part B.2. Is ammonia or ethylenediamine a stronger ligand? Explain.
4. Parts A–D. Of the five ligands, Cl^-, NH_3, $H_2NCH_2CH_2NH_2$, SCN^-, and H_2O, studied in this experiment, which ligand appeared to be the strongest ligand? Why? Which ligand appeared to be the weakest? Why?
5. Part E.2. a. A solution of potassium cyanide, KCN, instead of conc NH_3 is added. Write the formula of the expected complex ion.
   b. A solution of potassium chloride, KCl, instead of conc NH_3 is added. Write the formula of the expected complex ion.
6. Part E.2. Identify the precipitate that forms *before* the addition of excess *conc* NH_3.
7. Part E.3. Why is 95% ethanol used to wash the solid [the tetraamminecopper(II) sulfate monohydrate product] on the filter paper instead of deionized water?

Appendix A

Conversion Factors

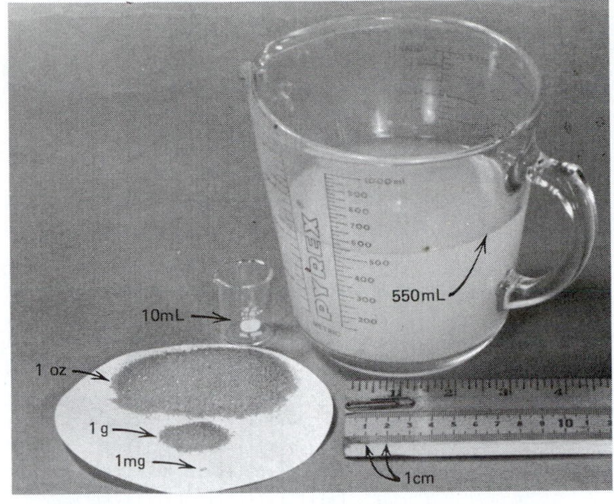

The magnitude of a measurement must be familiar to a chemist.

Length

1 meter (m) = 39.37 in. = 3.281 ft = distance light travels in 1/299,792,548th of a second
1 inch (in.) = 2.54 cm (exactly) = 0.0254 m
1 kilometer (km) = 0.6214 (statute) mile
1 angstrom (Å) = 1×10^{-10} m = 0.1 nm
1 micrometer or micron (μm) 1×10^{-6} m

Mass

1 gram (g) = 0.03527 oz = 15.43 grains
1 kilogram (kg) = 2.205 lb = 35.27 oz
1 metric ton = 1×10^{6} g = 1.102 short ton = 0.9843 long ton
1 pound (lb) = 453.6 g = 7000 grains
1 ounce (oz) = 28.35 g = 437.5 grains

Temperature

°F = 1.8°C + 32
K = °C + 273.15

Volume

1 liter (L) = 1 dm^3 = 1.057 fl qt = 1×10^{3} mL = 1×10^{3} cm^3 = 61.02 $in.^3$ = 0.2642 gal
1 fluid quart (fl qt) = 946.4 mL = 0.250 gal = 0.00595 bbl (oil)
1 fluid ounce (fl oz) = 29.57 mL
1 cubic foot (ft^3) = 28.32 L = 0.02832 m^3

Pressure

1 atmosphere (atm) = 760 torr (exactly) = 760 mm Hg = 29.92 in. Hg = 14.696 lb/$in.^2$ = 1.013 bar = 101.325 kPa
1 pascal (Pa) = 1 kg/(m • s^2) = 1 N/m^2
1 torr = 1 mm Hg = 133.3 N/m^2

Energy

1 joule (J) = 1 kg • m^2/s^2 = 0.2389 cal = 9.48×10^{-4} Btu 1 = 10^7 ergs
1 calorie (cal) = 4.184 J = 3.087 ft•lb
1 British thermal unit (Btu) = 252.0 cal = 1054 J = 3.93×10^{-4} hp • hr = 2.93×10^{-4} kW • hr
1 liter atmosphere (L • atm) = 24.2 cal = 101.3 J
1 electron volt (eV) = 1.602×10^{-19} J
1 kW • hr (kWh) = 3413 Btu = 8.606×10^{5} cal = 3.601×10^{6} J = 1.341 hp•hr

Constants and Other Conversion Data

velocity of light (c) = 2.9979×10^{8} m/s = 186,272 mi/s
gas constant (R) = 0.08206 L • atm/(mol • K) = 8.314 J/(mol•K) = 1.986 cal/(mol • K) = 62.37 L • torr/(mol • K)
Avogadro's number (N_o) = 6.0221×10^{23}/mol
Planck's constant (h) = 6.6261×10^{-34} J • s/photon
Faraday's constant ($\mathfrak{F}$) = 96,485 C/mol e^-

Appendix B

Treatment of Data

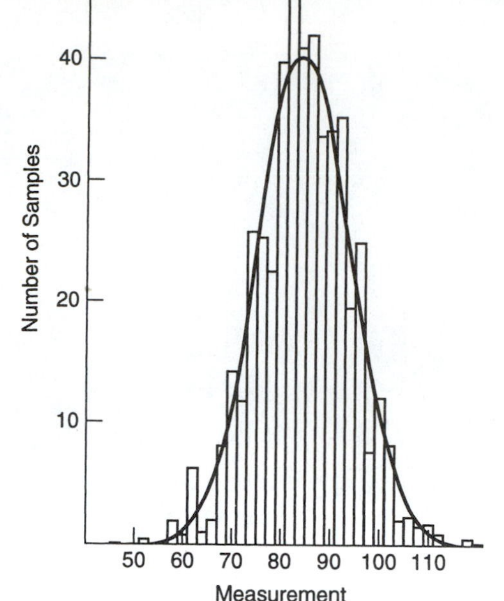

Bar graph and standard error curve for the multiple analysis of a sample.

Confidence in a scientific theory depends on the reliability of the experimental data on which the theory is based. For this reason, a scientist must be concerned about the quality of the data he or she collects. Of prime importance is the **accuracy** of the data—how closely the measured values lie to the true values.

To obtain accurate data we must use instruments that are carefully calibrated for a properly designed experimental procedure. Miscalibrated equipment, such as a balance or buret, may result in reproducible, but erred data. Flawed instruments or experimental procedures result in **systematic errors**—errors that can be detected and corrected. As a result of systematic errors, the data may have good **precision**, but not necessarily have good **accuracy**. To have good accuracy of data, the systematic errors must be minimized.

Systematic errors: Determinate errors that arise from flawed equipment or experimental design

Precision: Data with small deviations from an average value have high precision

Accuracy: Data with small deviations from an accepted or accurate value have good accuracy

Random errors: Indeterminate errors that arise from the bias of a chemist in observing and recording measurements

Because scientists collect data, **random errors** may also occur in measurements. Random errors are a result of reading/interpreting the value from the measuring instrument. For example, reading the volume of a liquid in a graduated cylinder to the nearest milliliter depends on the "best view" of the bottom of the meniscus, the judgment of the bottom of the meniscus relative to the volume scale, and even the temperature of the liquid. A volume reading of 10.2 mL may be read as 10.1 or 10.3, depending upon the chemist and the laboratory conditions. When the random errors are small, all measurements are close to one another and we say the data are of high precision. When the random errors are large, the values cover a much broader range and the data are of low precision. *Generally,* data of high precision are also of high accuracy, especially if the measuring device is properly calibrated. Remember that all measurements are to be expressed with the correct number of significant figures, the number being reflective of the measuring instrument.

Average (or Mean) Value, $\bar{x}$

Methods of analyzing experimental data, based on statistics, provide information on the degree of precision of the measured values. Applying the methods is simple, as you will see, but to understand their significance, examine briefly the **standard error curve** (Figure B.1a).

If we make a large number of measurements of a quantity, the values would fluctuate about the average value (also called the mean value). The **average, or mean, value** is obtained by dividing the sum of all the measured values by the total number of

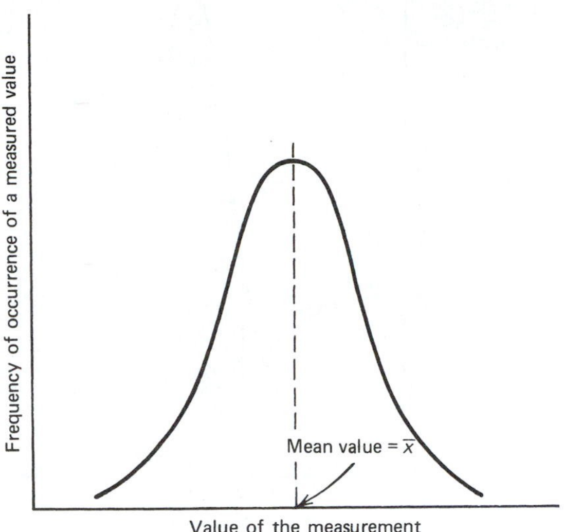

Figure B.1a The standard error curve.

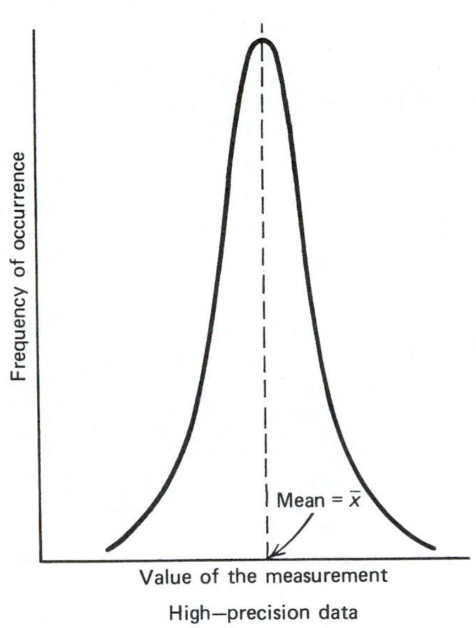

Figure B.1b Error curve for high-precision data.

values. If x_1, x_2, x_3, etc., are measured values, and there are n of them, then the average value, $\bar{x}$, is computed as

$$\text{average (or mean) value, } \bar{x} = \frac{x_1 + x_2 + x_3 + \cdots + x_n}{n} \tag{B.1}$$

Most values lie close to the average, but some lie further away. If we plot the frequency with which a measured value occurs versus the value of the measurement, we obtain the curve in Figure B.1a. When the random errors are small (high-precision data, Figure B.1b), the curve is very narrow and the peak is sharp. When the random errors are large (low-precision data, Figure B.1c), the data are more "spread out," and the error curve is broader and less sharp.

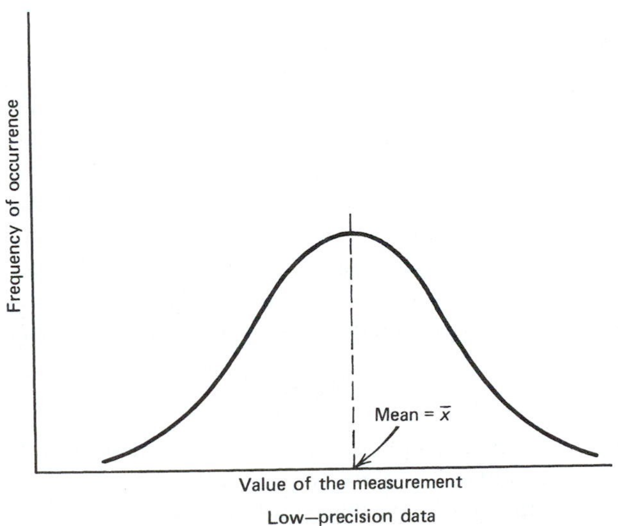

Figure B.1c Error curve for low-precision data.

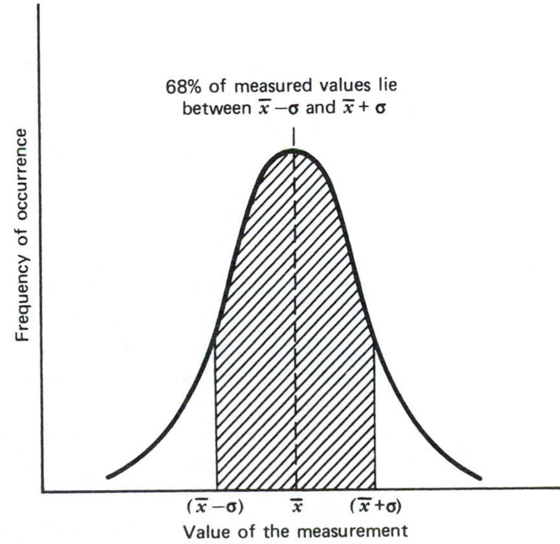

Figure B.2a Relationship of the standard deviation to the error curve.

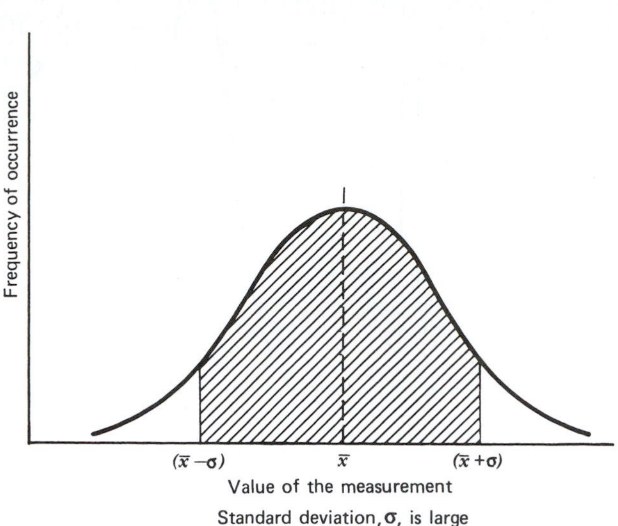

Figure B.2b Data set with a large standard deviation.

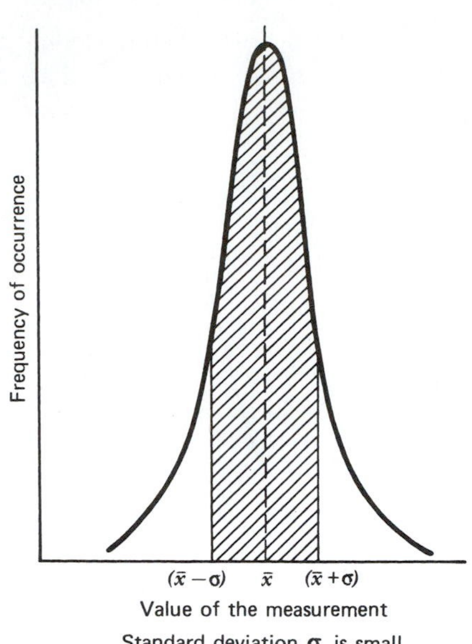

Figure B.2c Data set with a small standard deviation.

Standard Deviation, σ

Statistics gives us methods for computing quantities that tell us about the width of the error curve for our data and, therefore, about the precision of the data, even when the amount of data is relatively small. One of the most important statistical measures of precision is the **standard deviation,** σ (Greek letter sigma). To calculate the standard deviation, we first compute the average value, $\bar{x}$. The next step is to compute the *deviation, d,* from the average value for *each* measurement—the difference between the average and each measured value:

$$\text{deviation, } d_i = \bar{x} - x_i \tag{B.2}$$

d_i is the deviation for the measured value, x_i. The standard deviation is obtained by squaring these deviations, adding the squared values together, dividing this sum by $n - 1$ (where n is the number of measurements), and then taking the square root.

$$\text{standard deviation } \sigma = \sqrt{\frac{d_1^2 + d_2^2 + \cdots + d_n^2}{(n-1)}} \tag{B.3}$$

The standard deviation means that if we make yet another measurement, the probability that its value will lie within $\pm\sigma$ of the average value is 0.68. In other words, 68% of the measurements lie within $\pm\sigma$ of the average value (i.e., within the range $\bar{x} - \sigma$ to $\bar{x} + \sigma$). On the error curve in Figure B.2a, this represents the measurements falling within the shaded area. If we obtain a large calculated σ from a set of measured values, it means that the error curve for our data is broad and the precision of the data is low (Figure B.2b); a small value of σ for a set of data means that the error curve is narrow and the precision of the data is high (see Figure B.2c). Thus, σ is a statistical measure of the precision of the data.

Average Deviation, *a*

Another quantity that serves as a statistical measure of precision is the **average deviation, *a*.** To compute it we simply add all the deviations (Equation B.2) together, take their absolute values, and then divide this sum by the number of measurements.

$$\text{average deviation, } a = \frac{|d_1| + |d_2| + |d_3| + \cdots + |d_n|}{n} \tag{B.4}$$

Although simpler to compute than the standard deviation, the average deviation is less frequently used. In terms of the error curve, the probability that yet another measurement lies within $\pm a$ of the average value is 0.57. Thus, 57% of the measurements fall within the range of $\bar{x} + a$ to $\bar{x} - a$.

Let us now look at an example that illustrates how these statistical methods are applied. Suppose that four analyses of an ion ore sample give the following data with four significant figures:

Trial	% Iron in the Ore
1	39.74
2	40.06
3	39.06
4	40.92

The average value for the percent iron is

$$\text{average (or mean) value, } \bar{x} = \frac{39.74 + 40.06 + 39.06 + 40.92}{4} = 39.94\%$$

To calculate the average deviation and the standard deviation, compute the deviations and their squares. Let's set up a table.

Trial	Measured Values	$d_i = \bar{x} - x_i$	$\lvert d_i \rvert$	d_i^2
1	39.74	0.20	0.20	0.040
2	40.06	−0.12	0.12	0.014
3	39.06	0.88	0.88	0.77
4	40.92	−0.98	0.98	0.96
	$\bar{x}$ = 39.94		Sum = 2.18	Sum = 1.78

$$\text{average deviation, } a = \frac{2.18}{4} = 0.54$$

$$\text{standard deviation, } \sigma = \sqrt{\frac{1.78}{4 - 1}} = 0.77$$

If we wish to express the precision of our analysis in terms of an average deviation, the percent iron is reported as 39.94 ± 0.54%; this means that 57% of all subsequent analyses of the iron ore sample should be in the range of 39.94 ± 0.54% iron. If we choose to express our precision in terms of a standard deviation, the percent iron is reported as 39.94 ± 0.77%, meaning that 68% of all subsequent analyses should be in the range of 39.94 ± 0.77% iron.

Relative Error

Scientists check the *accuracy* of their measurements by comparing their results with values that are well established and are considered "accepted values." Many reference books, such as the Chemical Rubber Company's (CRC) *Handbook of Chemistry and Physics,* are used to check a result against an accepted value. To report the relative error in *your* result, take the absolute value of the difference between your value and the accepted value, divide this difference by the accepted value. Taking x to be your measured value and y to be the accepted value,

$$\text{relative error} = \frac{\lvert x - y \rvert}{y}$$

Relative error may be expressed as percent or parts per thousand, multiplying the relative error by 100 or 1000.

Appendix C

Graphing Data

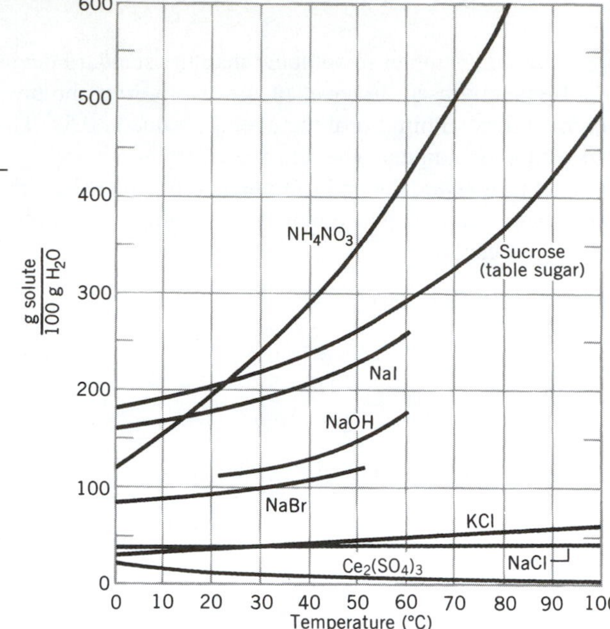

Plotted data show how the solubilities of salts vary with temperature.

A well-designed graph of experimental data is a very effective organization of the data for observing trends, discovering relationships, or predicting information. It is therefore worthwhile to learn how to effectively construct and present a graph and how to extract information from it.

Graph Construction

In general, a graph is constructed on a set of perpendicular axes (Figure C.1); the vertical axis (the y axis) is the **ordinate,** and the horizontal axis (the x axis) is the **abscissa.** Constructing a graph involves the following steps.

1. **Select the Axes.** First choose which variable corresponds to the ordinate and which one corresponds to the abscissa. Usually the dependent variable is plotted along the ordinate and values of the independent variable along the abscissa. For

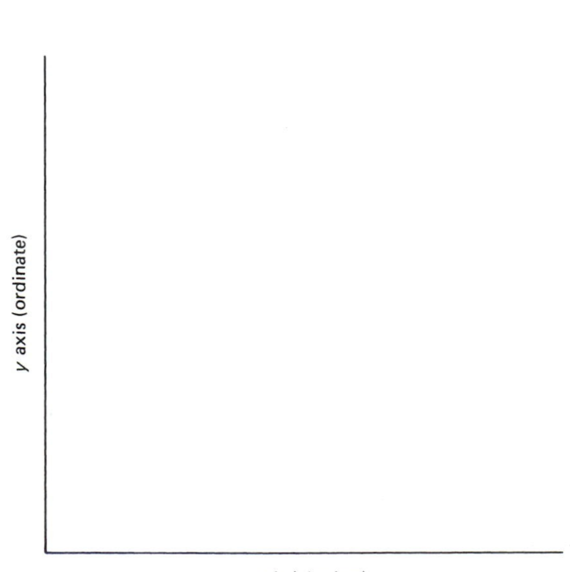

Figure C.1 A graph is usually constructed on a set of perpendicular axes.

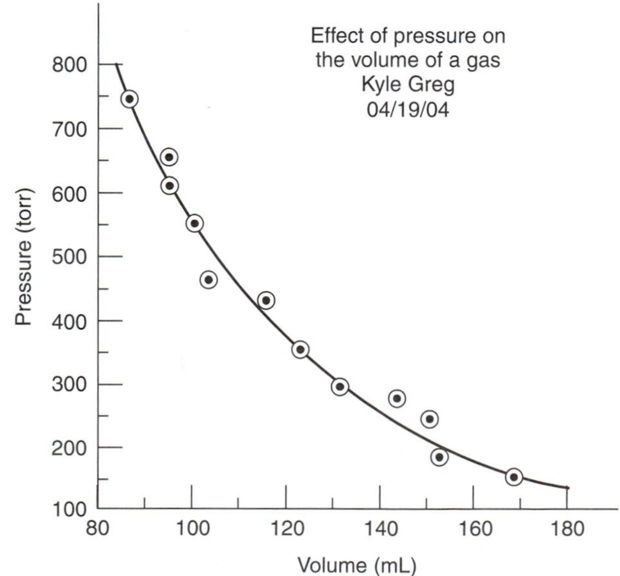

Figure C.2 An example of a properly drawn and labeled graph showing how the pressure of a gas depends on the volume of that gas.

example, if we observe how the pressure of a gas responds to a change in volume, pressure is the dependent variable. We therefore assign pressure to the vertical axis and volume to the horizontal axis; we say we are plotting pressure versus volume. Be sure to "label each axis" by indicating the units that correspond to the variables being plotted (Figure C.2).

2. **Set the Scales for the Axes.** Construct the graph so that the data fill as much of the space of the graph paper as possible. Therefore, choose scales for the x and y axes that cover the range of the experimental data. For example, if the measured pressure range is 150 to 740 torr, choose a pressure scale that ranges from 100 to 800 torr. This covers the entire data range and allows us to mark the major divisions at intervals of 100 torr (Figure C.2). When choosing the scale, always choose values for the major divisions that make the smaller subdivisions easy to interpret. With major divisions at every 100 torr, minor divisions occur at every 50 torr. This makes plotting values such as 525 torr very simple.

   Construct the scale for the x axis in the same manner that we've described for the y axis. In Figure C.2, the volumes range from 170 mL at a pressure of 150 torr to 85 mL at a pressure of 750 torr. The scale on the x axis ranges from 80 to 180 mL and is marked off in 20-mL intervals. Label each axis with the appropriate units.

   There are a few additional points to note about marking the scales of a graph:

   - The values do not have to begin at zero at the origin; in fact, they seldom do.
   - The size of the minor subdivisions should permit estimation of all the significant figures used in obtaining the data (if pressure measurements are made to the nearest torr, then the pressure scale should be interpreted to read to the nearest torr).
   - If the graph is used for extrapolation, be sure that the range of scales covers the range of the extrapolation.

3. **Plot the Data.** Place a dot for each data point at the appropriate place on the graph. Draw a small circle around the dot. *Ideally,* the size of the circle should approximate the estimated error in the measurement. If you plot two or more different data sets on the same graph, use different-shaped symbols (triangle, square, diamond, etc.) around the data points to distinguish one set of data from another.

4. **Draw a Curve for the Best Fit.** Draw a *smooth* curve that best fits your data. This line does not have to pass through the centers of all the data points, or even through any of them, but it should pass as closely as possible to all of them.

   Note that the line in Figure C.2 is not drawn through the circles. It stops at the edge of the circle, passes "undrawn" through it, and then emerges from the other side.

5. **Title Your Graph.** Place a descriptive title in the upper portion of the graph, well away from the data points and the smooth curve. Include your name and date under the title.

Straight-Line Graphs

Often, the graphical relationship between measured quantities produces a straight line. This is the case, for example, when we plot pressure versus temperature for a fixed volume of gas. Such linear relationships are useful because the line corresponding to the best fit of the data points can be drawn with a straight edge, and because quantitative (extrapolated) information about the relationship is easily obtained directly from the graph.

Algebraically, a straight line is described by the equation

$$y = mx + b \quad \text{(C.1)}$$

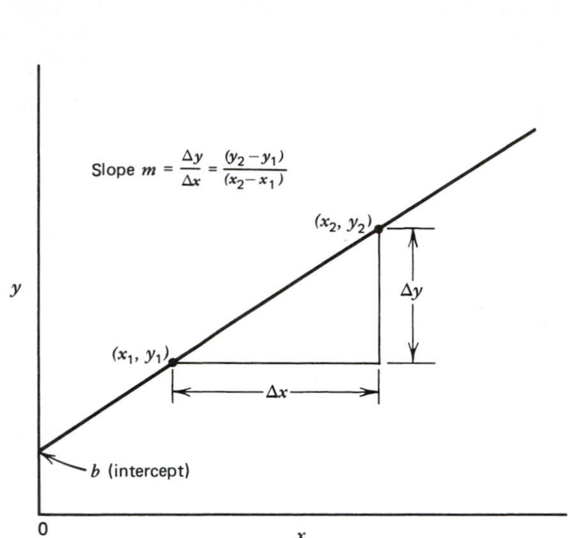

Figure C.3 The slope and intercept for a straight line, $y = mx + b$.

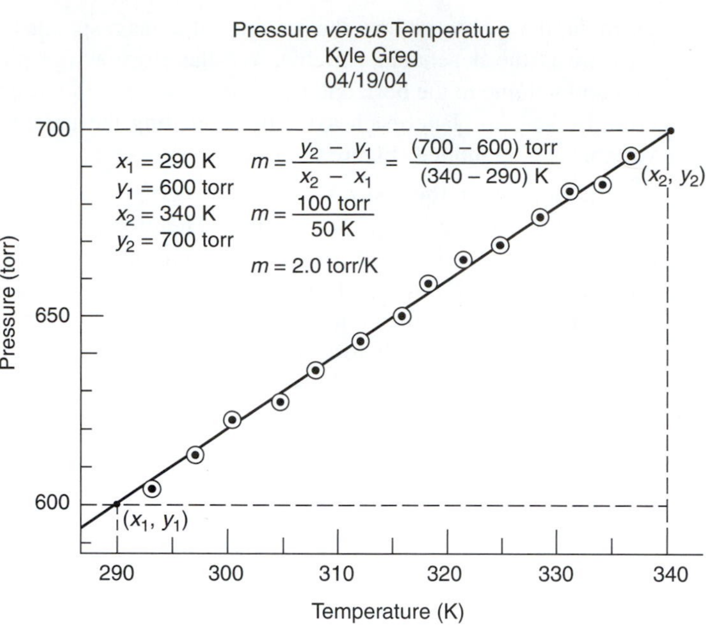

Figure C.4 Determination of the slope of a straight line drawn for a plot of pressure vs. temperature for a gas.

m is the slope of the straight line and b is the point of intersection of the line with the y axis when $x = 0$ (Figure C.3). The slope of the line, which is usually of greatest interest, is determined from the relationship

$$m = \frac{y_2 - y_1}{x_2 - x_1} = \frac{\Delta y}{\Delta x} \qquad \text{(C.2)}$$

Figure C.4 illustrates the determination of the slope for a typical plot of pressure versus temperature. First plot the data and then draw the *best straight line*. Next, choose points *on the drawn line* corresponding to the easily readable values along the x axis. Read corresponding y values along the y axis, and then compute the slope.

Appendix D

Familiar Names of Common Chemicals

Sodium bicarbonate is commonly called baking soda or bicarbonate of soda.

Familiar Name	Chemical Name	Formula
alcohol	ethanol (ethyl alcohol)	C_2H_5OH
aqua regia	mixture of conc nitric and hydrochloric acids	HNO_3 + 3 HCl by volume
aspirin	acetylsalicylic acid	$CH_3COOC_6H_4COOH$
baking soda	sodium bicarbonate	$NaHCO_3$
banana oil	amyl acetate	$CH_3COOC_5H_{11}$
bauxite	hydrated aluminum oxide	$Al_2O_3 \cdot xH_2O$
bleaching powder	calcium chloride hypochlorite	$Ca(ClO)_2$, $Ca(ClO)Cl$
blue vitriol	copper(II) sulfate pentahydrate	$CuSO_4 \cdot 5H_2O$
borax	sodium tetraborate decahydrate	$Na_2B_4O_7 \cdot 10H_2O$
brimstone	sulfur	S_8
calamine	zinc oxide	ZnO
calcite	calcium carbonate	$CaCO_3$
Calgon	polymer of sodium metaphosphate	$(NaPO_3)x$
calomel	mercury(I) chloride	Hg_2Cl_2
carborundum	silicon carbide	SiC
caustic soda	sodium hydroxide	$NaOH$
chalk	calcium carbonate	$CaCO_3$
Chile saltpeter	sodium nitrate	$NaNO_3$
copperas	iron(II) sulfate heptahydrate	$FeSO_4 \cdot 7H_2O$
cream of tartar	potassium hydrogen tartrate	$KHC_4H_4O_6$
DDT	dichlorodiphenyltrichloroethane	$(C_6H_4Cl)_2CHCCl_3$
dextrose	glucose	$C_6H_{12}O_6$
Epsom salt	magnesium sulfate heptahydrate	$MgSO_4 \cdot 7H_2O$
fool's gold	iron pyrite	FeS_2
Freon	dichlorodifluoromethane	CCl_2F_2
Glauber's salt	sodium sulfate decahydrate	$Na_2SO_4 \cdot 10H_2O$
glycerin	glycerol	$C_3H_5(OH)_3$
green vitriol	iron(II) sulfate heptahydrate	$FeSO_4 \cdot 7H_2O$
gypsum	calcium sulfate dihydrate	$CaSO_4 \cdot 2H_2O$
hypo	sodium thiosulfate pentahydrate	$Na_2S_2O_3 \cdot 5H_2O$
invert sugar	mixture of glucose and fructose	$C_6H_{12}O_6 + C_6H_{12}O_6$
laughing gas	nitrous oxide	N_2O
levulose	fructose	$C_6H_{12}O_6$
lye	sodium hydroxide	$NaOH$
magnesia	magnesium oxide	MgO
marble	calcium carbonate	$CaCO_3$
marsh gas	methane	CH_4
milk of lime (limewater)	calcium hydroxide	$Ca(OH)_2$
milk of magnesia	magnesium hydroxide	$Mg(OH)_2$
milk sugar	lactose	$C_{12}H_{22}O_{11}$
Mohr's salt	iron(II) ammonium sulfate hexahydrate	$Fe(NH_4)_2(SO_4)_2 \cdot 6H_2O$
moth balls	naphthalene	$C_{10}H_8$
muriatic acid	hydrochloric acid	$HCl(aq)$
oil of vitriol	sulfuric acid	$H_2SO_4(aq)$

Familiar Name	Chemical Name	Formula
oil of wintergreen	methyl salicylate	$C_6H_4(OH)COOCH_3$
oleum	fuming sulfuric acid	$H_2S_2O_7$
Paris green	double salt of copper(II) acetate and copper(II) arsenite	$Cu(CH_3CO_2)_2 \cdot Cu_3(AsO_3)_2$
plaster of Paris	calcium sulfate hemihydrate	$CaSO_4 \cdot \frac{1}{2}H_2O$
potash	potassium carbonate	K_2CO_3
quartz	silicon dioxide	SiO_2
quicklime	calcium oxide	CaO
Rochelle salt	potassium sodium tartrate	$KNaC_4H_4O_6$
rouge	iron(III) oxide	Fe_2O_3
sal ammoniac	ammonium chloride	NH_4Cl
salt (table salt)	sodium chloride	$NaCl$
saltpeter	potassium nitrate	KNO_3
silica	silicon dioxide	SiO_2
sugar (table sugar)	sucrose	$C_{12}H_{22}O_{11}$
Teflon	polymer of tetrafluoroethylene	$(C_2F_4)x$
washing soda	sodium carbonate decahydrate	$Na_2CO_3 \cdot 10H_2O$
white lead	basic lead carbonate	$PbCO_3 \cdot Pb(OH)_2$
wood alcohol	methanol (methyl alcohol)	CH_3OH

Appendix E

Vapor Pressure of Water

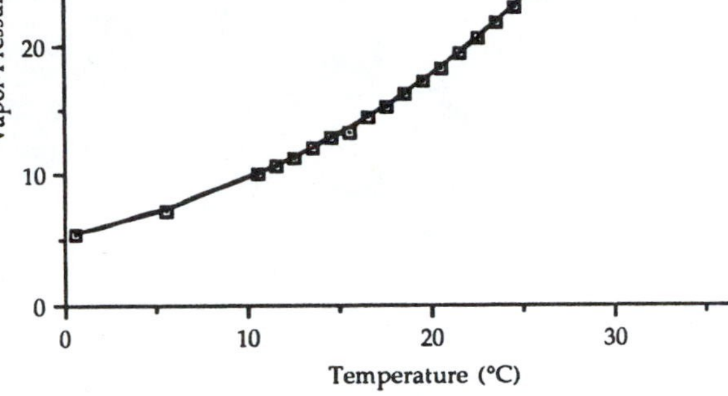

Vapor pressure of water as a function of temperature.

Temperature (°C)	Pressure (Torr)
0	4.6
5	6.5
10	9.2
11	9.8
12	10.5
13	11.2
14	12.0
15	12.5
16	13.6
17	14.5
18	15.5
19	16.5
20	17.5
21	18.6
22	19.8
23	21.0
24	22.3
25	23.8
26	25.2
27	26.7
28	28.3
29	30.0
30	31.8
31	33.7
32	35.7
33	37.7
34	39.9
35	42.2
37*	47.1
—	—
100	760

*Body temperature.

Appendix F

Concentrations of Acids and Bases

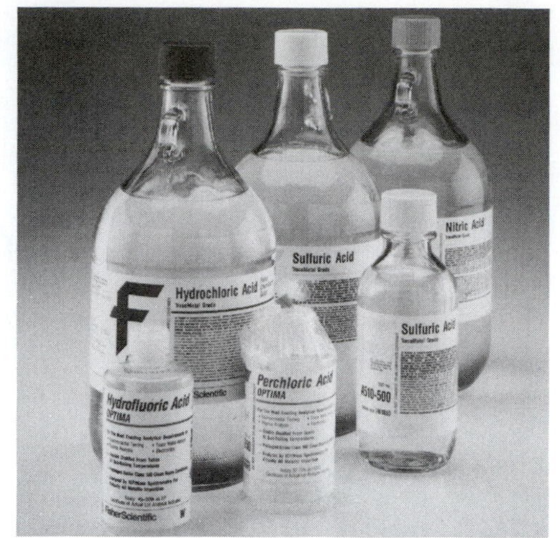

Concentrated laboratory acids and bases.

Reagent	Approximate Molar Concentration	Approximate Mass Percent	Specific Gravity	mL to Dilute to 1 L for a 1.0 *M* Solution
Acetic acid	17.4	99.5	1.05	57.5
Hydrochloric acid	11.6	36	1.18	86.2
Nitric acid	16.0	71	1.42	62.5
Phosphoric acid	14.7	85	1.70	68.0
Sulfuric acid	18.0	96	1.84	55.6
Ammonia (*aq*) (ammonium hydroxide)	14.8	28%(NH_3)	0.90	67.6

Caution: *When diluting reagents, add the more concentrated reagent to the more dilute reagent (or solvent).* ***Never*** *add water to a concentrated acid!*

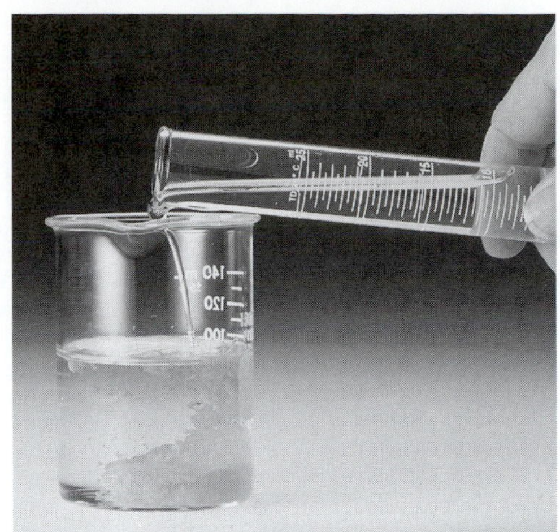

Appendix G

Water Solubility of Inorganic Salts

Many salts, such as cadmium sulfide, have very low solubilities.

Water-Soluble Salts

1. All salts of the chloride ion, Cl^-, bromide ion, Br^-, and iodide ion, I^-, are soluble *except* those of Ag^+, Hg_2^{2+}, Pb^{2+}, Cu^+, and Tl^+. BiI_3 and SnI_4 are insoluble. $PbCl_2$ is three to five times more soluble in hot water than in cold water.
2. All salts of the acetate ion, $CH_3CO_2^-$, nitrate ion, NO_3^-, chlorate ion, ClO_3^-, perchlorate ion, ClO_4^-, and permanganate ion, MnO_4^-, are soluble.
3. All common salts of the Group 1A cations and ammonium ion, NH_4^+, are soluble.
4. All common salts of the sulfate ion, SO_4^{2-}, are soluble *except* those of Ba^{2+}, Sr^{2+}, Pb^{2+}, and Hg^{2+}.
5. All Group 1A and 2A salts of the bicarbonate ion, HCO_3^-, are soluble.
6. *Most* salts of the fluorosilicate ion, SiF_6^{2-}, thiocyanate ion, SCN^-, and thiosulfate ion, $S_2O_3^{2-}$, are soluble. *Exceptions* are the Ba^{2+} and Group 1A fluorosilicates, the Ag^+, Hg_2^{2+}, and Pb^{2+} thiocyanates, and the Ag^+ and Pb^{2+} thiosulfates.

Water-Insoluble Salts

1. All common salts of the fluoride ion, F^-, are insoluble *except* those of Ag^+, NH_4^+, and Group 1A cations.
2. In general, all common salts of the carbonate ion, CO_3^{2-}, phosphate ion, PO_4^{3-}, borate ion, BO_3^{3-}, arsenate ion, AsO_4^{3-}, arsenite ion, AsO_3^{3-}, cyanide ion, CN^-, ferricyanide ion, $[Fe(CN)_6]^{3-}$, ferrocyanide ion, $[Fe(CN)_6]^{4-}$, oxalate ion, $C_2O_4^{2-}$, and the sulfite ion, SO_3^{2-}, are insoluble, *except* those of NH_4^+ and the Group 1A cations.
3. All common salts of the oxide ion, O^{2-}, and the hydroxide ion, OH^-, are insoluble *except* those of the Group 1A cations, Ba^{2+}, Sr^{2+}, and NH_4^+. $Ca(OH)_2$ is slightly soluble. Soluble oxides produce the corresponding hydroxides in water.
4. All common salts of the sulfide ion, S^{2-}, are insoluble *except* those of NH_4^+ and the cations that are isoelectronic with a noble gas (e.g., the Group 1A cations, the Group 2A cations, Al^{3+}, etc.).
5. Most common salts of the chromate ion, CrO_4^{2-}, are insoluble *except* those of NH_4^+, Ca^{2+}, Cu^{2+}, Mg^{2+}, and the Group 1A cations.
6. All common salts of the silicate ion, SiO_3^{2-}, are insoluble *except* those of the Group 1A cations.

Table G.1 Summary of the Solubility of Salts

Anion	Soluble Salts with These Cations	"Insoluble" Salts with These Cations
acetate, $CH_3CO_2^-$	most cations	none
arsenate, AsO_4^{3-}	NH_4^+, Group 1A (except Li^+)	most cations
arsenite, AsO_3^{3-}	NH_4^+, Group 1A (except Li^+)	most cations
borate, BO_3^{3-}	NH_4^+, Group 1A (except Li^+)	most cations
bromide, Br^-	most cations	Ag^+, Hg_2^{2+}, Pb^{2+}, Cu^+, Tl^+
carbonate, CO_3^{2-}	NH_4^+, Group 1A (except Li^+)	most cations
chlorate, ClO_3^-	most cations	none
chloride, Cl^-	most cations	Ag^+, Hg_2^{2+}, Pb^{2+}, Cu^+, Tl^+
chromate, CrO_4^{2-}	NH_4^+, Ca^{2+}, Cu^{2+}, Mg^{2+}, Group 1A	most cations
cyanide, CN^-	NH_4^+, Group 1A (except Li^+)	most cations
ferricyanide, $[Fe(CN)_6]^{3-}$	NH_4^+, Group 1A (except Li^+)	most cations
ferrocyanide, $[Fe(CN)_6]^{4-}$	NH_4^+, Group 1A (except Li^+)	most cations
fluoride, F^-	Ag^+, NH_4^+, Group 1A	most cations
fluorosilicate, SiF_6^{2-}	most cations	Ba^{2+}, Group 1A
hydroxide, OH^-	NH_4^+, Sr^{2+}, Ba^{2+}, Group 1A	most cations
iodide, I^-	most cations	Ag^+, Hg_2^{2+}, Pb^{2+}, Cu^+, Tl^+, Br^{3+}, Sn^{4+}
nitrate, NO_3^-	most cations	none
nitrite, NO_2^-	most cations	none
oxalate, $C_2O_4^{2-}$	NH_4^+, Group 1A (except Li^+)	most cations
oxide, O^{2-}	NH_4^+, Sr^{2+}, Ba^{2+}, Group 1A	most cations
perchlorate, ClO_4^-	most cations	none
permanganate, MnO_4^-	most cations	none
phosphate, PO_4^{3-}	NH_4^+, Group 1A (except Li^+)	most cations
silicate, SiO_3^{2-}	Group 1A	most cations
sulfate, SO_4^{2-}	most cations	Sr^{2+}, Ba^{2+}, Pb^{2+}, Hg^{2+}
sulfide, S^{2-}	NH_4^+, Groups 1A and 2A	most cations
sulfite, SO_3^{2-}	NH_4^+, Group 1A (except Li^+)	most cations
thiocyanate, SCN^-	most cations	Ag^+, Hg_2^{2+}, Pb^{2+}
thiosulfate, $S_2O_3^{2-}$	most cations	Ag^+, Pb^{2+}

Cations	Soluble Salts with These Anions	"Insoluble" Salts with These Anions
ammonium, NH_4^+	most anions	no common anions
Group 1A	most anions	no common anions

Photo Credits

Preface
Page V: Granger Collection.

Lab Safety
Page 1: Courtesy Fisher Scientific. Page 1 (bottom): Courtesy Fisher Scientific. Page 2 (top): Courtesy Fisher Scientific. Page 2 (bottom): Courtesy Fisher Scientific. Page 3: Courtesy Flow Sciences, Inc.

Lab Data
Page 5: Yoav Levy/Phototake/ PictureQuest. Page 7: Courtesy Fisher Scientific. Page 8 (top left): Courtesy VWR Scientific. Page 8 (top center): Courtesy Fisher Scientific. Page 8 (top right): Courtesy Fisher Scientific. Page 8 (bottom left): Courtesy VWR Scientific. Page 8 (bottom center): Kristen Brochmann/ Fundamental Photographs. Page 8 (bottom right): Courtesy VWR Scientific. Page 9: Yoav Levy/Phototake. Page 10: Yoav Levy/Phototake.

Lab Techniques
Page 11: Courtesy Mettler Instrument Corp. Page 13 (top): Ken Karp. Page 13 (bottom): Courtesy Fisher Scientific. Page 13 (center): Courtesy Fisher Scientific. Page 14 (top): Michael Watson. Page 14 (bottom left): Courtesy Fisher Scientific. Page 14 (bottom right): Courtesy VWR Scientific. Page 15: Peter Lerman. Page 16 (bottom left): Courtesy Fisher Scientific. Page 16 (bottom right): Courtesy Scientech, Inc. Page 16 (top): Courtesy VWR Scientific. Page 17 (top left): Courtesy VWR Scientific. Page 17 (top right): Courtesy Sartorius Co. Page 17 (bottom center): Courtesy Corning Glass Works. Page 17 (bottom right): Courtesy Fisher Scientific. Page 18 (top left): Ken Karp. Page 18 (top right): Ken Karp. Page 18 (bottom left): Ken Karp. Page 19 (left): Courtesy Professor Jo A. Beran. Page 19 (right): Courtesy Professor Jo A. Beran. Page 20 (top): Courtesy Professor Jo A. Beran. Page 20 (bottom left): Courtesy Professor Jo A. Beran. Page 20 (bottom right): Courtesy Professor Jo A. Beran. Page 21 (left): Ken Karp. Page 21 (right): Ken Karp. Page 22 (top right): Courtesy Professor Jo A. Beran. Page 22 (bottom left): Courtesy Professor Jo A. Beran. Page 22 (bottom right): Courtesy Professor Jo A. Beran. Page 23 (top left): Courtesy VWR Scientific. Page 24 (left): Courtesy VWR Scientific. Page 24 (right): Courtesy Professor Jo A. Beran. Page 25 (left): Courtesy Professor Jo A. Beran. Page 25 (center): Courtesy Professor Jo A. Beran. Page 25 (right): Ken Karp. Page 26 (top left): Ken Karp. Page 26 (top right): Courtesy Fisher Scientific. Page 26 (bottom): Courtesy Fisher Scientific. Page 27 (top left): Ken Karp. Page 27 (top center): Ken Karp. Page 27 (top right): Ken Karp. Page 28 (bottom left): Courtesy VWR Scientific. Page 28 (bottom center): Courtesy Fisher Scientific. Page 28 (bottom right): Courtesy Fisher Scientific. Page 29 (top left): Ken Karp. Page 29 (top center): Ken Karp. Page 29 (top right): Ken Karp. Page 29 (bottom left): Courtesy Professor Jo A. Beran. Page 29 (bottom right): Courtesy Professor Jo A. Beran. Page 30: Courtesy Fisher Scientific. Page 31 (top left): Courtesy VWR Scientific. Page 31 (top center): Courtesy Professor Jo A. Beran. Page 31 (top right): Courtesy Professor Jo A. Beran. Page 32 (top left): Ken Karp. Page 32 (top center): Courtesy Fisher Scientific. Page 32 (top right): Courtesy Professor Jo A. Beran. Page 32 (bottom left): Courtesy Professor Jo A. Beran. Page 32 (bottom center): Courtesy Professor Jo A. Beran. Page 32 (bottom right): Courtesy Professor Jo A. Beran. Page 33 (top right): Ken Karp. Page 33 (bottom): Courtesy Micro Essential Labs.

Dry Lab 1
Page 37: Courtesy Fisher Scientific. Page 38: Courtesy Fisher Scientific. Page 39: Courtesy Professor Jo A. Beran. Page 40: /Visuals Unlimited. Page 43 (bottom): /Fundamental Photographs. Page 43 (center): Courtesy VWR Scientific. Page 43 (top): Jo Beran. Page 44 (top): Stone/Getty Images. Page 44 (center): Yoav Levy/Phototake. Page 44 (bottom): Courtesy Professor Jo A. Beran.

Experiment 1
Page 45: /Fundamental Photographs. Page 46 (left): Courtesy Fisher Scientific. Page 46 (right): Courtesy VWR Scientific. Page 47: Courtesy Professor Jo A. Beran. Page 49: /Stock Boston. Page 50: /Fundamental Photographs.

Experiment 2
Page 53: Michael Watson. Page 54: OPC, Inc. Page 56: Ken Karp.

Experiment 3
Page 61: Courtesy USDA. Page 63: Courtesy Professor Jo A. Beran.

Experiment 4
Page 69: /Fundamental Photographs. Page 71: Courtesy Norton Seal View. Page 72: Courtesy Fisher Scientific.

Dry Lab 4
Page 377: Courtesy VWR Scientific.

Experiment 5
Page 79: Michael Watson. Page 79 (bottom): OPC, Inc.

Experiment 6
Page 87: Michael Watson. Page 87 (bottom): Courtesy Professor Jo A. Beran. Page 88: Courtesy Fisher Scientific. Page 89: Courtesy National Gypsum Company.

Experiment 7
Page 93: Ken Karp. Page 95: Courtesy Professor Jo A. Beran. Page 97: /Fundamental Photographs.

Experiment 8
Page 101: Ken Karp. Page 103: Ken Karp. Page 106: L.S. Stepanowicz/Visuals Unlimited.

Experiment 9
Page 109: Michael Watson. Page 112 (left): Courtesy VWR Scientific. Page 112 (right): Courtesy VWR Scientific. Page 113: Courtesy Fisher Scientific.

Experiment 10
Page 119: /Fundamental Photographs. Page 120: Courtesy Fisher Scientific. Page 122: Courtesy Borden Corporation.

Dry Lab 2A
Page 125: Keith Stone. Page 125 (bottom): Courtesy Fisher Scientific.

Dry Lab 2B
Page 128: /Fundamental Photographs. Page 129: Andy Washnik. Page 131: Peter Lerman.

Dry Lab 2C
Page 132: Peter Lerman. Page 136: Kathy Bendo.

Experiment 11
Page 137: /Fundamental Photographs. Page 137 (bottom): Courtesy New York Public Library.

Dry Lab 3
Page 149: /Fundamental Photographs. Page 152: Courtesy Library of Congress.

Experiment 12
Page 161: Yoav Levy/Phototake. Page 162: Peter Lerman. Page 163: Courtesy VWR Scientific.

Experiment 13
Page 169: Peter Lerman. Page 170 (top): Andy Washnik. Page 170 (center): Kathy Bendo and Jim Brady. Page 170 (bottom): Kathy Bendo and Jim Brady. Page 172: Courtesy VWR Scientific. Page 173 (left): Peter Lerman. Page 173 (right): Peter Lerman. Page 175: Ken Karp.

Experiment 14
Page 183: Yoav Levy/Phototake. Page 185: Fundamental Photographs. Page 186: Peter Lerman. Page 188: Michael Watson.

Experiment 15
Page 191: OPC, Inc. Page 193 (top left): Andy Washnik. Page 193 (top right): Andy Washnik. Page 193 (bottom): Courtesy Professor Jo A. Beran. Page 194 (right): Courtesy VWR Scientific. Page 194 (left): Courtesy VWR Scientific. Page 195 (top left): Courtesy Professor Jo A. Beran. Page 195 (bottom left): Hugh Lieck. Page 195 (bottom right): Courtesy VWR Scientific. Page 198: Ken Karp.

Experiment 16
Page 201: Ken Karp. Page 201 (bottom): Michael Siluk/The Image Works. Page 202: Courtesy Professor Jo A. Beran. Page 203: Courtesy VWR Scientific.

Experiment 17
Page 209: Michael Watson. Page 209 (bottom): Courtesy Professor Jo A. Beran. Page 212 (left): Courtesy Professor Jo A. Beran. Page 212 (right): Courtesy VWR Scientific. Page 213: Ken Karp.

Experiment 18
Page 219: Courtesy VWR Scientific. Page 221 (left): Ken Karp. Page 221 (right): Ken Karp.

Experiment 19
Page 227: Andy Washnik. Page 227 (bottom): Bruce Roberts/Photo Researchers. Page 230: Hugh Lieck.

Experiment 20
Page 235: Courtesy Fisher Scientific. Page 235 (bottom): Hugh Lieck. Page 238 (left): Hugh Lieck. Page 241: Michael Dalton/Fundamental Photographs.

Experiment 21
Page 245: Andy Washnik. Page 246: Courtesy Fisher Scientific. Page 247: Ken Karp. Page 249 (left): Courtesy Professor Jo A. Beran.

Experiment 22
Page 257: Ken Karp. Page 257 (bottom): Courtesy Fisher Scientific. Page 259 (left): Courtesy OPC, Inc. Page 259 (center): Courtesy OPC, Inc. Page 259 (right): Courtesy OPC, Inc. Page 260: Courtesy VWR Scientific. Page 263: John Dudak/Phototake.

Experiment 23
Page 267: Ken Karp. Page 271: Courtesy VWR Scientific.

Experiment 24
Page 281: Peter Lerman. Page 282: Michael Watson. Page 283: Ken Karp. Page 284: Courtesy Center for Disease Control. Page 285: Ken Karp.

Experiment 25
Page 293: Ken Karp. Page 294 (top): Courtesy VWR Scientific. Page 294 (bottom): Courtesy VWR Scientific. Page 297: Courtesy Fisher Scientific.

Experiment 26
Page 305: Ken Karp. Page 305 (bottom): Kathy Bendo. Page 307: Courtesy Professor Jo A. Beran.

Experiment 27
Page 313: Richard Megna/Fundamental Photographs. Page 314: Courtesy VWR Scientific. Page 316: Courtesy Fisher Scientific. Page 317: Richard Megna/Fundamental Photographs.

Experiment 28
Page 323: Ken Karp. Page 325: Ken Karp. Page 326: Courtesy Professor Jo A. Beran.

Experiment 29
Page 331: Michael Watson. Page 332: Hugh Lieck. Page 335: Richard Megna/Fundamental Photographs.

Experiment 30
Page 339: Courtesy The Permutit Co., a division of Sybron Corporation. Page 340: /Photo Researchers.

Experiment 31
Page 347: Andy Washnik. Page 350: Courtesy Fisher Scientific. Page 351 (top): Courtesy Fisher Scientific.

Experiment 32
Page 357: Michael Watson. Page 362: Michael Watson.

Experiment 33
Page 369: Ken Kay/Fundamental Photographs. Page 370: Michael Watson. Page 372: Ken Karp.

Experiment 34
Page 383: Peter Lerman. Page 385: Peter Lerman.

Experiment 35
Page 393: Peter Lerman. Page 396: Andy Washnik. Page 398: Ken Karp. Page 399 (left): Ken Karp. Page 399 (right): Courtesy Professor Jo A. Beran. Page 400: Ken Karp.

Experiment 36
Page 405: OPC, Inc. Page 405 (bottom): OPC, Inc.. Page 409 (left): Courtesy Professor Jo A. Beran. Page 409 (right): Courtesy Professor Jo A. Beran.

Experiment 37
Page 415: Yoav Levy/Phototake.

Experiment 38
Page 423: OPC, Inc. Page 424 (top): National Audobon Society. Page 424 (center): Courtesy Fisher Scientific. Page 428: Ken Karp. Page 429: Courtesy Professor Jo A. Beran.

Appendix A
Page 435: Courtesy Professor Jo A. Beran.

Appendix D
Page 443: Kathy Bendo.

Appendix F
Page 446: Courtesy Fisher Scientific.

Appendix G
Page 447: Michael Watson.